Emergent Research on Polymeric and Composite Materials

R. Somashekar
University of Mysore, India

Thejas Urs G.
University of Mysore, India

A volume in the Advances in Chemical and Materials Engineering (ACME) Book Series

IGI Global
DISSEMINATOR OF KNOWLEDGE

Published in the United States of America by
IGI Global
Engineering Science Reference (an imprint of IGI Global)
701 E. Chocolate Avenue
Hershey PA, USA 17033
Tel: 717-533-8845
Fax: 717-533-8661
E-mail: cust@igi-global.com
Web site: http://www.igi-global.com

Library of Congress Cataloging-in-Publication Data

Names: Somashekar, R. (Rudrappa), 1953- editor. | Urs G., Thejas (Urs Gopal), 1989- editor.
Title: Emergent research on polymeric and composite materials / R. Somashekar and Thejas Urs G., editors.
Description: Hershey, PA : Engineering Science Reference, [2018] | Includes bibliographical references.
Identifiers: LCCN 2017012022| ISBN 9781522530237 (hardcover) | ISBN 9781522530244 (ebook)
Subjects: LCSH: Polymeric composites--Industrial applications.
Classification: LCC TA418.9.C6 E4255 2018 | DDC 620.1/92--dc23 LC record available at https://lccn.loc.gov/2017012022

This book is published in the IGI Global book series Advances in Chemical and Materials Engineering (ACME) (ISSN: 2327-5448; eISSN: 2327-5456)

British Cataloguing in Publication Data
A Cataloguing in Publication record for this book is available from the British Library.

All work contributed to this book is new, previously-unpublished material. The views expressed in this book are those of the authors, but not necessarily of the publisher.

For electronic access to this publication, please contact: eresources@igi-global.com.

Advances in Chemical and Materials Engineering (ACME) Book Series

J. Paulo Davim
University of Aveiro, Portugal

ISSN:2327-5448
EISSN:2327-5456

Mission

The cross disciplinary approach of chemical and materials engineering is rapidly growing as it applies to the study of educational, scientific and industrial research activities by solving complex chemical problems using computational techniques and statistical methods.

The **Advances in Chemical and Materials Engineering (ACME) Book Series** provides research on the recent advances throughout computational and statistical methods of analysis and modeling. This series brings together collaboration between chemists, engineers, statisticians, and computer scientists and offers a wealth of knowledge and useful tools to academics, practitioners, and professionals through high quality publications.

Coverage

- Heat Treatments
- Biomaterials
- Electrochemical and Corrosion
- Artificial Intelligence Methods
- Composites
- Sustainable Materials
- Thermo-Chemical Treatments
- Computational methods
- Ceramics and glasses
- Fatigue and Creep

IGI Global is currently accepting manuscripts for publication within this series. To submit a proposal for a volume in this series, please contact our Acquisition Editors at Acquisitions@igi-global.com or visit: http://www.igi-global.com/publish/.

The Advances in Chemical and Materials Engineering (ACME) Book Series (ISSN 2327-5448) is published by IGI Global, 701 E. Chocolate Avenue, Hershey, PA 17033-1240, USA, www.igi-global.com. This series is composed of titles available for purchase individually; each title is edited to be contextually exclusive from any other title within the series. For pricing and ordering information please visit http://www.igi-global.com/book-series/advances-chemical-materials-engineering/73687. Postmaster: Send all address changes to above address.

Titles in this Series

For a list of additional titles in this series, please visit: www.igi-global.com/book-series

3D Printing and Its Impact on the Production of Fully Functional Components...
Petar Kocovic (Union – Nikola Tesla University, Serbia)
Engineering Science Reference • © 2017 • 115pp • H/C (ISBN: 9781522522898) • US $150.00 (our price)

Advanced Applications of Supercritical Fluids in Energy Systems
Lin Chen (Tohoku University, Japan & Japan Society for the Promotion of Science (JSPS), Japan) and Yuhiro Iwamoto (Nagoya Institute of Technology, Japan)
Engineering Science Reference • © 2017 • 682pp • H/C (ISBN: 9781522520474) • US $235.00 (our price)

Innovative Applications of Mo(W)-Based Catalysts in the Petroleum and Chemical Industry...
Hui Ge (Chinese Academy of Sciences, China) Xingchen Liu (Chinese Academy of Sciences, China) Shanmin Wang (Oak Ridge National Laboratory, USA) Tao Yang (China University of Petroleum, China) and Xiaodong Wen (Synfuels China, China)
Engineering Science Reference • © 2017 • 162pp • H/C (ISBN: 9781522522744) • US $135.00 (our price)

Sustainable Nanosystems Development, Properties, and Applications
Mihai V. Putz (West University of Timişoara, Romania & Research and Development National Institute for Electrochemistry and Condensed Matter (INCEMC) Timişoara, Romania) and Marius Constantin Mirica (Research and Development National Institute for Electrochemistry and Condensed Matter (INCEMC) Timişoara, Romania)
Engineering Science Reference • © 2017 • 794pp • H/C (ISBN: 9781522504924) • US $245.00 (our price)

Computational Approaches to Materials Design Theoretical and Practical Aspects
Shubhabrata Datta (Calcutta Institute of Engineering and Management, India) and J. Paulo Davim (University of Aveiro, Portugal)
Engineering Science Reference • © 2016 • 475pp • H/C (ISBN: 9781522502906) • US $215.00 (our price)

Green Approaches to Biocomposite Materials Science and Engineering
Deepak Verma (Graphic Era Hill University, Dehradun, India) Siddharth Jain (College of Engineering Roorkee, India and University of Alberta, Canada) Xiaolei Zhang (Queens University, Belfast, UK) and Prakash Chandra Gope (College of Technology, G.B.Pant University of Agriculure and Technology, Pantnagar, India)
Engineering Science Reference • © 2016 • 322pp • H/C (ISBN: 9781522504245) • US $165.00 (our price)

Position-Sensitive Gaseous Photomultipliers Research and Applications
Tom Francke (Myon, Sweden) and Vladimir Peskov (CERN, Switzerland)
Engineering Science Reference • © 2016 • 562pp • H/C (ISBN: 9781522502425) • US $240.00 (our price)

IGI Global
DISSEMINATOR OF KNOWLEDGE

701 East Chocolate Avenue, Hershey, PA 17033, USA
Tel: 717-533-8845 x100 • Fax: 717-533-8661
E-Mail: cust@igi-global.com • www.igi-global.com

Editorial Advisory Board

Table of Contents

Preface xiv

Acknowledgment xv

Chapter 1
Fibre Diffraction and Whole Powder Pattern Fitting in Polymers 1
Thejas Urs G., University of Mysore, India
R. Somashekar, University of Mysore, India

Chapter 2
Cellulose Application in Food Industry: A Review 38
Diego Mauricio Sanchez Osorno, Pontificia Bolivariana University, Colombia
Cristina Castro, Pontificia Bolivariana University, Colombia

Chapter 3
Comparative Study of Mechanical and Tribological Behaviour of Thermoplast-Based Composites: With Inorganic and Natural Particulate Fillers 78
Kaushik Kumar, Birla Institute of Technology, India
J. Paulo Davim, University of Aveiro, Portugal

Chapter 4
Properties Enhancement of Polymer/Ceramic Nanocomposites: Polymer Nanocomposites 99
Noureddine Ramdani, Harbin Engineering University, China
Mehdi Azibi, Independent Researcher, Algeria

Chapter 5
Structure-Property Relations in Polymers and Polymer Blends Using X-Ray Diffraction Technique 122
Somashekarappa H, University of Mysore, India
Manjunath Alabur, Government Science College Chitradurga, India

Chapter 6
Advances in Polymer-Based Piezoelectric Systems 139
Nithin Kundachira Subramani, The National Institute of Engineering, India
Shilpa K. N., Sri Jayachamarajendra College of Engineering, India
Sachhidananda Shivanna, Sri Jayachamarajendra College of Engineering, India
Jagajeevan Raj B. M., The National Institute of Engineering, India
Siddaramaiah Hatna, Sri Jayachamarajendra College of Engineering, India

Chapter 7
Effect of Cenosphere on the Performance of Polyurethane/Polystyrene Interpenetrating Polymer Network Green Composites 158
Roopa S., Sri Jayachamarajendra College of Engineering, India & JSS Science and Technology University, India
Siddaramaiah, Sri Jayachamarajendra College of Engineering, India

Chapter 8
Physical Characteristics of CBD Synthesized CdZnS Thin Films: CdZnS Thin Film Preparation Techniques, Properties, and Characterization 186
Ayan Mukherjee, The University of Burdwan, India
Partha Mitra, The University of Burdwan, India

Chapter 9
Studies on Metal Oxide Nanoparticle Doped PVP Polymer Nanocomposites 209
Madhu B. J., Government Science College Chitradurga, India

Chapter 10
Preparation and Characterization of Biopolymeric Nanoparticles as Drug Delivery Vehicles: A Comparative Study 225
Sai S. Sagiri, National Institute of Technology Rourkela, India
Suraj K. Nayak, National Institute of Technology Rourkela, India
S. Lakshmi, Bharatiyar University, India
Kunal Pal, National Institute of Technology Rourkela, India

Chapter 11
Methods of Preparation of Nanoparticle Formulations for the Treatment of COPD: A Short Review 247
Suprio R. Paul, Amity University, India
Suraj K. Nayak, National Institute of Technology, India
Arfat Anis, King Saud University, Saudi Arabia
Kunal Pal, National Institute of Technology, India

Chapter 12
Efficacy of Composites for Fabrication of Orthotic Calipers 267
Nisha Kumari, Birla Institute of Technology, India
Kaushik Kumar, Birla Institute of Technology, India

Compilation of References 287

Index 340

Detailed Table of Contents

Preface xiv

Acknowledgment xv

Chapter 1
Fibre Diffraction and Whole Powder Pattern Fitting in Polymers 1
Thejas Urs G., University of Mysore, India
R. Somashekar, University of Mysore, India

Crystal structure determination using a single crystal of any material has become a routine exercise. However in the case of fibres, the micro-crystallites are randomly arranged with their axis parallel to the fibre axis giving rise to poor reflections which are due to diffuse scattering, crystal structure determination is not possible. Hence crystal structure determination using poor fibre diffraction data is a challenging one. Model based correlations of X-ray diffraction data from fibrous materials are extensively discussed in a book by Fraser and MacRue(1973). The crux of the problem in fibre diffraction analysis is that the number of X-ray diffraction data is less than what could have been obtained for a single crystal. This aspect has limited the use of existing well developed crystallographic method such as Fourier synthesis of electron density or least squares refinement of atomic parameters.

Chapter 2
Cellulose Application in Food Industry: A Review 38
Diego Mauricio Sanchez Osorno, Pontificia Bolivariana University, Colombia
Cristina Castro, Pontificia Bolivariana University, Colombia

Cellulose polysaccharide is the most important component in plants with a fascinating structure and properties. Despite the origin, cellulose is a linear homopolymer of β-(1-4)-linked D-glucopyranose units varying mainly on purity, degree of polymerization (DP) and crystallinity index. This linear stiff-chain homopolymer is characterized by its hydrophilicity, chirality, biodegradability, broad chemical modifying capacity, and its formation of versatile semicrystalline fiber morphologies. This chapter aims to show the most important applications of cellulose in food, presenting other cellulose derivatives as methylcellulose, carboxymethyl cellulose, and novel cellulose forms as bacterial cellulose. New frontiers, including environmentally friendly cellulose fiber technologies for food packaging, bacterial cellulose in foodstuff and other applications as thickening agent, stabilizing agent, gelling agent, suspending agent were highlighted with future aims, strategies, and perspectives of cellulose research and its applications.

Chapter 3
Comparative Study of Mechanical and Tribological Behaviour of Thermoplast-Based Composites: With Inorganic and Natural Particulate Fillers ... 78
Kaushik Kumar, Birla Institute of Technology, India
J. Paulo Davim, University of Aveiro, Portugal

The usage of composite material has been increasing day by day over the years due to the excellent properties being offered by them such as lower volume-to-weight ratio, improved toughness, recyclable, environment friendly and also due to their short cycle productivity, easiness in fabrication, long lasting life span etc. This work presents study of mechanical and tribological behaviour of ABS polymer matrix filled with micron-sized inorganic and Natural / Green fillers. The main purpose behind this work is to compare the performance of composites with different class of fillers. The experimental results unveil that the composite with inorganic fillers perform better than with natural based fillers.

Chapter 4
Properties Enhancement of Polymer/Ceramic Nanocomposites: Polymer Nanocomposites ... 99
Noureddine Ramdani, Harbin Engineering University, China
Mehdi Azibi, Independent Researcher, Algeria

This chapter reviews the various properties enhancement of polymer/ceramic nanocomposites. Ceramics nanofillers have attracted both academic and industrial interest as they can produce a significant improvement in the properties of polymers even at lower filler loadings. Recently, numerous kinds of polymeric matrices reinforced with ceramic nanoparticles have been reported. The surface-modification of ceramic nanoparticles was reported to provide extra-improvements in the thermal and mechanical properties of these materials. In addition, the type of the used ceramic nanofillers agent determines the final properties of the nanocomposites. Herein, the various effects of adding ceramic nanoparticles on the thermal, electrical, optical, and mechanical properties of polymer/ceramic nanocomposites as well as the reinforcing mechanism are discussed in general along with detailed examples drawn from the scientific literature.

Chapter 5
Structure-Property Relations in Polymers and Polymer Blends Using X-Ray Diffraction Technique ... 122
Somashekarappa H, University of Mysore, India
Manjunath Alabur, Government Science College Chitradurga, India

This chapter discusses about the synthesis and characterization of polymers, polymer blends, polymer composites, and polymer nanocomposites. Electrically conductive blends polyaniline dodecyl benzene sulphonic acid (Pani:DBSA)/Styrene Butadiene Styrene (SBS) block copolymer have been prepared by melt mixing and using in situ polymerization method. The microstructural parameters were computed using wide angle X-ray scattering (WAXS) and Small angle X-ray scattering (SAXS). Bimodal distribution method has been studied for SBS, DCH32 cotton fiber and Pure Mysore multi-voltine silk (PMS) samples using Line profile analysis.

Chapter 6
Advances in Polymer-Based Piezoelectric Systems 139
Nithin Kundachira Subramani, The National Institute of Engineering, India
Shilpa K. N., Sri Jayachamarajendra College of Engineering, India
Sachhidananda Shivanna, Sri Jayachamarajendra College of Engineering, India
Jagajeevan Raj B. M., The National Institute of Engineering, India
Siddaramaiah Hatna, Sri Jayachamarajendra College of Engineering, India

Lately, polymer based piezoelectric materials that harness energy from mechanical vibrations and/or impact are being increasingly investigated as radical alternates to conventional batteries that are hard to service once deployed. Nevertheless, the optimization of energy outputs of piezoelectric energy harvesters is one of the prime challenges faced by the scientific community. This chapter provides an overview of polymer based piezoelectric energy harvesters with special emphasis on current state of research on polymer composites/nanocomposites for vibrational energy harvesting. A detailed summary of piezoelectric phenomenon in polymers is also presented. An in-depth narration detailing the enhancement of piezoelectric behavior of one of the most commonly employed piezoelectric polymer (PVDF) is presented with special emphasis on some of the promising filler materials towards realizing high efficient piezoelectric modules. This chapter is intended to give an insight on the recent advances in the field of polymer based piezoelectric materials.

Chapter 7
Effect of Cenosphere on the Performance of Polyurethane/Polystyrene Interpenetrating Polymer Network Green Composites 158
Roopa S., Sri Jayachamarajendra College of Engineering, India & JSS Science and Technology University, India
Siddaramaiah, Sri Jayachamarajendra College of Engineering, India

The effect of cenosphere content on the performances of polyurethane/polystyrene (PU/PS, 90/10) interpenetrating polymer network (IPN) based green composites have been studied. The PU/PS IPNs have been prepared using castor oil, toluene diisocyanate and styrene. IPN/cenosphere composites have been prepared with different weight fractions viz., 0, 5, 10, 20 and 30 wt % of cenosphere. The prepared IPN composites have been characterized by physico – mechanical, chemical and thermal behavior. The tensile strength of unfilled IPN was 1.79 MPa and a significant improvement in tensile strength (34%) was noticed for 10% cenosphere loaded IPN composite. The swelling behavior of the composites has been studied in different organic solvents. Thermal characteristics of the composites have been measured using differential scanning calorimeter, thermogravimetric analysis and dynamic mechanical analysis (DMA). A slight improvement in thermal stability was noticed for filler loaded specimens. Morphological features of cryo-fractured IPN/cenosphere green composites have been analyzed using SEM.

Chapter 8
Physical Characteristics of CBD Synthesized CdZnS Thin Films: CdZnS Thin Film Preparation Techniques, Properties, and Characterization 186
Ayan Mukherjee, The University of Burdwan, India
Partha Mitra, The University of Burdwan, India

In recent years, ternary cadmium zinc sulfide (CdZnS) alloy compounds have been paid much attention in the fields of opto-electronics, particularly in photovoltaic devices. CdZnS thin films can be prepared by different techniques among which chemical methods have more advantages. Among different chemical method, Chemical Bath Deposition (CBD) is simple, low cost and widely applicable in industrial applications. In this chapter, we have discussed different methods of preparation of CdZnS thin film and their obtained properties. Also, the films are characterized by XRD, TEM, FESEM, EDAX, UV-Vis spectroscopy, etc. The properties of CdZnS gives insight of the properties of ternary thin film semiconductor and it will help to design semiconductor with tuneable properties for future applications in optoelectronic sector.

Chapter 9
Studies on Metal Oxide Nanoparticle Doped PVP Polymer Nanocomposites 209
Madhu B. J., Government Science College Chitradurga, India

Magnesium Oxide (MgO) nanoparticles have been synthesized by solution combustion technique using stoichiometric composition of magnesium nitrate as oxidizer and urea as fuel. Structure of the MgO was studied with the X-ray diffraction (XRD) using Cu-Kα radiation. MgO/polyvinylpyrrolidone (PVP) nanocomposites have been prepared by blending MgO nanoparticles with the polyvinylpyrrolidone. MgO/PVP nanocomposites were characterized by Fourier transform infrared (FTIR) spectroscopy. Frequency dependence of dielectric constant (ε′), dielectric loss tangent (tanδ) and AC conductivity studies have been undertaken on the MgO/PVP nanocomposites in the frequency range 50Hz-5MHz at room temperature. Dielectric properties such as dielectric constant (ε′) and dielectric loss tangent (tanδ) are found to decrease with the increase in the frequency. Further, AC conductivity of MgO/PVP nanocomposites was found to increase with an increase in the frequency. Observed variation in the a. c. conductivity with the frequency has been understood on the basis of electron hopping model.

Chapter 10
Preparation and Characterization of Biopolymeric Nanoparticles as Drug Delivery Vehicles: A Comparative Study 225
Sai S. Sagiri, National Institute of Technology Rourkela, India
Suraj K. Nayak, National Institute of Technology Rourkela, India
S. Lakshmi, Bharatiyar University, India
Kunal Pal, National Institute of Technology Rourkela, India

In recent years, the use of biopolymeric nanoparticles as vehicles for drug delivery has increased exponentially. In the present study, chitosan and gelatin nanoparticles were prepared by ionic gelation and desolvation methods, respectively. Salicylic acid was used as the model drug. The nanoparticles were characterized using SEM, XRD analysis and FTIR spectrophotometric studies. In vitro drug release experiments were carried out to understand the mechanism of drug release. SEM micrographs showed the formation of spherical nanoparticles. XRD studies indicated a higher crystalline nature of the chitosan nanoparticles as compared to the gelatin nanoparticles. FTIR studies indicated the presence of salicylic acid within the drug- loaded nanoparticles. Drug release studies indicated that the developed nanoparticles may be used as carriers for various bioactive agents.

Chapter 11
Methods of Preparation of Nanoparticle Formulations for the Treatment of COPD: A Short Review 247
Suprio R. Paul, Amity University, India
Suraj K. Nayak, National Institute of Technology, India
Arfat Anis, King Saud University, Saudi Arabia
Kunal Pal, National Institute of Technology, India

The incidence of chronic obstructive pulmonary disease (COPD) has increased many folds in the last couple of decades. This can be explained by the increased pollution levels and drastic changes in the lifestyle of the people. Taking a note of this, in this review, a conscious attempt was made to understand the physiological changes in the respiratory tract in a COPD patient. Nanoparticulate formulations play an extensive role in the current-day treatment regime of COPD patients. Hence, different methods for developing nanoparticulate formulations for the treatment of COPD are discussed in details.

Chapter 12
Efficacy of Composites for Fabrication of Orthotic Calipers 267
Nisha Kumari, Birla Institute of Technology, India
Kaushik Kumar, Birla Institute of Technology, India

Composites based materials are becoming more important in the field of aerospace, automobile, medicine, sports, energy etc. The principal quality of composite materials is their higher strength, firmness, resistant to corrosion and comparatively low weight when compared to the metals. The main aim of the work is to replace the presently used orthotic calipers with the proposed polymeric based composites as in last few years composites have generated broad research in engineering and research field due to its smaller density, low cost, recyclable, environment friendly and capable of being decomposed. Here, the authors have proposed two polymeric based composites (thermoplast and thermoset) as an alternative material of construction and hence their various properties were investigated. The authors after performing the experiment unveil that the strength and firmness of thermoset based composites (Epoxy + Carbon) is higher and have lower weight- to- volume ratio than that of the presently used (Aluminum based) orthotic calipers.

Compilation of References 287

Index 340

Preface

As we approach a new century, it is an era of materials, the prominent subject being polymers and composite materials. Polymers have remarkable interest in the field of research and development of various facets of science. The twenty first century has brought the preparation to a state of substantial sophistication and here is an attempt to provide researchers a handbook for fundamentals in this arena.

There has not been book devoted to preparation and characterization of polymers and materials for many years and so this book was designed by asking prominent international and national experts in this field to contribute chapters. We have included chapters on “Fibre Diffraction and Whole Powder Pattern Fitting in Polymers”, “Cellulose Application in Food Industry: A Review”, “Comparative Study of Mechanical and Tribological Behaviour of Thermoplast-Based Composites: With Inorganic and Natural Particulate Fillers”, “Structure-Property Relations in Polymers and Polymer Blends Using X-Ray Diffraction Technique”, “Properties Enhancement of Polymer/Ceramic Nanocomposites: Polymer Nanocomposites”, “Effect of Cenosphere on the Performance of Polyurethane/Polystyrene Interpenetrating Polymer Network Green Composites”, "Synthesis, Characterization, Dielectric, and AC Conductivity Behavior of MgO/PVP Polymer Nanocomposites”, and various other related materials such as “Advances in Polymer-Based Piezoelectric Systems”, “Physical Characteristics of CBD Synthesized CdZnS Thin Films: CdZnS Thin Film Preparation Techniques, Properties, and Characterization”, “Preparation and Characterization of Biopolymeric Nanoparticles as Drug Delivery Vehicles: A Comparative Study”, and “Efficacy of Composites for Fabrication of Orthotic Calipers”. Both preparation and characterization procedures are discussed in detail and a significant portion is devoted to various applications. Because of different backgrounds of various authors, it was decided to not rigidly enforce a uniform notation but to use all of the most common notations found in the modern literature.

The production of this book involved a huge effort by Mr. Gowtham G K, Mr. Manju V V, Mr. Nandaprakash M B, Dr. Mahadevaiah, and a host of editors which we have listed in this book. We are deeply indebted to them for their efforts.

The editors and authors hope that this book will be a milepost along the way to the development of a complete polymer based materials before the end of next century.

Acknowledgment

We thank all the editorial advisory board members and the potential reviewers who have contributed significantly to upgrade the articles submitted to this book with their scientific knowledge. In this, we would wish to name the team of reviewers who have spent their valuable time in evaluating the scientific quality of the articles and in improvising the same.

Dr. T Demappa, Professor, Dept. of Polymer Science, University of Mysore; Dr. Gopal Krishna Naik and Dr. Y Sangappa, Professors at the Dept. of Physics, Mangalore University; Dr. Santosh Kumar Biswas, Professor of Textile Physics, University of Calcutta; Dr. Kaushik Kumar, BITS, Mesra and Dr. P Parameshwara, Asst. Prof. of Physics, NIE College, Mysore. They all have been very helpful in editing this book by sharing their useful suggestions and concerns in the development of chapters. Dr. Mahadevaiah, Research Associate; Nandaprakash M B; Gowtham G K; Vinayakprasanna N. Hegde and Manju V V, fellow colleagues, were very supportive in scrutinizing the chapters based on their scientific and grammatical content. We appreciate the support and assistance provided by all our authors who have contributed chapters for the successful completion of this book project, and we are thankful to them.

Chapter 1
Fibre Diffraction and Whole Powder Pattern Fitting in Polymers

Thejas Urs G.
University of Mysore, India

R. Somashekar
University of Mysore, India

ABSTRACT

Crystal structure determination using a single crystal of any material has become a routine exercise. However in the case of fibres, the micro-crystallites are randomly arranged with their axis parallel to the fibre axis giving rise to poor reflections which are due to diffuse scattering, crystal structure determination is not possible. Hence crystal structure determination using poor fibre diffraction data is a challenging one. Model based correlations of X-ray diffraction data from fibrous materials are extensively discussed in a book by Fraser and MacRue(1973). The crux of the problem in fibre diffraction analysis is that the number of X-ray diffraction data is less than what could have been obtained for a single crystal. This aspect has limited the use of existing well developed crystallographic method such as Fourier synthesis of electron density or least squares refinement of atomic parameters.

INTRODUCTION

Crystal structure determination using a single crystal of any material has become a routine exercise. However in the case of fibres, the micro-crystallites are randomly arranged with their axis parallel to the fibre axis giving rise to poor reflections which are due to diffuse scattering, crystal structure determination is not possible. Hence crystal structure determination using poor fibre diffraction data is a challenging one.

Model based correlations of X-ray diffraction data from fibrous materials are extensively discussed in a book by Fraser and Mac Rue (1973). The crux of the problem in fibre diffraction analysis is that the number of X-ray diffraction data is less than what could have been obtained for a single crystal. This aspect has limited the use of existing well developed crystallographic method such as Fourier synthesis

DOI: 10.4018/978-1-5225-3023-7.ch001

of electron density or least squares refinement of atomic parameters. To get over this difficulty, Arnott and Wonacott(1966), Smith and Arnott (1978) have developed a method wherein the approach involves to incorporate the knowledge of chemical sequence like bond lengths and bond angles in molecular models and then by systematic trial and error, to adjust their conformations and positions in a unit cell to obtain a good agreement between observed and calculated amplitudes from the model parameters. This technique has been used to solve the crystal structure of collagen (Okuyama(2008)), silk-I (Okuyama et al) and silk-II modifications (Sangappa (2005)) and cotton fibres (Samir & Somashekar (2007)) and also of polymer (Hall & Somashekar 1991). We apologize, in advance, for the omission of some structures to which our survey could not reach.

In addition to the problem mentioned above, the parallelism of the crystallites is invariably imperfect and the diffracted intensities are distributed not in spot but along arcs the length of which increases with distance from the pattern center. This dissipation of diffracted intensity over increasing lengths of arc implies a similar rise in the threshold for observation and even at moderate distances from the pattern center all but the strongest intensities are merged in the background. This leads to the fact that the resolving power of the data is low since it is the diffraction furthest from center that defines the finer periodicities in crystals. In the following we describe a method for refining the positional and conformational parameters of model systems while retaining given sterio chemical constraints. It is readily adaptable for automatic digital computers and is related to one studied for DNA fibres and to others described for single crystals.

Mathematical Analysis of the Refinement Method

Essence of mathematics that goes into the methodology of refinement are reproduced from the earlier work of Arnott & Wonacott (1966). The 'best fit' considered to be that which minimizes.

$$¦ = \sum_{1}^{M} É_{m}\left(F_{o} - F_{m}\right)^{2} = \sum_{1}^{M} É_{m}\text{”}F_{m}^{2} \tag{1}$$

where there are M observed diffraction amplitudes, *Fo*, and *Fm* are the corresponding amplitudes calculated for the model.

$$F_{m} = \left(A_{m}^{2} + B_{m}^{2}\right)^{\frac{1}{2}} . \, {}^{1}\!/_{K} . \exp\left({}^{-BÁ^{2}}\!/_{4}\right) \tag{2}$$

Where

$$A_{m} = \sum_{1}^{P} f_{p}\left(h_{m} k_{m} l_{m}\right) \cos 2\pi\left(h_{m} x_{p} + k_{m} y_{p} + l_{m} z_{p}\right) \tag{3}$$

and

$$B_m = \sum_{1}^{P} f_p\left(h_m k_m l_m\right) \sin 2\pi\left(h_m x_p + k_m y_p + l_m z_p\right) \tag{4}$$

when there are P atoms with functional Cartesian coordinates $\left(x_p y_p z_p\right)$ in the crystal unit; $fp\left(h_m k_m l_m\right)$ is the scattering power of the pth atom and $h_m k_m l_m$ are the indeces of the X-ray reflection; *K* is the factor required to put the observations on an absolute scale; *B* is the average isotropic thermal vibration parameter which results in a radial fall-off of diffraction intensity; ρ=2(sin θ)/λ where θ is the diffraction angle. w_m theweight of each term in Φ, should be proportional to the reliability of F_o. Usually an analytic function $\left\{\left(1 + a.F_o + F_o^2\right).\rho\right\}^{-1}$ is satisfactory. *a* and *b* are chosen constants which results in the very intense reflections having low weights while $\rho - 1$ reduces the weight of the reflections in the outer part of the diffraction pattern where observed intensities have large geometrical factors to convert them to $F_o s$. When a reflection is composite

$$F_m = \left(\sum_{1}^{R} F_{mr}^2\right)^{1/2} .\left(1 / K\right).exp\left(-B\rho^2 / 4\right) \tag{5}$$

where the F_{mr}s are the amplitudes of the reflections in the group. If we begin with a model with *N* parameters [*un*] we can obviously calculate the value of Φ. If we change the parameters to [$u_n + \Delta u_n$] the new value of Φ will be by Taylor'sexpansion

$$\Phi = \sum_{1}^{M} w_m \left\{\Delta F_m\left(\left[u_n\right]\right) - \sum_{1}^{N} \Delta u_n \frac{\partial F_m\left[u_n\right]}{\partial u_n}\right\}^2 \tag{6}$$

If the terms in $(un)^2$ are small, which will be so if the position represented by [$u_n + \Delta u_n$] is close enough to [*un*]. The *N* values of [Δ*un*] which will minimize Φ are given by the *N* linear equations ∂Φ/∂(Δ*un*)=0, i.e. by *N* equations like

$$\frac{\partial \Phi}{\partial\left(\Delta u_q\right)} = -2\sum_{1}^{M} w_m \left\{\Delta F_m - \sum_{1}^{N} \Delta u_n \frac{\partial F_m}{\partial u_n}\right\} \frac{\partial F_m}{\partial u_q} = 0 \tag{7}$$

They may be conveniently written in matrix notation (which also provides an indication of how they may be solved using standard computer routines)

$$F.M - U.M^T.M = 0$$

i.e

$$U = F.M.\left(M^T.M\right)^{-1} \tag{8}$$

Where

$$F \equiv \left[\sqrt{w_1}\,\Delta F_1 \ldots\ldots\ldots \sqrt{w_m}\,\Delta F_m\right] \tag{9}$$

$$M \equiv \begin{bmatrix} \sqrt{w_1}\dfrac{\partial F_1}{\partial u_1} & \cdots & \cdots & \sqrt{w_1}\dfrac{\partial F_1}{\partial u_N} \\ \vdots & & \ddots & \vdots \\ \sqrt{w_m}\dfrac{\partial F_M}{\partial u_1} & \cdots & \cdots & \sqrt{w_m}\dfrac{\partial F_M}{\partial u_N} \end{bmatrix} \tag{10}$$

$$U = \left[\Delta u_1 \ldots\ldots\ldots\ldots \Delta u_N\right] \tag{11}$$

In general, equation 11 is an inadequate approximation (the $[u_n]$ are not small enough) and the parameters $[u_n + \Delta u_n]$ obtained after solving (8) will not give minimum Φ and the process has to be repeated with the new parameters and iterated until the $[u_n]$ are so small that Φ does not change significantly in successive cycles.

APPLICATION TO TERYLENE

The scheme devised allows a set of standard bond lengths and angle to be included explicitly in the least-squares analysis. Terylene is the example chosen and indicated in Figure

The unit in the Terylene molecule consists of four main chain atoms 1,2,3,4 and three substituent's, 2a,3a,3b (for the moment hydrogen atoms will be ignored). Sets of axes may be set up at each atom of the main chain after the first in a systematic way such that in each case the x-axis lies along the bond between atom n and atom (n-1); the y axis is perpendicular to the x-axis and lies within the obtuse angle formed by atoms (n-1), n, and (n+1) and is coplanar with them; the z-axis completes a right-handed orthogonal set. The coordinates of atom 1 with respect to the set of axes at 2 are immediately known and are given by

$$X_2 = L_{12} = \begin{bmatrix} l_{12} \\ 0 \\ 0 \end{bmatrix} \tag{12}$$

Figure 1. The chain geometry of Terylene. φ1, φ2 and φ3 are bond angles along the chain

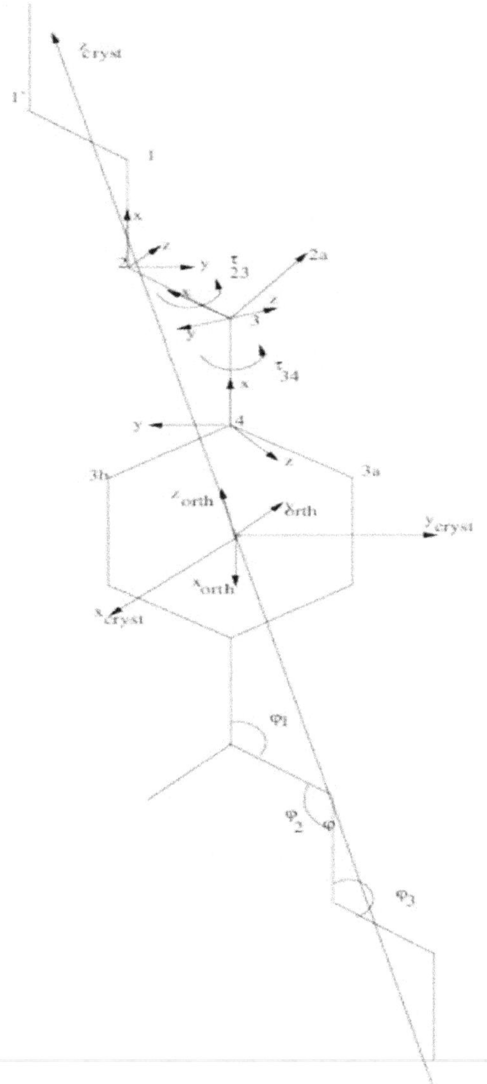

where l_{12} is the bond length between atoms 1 and 2. Its coordinates with respect to the next axis system at atom 3 are

$$X_3 = \begin{bmatrix} -\cos\phi_2 & -\sin\phi_2 & 0 \\ \sin\phi_2\cos\tau_{23} & -\cos\phi_2\sin\tau_{23} & \sin\tau_{23} \\ \sin\phi_2\cos\tau_{23} & \cos\phi_2\sin\tau_{23} & \cos\tau_{23} \end{bmatrix} . X_2 + \begin{bmatrix} l_{23} \\ 0 \\ 0 \end{bmatrix} \tag{13}$$

i.e, $X3 = A_{23}X_2 + L_{23} = A_{23}L_{12} + L_{23}$. This transformation involves the stereo chemical constants φ_2 and l_{23} together with angle τ_{23}, the dihedral angle between the plane formed by atoms 1, 2 and 3 and that formed by 2,3 and 4. It is this angle which is regarded as a refinable parameter of the chain. A further similar transformation involving φ_3, l_{31} with yield the coordinate of atoms 1 with respect to the axial set at 4.

$$X_4 = A_{34}X_3 + L_{34} = A_{34}A_{23}A_{12} + A_{34}L_{23} + L_{34} \tag{14}$$

It is clear that the coordinates of all the atoms in the asymmetric unit of Terylene with respect to the axial set at 4 can be derived in a similar way as functions of the chain parameters τ_{23} and τ_{34} and the various bond lengths and angles. It remains only to relate them to the unit cell axes. Since crystal axes are not generally orthogonal it is convenient to define an orthogonal set which has $Z_{orthogonal}$ coincident with

$Z_{crystal}$, $Y_{orthogonal}$ in the same phase as $Y_{crystal}$ and $Z_{crystal}$.

Coordinates can then be transformed to a set at 4 parallel to the crystal set by

$$X_{orthogonal} = R \cdot X$$

Where

$$R = \begin{bmatrix} 1 & 0 & 0 \\ 0 & \cos\theta_1 & -\sin\theta_1 \\ 0 & \sin\theta_1 & \cos\theta_1 \end{bmatrix} \begin{bmatrix} \cos\theta_2 & 0 & -\sin\theta_2 \\ 0 & 1 & 0 \\ \sin\theta_2 & 0 & \cos\theta_2 \end{bmatrix} \begin{bmatrix} \cos\theta_3 & \sin\theta_3 & 0 \\ \sin\theta_3 & \cos\theta_3 & 0 \\ 0 & 0 & 1 \end{bmatrix} \tag{16}$$

θ_1, θ_2 and θ_3 are the Eulerian angles relating the orientation of the axial set 4 to the orthogonal crystal set. A final transformation

$$X_{crystal} = C \cdot X_{orthogonal} + T \tag{17}$$

yields fractional crystal coordinates

$$C = \begin{bmatrix} \dfrac{1}{a\,\sin\beta\sin\gamma} & 0 & 0 \\ \dfrac{1}{b\sin\alpha\tan\gamma\,*} & \dfrac{1}{b\sin\alpha} & 0 \\ \dfrac{-\sin\beta\cos\gamma\,* + \tan\alpha\cos\beta}{c\tan\alpha\sin\beta\sin\gamma\,*} & \dfrac{-1}{c\tan\alpha} & \dfrac{1}{c} \end{bmatrix} \tag{18}$$

and

$$T = \begin{bmatrix} u_1 \\ u_2 \\ u_3 \end{bmatrix} \tag{19}$$

u_1, u_2, u_3 are the fractional crystal coordinates of atom 4 and are the parameters necessary to define the position of the chain. This description of a chain molecule is entirely general and since it involves a set of systematic transformations in matrix form it is readily adaptable for automatic computations. The immediate advantages achieved apart from the important one of maintaining predetermined bonded stereochemistry, are twofold: the number of parameters is reduced by a factor of two, at least in the case of Terylene, and since changes in these parameters will, in general, result in movements of large groups of atoms the lack of resolving power in the diffraction data becomes irrelevant. When the structure parameters refined are the fractional Cartesian co-ordinates of the atoms in the crystal unit we have the case similar to single crystal analysis. This is illustrated in refining the structure of Terysene, described in Arnott and Wonacott(1966). For details of the refinement procedure one can refer to earlier papers by Arnot & Wonacott (1966), Smith & Arnott(1978). Here we re-produce two methods of calculating intensities of Bragg reflections: first one being the calculation of intensities using the supplied software of the instruments (silk-I structure) and another using CCP13, a fibre software site (Collagen, cotton, polymer, silk-II structure) CCP13.

APPLICATION OF LALS TECHNIQUE: EARLIER REPORTS

Heteronomous DNA structure was reported using the above technique by Arnott et al 1983. The corresponding X-ray fibre diffraction patterns along crystal packing are reproduced here.

Further, this group also have reported the crystal packing of B forma of poly d(GC):ply d(GC) in oriented fibers using LALS technique (Arnott et al, (1983)). X-ray diffraction patterns and crystal packings are reproduced here.

Desborough & Hall (1977) have compared the structures of poly (tetramethyleneterephthalate) using LALS technique. Hall & Pass have also reported crystal structure packing in the case of poly (tetramethylene terephthalate) and further studied the changes with strain using LALS technique. Here we reproduce X-ray diffraction patterns and crystal packing for this sample.

Arnott et al (1983) have investigated DNA secondary structures using LALS technique using X-ray diffraction patterns of DNA fibers like Calf thymus B-DNA, B forma of Poly[d(GC)]*poly[d(GC)], screw-disordered form of D-DNA and oriented polycrystalline form of D-DNA.

CRYSTAL AND MOLECULAR STRUCTURE OF SILK-I MODIFICATION

Introduction

Silk has two modifications, silk-I and silk-II. Silk-I can be obtained from the silk gland content by quiescent drying without mechanical stress and crystalline dimorphism of the silk. This was first investigated in detail by Shimizu(1941). Later onsilk-I crystal structure has also been studied by Okuyama et al

Figure 2. X-ray diffraction patterns of Poly d(A): poly d(T); (a) alpha-form (b) beta-form

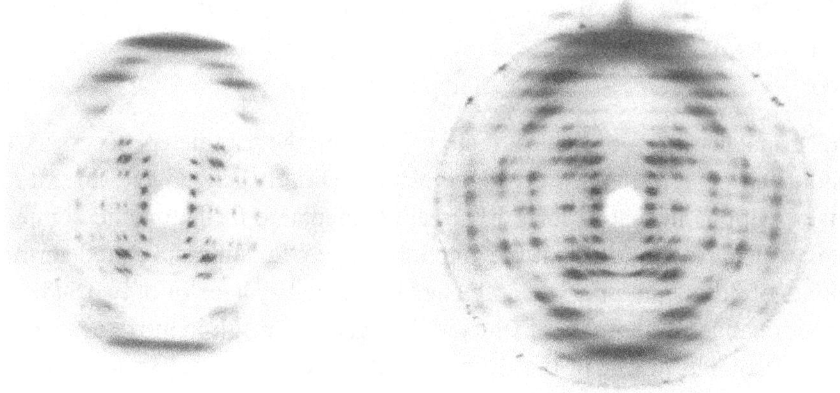

Figure 3. The molecular packings of poly d(A): poly d(T) (a) Alpha-forma and (b) Beta-form

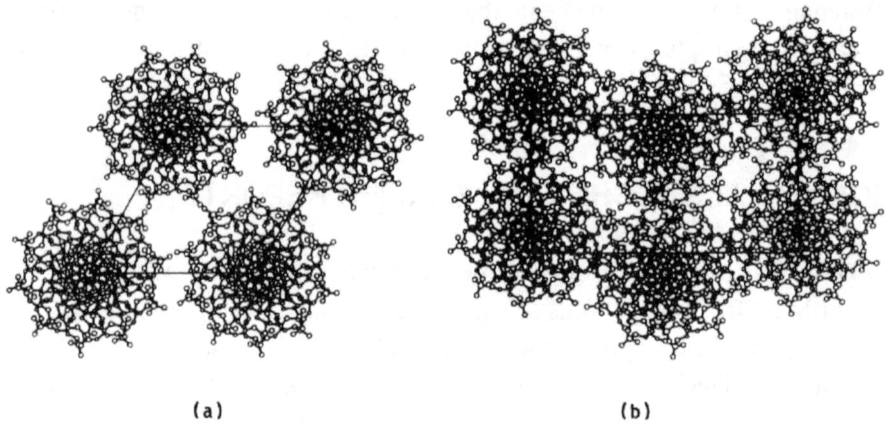

Figure 4. X-ray diffraction patterns of (a) poorly crystalline B poly d(GC): poly d(GC)and (b) polycrystalline poly d(GC): poly d(GC)

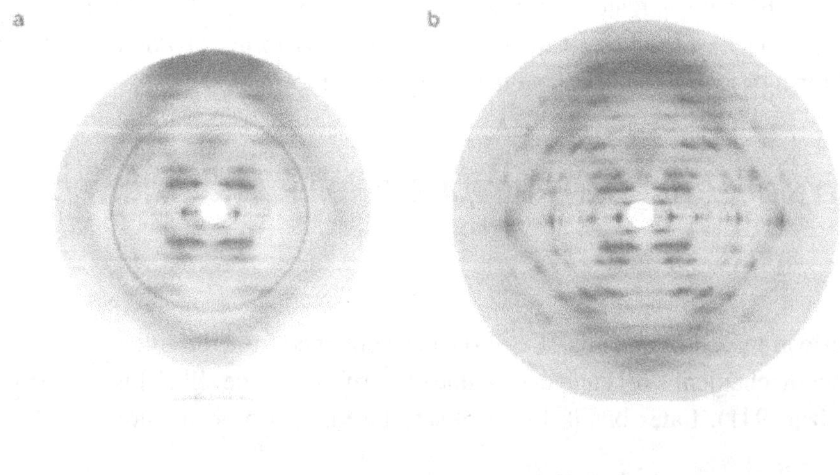

Figure 5. The molecular packings of (a) Orthorhombic B poly d(GC): poly d(GC) and tetragonal D poly d(GC): poly d(GC)

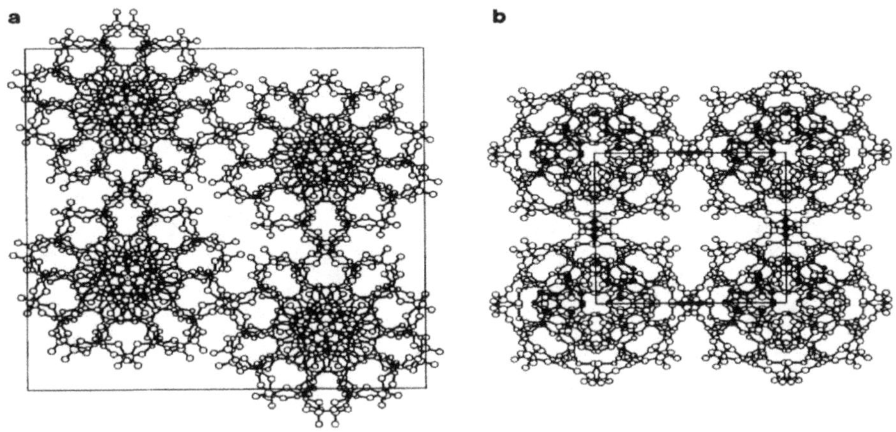

Figure 6. X-ray diffraction pattern of 4GT fibre. (a) Relaxed

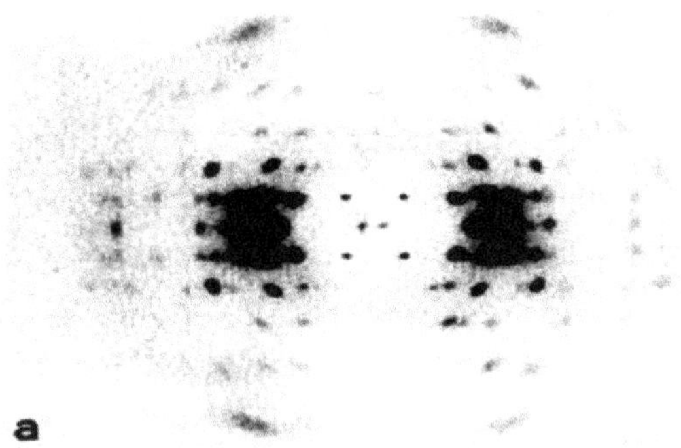

Figure 7. X-ray diffraction pattern of 4GT fibre. (b) Strained at 10%

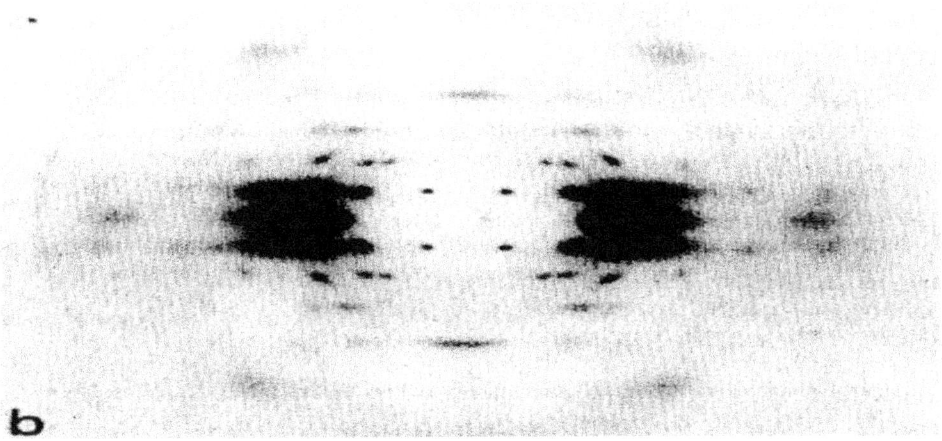

Figure 8. Projection of unit cell and contents onto ac plane (a) Relaxed; (b) 10% strain

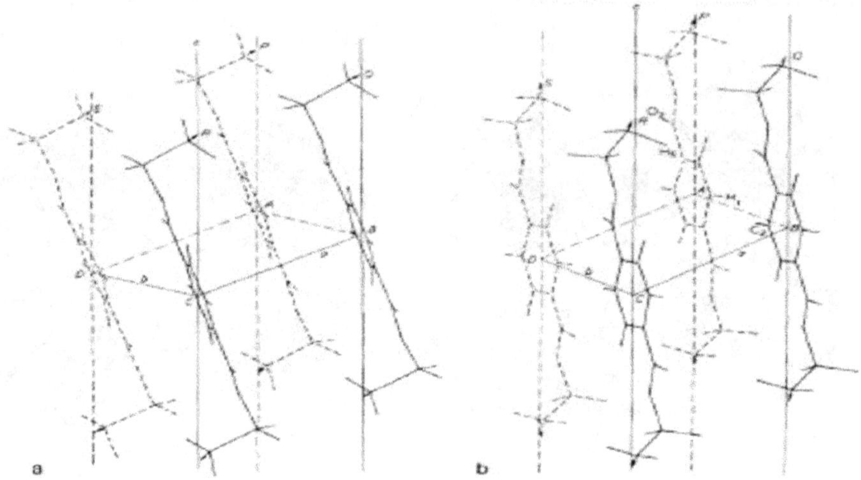

(1988), Asakura et al(1999). In the following sections, we present the procedure for the determination of structure of Silk-I based on our earlier publications (Okuyama et al (2001)). On the other hand, silk-II can be obtained from cocoons spun by *Bombyx mori* silk worm sand its structure has been reported by Marsh et al (1955) to be a β-sheet structure. A detailed discussion of silk-II structure can be seen in relatively recent report by Takahashi et al (1999).

Specimen Preparation

For our study we have used the cocoon of *Bombyx mori* and it was degummed with 0.6% Marsciltes soap solution (300 ml) plus 0.3 gm of sodium carbonate at 100°*C* for 0.5 hour. It was washed with hot, distilled water. Then this cocoon was again soaked in the above solution at 100°*C* for 1 hour and washed with a large quantity of hot aqueous solution of sodium carbonate. This is associated with gradually changing its concentrations from 0.3 to 0.1% and then washing with 0.2% acetic acid (1L), and finally, washing by using hot, distilled water until no bubble was observed. This degummed cocoon was defatted with benzene and methanol solution (2/1 by volume) for 100 hour in a Soxhlet extractor. Then the silk fibroin (1gm) so prepared was dissolved in aqueous solution of 8M LiBr (20ml). To remove LiBr salt, the solution was dialyzed against running water for 6 hour and then renewed against distilled water every 3 hour until no bromide ion was observed in the solution (usually more than 40 hour). To make the fibroin concentration of about 1% distilled water (100ml) was added to the dialyzed solution (20 ml). After being prepared at pH 7.8 by the phosphate buffer, the sample fibroin was digested by $\alpha-$drymotripsin (10mg) with $CaCl_2$ (40 mg) in an incubator at 37°*C* for 24 hours. The precipitate of *Cp* fraction (crystalline fraction of silk fibroin) was washed several times with distilled water. This was again defatted by a diethyl ether solution and then by chloroform and methanol solution for 20 hours for each. After washing several times with distilled water, the *Cp* fraction was lyophilized. Because the crystal form of the *Cp* fraction at this stage is usually the silk II modification, the specimen (0.3g) was again dissolved in aqueous 8m LiBr solution (18ml) at 37°*C*. And after filtering the solution, distilled water (18ml) was added and the solution was prepared at pH 4.2 by 10% acetic acid (Nakajima et al (1987)).

This solution was dialyzed against distilled water prepared at pH 4.2 for about 60 hour at room temperature. After washing several times with distilled water, the precipitate was lyophilized. This work was done in Kenji Okuyama's laboratory, Department of Biotechnology, Tokyo University of Agricultural and Technology, Tokyo, Japan.med with 0.6% Marsciltes soap solution (300 ml) plus 0.3 gm of sodium carbonate at 100°*C* for 0.5 hour and it was washed with hot, distilled water. Then this cocoon was again soaked in the above solution at 100°*C* for 1 hour and washed with a large quantity of hot aqueous solution of sodium carbonate. This is associated with gradually changing its concentrations from 0.3 to 0.1% and then washing with 0.2% acetic acid (1L), and finally, washing by using hot, distilled water until no bubble was observed. This degummed cocoon was defatted with benzene and methanol solution (2/1 by volume) for 100 hour in a Soxhlet extractor. Then the silk fibroin (1gm) so prepared was dissolved in aqueous solution of 8M LiBr (20ml). To remove LiBr salt, the solution was dialyzed against running water for 6 hour and then renewed against distilled water every 3 hour until no bromide ion was observed in the solution (usually more than 40 hour). To make the fibroin concentration of about 1% distilled water (100ml) was added to the dialyzed solution (20 ml). After being prepared at pH 7.8 by the phosphate buffer, the sample fibroin was digested by α−dry-motripsin (10mg) with $CaCl_2$ (40 mg) in an incubator at 37°*C* for 24 hours. The precipitate of *Cp* fraction (crystalline fraction of silk fibroin) was washed several times with distilled water. This was again defatted by a diethyl ether solution and then by chloroform and methanol solution for 20 hours each. After washing several times with distilled water, the *Cp* fraction was lyophilized. Because the crystal form of the *Cp* fraction at this stage is usually the silk II modification, the specimen (0.3g) was again dissolved in aqueous 8M LiBr solution (18ml) at 37°*C*. And after filtering the solution, distilled water (18ml) was added and the solution was prepared at pH 4.2 by 10% acetic acid (Nakajima et al (1987)). This solution was dialyzed against distilled water prepared at pH 4.2 for about 60 hour at room temperature. After washing several times with distilled water, the precipitate was lyophilized. This work was done in Kenji Okuyama's laboratory, Department of Biotechnology, Tokyo University of Agricultural and Technology, Tokyo, Japan.

X-Ray Recording

X-ray diffraction pattern from the *Bombyx mori* in silk I modification was recorded using a Rigaku co. X-ray diffractometer system Rota FLEX RU-200 equipped with a monochromator with nickel filtered Cu$K\alpha$ radiation using the cylindrical camera. X-ray diffraction pattern was obtained from powder specimens. It is shown in Figure 9. Here previously reported cell parameters were used to index the observed reflections (Okuyama et al (1988)). Intensities were obtained by micro-densitometer scanning and were corrected for the Lorentz and polarization factors (LP) but not for the absorption effect.

Molecular Model Building

The amino acid composition of the crystalline fraction of *Bombyx mori* silk fibroin is known as Gly:0.48, Ala:0.33, Ser: 0.15 and Tyr:0.01 by fraction (Lucas et al (1957); Strydomet al (1977)). Since Tyr residues locate only at the end of the Cp-fraction and their amount is very small, the ratio of the amino acid in the Cp-fraction may be expressed as Gly:Ala:Ser = 3:2:1. On the other hand, from the sequence analysis of oligopeptides obtained by the partial hydrolysis of the crystalline part, the amino acid sequence was proposed as Gly-L-Ala-Gly-L-Ala-Gly-[L-Ser-Gly-(L-Ala-Gly)*n*]8-L-Ser-Gly-L-ala-L-Ala-Gly-L-Tyr (Lucas et al (1957)), where n is usually 2. Therefore, in this study, a dipeptide, L-Ala-Gly, was used as a

Figure 9. X-ray diffraction pattern of silk-I

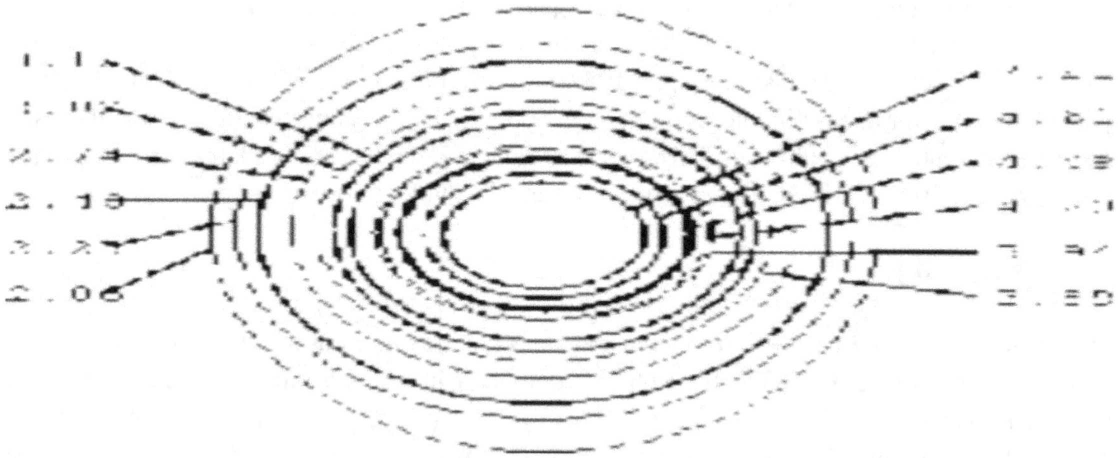

chemical repeating unit of the Cp-fraction at the initial stage of the analysis and a hexa peptide, L-Ala-Gly-L-Ala-Gly-L-Ser-Gly, was used at the final stage. Molecular models having the appropriate helical symmetry and fibre repeating period were generated using a linked-atom description (Smith & Arnott (1978)) with all bond lengths and angles held constant (Momany et al (1975)). These values are shown in Figure 11. As shown in the following section the molecular structure of the silk fibroin has 2/1-helical symmetry. That is, two chemical repeating units of L-Ala-Gly are contained in the fibre repeating period. Since the dihedral angle at the peptide bond (ω) was fixed at 180° in the early stage of the analysis, four dihedral angles, φ and ψ of the glycine and alanine residues, have to be defined in order to construct a molecular conformation. Out of four, however, only two are independent, since two constraints of the helical parameters (unit height, h=4.4 Å and unit twist, θ=180°) reduce the degree of freedom for the

Figure 10. Starting model parameters for Silk-1

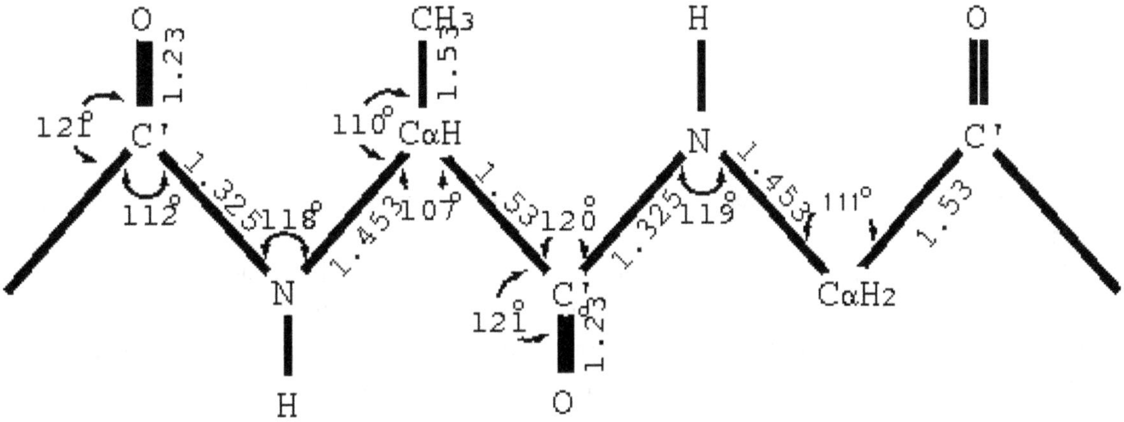

Figure 11. Crystal and molecular structure silk-I

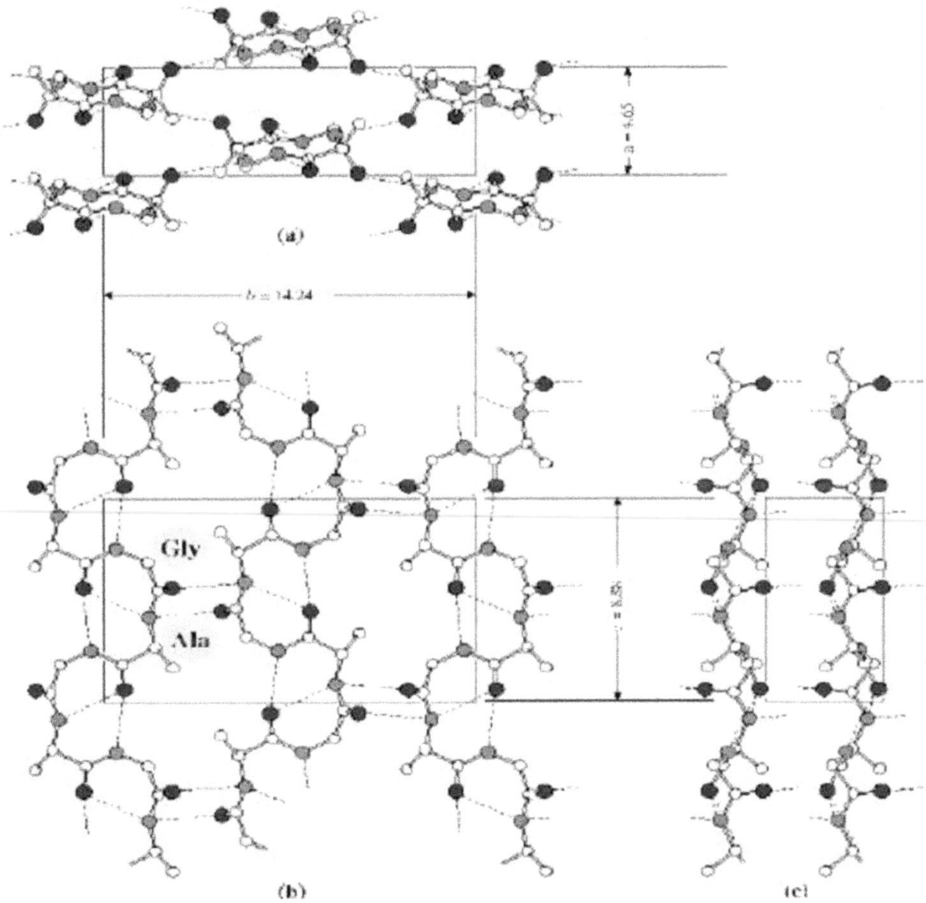

molecular conformation by two. In order to obtain the starting values of φ and ψ, those of the alanine residue were chosen as independent variables and those of the glycine were determined from those of the alanine residue and the helical parameters by using numerical calculation method (Go & Okuyama, (1976)). The molecular conformation must satisfy both sterical and mathematical requirements. The former was fulfilled by choosing φ and ψ within the allowed region of the $\varphi - \psi$-map (Ramachandran et al (1963)) for the alanine residue. The latter requirement was fulfilled by choosing φ and ψ within the region where the main chain could be constructed with an appropriate helical parameters together with bond lengths and angles shown in Figure 11. The overlapping regions with these two requirements were divided by 20° (in the fully allowed region) or 30° (in the allowed region) intervals, which makes 37 grid points. Since two sets of (φ, ψ) values of the glycine residue were usually obtained from each point of the alanine residue, 73 molecular models were built from the above grid points. The investigation was carried out for each of these 73 molecular models. In the analysis of hexapeptide, we have optimized the position of the oxygen atom in serin residue in terms of *R* and *Rw* factors by changing the torsion angle ($O - C\beta - C\alpha - N$) intervals from 0° *to* 360°.

Packing Models and Their Refinement

In the following section, the most plausible space group and the number of a chemical repeating unit of L-Ala-Gly in a unit cell were determined to be $P2_12_12_1$ and four (z=4), respectively. In order to pack four chemical repeating units in a unit cell, the 2/1-helical axis of the molecule must coincide with the crystallographic 2_1- axis parallel to the *c*-direction. For the positioning the molecular model in the unit cell, two additional parameters were used to define the relative axial rotation (μ) and translation (w) along the *c*-axis. In order to obtain the appropriate values for each packing model, μ and w were systematically changed by 20° intervals from 0° to 180°, and 0.05 intervals from 0 to 0.5 respectively. At each stage in the modeling and refinement of the structure, we sought to minimize the quantity Ω in the following least-squares fashion (Smith & Arnott, (1978)).

$$\Omega = \sum w\left(\left|F_o\right| - \left|F_c\right|\right)^2 + S\sum_j \in_j + \sum_h \lambda_h G_h \tag{20}$$

The first summation in Ω ensures the optimum agreement between the observed (*Fo*) and the calculated (*Fc*) X-ray structure amplitudes. *w* is the weight of each reflection. The second ensures the optimization of non-covalent interactions (*j*) S is the scale factor used to adjust the overall weight of the second term with respect to the first. The third imposes, by the method of Lagrange undetermined multipliers (λh), the exact values of constraints (*Gh*) we have chosen.

Atomic scattering factors for calculating structure factors were obtained using this method and values given in International Tables for X-ray Crystallography International Crystallography Table (1974)). Computations were carried out using SUN workstation.

Structure Determination

Figure 9 shows an X-ray powder diffraction pattern from the Cp-fraction of the Bombyxmori silk fibroin. A total of 12 observed diffraction rings were indexed by the rectangular unit cell with the dimensions of a=4.65, b=14.24 and c(fibre axis)=8.88 Å which are determined by using the oriented specimen (Asakuara et al (1999)). The reflections below the observational threshold were also included in the structure analysis. That is, un observed reflections within 2.0Å resolution were assumed to have half of the intensity of the observational threshold. These unobserved reflection data were used only when the magnitude of the calculated structure amplitude became larger than that of the estimated structure amplitude (|Fo|_|Fc|) during refinement. Assuming that the unit cell contains four chemical repeating units of L-Ala-Gly, the calculated density 1.493 gcm^{-3} agrees with the observed value (1.39 gcm^{-3} (Okuyama et al, (1988)). Since the extended length of a chemical repeating unit would be less than 7.5 Å, only two packing models are possible. One is a two-chain model in which the unit cell contains two peptide chains plausibly with a 2-fold helical symmetry, the other is a one-chain model in which one peptide chain is contained four chemical repeating units. The silk-I structure transforms easily to the silk-II structure by the elongation in the solid state. The unit cell of the resultant silk-II crystal contains two antiparallel peptide molecules. Therefore, in the silk-I crystal, molecules should be arranged in an antiparallel fashion, which is ruled out in the one-chain model. Furthermore, the long dimension of b also suggests the two chain model in which two chains parallel to the c-axis are arranged along the b–direction.

Since odd reflections on 0k0 were systematically absent up to 060(Okuyama et al, (1988)), the 21-symmetry along the b– axis was assumed, which is consistent with antiparallel packing of peptide chains of silk-II crystal. On the other hand, only 004 reflection was observed on 00l and no reflection was observed on h00. Therefore, we performed structure analysis based on both P21 as the lowest symmetry space group and P212121 as the highest symmetry space group. Since the detailed refinement calculation for P21 resulted in a very similar structure to that for P212121, we have decided that the most plausible space group of silk-I form is P212121 and only the latter case will be shown in this article. At first, packing parameters μ and w were divided by 200 intervals from 0o to 180o, and by 0.05. In order to get appropriate packing parameters, the discrepancy factors R (conventional R-factor) and Rw (weighed R-factor) and the shortest contact between non-bonded atoms were calculated for every 73 molecular models at each grid point of μ and w. Here, Rw were defined by R =

$$R = \Sigma\left(\left|F_0\right| - \left|f_c\right|\right) / \Sigma\left|F_0\right| and\ R_w = \Sigma w\left(\left|F_0\right| - \left|F_c\right|\right)^2 / \Sigma w\left|F_0\right|^2$$

In this study, the weight of each reflections, w was fixed to 1.0. Out of 73 molecular models, 11 models with R less than 0.3 and the shortest contact larger than 1.6Å, were selected as the candidates for the further calculations. After several refinement cycles, these 11 models could be classified on the basis of φ and ψ angles. The mode l which has fairly good agreement between observed and calculated structure amplitudes(R=0.12, Rw=0.14) was chosen as the most suitable model for silk-I structure. With hexapeptide, the repetitive part of the amino acid sequence in the crystalline part of Bombyx mori silk fibroin is well represented by (L-Ala-Gly)2-L-Ser-Gly. Therefore, the refinement calculations were also carried out for the above classified models with this hexa peptide repeating unit. In these calculations, the hydroxyl moiety with 1/3 occupancies was attached to the Cβ atom of alanine with an appropriate geometry. Refined structures after several cycles are summarized in Table 1.

CRYSTAL AND MOLECULAR STRUCTURE OF SILK-II

In this section, we look at the crystal and molecular structure of a variety of silk fibre (silk-II) known as Bivoltine which belongs to Bombyx mori family.

Table 1. Refinement parameters with (Ala-Gly)2-Ser-Gly repeating unit

RefinedParameters	Final Values
Torsional angle /°	
φAla & Ser	46.94
ψAla&Ser	100.85
ωAla&Ser	179.15
χSer	97.32
φGly	88.14
ψGly	-48.69
ωGly	171.06
Eulerian angle /o	
ϵ_x	91.75
ϵ_y	32.19
ϵ_z	126.79
Other parameters	
S/ Å-	1.3253
μ/°	115.04
w	0.2260
Scale factor	1.588
Attenuation factor	4.488

Materials and Methods

For our study we have used silk fibre belonging to *Bombyx mori* family, which comes under the classification Bivoltine on the basis of shape, colour, denier(thickness of the fibre) and life cycle of the fibres/ cocoons. Bivoltine races are well suited to temperate regions. They produce in majority white cocoons. The cocoons are either oval, ellipsoidal or dumbbell in shape. It is well known that they produce well built cocoons. Generally cocoon weight of Bivoltine races are more. Hence Bivoltines are called as races producing heavy cocoons.

When Bivoltine cocoons are reeled using standard mono-cocoon reeling devices (EPPROVITE) the filament length is longer(1800-2000 meters) and there are bivoltines which produces silk filaments measuring a length 1800-2000 meters.

The denier of bivoltine races ranges from 2.2 to 3.5 denier. Regarding other fibre qualities of bivoltine races, the reelability is greater. The number of knots observed in silk fibre is very less (cleanness test). When white coloured Bivoltine silk fibre is exposed to Ultra violet light it gradually changes to yellowish colour on prolonged exposure. Cocoons were collected from the germ plasma stock of the Department of Sericulture which were then cooked in boiling water (100 machine reeled Bivoltine mulberry raw silk fibre of 25.2 d tex 15tpcm with zero twist) and were degummed with sodium silicate solution having a pH 10 for 1hour at boil and the cocoons were reeled in warm water with the help of mono cocoon reeling equipment EPPROUVITE. The characteristics features of these fibres are that they are white in color with average filament length of 1050 meter and denier being in the range 2.4-2.8. These fibres were mounted on rectangular frame in just *taut* condition, which does not involve any mechanical stretching of fibres. The whole process, starting from reeling to mounting of fibres, does not involve any type of mechanical deformation.

X-Ray Recording

X-ray diffraction patterns of silk fibres were recorded on an imaging plate. In this method, a rotating anode X-ray generator (ULTRA-X, RIGAKU) was operated in a normal focus mode to provide a monochromatic (Mo) beam of wavelength (λ= 0.07107nm), at 50 kV and 250 mA. Diffraction data were recorded on a disk shaped imaging plate with sample to plate distance of 150 mm. The measurement of X-ray diffraction data was implemented by the hardware system DIP100S (MACSCIENCE). The digitized X-ray diffraction pattern was obtained from the imaging plate with 100μm intervals in both x and y directions. The centre of the diffraction pattern, fibre tilt and rotation angle around the X-ray beam were determined by using XFIX software of CCP13 package (http://www.dl.ac.uk/SRS/CCP13). Then, the background subtraction based on the method of Ivanova & Makowski (1998) was proceeded.

This pattern was transformed to a diffraction image in a Cartesian reciprocal space, and then continuous X-ray intensities were sampled at every 0.001Å^{-1} on each layer line according to the corresponding fibre repeating periods of 6.97 Å. The background subtracted X-ray pattern from raw Bivoltine fibre is shown in Figure 12. Using XCONV, XFIX, and FTOREC, we have performed all data processing cited above and LSQINT programs in the CCP134 package. After correcting Lorentz and polarization effect, structure amplitudes were used as continuous X-ray data in the Linked-Atom-Least-Squares calculation. By inputting unit cell parameters, the space group symmetry and the profile parameters, further processing of the pattern is carried out. In fact there are six profile parameters. This includes Hosemann's

Figure 12. Background-corrected X-ray diffraction pattern of raw Bivoltine silk fibre

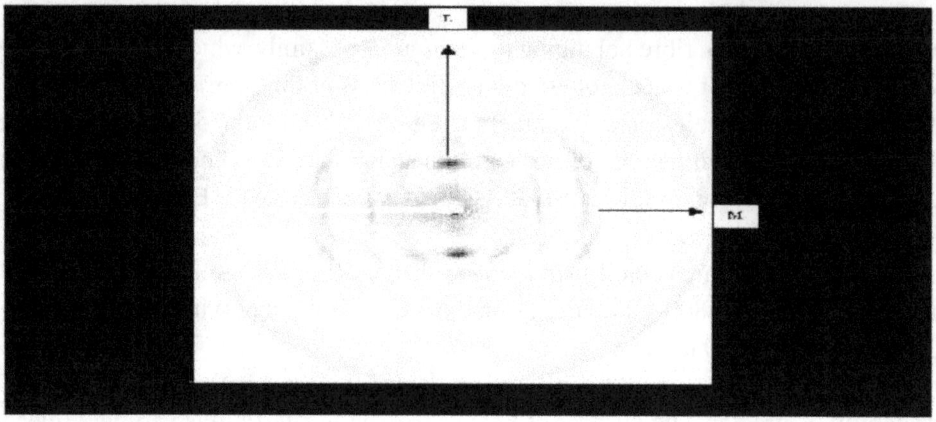

paracrystallinity, which is defined as the extent of disorder of second kind present in the lattice. The six parameters are varied to generate the profiles and then to fit these profiles to the observed pixel intensities by the maximum entropy of Skilling and Bryan(1984) where the R-factor is given by

$$R = \sqrt{(\pounds\left(Po - Pc\right)^2 / \pounds\left(P_0^2\right))}$$

Where *Po* and *Pc* are the observed and calculated pixels respectively. Here, *Po* is the background corrected observed pixel values. The R-factor for the pattern fitting in this case was less than 20%. Here it should be noted that the procedure is used to simulate equatorial plot do separate the overlapping reflections.

Molecular Model

The amino acid sequence in Silk-II was *Gly–L–Ala–Gly–L–Ala–Gly–[L–Ser –Gly –(L–Ala–Gly) n]*$_8$ *–L–Ser–Gly –L–Ala–L–Ala–Gly–L–Tyr* (Strydom et al (1977)) where n is usually 2. Therefore in this study a hexa-peptide,L-Ala-Gly-L-Ala-Gly-L-Ser-Gly was used. Molecular models having the appropriate helical symmetry and fibre-repeating period were generated using a Linked – atom description with all bond lengths and angles held constant (Momany et al (1975) & Smith & Arnott (1978)). These values are shown in Table 2. The molecular structure of the silk fibroin, with three amino acids Ala, Gly and Ser has a 1/1-helical symmetry. That is, two chemical repeating units of Ala-Gly are constrained in the fibre-repeating period. Actually there is an option in Linked-Atom-Least-Squares(LALS) program to indicate two molecular chains by defining the constraints for only one of them. Here, the geometry for Ala is for the D-Ala and not L-Ala. The molecular conformation must satisfy the both sterical and mathematical requirements. Hence, we have chosen initial φ and ψ for Ala and Gly residues to be the same as that of Sangappa et al (2004) and the main chain was constructed with an appropriate helical parameters together with bond lengths and angles shown in Figure 3.

Packing Models and Their Refinement

In the following section, the most plausible space group and the number of a chemical repeating unit of (Ala-Gly)2-Ser-Gly in a unit cell were determined to be $P2_1$ and four (Z=4), respectively. The symmetries and four molecular sites in the unit cell with the space group $P2_1$ are shown in Figure 17. The molecules at 1(1') and at site 2(2') are symmetrically independent of each other and form a sheet structure parallel to the ac-plane by hydrogen bonds. Two molecules at sites 1 and 2 can be related to the molecules at sites 1' and 2' by the 2-fold screw symmetry, respectively. Takahashiet al. (1999) have shown that there are four models for the sheet structure formed by hydrogen bonds. They are: (i) polar-antiparallel [PA (1)]; (ii) polar-parallel(PP);(iii) antipolar-antiparallel (AA); and (iv) antipolar-parallel (AP). In the polar model, the methyl groups of alanine residues are on one side of the sheet only, while in the antipolar model, the methyl groups alternately point to both sides of the sheet along the hydrogen bonding direction (Takahashi et al (1999)). The crystal structure proposed by Marsh et al (1955) corresponds to PA sheets. In order to pack four chemical repeating units in a unit cell the helical axis of the molecule must coincide with the c-axis, which is parallel to 21-axis. For the positioning the molecular model in the unit cell, two additional parameters were used to define the relative axial rotation μ and translation ω along the c-axis. At each stage in the modeling and refinement of the structure, we minimized the quantity Ω in the following least-squares fashion Smith and Arnott (1978).

The first summation in Ω ensures the optimum agreement between the observed (F*o*) and the calculated (F*c*) X-ray structure amplitudes given in the first section. The *w* is the weight of each reflection. Computations were carried out using a Linux based PC. We have compiled the software LALS for execution using LINUX operating system.

Molecular and Crystal Structure for the Chemical Repeating Unit of (Ala-Gly)2-Ser-Gly

The refinement was carried out for the crystal structure in which two antiparallel molecules related by a 2-fold rotation axis parallel to c-axis. In order to get appropriate packing parameters, the discrepancy factors R*c*(conventional R factor) and R*w* (weighted R*c* factor) and shortest contact between non-bonded atoms were calculated. Here *Rc* and *Rw* are as defined earlier. Initially a di-peptide Ala-Gly was used as chemical repeating unit for refinement and later on a hexa-peptide (Ala-Gly)2-Ser-Gly was used at a later stage. In this study, the weight of each reflection, *w* was fixed at 1.0. The refinement gave far better agreement between the observed and calculated structure factors with the consideration of serine residue by replacing the 1/3 of alanine in the peptide chain. The R-factor of bivoltine turns out to be 21.2% . The refined parameters and the final values are summarized in Table 4.1. Here, the azimuthal angles of the two molecules forming a sheet were also refined independently. The internal rotation angle C(=O)-C(α)-C(β)-O(H) of serine residue was refined. Table gives the comparison between the observed and calculated structure factors. The crystal structure of Bivoltine silk fibre in raw form is shown in Figure 15.

Result and Discussion

The internal rotation angles ω of the glycine and alanine residues about N-C (=O) are 169.4^o and 177.4^o respectively, which are nearly trans conformation. The internal rotation angles φ and ψ for the glycine

Figure 13. (a) Fibre axis projection of crystal structure of stretched silk fibre obtained by Marsh et al (1955). (b) Down the fibre axis projection of untreated and unstretched pure Mysore silk fibre obtained by Sangappa et al (2003). (c)Down the fibre axis projection of untreated and unstretched Bivoltine silk fibre obtained using the method described in the text.

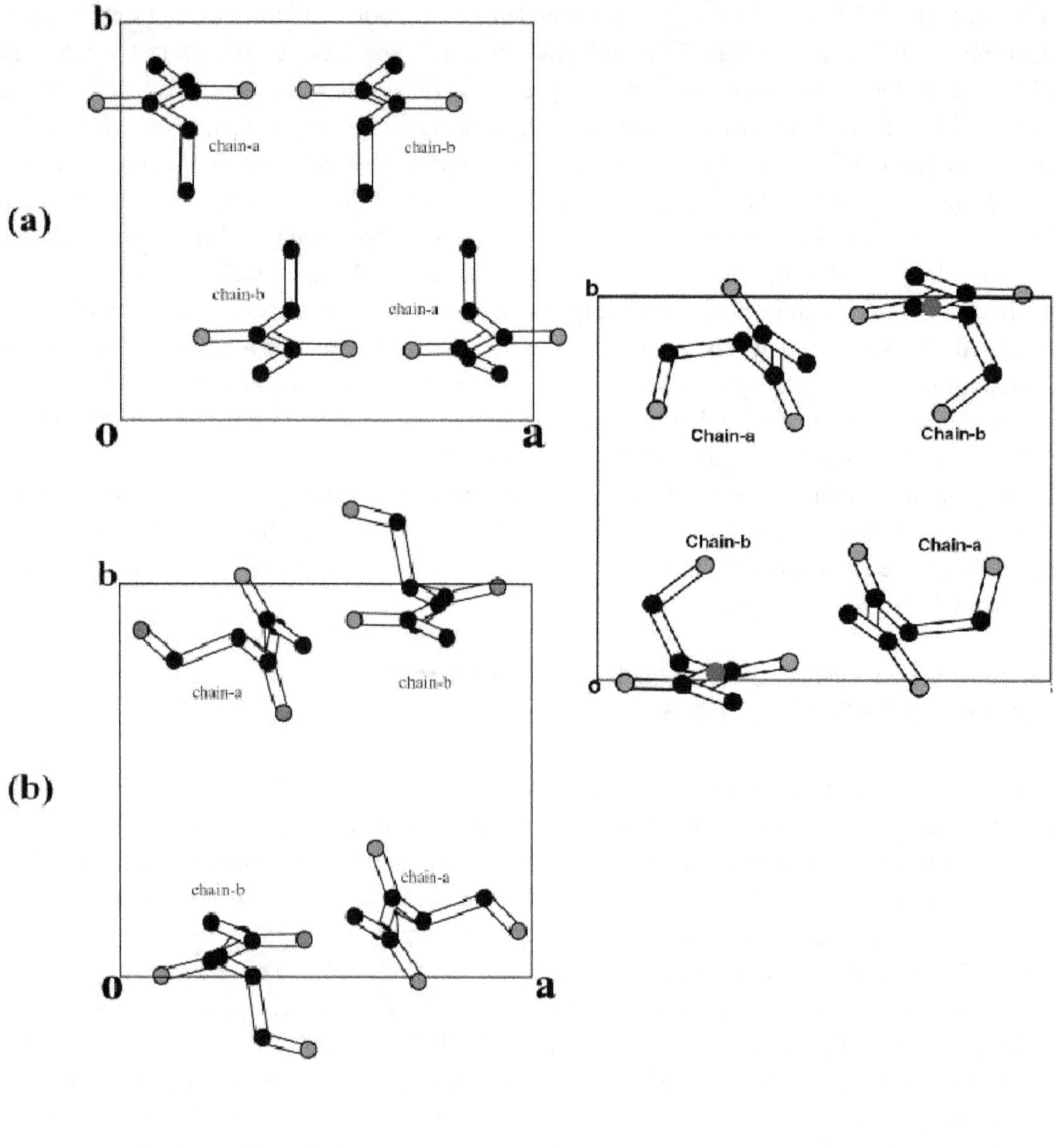

residue are -144.4*o* and 155.9*o* respectively which are between skew and trans conformations. The internal rotation angles φ and ψ for the alanine residue are -158.9° and 151.8° respectively which are also between skew and trans conformations. We have observed the torsion angles for Gly (-144.64° and 146.48° and Ala (-149.99^{o}) and 143.68^{o}) for pure Mysore silk fibre of unstretched one. Marsh et al(1955) have reported torsion angles for Gly (-139,140) and Ala (-139) for the Silk II with mechanically oriented and treated samples. Torsion angles (φ and ψ) of Silk II structure using energy calculations of Fossey's model (1991) have been reported for Ala (-149,148) and for Gly (-150,146) whereas on the basis of solid state

Table 2. Refined details of Silk-II with (Ala – Gly)2 – Ser – Gly repeating unit

Refined Parameter	Pure Mysore Silk Fibre	Bivoltine Silk Fibre
φAla	-147.99	-158.9(-139)
ψAla	143.68	151.8(140)
ωAla	178.62	177.4(176)
φGly	-144.64	-144.4(-139)
ψGly	146.48	155.9(140)
ωGly	178.19	169.4 (176)
χser	79.6	74.3
Eulerian angle (*o*)		
ϵ_x	-88.06	-90.1
ϵ_y	128.81	134.6
ϵ_z	-167.87	-170.9
Other parameters		
S(Å)	42.10	42.10
μ(°)chain-a	167.31	167.30
μ(°)chain-b	133.07	133.3
u(a);v(a);w(a)	0.1657;0.0042;0.2113 0.181;0.004;0.21	
u(b);v(b);w(b)	0.7018;0.7712;0.3018	0.715;0.771;0.302
Scale factor	2.783	1.28
Attenuation factor	-8.755	-6.245
Hydrogen Bonds		
Intra-molecular		
Distance (Å); angle(°)	N(Ala)... O(Ala)	N(Ala)... O(Ala)
	(2.74,165.2)	(2.70,170.81)
	N(Ala)... O(Ser)	N(Ala)... O(Ser)
	(3.1,136.0)	(2.99,126.6)
	O(Gly)...N(Gly)	O(Gly)...N(Gly)
	(2.7,161.5)	(2.65,156.7)

continued on following page

Table 2. Continued

Refined Parameter		Pure Mysore Silk Fibre		Bivoltine Silk Fibre
		O(ser)...O(Ala)	O(ser)...O(Ala)	
		(2.8, 89.5)	(2.98,86.89)	
R-factor				
Including unobserved	R*c* 0.168		0.212	
Reflections	R*w* 0.172		0.205	
Contact sigma(σ)	2.35E+03		2.59E+03	

NMR for Ala (-140,142) and for Gly (-139,135) (Demura et al (1998)). In NMR studies, the samples were not subjected to any mechanical treatment. The stereo chemical energy which is represented by σ of Bivoltine silk fibre turns out to be 2.59E+03 is little high compared to the pure Mysore silk fibre, 2.35E+03 .

Here σ is given by the sum of the second term of the equation (Okuyama et al (2001)). Essentially, this justifies that pure Mysore silk fibre is more stable compared to Bivoltine in respect of durability observed in textile industry. The torsion angles and Eulerian angles are the same for chain-a and chain-b as they are symmetric. In the network, the O and N atoms of glycine are at 2.65Å and the bond angles of 156.7°, whereas the intra chain hydrogen bonding, N – H···O and N-H···O_{ser} in alanine are at bond distances of 0.270 nm and 0.299 nm respectively. The corresponding bond angles are 170.81° and 126.6°. Further the intermolecular distances between O atom of serine residue and O atom of alanine residue is 0.298 nm and the bond angle is 86.89°. This may suggest that the OH groups of the serine residues are associated with the hydrogen-bonding network forming the sheet structure. In crystal structure down the c-axis (i.e. fibre axis) of Bivoltine fibre, shown in Figure 21 (Spek, 2003), indicates that chains 2 and 2' are shared and twisted along the fibre axis. In fact, for PMS it is around 22° and for Bivoltine it is 7°. In a unit cell, the position of the C=O group of chain along C-axis is oriented in one direction at one position and in opposite direction in another position. Such an arrangement results in a dipole, which stabilize the crystal lattice of bivoltine fibre in this study.

CRYSTAL STRUCTURE OF COLLAGEN

Crystal structure of collagen was re-visited by Okuyama (1978,2008) and the work which uses the software's available at CCP13 site, has been briefly reproduced here. The Rich and Crick model is widely accepted as the molecular structure of collagen. This model was first proposed in 1955 and its detailed structure was presented in 1961, based on the X-ray fibre diffraction patterns from native collagen (Ramachandran et al (1961)). It is made up of a triple helical structure in which each peptide strained forms a right-handed 10/1-helical conformation and a pitch length of 85.8Å. This triple helical structure is considered to be a single left-handed 10/3-helix with a 28.6Å axial repeat unit. With a model peptide Okuyama (1978,2008) found a left-handed 7/2-helical conformation with 20Å axial repeat in the single crystal structure of a collagen model peptide. There are many such single crystal structures of collagen model peptides which have been analyzed at high resolution using the method of protein crystallography.

Such studies have provided valuable information about collagen structure. It is significant to note that all the main chain conformation found in these peptide crystals generally maintain left-handed 7/2-helical symmetry. To demonstrate that both the 10/3 and 7/2 helical models can explain collagen fibre diffraction data, both models have been refined against the same diffraction data from collagen to compare their structural viability. For this purpose, the process of digitized X-ray data was carried out with software available at CCP13 package (http://www.dl.ac.uk/SRS/CCP13) as described earlier while discussing the structure of Silk-II.

Using fibre diffraction method, it has been found that the R-factor with 7/2-helicalmodel was somewhat lower than or at least the same as, that of the 10/3-helical model. This approach gave quantitative data by refining both the helical models against continuous data from the same X-ray diffraction pattern from the native collagen.

Figure 14. X-ray diffraction pattern of Collagen

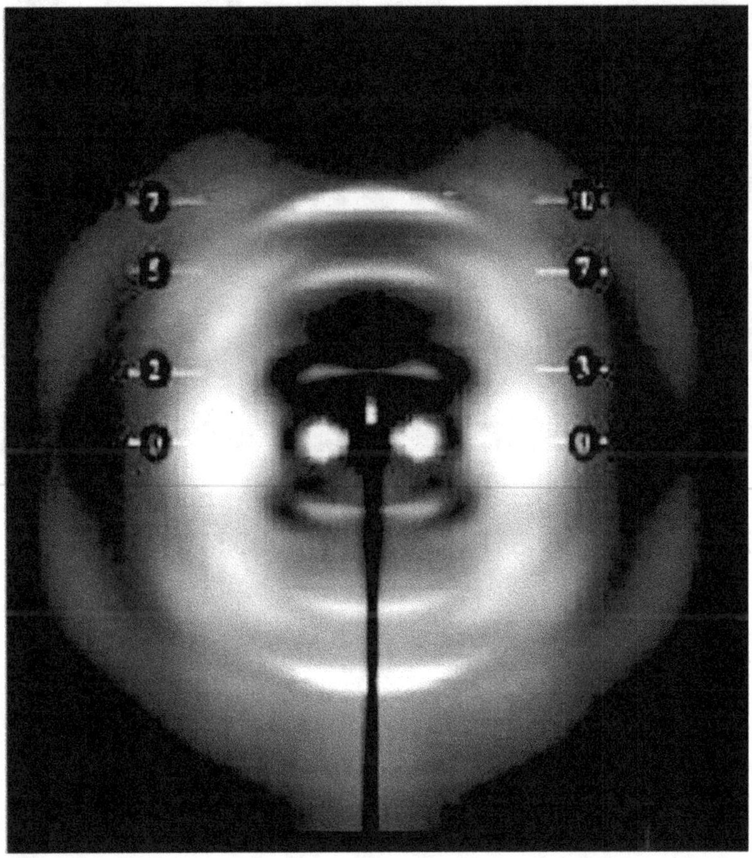

Figure 15. Refined peptide conformations of the 7/2- (a) and 10/3- helical models (b). final dihedral angles in degree of Gly-X-Y Triplet of the 7/2- and 10/3- helix are given in Table 3.

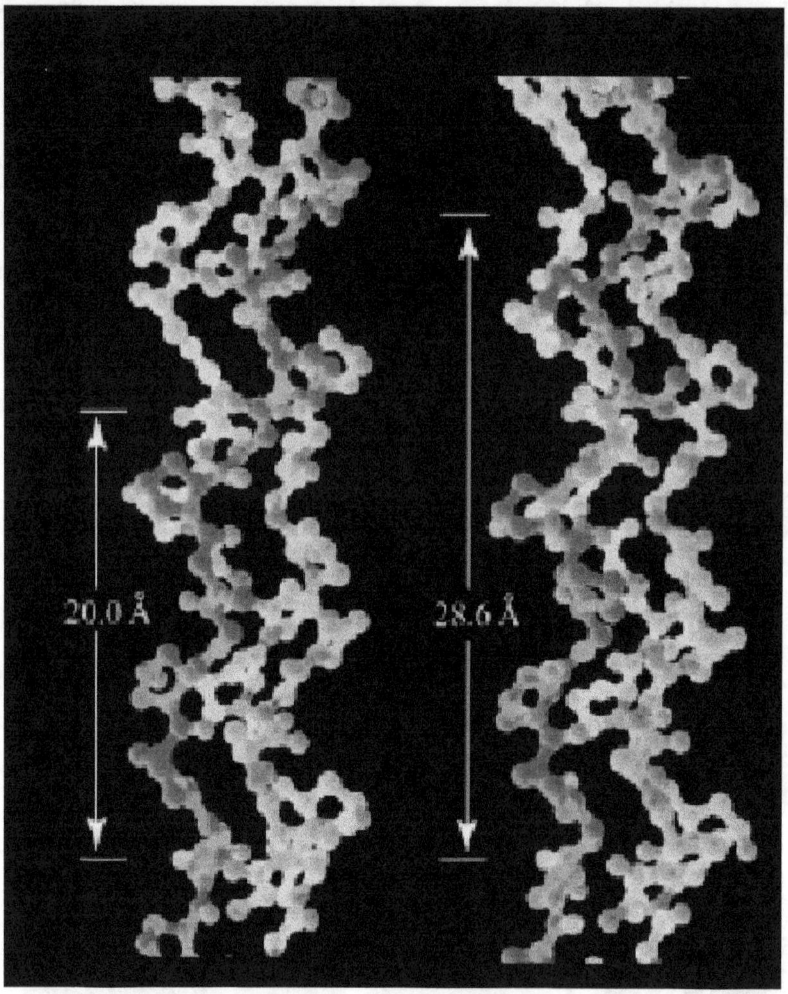

CRYSTAL AND MOLECULAR STRUCTURE OF COTTON FIBRES

Cotton fibre belongs to a family of "Gossypium" (Brubaker (1999)). It has been used for more than 3000 years because of its versatility and needs less cost to produce finished products. Cotton is essentially pure cellulose. The structure-property relation of cotton is considered to be important in textile industry. The elastic constants of cellulose were computed by assuming molecular models of Jaswon (1968). In

Table 3. Dihedral angles of Gly-X-Y Triplet

	φ(gly)	ψ(gly)	ω(gly)	φ(X)	ψ(X)	ω(X)	φ(Y)	Ψ(Y)	ω(Y)
7/2	-75.71	176.28	173.50	-77.94	166.13	175.83	-60.26	163.44	179.99
10/3	-78.48	147.06	-165.60	-67.56	147.70	-171.54	-69.04	155.44	-166.9

this model, the strains present along the chain depend on the deformation of primary (covalent) bonds and hydrogen bonds. A comparative study of this model has been carried out by Eichhorn and Davies (2006). We have investigated the crystal structural features of these fibres, using LALS technique.

Material and Methods

Raw cotton is plucked manually after ripening. This cotton is freed from the buds and taken for ginning. Ginning is a process where raw cotton is separated from seed, seed hulls and other small objects. This process must be done before cotton fibres can be used for studying. Depending on the varieties of cotton, ginning machine is employed. After removing the seed it is taken for baling till it became as lint. The lint raw cotton samples are used for our study (Dch32 and Dharwar) without any additional treatment.

X-Ray Diffraction Measurement

The sample is taken as small bundles and clamped into a sample holder. The sample holder is then mounted on a goniometer such that the rotational axis is parallel to fibre axis and perpendicular to X-ray beam. Here, we have used an imaging plate system (Dip-3200) with dimensions (440/240)mm^2, which is moved parallel to the axis of rotation. The wavelength of X-ray radiation is $\lambda = 0.71073Å$ (Mo target).

The X-ray generator (RIGAKU) settings are 50kV and 32mA. The time of exposure is kept at 400s and X-ray diffraction pattern of cotton fibre (Dharwar and Dch32) is recorded using Weissenberg geometry. The diffraction pattern is given in Figure 16. Equatorial scan of intensity versus 2θ was carried using the "MOSFLM" software supplied with the instrument. This data is analyzed using Peak-Fit software(Chen et al 2004)). Background intensity is subtracted during peak fitting procedures. For peak fitting purpose, a Gaussian deconvolution is selected as peak profile of X-ray diffraction line of cotton fibres. The position of each diffraction line is calculated by resolving the symmetric peak position and is identified from residual peak-fitting procedure of Peak-Fit. The 2θ positions are computed by calculating the mean and standard deviation of each diffraction profile Cell parameters and space group of crystal system of cotton fibre are determined by using a "CHEKCELL" a freely available program. "CHEK-

Figure 16. X-ray diffraction photograph of cotton

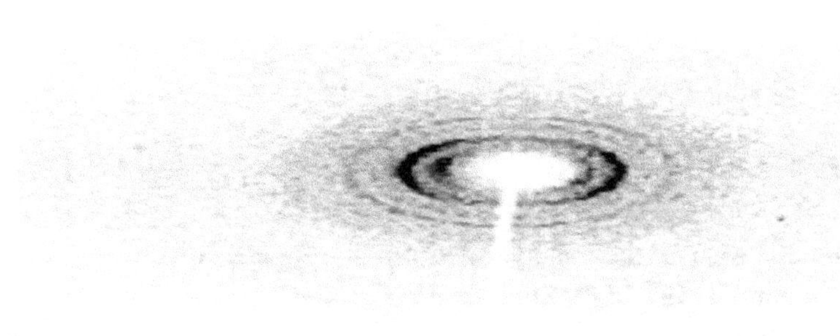

Figure 17. Standard starting bond angles and bond lengths for cotton fibres

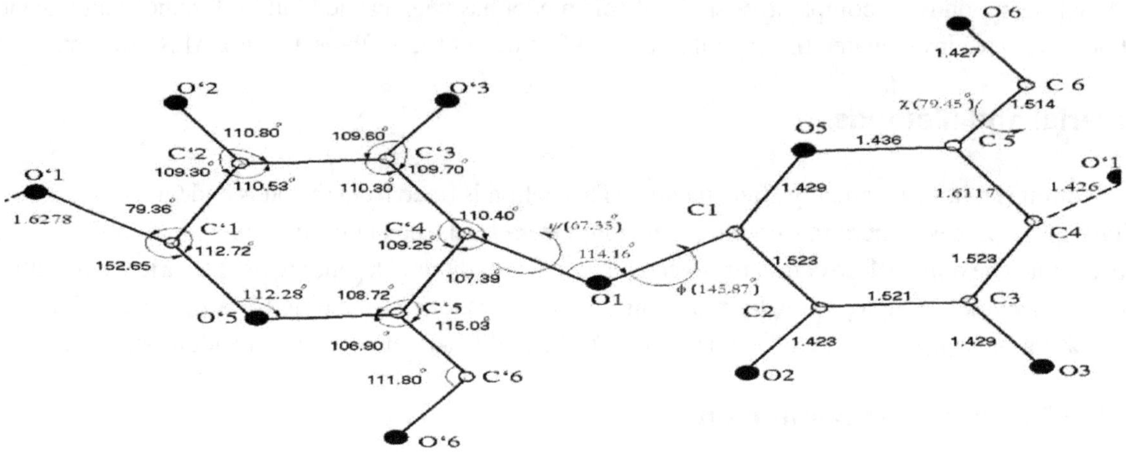

CELL" program is executed until a set minimum standard deviation is achieved. We could identify 11 reflections. Observed reflections are indexed by a unit cell with dimensions $a = 4.49$Å; $b = 6.19$Å; $c = 5.36$Å and the space group is $P2_1$.

The Molecular Model

It is well known that cellulose is a linear polymer of $\beta(1-4)-D-$glucopyranose units in 4C1 conformation (Joel (2002)). There are two types of cellulose I which are cellulose Iα and cellulose Iβ, but for higher plants such as cotton fibre the dominant is cellulose Iβ . The chair conformation of 4C1-linked glucopyranose residues stabilizes the structure. The flexibility of polysaccharide chains depends on the ease of rotation around this glycosidic linkage. To find molecular structure of cotton fibres, we have considered molecular models having the 2/1 -helical symmetry and the fibre-repeating period of 10.34Å together with the pyranose ring in the standard with all bonds lengths and angles. There are two dihedral angles (φ, ψ) between two adjacent pyranoserings, which are called glycosidic linkage angles, and there is a dihedral angle (χ) to define the orientation of O6 oxygen atom (French and Johnson, 2004). There are three possible values of O6, depending on the orientation of this atom related to pyranose ring, *gauche-gauche*, *gauche-trans* and *trans-gauche*. The molecular model of cotton fibres with standard bond lengths, bond angles and dihedral angles is shown in Figure.

The Crystal Structure Packing

There are two molecular chains with 2/1 helical symmetry in a unit cell with space group $P2_1$ (French (2004)). These two independent polymer chains are taken in parallel positions and pointing in the same direction in the unit cell. To obtain appropriate packing models, the fractional coordinates (u, w, v) of the origin atom (C1) were refined with stepwise procedures and the one which gives minimum standard deviation was used (Okuyama et al (1997)). Here, u and v are the positional parameters of polymer chain in the ac-plane; *w* is the height of the origin along the *b*-direction and the azimuth angle (μ) around the molecular axis. The Eulerian angles (*_x, _y, _z*), the three angular parameters that specify the orientation of a molecule with respect to reference axes, were also determined using stepwise method. Other required packing parameters were also determined.

Linked Atom Least Squares (LALS) Program

Molecular and crystal structure was generated by using LALS (Linked Atom Least Squares) method and the computations were carried out using a Linux based PC(Smith & Arnott (1978)). We have compiled the software LALS for execution using LINUX operating system. Actually there is a method to pack the two polymer chains with 2/1-helical symmetry in a unit cell with space group $P2_1$, that is by locating the polymer chains at special positions on the crystallographic axes at (u=0.25, v=0.0) and(u=0.75, v=0.0). The azimuthal angles $\mu 1$ and $\mu 2$ are searched within the asymmetric unit by calculating R-values for the equation 1.1.2.12. The two independent polymer chains are in the same directions. The next step, we have searched the positional parameter *w* of the polymer chain along the *c*-direction. Then, this model was refined against all reflections, including unobserved ones. We have observed 15 reflections, of which 13 reflections have been identified conclusions in LALS program. The weight of the reflection (w) was fixed to 1.0 for observed reflections. R and *Rw* were used to evaluate the agreement between observed and calculated structure amplitudes. We have carried out refinement of molecular and crystal structure parameters to minimize the difference in structure amplitudes of experimental *Fobs* and simulated *Fcal*

Figure 18. Cotton crystal structure showing weak hydrogen bonds

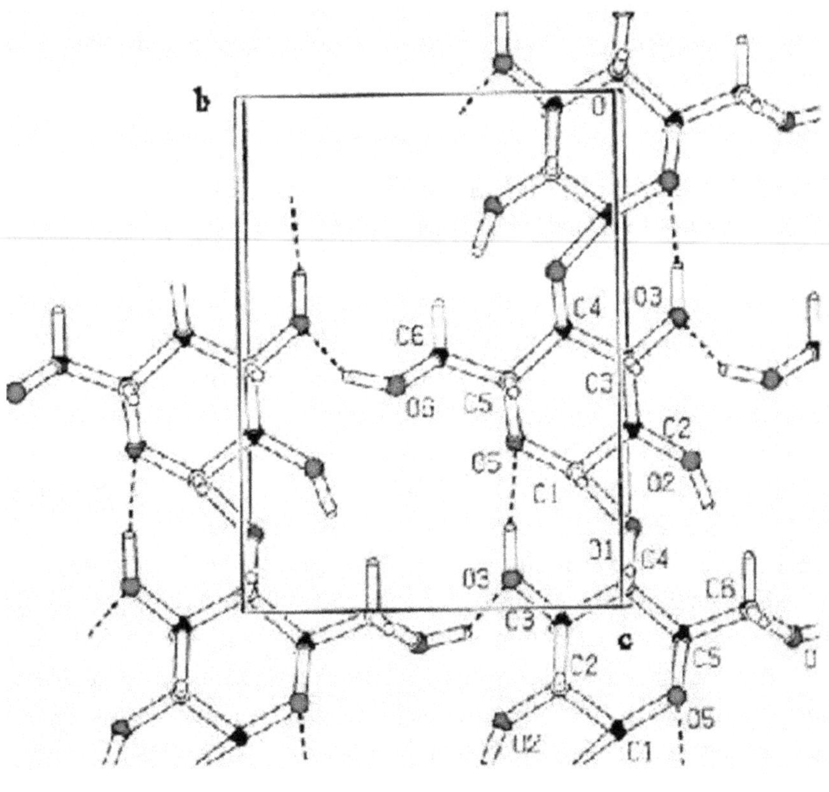

Table 4. Final model parameters of cotton fibre after several refinements using experimental

Intensities	Torsion angles (in deg.)
φ(C2-C1-O1-C'4)	144.52
ψ(C1-O1-C'4-C'3)	62.81
Bond angle at Glycosidic Linkage,(deg.)	
τ (C1-O1-C'4) 116.20	
Orientation of O6 (in deg.)	
χ(O5-C5-C6-O6) 8.73	
Packing parameters of cellulose chains	
$\mu 1$ (in deg.)	127.89
$\mu 2$ (in deg.)	96.04
$w1$	0.832
$w2$	0.434
scale factor	0.664
attenuation factor	-31.43
R- factors (in %)	
R	22.21
Rw	22.68

in terms of R and *Rw* factors as given earlier. Atomic scattering factors for calculating structure factors were obtained by using the method and values given in the literature mentioned in the earlier sections. Final values of refined parameters are listed in Table 4. The packing structure is shown in Figure 17 in which hydrogen bonds are denoted by dashed lines.

WHOLE POWDER PATTERN FITTING METHOD FOR THE ANALYSIS OF X-RAY PATTERNS FROM POLYMERS

Line Profile Analysis is a powerful method to investigate the microstructure parameters like crystal size ($<N>$), lattice strain (g) and stacking faults in polymer materials (Mittemeijer (2003); Snyder et al (1999); Scardi & Leoni (2006); Welberry and Butler (1994); Warren & Averbach (1954); Langford (1964, 1968)). Currently there are several software packages available commercially to extract microstructural parameters by whole powder pattern fitting programs (Ribarik et al (2001); Dong & Scardi (2000); Scardi & Leoni (2002)). Recently IUCr conducted Round Robin test to evaluate the different procedures used to determine the crystallite size and lattice strain present in the material (Balzar et al (2004)). All these approaches, in principle, are based on multiple order method proposed by Warren and Averbach (1954). A majority of these methods employ simple single distribution function for whole pattern and average set of microstructural parameters are computed. In most of the materials we have observed in our investigation, the crystal size parameters change with directions. We have been emphasizing in all our investigations that individual line profile analysis or single order method (Hall & Somashekar (1991); Somashekar et al (1989)) is a reasonably good approach to obtain microstructural parameters. The Keijser et al (1982) have suggested a single order method employing Voigt function and using integral Breadth method. For silk fibres or for that matter, any natural fibres, it is very rare to find multiple reflections. So, to find the finer details of microstructure, we have reported methods by approximating the size profile by simple analytical asymmetric functions (Hall & Somashekar (1991)). Another advantage of this method is that the distribution function is different along different directions. whereas (Scardi & Leoni(2001); Pop & Balzar (2002)) single crystal size distribution function is used for the whole pattern fitting which we feel, may be inadequate to describe silk fibre diffraction patterns, but adequate enough to quantify the stacking faults in metal oxide compounds. Here, we would like to emphasize that the Fourier method of profile analysis(single order method used here) is quite a reliable one as per the recent survey and results of Round Robin test conducted by IUCr (Balzar, (2002)). In fact for refinement, we have also considered the effect of background by introducing a parameter, in a manner similar to what has been reported in our earlier paper (Somashekar et al(1989).In this article, we have computed crystallite size, lattice strain and stacking faults present in silk fibres using single order method. Essentially the peak broadening and peak shifts in these fibres are due to deformation (stacking) and twin fault probabilities along with crystallite size parameter. Using these computed parameters we have simulated the whole pattern observed in three non-mulberry silk fibres Tassar, Muga and Eri, and the necessary procedure for doing this is given in this paper. The extent of variation in the parameters obtained from single order method while refining against whole pattern is also discussed.

The Theory

The contribution of crystallite size, lattice strain and stacking faults to a Bragg reflection profile can be written as (Scardi & Leoni (2002)),

$$Ihkl\left(shkl\right) = \int_{\infty}^{-\infty} T^{IP}\left(nd\right) e^{\left[2\pi i\zeta(nd)\right]} \mathrm{e}^{\left[2\pi i\varphi(nd)\right]} \mathrm{e}^{\left[2\pi indshkl\right]} d\left(nd\right) \tag{21}$$

where, *TIP* is the Fourier transform of instrument profile, $e[2\pi i\zeta(n)]$ is the average phase factors due to crystallite size and $e[2\pi i\varphi(n)]$ is due to lattice distortion and stacking faults. $L = nd$ (with $d = dhkl$) is the column length. This equation can also be written in the form of Fourier series as

$$Ihkl\left(s\right) = \sum_{\mathrm{n}=-\infty}^{\infty} A_{hkl}\left(n\right) \cos\left\{2\pi n d_{hkl}\left(s - s_{o}\right)\right\} \tag{22}$$

where *Ahkl*(*n*) are corrected Fourier coefficients with the Fourier coefficients of instrumental profile function *TIP* (*nd*) using the method of Stokes (1948). Here afterwards, we refer to crystallite size in terms of the number of unit cells measured in a direction perpendicular to the Bragg plane with a notation _N_ and the crystallite size in Å is given by $Dhkl = \langle N \rangle d_{hkl}$. These Fourier coefficients *Ahkl*(*n*) are functions of the size of the crystallite, the disorder of the lattice and stacking faults coefficients,i.e.

$$A_{hkl}\left(n\right) = A^{s}{}_{hkl}\left(n\right).A^{d}{}_{hkl}\left(n\right).A^{F}{}_{hkl}\left(n\right) \tag{23}$$

Fourier analysis of a Bragg reflection profile must always be performed (Warren, (1969)) over the complete cycle of the fundamental from $d(s-s0)$=-1/2 to +1/2, which is rarely possible, experimentally. We do this analysis with the available truncated range by introducing truncated correction (Somashekar et al (1989)). For a paracrystalline material like silk fibre, with Gaussian distribution the expression *Adhkl*(*n*) has been given in earlier references (Warren & Averbach (1954); Hall & Somashekar (1991); Somashekar et al (1989); Warren (1969); Warren (1959)). Using the exponential distribution function for column length, we have reported an expression for *As hkl*(*n*) in our earlier paper (Hall & Somashekar (1991); Somashekar et al (1989)).

Warren (1969) has given an integral analysis for deformation faults and twin faults in various crystal systems. According to this paper, the shift, broadening and asymmetry of the profile are proportional to these fault densities. Here stacking fault density is defined as the chance of finding a stacking fault between any two adjacent layers causing a Bragg reflection and is denoted by αd. αd is normally expressed in percentage and the average number of Bragg planes between stacking faults is given by $1/\alpha d$.

The twin fault probability β is defined as the chance of finding a twin fault between any two adjacent (hkl) layers and the average number of (hkl) layers between twin faults is $1/\beta$. With these aspects, Veltrop et al (2000) has obtained an equation for Fourier coefficients *AF hkl*(*n*) in terms of the deformation faults (αd) and twin faults (β) probabilities as

$$A^{F}{}_{hkl}\left(n\right) = (1 - 3\pm^{\mathrm{d}} - 2\beta + 3\left(\alpha^{d}\right)^{2} \left|\left(\frac{1}{2}\right) n d_{hkl} s \left(\frac{L_0}{h_0^2}\right) \tilde{\mathrm{A}}\mathrm{L}_0\right| \tag{24}$$

where $L_0 = \mathrm{h} + \mathrm{k} + \mathrm{l}$, $h^2{}_0 = h^2 + k^2 + l^2$.

We have assumed σL_0 to be positive for all reflections studied here.

The whole powder pattern of silk fibre was simulated using individual Bragg reflections represented by the above equations using

$$I\left(s\right) = \sum_{\mathrm{hkl}} \left(\acute{\mathrm{E}}_{\mathrm{hkl}} \mathrm{I}_{\mathrm{hkl}} - \mathrm{BG}\right) \tag{25}$$

where ω_{hkl} are the appropriate weight functions for the hkl Bragg reflection. Here s takes the whole range of X-ray diffraction of the sample. BG is an error parameter introduced to correct the background estimation.

We have used an exponential function for the crystallite size distribution and evaluated the microstructural parameters using the equations (1), (2) and (3). Using these equations we have initially computed the parameters N, α, background correction factor and g for individual Bragg reflections. For this purpose we have utilized SIMPLEX program (Press et al (1986)). Then, we have carried out the whole pattern fitting by introducing weight factors for the individual profiles and also taking into effect the average stacking faults by considering equations (4) and (5) in the final stage of refinement with the whole experimental diffraction data of the sample. Computational procedure is as given in the flow chart (see figure). The reasons for the observed variation in the microstructural parameters given in Table 1 and Table 2 are due to a two-fold refinement. First, we have carried out the line profile analysis of extracted profiles from overlapping regions, which is the standard procedure to compute microstructural parameters.

Figure 19. Flow chart for the computation of microstructural parameters

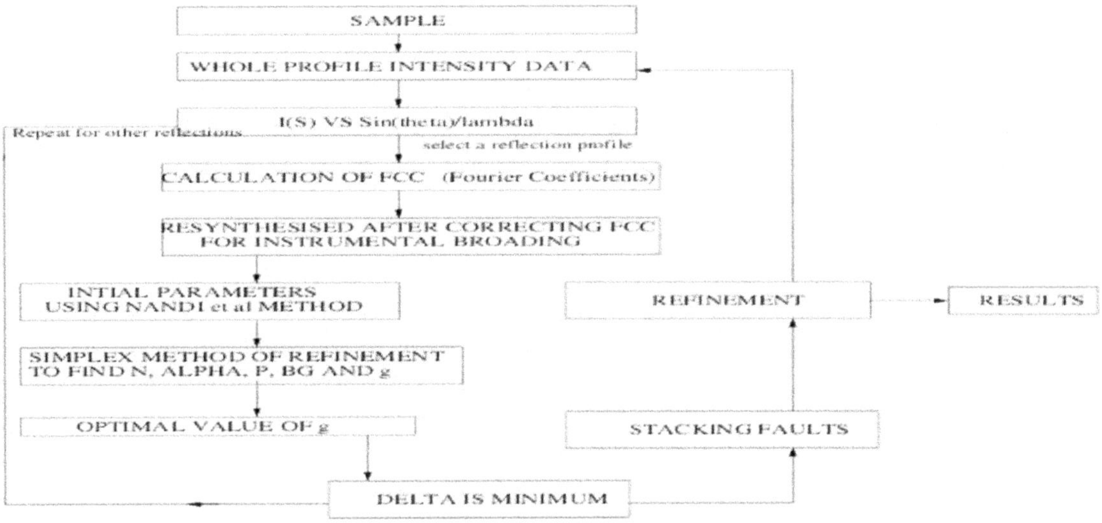

Figure 20. Simulated and experimental profiles for Tassar, Muga and Eri silk fibres along with the differences

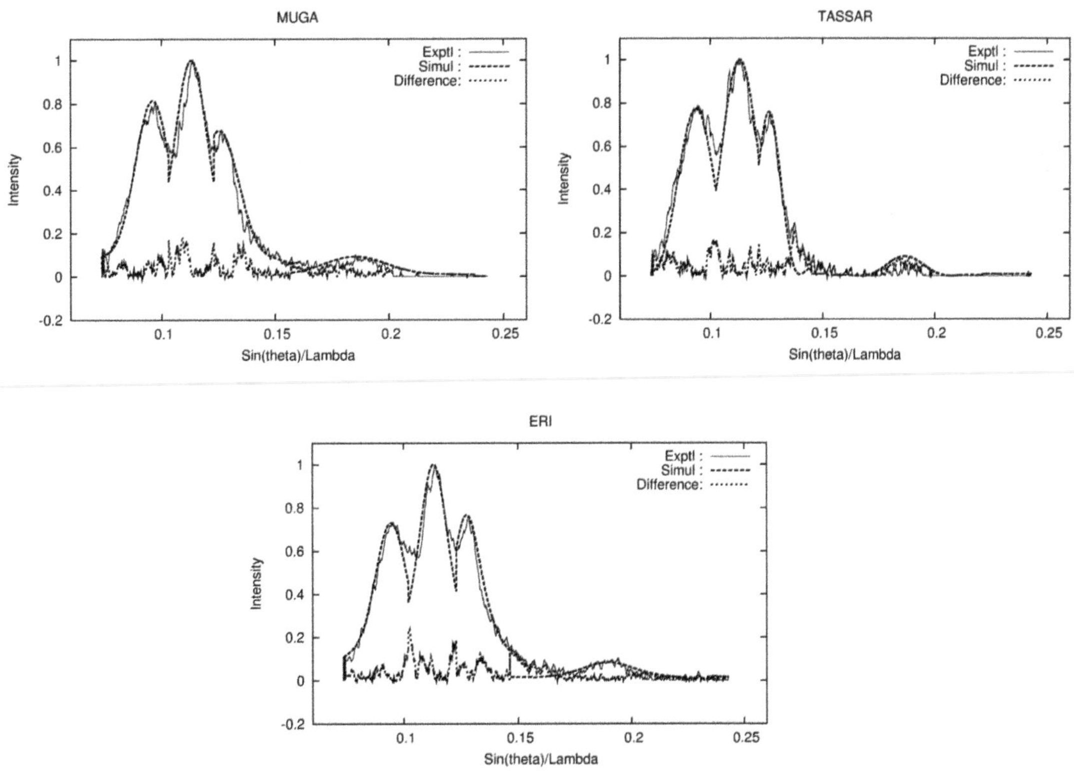

Secondly, the range of overlapping regions, do determine the extent of broadening of the reflections. In fact, the broadening may decrease, if the reflections are close by and hence result in increase in the crystallite size values. A closer look at the results in Table 1 and Table 2 and also the whole powder pattern indicates such a problem. It is also worth noting that none of the other parameters like lattice strain, stacking fault probabilities varied much during the refinement against the whole pattern data of silk fibres.

Projecting the results on to two-dimensions we have crystallite shape ellipsoids and these are shown in Figures 21 for cotton and dry and wet silk fibres.

Table 5. Microstructural parameters computed from whole pattern refinement

SAMPLE	2Õ	<N>	ω	BG	g
TASSAR SILK FIBRE	16.62 20.08 22.42 33.47	5.7±0.6 6.2±0.6 8.7±0.9 6.2±0.6	0.85±0.09 1.09±0.11 0.83±0.08 0.01±0.01	0.01±0.001	0.02±0.002
MUGA SILK FIBRE	17.04 20.08 22.28 33.19	6.0±0.6 5.1±0.5 7.1±0.7 4.4±0.4	0.88±0.09 1.08±0.11 0.73±0.07 0.01±0.01	0.001±0.0001	0.11±0.011
FIRI SILK FIBRE	16.76 20.08 22.70 34.02	5.2±0.5 6.1±0.6 8.4±0.8 5.2±0.5	0.87±0.09 1.20±0.12 0.92±0.09 0.01±0.01	0.020±0.002	0.13±0.013

Figure 21. Variation of crystallite shape ellipsoids in (a) dry Bivoltine silk fibres and (b)wet Bivoltine silk fibres.
Source: Divakara et al., 2010

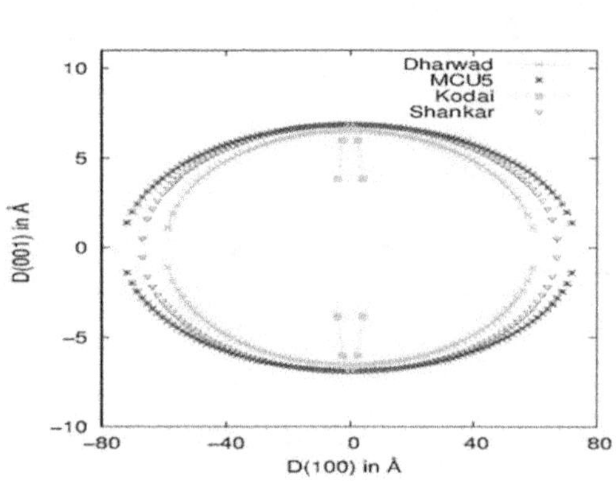

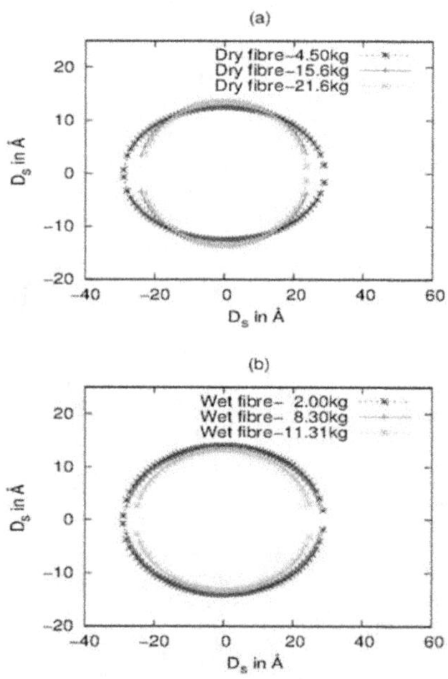

REFERENCES

Arnott, S., Chandrasekaran, R., Hall, I. H., & Puigjaner, L. C. (1983). Heteronomous DNA. *Nucleic Acids Research, 11*(12), 4141–4155. doi:10.1093/nar/11.12.4141 PMID:6866768

Arnott, S., Chandrasekaran, R., Hall, I. H., Puigjaner, L. C., Walker, J. K., & Wang, M. (1983, January). DNA secondary structures: Helices, wrinkles, and junctions. *Cold Spring Harbor Symposia on Quantitative Biology*, *47*(0), 53–65. doi:10.1101/SQB.1983.047.01.008 PMID:6574857

Arnott, S., Chandrasekaran, R., Puigjaner, L. C., Walker, J. K., Hall, I. H., Birdsall, D. L., & Ratliff, R. L. (1983). Wrinkled DNA. *Nucleic Acids Research, 11*(5), 1457–1474. doi:10.1093/nar/11.5.1457 PMID:6572358

Arnott, S., & Wonacott, A. J. (1966). The refinement of the crystal and molecular structures of polymers using X-ray data and stereochemical constraints. *Polymer*, *7*(4), 157–166. doi:10.1016/0032-3861(66)90009-7

Asakura, T., Iwadate, M., Demura, M., & Williamson, M. P. (1999). Structural analysis of silk with 13 C NMR chemical shift contour plots. *International Journal of Biological Macromolecules*, *24*(2), 167–171. doi:10.1016/S0141-8130(98)00082-8 PMID:10342761

Balzar, D. (2002). Report on the size-strain Round Robin. *IUCr-Newsletter*, *14*, 228.

Balzar, D., Audebrand, N., Daymond, M. R., Fitch, A., Hewat, A., Langford, J. I., & Toby, B. H. et al. (2004). Size–strain line-broadening analysis of the ceria round-robin sample. *Journal of Applied Crystallography*, *37*(6), 911–924. doi:10.1107/S0021889804022551

Bonham, R. A., & Schafer, L. (1974). *International Tables for X-Ray Crystallography*. Birmingham: Kynoch Press.

Brubaker, C. L., Bourland, F. M., & Wendel, J. E. (1999). The origin and domestication of cotton. In C. W. Smith & J. T. Cothren (Eds.), *Cotton: Origin, History, Technology, and Production* (p. 3). New York: John Wiley and Sons, Inc.

Chen, R., Jakes, K. A., & Foreman, D. W. (2004). Peak-fitting analysis of cotton fiber powder X-ray diffraction spectra. *Journal of Applied Polymer Science*, *93*(5), 2019–2024. doi:10.1002/app.20666

De Keijser, T. H., Langford, J. I., Mittemeijer, E. J., & Vogels, A. B. P. (1982). Use of the Voigt function in a single-line method for the analysis of X-ray diffraction line broadening. *Journal of Applied Crystallography*, *15*(3), 308–314. doi:10.1107/S0021889882012035

Demura, M., Minami, M., Asakura, T., & Cross, T. A. (1998). Structure of Bombyx mori silk fibroin based on solid-state NMR orientational constraints and fiber diffraction unit cell parameters. *Journal of the American Chemical Society*, *120*(6), 1300–1308. doi:10.1021/ja9720691

Desborough, I. J., & Hall, I. H. (1977). A comparison of published crystalline structures of poly (tetramethylene terephthalate). *Polymer*, *18*(8), 825–830. doi:10.1016/0032-3861(77)90189-6

Divakara, S., Niranjana, A. R., & Somashekar, R. (2009). Computation of stacking and twin faults in varieties of cotton fibers using whole powder pattern fitting technique. *Cellulose (London, England)*, *16*(6), 1187–1200. doi:10.1007/s10570-009-9354-5

Divakara, S., Reddy, T., Roy, S., Siddaraju, G. N., & Somashekar, R. (2010). Changes in Stacking Faults and Microstructural Parameters of Dry and Wet Bivoltine Silk Fibers using Whole Powder Pattern Fitting Technique. *Journal of Polymer Materials*, *27*(2), 135–148.

Dong, Y. H., & Scardi, P. (2000). MarqX: A new program for whole-powder-pattern fitting. *Journal of Applied Crystallography*, *33*(1), 184–189. doi:10.1107/S002188989901434X

Eichhorn, S. J., & Davies, G. R. (2006). Modelling the crystalline deformation of native and regenerated cellulose. *Cellulose (London, England)*, *13*(3), 291–307. doi:10.1007/s10570-006-9046-3

Fossey, S. A., Némethy, G., Gibson, K. D., & Scheraga, H. A. (1991). Conformational energy studies of β-sheets of model silk fibroin peptides. I. Sheets of poly (Ala-Gly) chains. *Biopolymers*, *31*(13), 1529–1541. doi:10.1002/bip.360311309 PMID:1814502

Fraser, R. D. B., & MacRue, T. P. (1973). *Conformations in fibrous proteins*. New York: Academic Press.

French, A. D., & Johnson, G. P. (2004). What crystals of small analogs are trying to tell us about cellulose structure. *Cellulose (London, England)*, *11*(1), 5–22. doi:10.1023/B:CELL.0000014765.94239.fe

Fried Joel, R. (2002). *Polymer science and technology*. Academic Press.

Gō, N., & Okuyama, K. (1976). A Method for Calculating Dihedral Angles in Helical Polymers with Given Values of Unit Height and Unit Rotation. *Macromolecules*, *9*(5), 867–868. doi:10.1021/ma60053a032

Hall, I. H., & Pass, M. G. (1976). Chain conformation of poly (tetramethylene terephthalate) and its change with strain. *Polymer*, *17*(9), 807–816. doi:10.1016/0032-3861(76)90036-7

Hall, I. H., & Somashekar, R. (1991). The determination of crystal size and disorder from the X-ray diffraction photograph of polymer fibres. 2. Modelling intensity profiles. *Journal of Applied Crystallography*, *24*(6), 1051–1059. doi:10.1107/S0021889891007707

He, S. J., Valluzzi, R., & Gido, S. P. (1999). Silk I structure in Bombyx mori silk foams. *International Journal of Biological Macromolecules*, *24*(2), 187–195. doi:10.1016/S0141-8130(99)00004-5 PMID:10342764

Ivanova, M. I., & Makowski, L. (1998). Iterative low-pass filtering for estimation of the background in fiber diffraction patterns. *Acta Crystallographica. Section A, Foundations of Crystallography*, *54*(5), 626–631. doi:10.1107/S0108767398003213

Jaswon, M. A., Gillis, P. P., & Mark, R. E. (1968, September). The elastic constants of crystalline native cellulose. In *Proceedings of the Royal Society of London A: Mathematical, Physical and Engineering Sciences* (Vol. 306, No. 1486, pp. 389-412). The Royal Society. doi:10.1098/rspa.1968.0157

Konishi, T., Kelsey, E., & Singleton, G. T. (1967). Negative potential in scala media during early stage of anoxia. *Acta Oto-Laryngologica*, *64*(1-6), 107–118. doi:10.3109/00016486709139097 PMID:6055793

Langford, J. I. (1965). Variance as a Measure of Line Broadening: Particle-size Determination. *Nature, 207*(5000), 966–967. doi:10.1038/207966a0

Langford, J. I. (1968). The variance and other measures of line broadening in powder diffractometry. II. Determination of particle size. *Journal of Applied Crystallography, 1*(3), 131–138. doi:10.1107/S0021889868005182

Lotz, B., & Cesari, F. C. (1979). The chemical structure and the crystalline structures of Bombyx mori silk fibroin. *Biochimie, 61*(2), 205–214. doi:10.1016/S0300-9084(79)80067-X PMID:465571

Lotz, B., & Keith, H. D. (1971). Crystal structure of poly (l-Ala-Gly) II: A model for silk I. *Journal of Molecular Biology, 61*(1), 201IN7203-202IN9215.

Lucas, F., Shaw, J. T. B., & Smith, S. G. (1957). The amino acid sequence in a fraction of the fibroin of Bombyx mori. *The Biochemical Journal, 66*(3), 468–479. doi:10.1042/bj0660468 PMID:13459882

Marsh, R. E., Corey, R. B., & Pauling, L. (1955). An investigation of the structure of silk fibroin. *Biochimica et Biophysica Acta, 16*, 1–34. doi:10.1016/0006-3002(55)90178-5 PMID:14363226

Mittemeijer, E. J., & Scardi, P. (Eds.). (2013). *Diffraction analysis of the microstructure of materials* (Vol. 68). Springer Science & Business Media.

Momany, F. A., McGuire, R. F., Burgess, A. W., & Scheraga, H. A. (1975). Energy parameters in polypeptides. VII. Geometric parameters, partial atomic charges, nonbonded interactions, hydrogen bond interactions, and intrinsic torsional potentials for the naturally occurring amino acids. *Journal of Physical Chemistry, 79*(22), 2361–2381. doi:10.1021/j100589a006

Nakajima, Y., Qu, X. M., & Natori, S. (1987). Interaction between liposomes and sarcotoxin IA, a potent antibacterial protein of Sarcophaga peregrina (flesh fly). *The Journal of Biological Chemistry, 262*(4), 1665–1669. PMID:3543008

Okuyama, K. (2008). Revisiting the molecular structure of collagen. *Connective Tissue Research, 49*(5), 299–310. doi:10.1080/03008200802325110 PMID:18991083

Okuyama, K., Noguchi, K., Miyazawa, T., Yui, T., & Ogawa, K. (1997). Molecular and crystal structure of hydrated chitosan. *Macromolecules, 30*(19), 5849–5855. doi:10.1021/ma970509n

Okuyama, K., Somashekar, R., Noguchi, K., & Ichimura, S. (2001). Refined molecular and crystal structure of silk I based on Ala–Gly and (Ala–Gly) 2—Ser–Gly peptide sequence. *Biopolymers, 59*(5), 310–319. doi:10.1002/1097-0282(20011015)59:5<310::AID-BIP1028>3.0.CO;2-5 PMID:11514934

Okuyama, K., Takanashi, K., Nakajima, Y., Hasegawa, Y., Hirabayashi, K., & Nishi, N. (1988). Analysis of silk I structure by X-ray and electron diffraction methods. *The Journal of Sericultural Science of Japan, 57*(1), 23–30.

Popa, N. C., & Balzar, D. (2002). An analytical approximation for a size-broadened profile given by the lognormal and gamma distributions. *Journal of Applied Crystallography, 35*(3), 338–346. doi:10.1107/S0021889802004156

Press, W. H., Flannery, B. P., Teukolsky, S. A., & Vetterling, W. T. (1986). *Numerical Recipes*. Cambridge, UK: Cambridge Univ. Press.

Ramachandran, G. N., Ramakrishnan, C., & Sasisekharan, V. (1963). Stereochemistry of polypeptide chain configurations. *Journal of Molecular Biology*, *7*(1), 95–99. doi:10.1016/S0022-2836(63)80023-6 PMID:13990617

Ribárik, G., Ungár, T., & Gubicza, J. (2001). MWP-fit: A program for multiple whole-profile fitting of diffraction peak profiles by ab initio theoretical functions. *Journal of Applied Crystallography*, *34*(5), 669–676. doi:10.1107/S0021889801011451

Samir, O. M., & Somashekar, R. (2007). Intrinsic strain effect on crystal and molecular structure of (dch32) cotton fiber. *Powder Diffraction*, *22*(1), 20-26.

Sangappa, E. Y., Mahesh, S. S., & Somashekar, R. (2005). Crystal structure of raw pure Mysore silk fibre based on (Ala-Gly) 2-Ser-Gly peptide sequence using Linked-Atom-Least-Squares method. *Journal of Biosciences*, *30*(2), 259–268. doi:10.1007/BF02703707 PMID:15933415

Scardi, P., & Leoni, M. (2001). Diffraction line profiles from polydisperse crystalline systems. *Acta Crystallographica. Section A, Foundations of Crystallography*, *57*(5), 604–613. doi:10.1107/S0108767301008881 PMID:11526309

Scardi, P., & Leoni, M. (2002). Whole powder pattern modelling. *Acta Crystallographica. Section A, Foundations of Crystallography*, *58*(2), 190–200. doi:10.1107/S0108767301021298 PMID:11832590

Scardi, P., & Leoni, M. (2006). Line profile analysis: Pattern modelling versus profile fitting. *Journal of Applied Crystallography*, *39*(1), 24–31. doi:10.1107/S0021889805032978

Shimizu, M. (1941). Eine röntgenographische Untersuchung des Sericins. *Bull. Sericult. Exp. Stn*, *10*, 441–474.

Skilling, J., & Bryan, R. K. (1984). Maximum entropy image reconstruction: General algorithm. *Monthly Notices of the Royal Astronomical Society*, *211*(1), 111–124. doi:10.1093/mnras/211.1.111

Smith, P. C., & Arnott, S. (1978). LALS: A linked-atom least-squares reciprocal-space refinement system incorporating stereochemical restraints to supplement sparse diffraction data. *Acta Crystallographica. Section A, Crystal Physics, Diffraction, Theoretical and General Crystallography*, *34*(1), 3–11. doi:10.1107/S0567739478000029

Smith, P. C., & Arnott, S. (1978). LALS: A linked-atom least-squares reciprocal-space refinement system incorporating stereochemical restraints to supplement sparse diffraction data. *Acta Crystallographica. Section A, Crystal Physics, Diffraction, Theoretical and General Crystallography*, *34*(1), 3–11. doi:10.1107/S0567739478000029

Snyder, R. L., Bunge, H. J., & Fiala, J. (1999). *Defect and microstructure analysis by diffraction*. Oxford Univ. Press.

Somashekar, R., Hall, I. H., & Carr, P. D. (1989). The determination of crystal size and disorder from X-ray diffraction photographs of polymer fibres. 1. The accuracy of determination of Fourier coefficients of the intensity profile of a reflection. *Journal of Applied Crystallography*, *22*(4), 363–371. doi:10.1107/S0021889889004085

Spek, A. L. (2003). Single-crystal structure validation with the program PLATON. *Journal of Applied Crystallography*, *36*(1), 7–13. doi:10.1107/S0021889802022112

Stokes, A. R. (1948). A numerical Fourier-analysis method for the correction of widths and shapes of lines on X-ray powder photographs. *Proceedings of the Physical Society*, *61*(4), 382–391. doi:10.1088/0959-5309/61/4/311

Strydom, D. J., Haylett, T., & Stead, R. H. (1977). The amino-terminal sequence of silk fibroin peptide Cp—a reinvestigation. *Biochemical and Biophysical Research Communications*, *79*(3), 932–938. doi:10.1016/0006-291X(77)91200-1 PMID:597316

Takahashi, Y., Gehoh, M., & Yuzuriha, K. (1999). Structure refinement and diffuse streak scattering of silk (Bombyx mori). *International Journal of Biological Macromolecules*, *24*(2), 127–138. doi:10.1016/S0141-8130(98)00080-4 PMID:10342756

Velterop, L., Delhez, R., Keijser, T. H. D., Mittemeijer, E. J., & Reefman, D. (2000). X-ray diffraction analysis of stacking and twin faults in fcc metals: A revision and allowance for texture and non-uniform fault probabilities. *Journal of Applied Crystallography*, *33*(2), 296–306. doi:10.1107/S0021889800000133

Warren, B. E. (1959). X-ray studies of deformed metals. *Progress in Metal Physics*, *8*, 147–202. doi:10.1016/0502-8205(59)90015-2

Warren, B. E., & Averbach, B. L. (1950). The effect of cold-work distortion on X-ray patterns. *Journal of Applied Physics*, *21*(6), 595–599. doi:10.1063/1.1699713

Welberry, T. R., & Butler, B. D. (1994). Interpretation of diffuse X-ray scattering via models of disorder. *Journal of Applied Crystallography*, *27*(3), 205–231. doi:10.1107/S0021889893011392

Chapter 2
Cellulose Application in Food Industry:
A Review

Diego Mauricio Sanchez Osorno
Pontificia Bolivariana University, Colombia

Cristina Castro
Pontificia Bolivariana University, Colombia

ABSTRACT

Cellulose polysaccharide is the most important component in plants with a fascinating structure and properties. Despite the origin, cellulose is a linear homopolymer of β-(1-4)-linked D-glucopyranose units varying mainly on purity, degree of polymerization (DP) and crystallinity index. This linear stiff-chain homopolymer is characterized by its hydrophilicity, chirality, biodegradability, broad chemical modifying capacity, and its formation of versatile semicrystalline fiber morphologies. This chapter aims to show the most important applications of cellulose in food, presenting other cellulose derivatives as methylcellulose, carboxymethyl cellulose, and novel cellulose forms as bacterial cellulose. New frontiers, including environmentally friendly cellulose fiber technologies for food packaging, bacterial cellulose in foodstuff and other applications as thickening agent, stabilizing agent, gelling agent, suspending agent were highlighted with future aims, strategies, and perspectives of cellulose research and its applications.

INTRODUCTION

In the last years, cellulose has become a popular food additive due to its unique chemical and physical properties. Cellulose is the most abundant natural polymer in the world and its properties have been exploited by mankind for millennia, especially to manufacture products for sectors such as paper, textile, construction, new materials, pharmaceutical, cosmetic, food, coatings and adhesives, gaskets and packing.Even other potential applications for this renewable resource are being increasingly hunted (Klemm et al., 2005)especially due to the lack of fossil resources (Martins et al., 2009).

DOI: 10.4018/978-1-5225-3023-7.ch002

Cellulose is a linear polymer composed of β-D-glucopyranose units linked by (1-4) glycosidic bonds forming chains(Klemm et al., 2005) with approximately 10.000 to 15.000 units (in native cellulose) (O'sullivan, 1997). The cellulose chain structure is indicated in Figure 1. Three hydroxyl groups in equatorial position in each glucose unit are responsible for its supramolecular structure, which determines many of its chemical and physical properties. The hydroxyl groups can be modified by chemical reactions generating esterified fibers with DS (degree of substitution) from 0.002 to 0.41; and in all instances, the cellulose surface acquires high hydrophobicity (Tomé et al., 2011).

By characterization techniques like infrared spectroscopy, X-ray diffraction and ^{13}C nuclear magnetic resonance, it has been possible to establish the formation of intramolecular hydrogen bonds between the hydroxyl group of C3 of a glucose unit and the oxygen of pyranose ring of contiguous glucose.Formation of intramolecular and intermolecular hydrogen bonds are shown in Figure 2. These are responsible for the relative rigidity of the cellulose molecule, which is reflected in the high tendency to crystallize in parallel arrangementsand the ability to form fibrillary structures (Krassig, 1993; Fengel& Wegener, 1984).

Figure 1. Cellulose chain structure

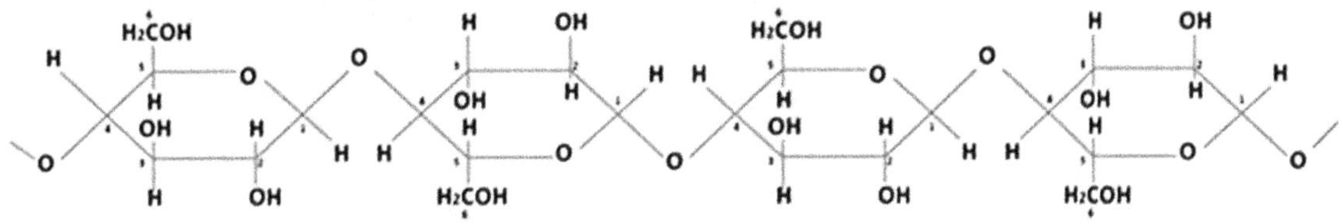

Figure 2. Formation of intramolecular and intermolecular hydrogen bonds between two adjacent cellulose molecules

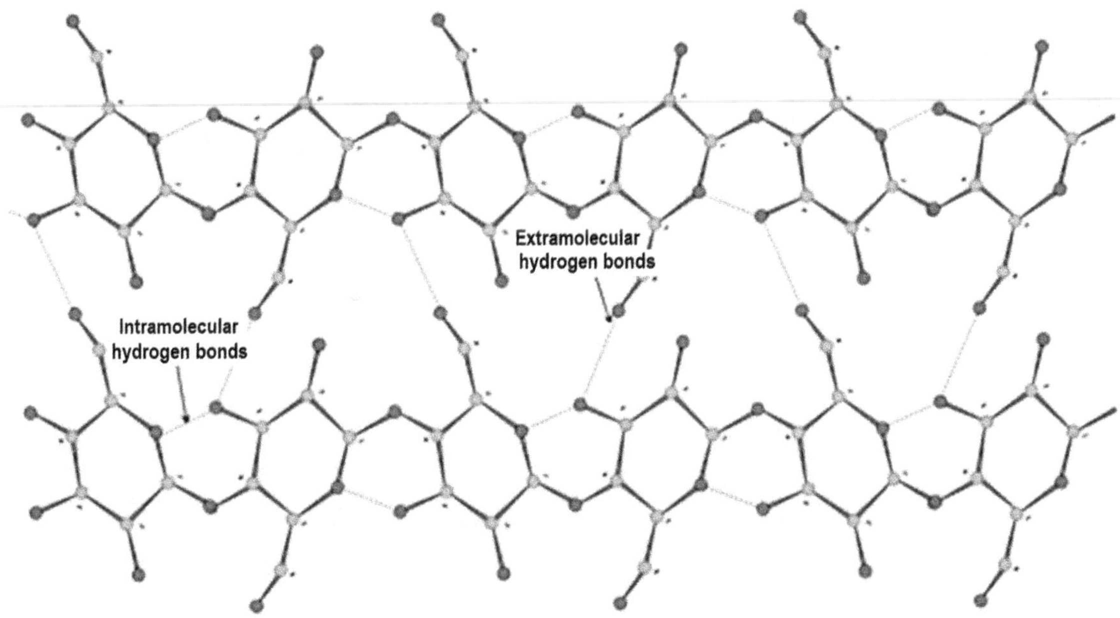

Therefore, molecules tend to organize into individual cellulose nanofibrils of 2 to 5 nm in widthwise in higher plants, 2 to 4 nm for bacterial cellulose and 15 to 30 nm for algae (Meshitsuke & Isogai, 1995). Within nanofibrils, chains form crystalline domains with different arrangements that give rise to several polymorphisms (I, II, III, and IV), in which each of the unit cell has a specific diffraction pattern. The most abundant polymorphism is cellulose I in native cellulose and it can be converted reversibly or irreversibly into the others polymorphisms by chemical, physical and thermal treatments. Transformation between different cellulose polymorphisms are shown in Figure 3.

As already mentioned above, cellulose-based fibers have a multilevel hierarchical organization ranging from nanometers to meters in nature, which provides mechanical resistance to the structure of many living speciessuch as plants, bacteria, fungi, algae, amoebas and animals (Kalia, S., Kaith, B. S., & Kaur, 2011). In nature, cellulose fiber biosynthesis results from combined action of polymerization and crystallization in specific enzymatic terminal complexes (TCs), followed by linear association of cellulose chains forming nanofibrils. General organization of terminal complexes are shown in Figure 4. In many organisms, TCs have been seen in two different morphologies, rosettes as for instance in high plants and linear as in algae and bacteria (Hayashi, 2006).

Cellulose chains synthesized in TCs are associated with adjacent ones forming a two-phase system consisting of amorphous and crystalline domains by secondary valence hydrogen bonds (Zugenmeier, 2008). These parallel arrangements constitute nanofiber, which in turn is joined to other ones by hydrogen bonds to form macrofibers (H. A. Krässig, 1993).

Figure 3. Transformations between different cellulose polymorphisms

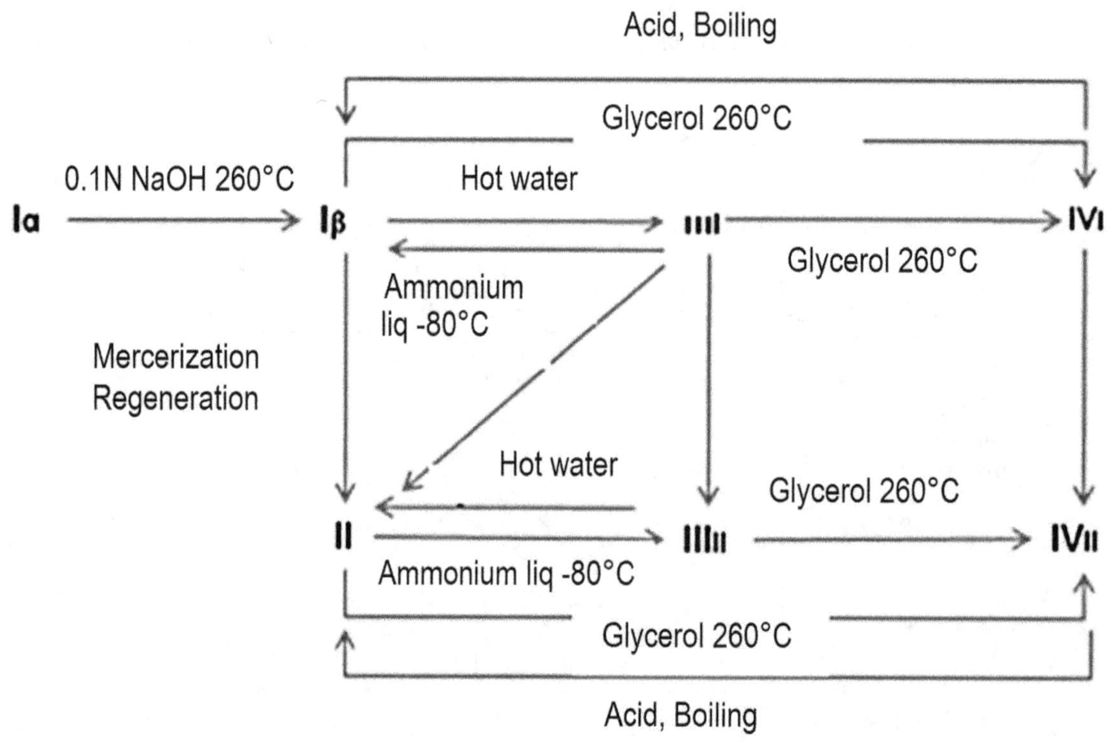

Figure 4. General organization of terminal complexes (TCs) in different organisms

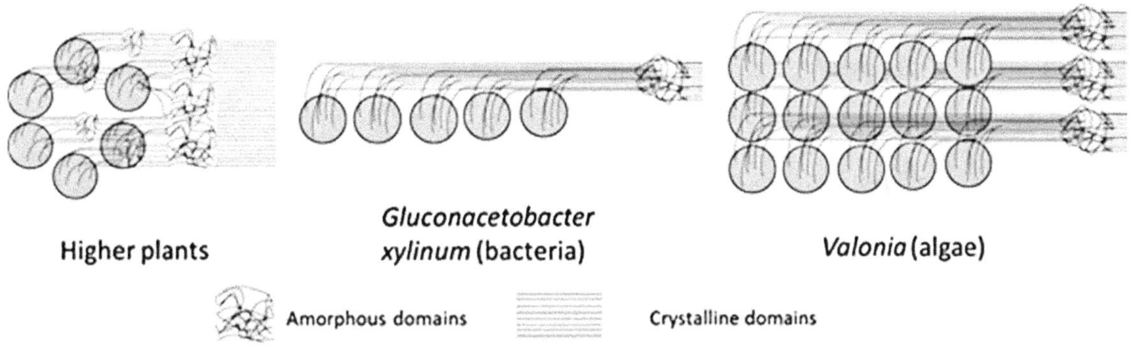

Cellulose properties are strongly related to its degree of crystallinity. Crystallinity influence the accessibility for chemical derivatization, swelling and water binding. The degree of crystallinity is important that to be taken in to account when considering the manufacturing and applications of cellulose and cellulose-based materials. Crystalline regions are resistant to solvent penetration and have low accessibility and reactivity when compared to disordered amorphous regions (Cybulska et al., 2003).

Looking at the history of science and technology of cellulosic materials, there are numerous technologies for cellulose modification by both chemical and mechanical methods for several industrial applications. Cellulose is of interest for these applications because of its characteristic properties/qualities such as high mechanical properties, hydrophilicity, insolubility in water, stability to chemicals, safe use in the body, recyclability and biodegradability. To reinforce these properties or add new functionalities, modification techniques are used on hydroxyl functional groups of cellulose. Chemical treatments are the most common methods and the typical modifications are esterification and etherification at hydroxyl groups of cellulose. Those reactions may occur over surface hydroxyl groups of the fiber or on the ones inside, affecting first the amorphous areas and continuing with crystalline zones (Cybulska et al., 2011; H. A. Krässig, 1993).

Food industry is not an exception, cellulose or modified cellulose is used in many processes as thickening or stabilizer hydrocolloid due to the facts that it is harmless, without odor or taste, absorbs water and other liquids, is 100% insoluble (the body cannot produce enzymes capable of cleaving the glycosidic bond) and therefore not absorbed by the body during intake. Cellulose is also used to control the ice crystal growth, replace fat, increase the fiber content and reduce the need to use ingredients like oil or flour, which are becoming more expensive.

Hydrocolloids are macromolecules which have a great affinity to water (the presence of a large number of hydroxyl groups (-OH) increases the affinity to water molecules by the compounds), which dissolves or disperses modifying its rheology, increasing the liquid viscosity or in many cases forming a gel with water by solvatation. Scheme of interaction of water with macromolecular components in hydrocolloids are shown in Figure 5. In case of gel, their association or crosslinking form a three-dimensional network that traps water and form a rigid structure that is resistant to flow (exhibiting characteristics of a liquid and a solid). Food hydrocolloids modify two basic properties in food, flow behavior (viscosity) and

Figure 5. Scheme of interaction of water with macromolecular components in hydrocolloids

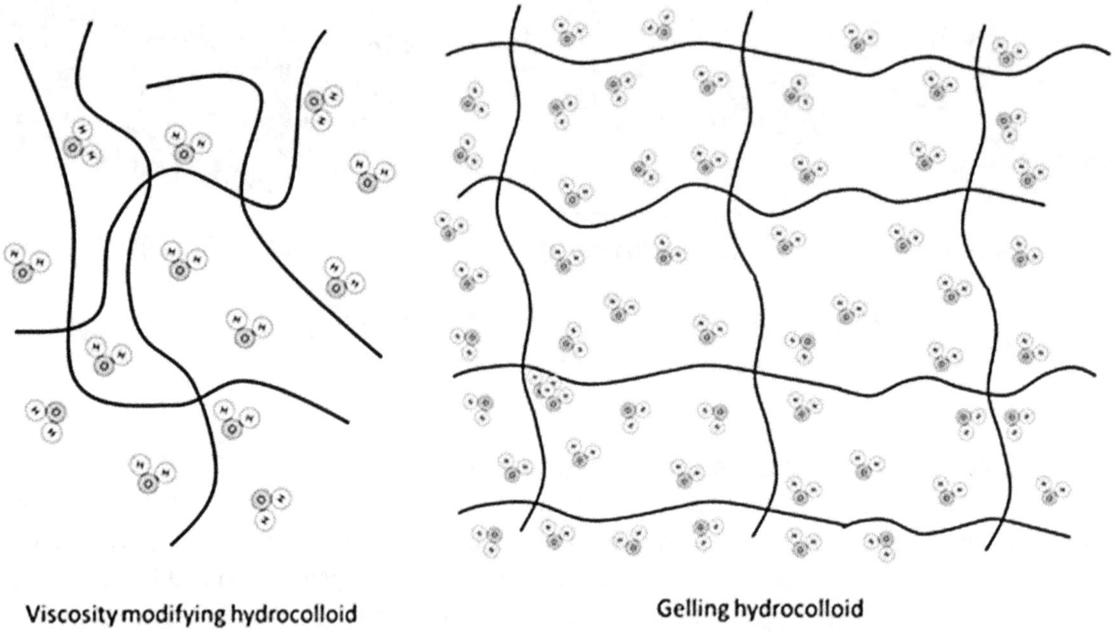

mechanical solid property (texture). The modification of the texture and/or viscosity of thefood system helps to modify its sensory properties. Various formulations of foods like soups, salad dressings and so on are using hydrocolloids as additives to achieve the desired viscosity and mouth feel. The most used colloids as thickening agent are locust bean gum, xanthan, gum tragacanth, gum Arabic and cellulose derivatives. Some gelling hydrocolloids used in food industry are alginate, pectin, carrageenan, gelatin, gellan and agar (Saha & Bhattacharya, 2010).

As mentioned above, cellulose is used as a thickening agent in food in two forms, without modification (natural) and partially or totally modified. Chemical structure of cellulose and partially or totally modified cellulose is shown in Figure 6. Most of cellulose used in food applications comes from vegetable origin, where fibers are purified by physico-chemical processes to remove other polysaccharides (hemicellulose and lignin), and then modified by chemical reactions (AQUALON, 1999; Keyser, W. L., & Kinney, 1983) on hydroxyl groups (-OH) with various reagents to produce derivatives with useful properties like mainly cellulose esters and ethers (-OR).Cellulose and cellulose derivatives are shown in Table 1.

The production of nano-scale cellulose fibers and their application in various fields has gained increasing attention due to their high strength and stiffness, low weight, biodegradability, renewability combined with a high surface area (Siró&Plackett, 2010). In the last few years the scientific community has centered their focus on different other forms of cellulose in nano-scale, in particular bacterial nanocelullose (BC), cellulose nanocrystals and nanofibrillated cellulose (Tomé et al., 2011).

The purpose of this paper is to show an updated overview of cellulose being used in the food industry and present new forms of cellulose, which are being investigated to enhance the skills of traditional forms.

Figure 6. Chemical structure of cellulose and partially or totally modified cellulose

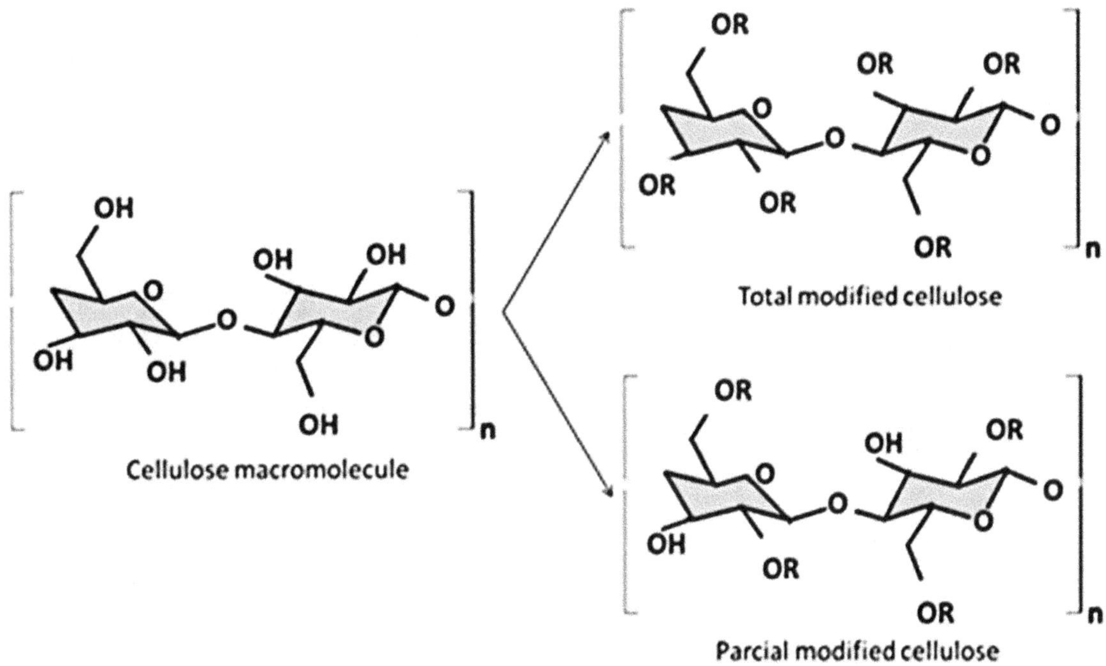

CARBOXYMETHYL CELLULOSE

Definition

Carboxymethylcellulose (CMC) is a white, water-soluble polymer derived from cellulose, used as a coating agent and sizing for paper and textiles, as stabilizer for various foods, and as appetite suppressor. Due to its insolubility in water, cellulose needs to be modified in order to increase the degree of solubility. The process consists in transforming cellulose into sodium carboxymethyl cellulose, which is soluble in hot and cold water. This is a highly effective additive for improving product and processing characteristics in various fields, including food, industrial additives and pharmaceuticals. Rheology profiles of CMC grades vary from smooth-flowing aqueous solutions to stable gels (Zorba & Ova, 1999).

Production

CMC is cellulose ether; which is possible to produce by reacting alkali cellulose with sodium monochloroacetate under controlled conditions. CMC is obtained by substituting some of the hydrogens of the hydroxyl groups for carboxymethyl groups(AQUALON, 1999; Keyser, W. L., & Kinney, 1983). Chemical structure of carboxymethyl cellulose is shown in Figure 7. The reaction of sodium carboxymethylcellulose is as follows:

NaOH + R-OH R-ONa + H_2O

Table 1. Cellulose and cellulose derivatives usedin food

Cellulose Form	Use	Reference
Carboxymethyl cellulose (water absorbing characteristics	Whipping agent in ice-cream, confectionery, jellies, etc.	Abughoush, M., Al-Mahasneh, M., Samhouri, M., Al-Holy, M., & Herald, 2008; Arbuckle, 1977; Diftis & Kiosseoglou, 2003; Flores & Goff, 1999; Kohajdová & Karovicová, 2008; Sun-Waterhouse et al., 2013
Methylcellulose (viscosity)	Thickener and emulsifier, and in foods formulated to be low in gluten	Debeaufort et al., 1994; Haque & Morris, 1993; Kamper & Fennema, 1985; Mallikarjunan et al., 1997; Psomiadou et al., 1996; Vojdani & Torres, 1990
Hydroxyethyl methylcellulose, hydroxypropyl-methylcellulose, and ethyl cellulose	Emulsifiers	(AQUALON, 2002; Larsen, J. C., Nørby, K. K., Reffstrup, T. K., & Beltoft, 2004; Organization, 2006; The Dow Chemical company, 2013; Valencia-Chamorro, S. A., Palou, L., Del Rio, M. A., & Pérez-Gago, 2008; Villalobos-Carvajal, Hernández-Muñoz, Albors, & Chiralt, 2009; Vojdani & Torres, 1990)
Hydroxyethyl methylcellulose, methylcellulose, hydroxypropyl methylcellulose, and cellulose gum	Cosmetic products at concentrations up to 10%. The cellulose derivatives pass essentially unchanged through the gastrointestinal tract following oral administration.	(Tarantino, 2006)
Sodium carboxymethyl cellulose; methyl cellulose; cellulose acetate; ethyl cellulose and hydroxypropylmethyl cellulose	Permitted for direct addition to food as food additives GRAS (Generally Recognized As Safe)	(The Dow Chemical company, 2013).

Figure 7. Chemical structure of carboxymethyl cellulose

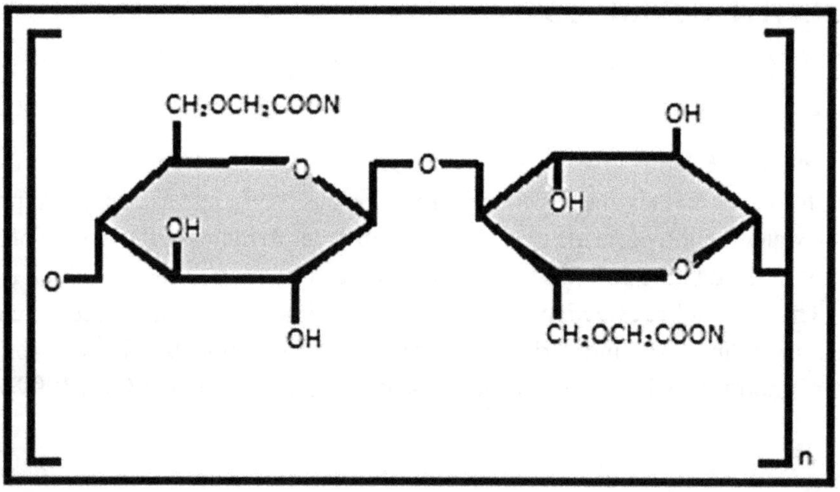

R-ONa + Cl – CH_2COONa R-O-CH_2COONa + NaCl

This condition has an impact on important CMC characteristics as solubilization and viscosity. After the reaction, the common pH of CMC is 8.25, which is generally adjusted to about 7-7.5. If the pH of CMC is 6.0 or less, the dried product does not have a good solubility in water; solutions are hazy and contain insoluble gel particles. If the pH is 4.0 or below, the dried product is insoluble in water (AQUALON@, 1999).

The desired degree of substitution, degree of polymerization and uniformity of substitution are controlled by monitoring the chemical reaction. Considering the highest degree of substitution to be obtained in cellulose is 3 (each anhydroglucose unit contains three available hydroxyl groups with which the monochloroacetate can react), better physical properties as for instance solubility and viscosity are obtained with CMC at a lower degree of substitution.

The degree of substitution of commercial material usually ranges from 0.4 to 1.2 (Klose & Glicksman, 1972), due to the fact that molecules of CMC with values up to 1.6 the molecule are markedly fragmented by mono component endoglucanase (Heinze & Koschella, 2005) decreasing the degree of polymerization and thus the viscosity. Characteristics of CMC are tabulated in Table 2.

Carboxymethylcellulose is regulated by European Economic Community (EEC) directive 74/329/EEC and its amendments of emulsifiers, stabilizers, thickeners, and gelling agents. Carboxymethylcelluloseis identified by E466.

Raw Materials for Food

Sodium carboxymethyl cellulose is used widely to modify the viscosity invarious food products such as dairy products, frozen desserts, jellies, cake mixes, salad dressings etc. Its chemical and physical properties (viscosity, dispersibility, and degree of substitution) have been studied considerably& Glicksman, 1972; Zecher, & Van Coilie, 1992)(AQUALON, 1999).

The following is the literature from various research articles, which reveals applications of CMC in food industry.

Table 2. Characteristics of carboxymethyl cellulose

<table>
<tr><th colspan="2">Degree of substitution</th><th>Solubility</th></tr>
<tr><td colspan="2">0.3 or less</td><td>Alkali</td></tr>
<tr><td colspan="2">0.45 or higher</td><td>Water</td></tr>
<tr><td colspan="2">0,65 -0,85</td><td>Most commonly used</td></tr>
<tr><th colspan="2">pH</th><th>Viscosity</th></tr>
<tr><td colspan="2">2-3</td><td>Insoluble free CMC may be precipitated</td></tr>
<tr><td colspan="2">Below 5</td><td>Reduction of viscosity</td></tr>
<tr><td colspan="2">7-9</td><td>Highest viscosity stability</td></tr>
<tr><td colspan="2">5 - 11</td><td>Fairly stable</td></tr>
<tr><th>Viscosity Type</th><th>Degree of Polymerization</th><th>Molecular Weight</th></tr>
<tr><td>High</td><td>3,200</td><td>700,000</td></tr>
<tr><td>Medium</td><td>1,100</td><td>250,000</td></tr>
<tr><td>Low</td><td>400</td><td>90,000</td></tr>
</table>

Source: Klose, R. E., & Glicksman, 1972.

Ice Cream

CMC has a high water-holding capacity and is easily dissolved in the mix where it act as an emulsifier (Josephson and Dahle, 1949). The amount of CMC used in ice cream varies from 0.15 to 0.20% (Panda, 2010)since this range allows an improvement of certain properties as the degree of coldness, sensory viscosity, sensory iciness and the degree of melting in the mouth. Moore and Shoemaker showed that that molecule migration is restricted by increasing viscosity, which limits the crystal nucleation and the size of ice crystals. This indicates that the iciness in ice cream is related to the mixed viscosity. Further, a positive correlation was observed between the duration of iciness and the concentration of CMC, probably due to the reduced rate of heat transfer from the mouth to the ice crystals through increased viscosity(Arbuckle, 1977; Josephson, Dahle and Patton, 1943; Moore & Shoemaker, 1981). Some non-gelling stabilizers (i.e. CMC) are more effective in retarding recrystallization than gelling stabilizers (i.e. gelatin), this suggests that steric blocking of the interface or inhibition of solute transport to and from the ice interface caused by gelation of the polymer is not the only mechanism of the stabilizer action. These differences in recrystallization rates can be attributed to the visco-elastic properties of the unfrozen phase imparted by stabilizers. Even molecular interactions between polysaccharides and proteins appear to be key factors for retarding ice recrystallization (Regand & Goff, 2003). This microstructural arrangement gives high flexibility to the unfrozen phase and it is expected that it would not be affected by the temperature cycling protocol, thus resulting in a more effective reduction in water mobility of the system and as a consequence lower recrystallization rates would be observed than in the gelling systems. Proteins may play a critical role in this process due to phase separation with polysaccharides, causing localized increases in concentration of each phase and a further element of structure formation (Regand & Goff, 2003).

Nevertheless, Flores (Flores & Goff, 1999; Wanga et al., 1998) affirms that stabilizers did not directly affect the initial ice crystal size distribution for any model solutions or ice cream and no effects on freezing properties of the mix, such as freezing point depression, latent heat of fusion of water and crystallization temperature. Most of the differences observed within each formulation were explained by the different amounts of air incorporated according to the type of stabilizer or the type of sweetener used. Additionally, in a sensorial panel the duration of the perception of iciness increased significantly ($r = 0.98$, $p < 0.01$) with increasing CMC concentration. Since smaller ice crystals would melt faster than an equivalent weight of larger ones, a negative correlation between duration of iciness and concentration of CMC in the ice cream was expected instead of the positive correlation which was observed. This positive correlation results probably from the reduced rate of heat transfer from the mouth to the ice crystals through an increased viscosity (Moore & Shoemaker, 1981).

CMC has gained good acceptance after it was introduced as an ice cream stabilizer, due to its water-holding capacity and its easy dissolubility in the mix where it acts as an emulsifier. Used in slightly smaller quantities than gelatin, it does not form a firm gel as gelatin and some of the vegetable stabilizers. CMC has been found to give the best results when used together with one or more other stabilizers, preferably carrageenan, locust bean gum or guar gum. The useful characteristics imparted to ice cream are a chewy body, smooth texture, and improved whipping properties (Arbuckle, 1977; Klose & Glicksman, 1972; Rao, Walter, & Cooley, 1981).

Edible Film

In edible films based on carboxymethyl cellulose and soy protein it was observed that mechanical properties increase and the water sensitivity is reduced with the CMC content. The results indicate that the structure and properties of edible films were modified and improved by blending with CMC (Su, Huang, Yuan, Wang, & Li, 2010) and those changes are attributed to Maillard reaction occurring between CMC and soy protein. The films developed in this work are found to be suitable for low moisture foods and pharmaceutical products (Su et al., 2010).

Improvement of Emulsify Properties

Proteins constitute one of the most important groups of emulsifying agents used in food; they stabilize the oil droplets against coalescence as well as reduce the oil-water interfacial tension and in this way contribute to the formation of oil droplets (Diftis & Kiosseoglou, 2003). Diftis, in 2003 studied the emulsifying behavior of the protein mixed with CMC, where emulsifying behavior was expected to be influenced by the pH of the solution. Soy protein may interact ionically with CMC at a pH below 6 and the interaction being stronger at pH values around the protein's isoelectric point (Diftis & Kiosseoglou, 2003). Protein-CMC mixtures result in emulsions of pH 6.5, with droplet size distribution shifted to lower values compared to those of non-treated samples. With regard to stability against droplet coalescence, the emulsion prepared with the protein-CMC mixtures were quite stable for a period of up to 50 days (Diftis & Kiosseoglou, 2003).

Baked Goods

In this kind of products, cellulose derivative improves moisture retention (from 60.5 to 68.3%) and texture. Stability of doughs was positively affected by cellulose derivative, while mixing tolerance index was practically not affected by the presence of additives (Kohajdová & Karovicová, 2008), the authors showed significant differences from control sample in: dough development time, degree of softening, dough stability, elasticity. In the sensory evaluation CMC showed lower crumb elasticity when compared to control samples. Baked goods with CMC were also presented as products with compact and close grain structures. In this case, cellulose reduces firmness and softens crumb than the fresh or stored samples. Baked goods shelf life evaluated during 72 h storage period through bread firmness values showed, that loaves prepared with hydrocolloids extent products contained CMC were softened when compared to control sample.

CMC are able to retard the migration of moisture to the bread surface and thus retard its staling process during storage, as showed by Armero in his study, conversely in the bread surface, where the addition of CMC increase the firmness of baked goods crumb, which could be the consequence of the thickening effect on the crumb walls surrounding air spaces (Armero & Collar, 1996; Bárcenas & Rosell, 2006). CMC showed positive effects on both crumb stability and elasticity for white breads. Bread having 1% CMC gave better sensory characteristics and golden brown color of bread crust, provides good product appeal (Armero & Collar, 1996; Bárcenas & Rosell, 2006). Others sensory characteristics were enhanced in bread during its shelf life with CMC presence, character of crust, grain of bread, aroma and taste (Ahmed, 2002).

Beverages

CMC polysaccharide is widely used in the beverage industry because it is inexpensive and has the ability to form complexes with whey proteins because of its anionic properties (Abughoush,, Al-Mahasneh,, Samhouri,, Al-Holy, & Herald, 2008; Dickinson, 1998; Rao, Walter, & Cooley, 1981).

CMC acts as suspending aid; rapid viscosifier; improves mouthfeel and body; protein stabilizer in acidified drinks (Abughoush, Al-Mahasneh, Samhouri, Al-Holy, & Herald, 2008; Adeyeye, Jain, Ghorab, & Reilly, 2002; Arancibia et al., 2011; Shirkhani, Asl, & Madadlou, n.d.).CMC concentration increases the stability of beverage, this could be due to the formation of a protein polysaccharides network. The interaction between CMC and whey proteins located at the droplet surface of the emulsions influences the behavior of the solutions and stabilizes the emulsion through adsorption of the secondary layer of the CMC. The emulsion stability index is increased in the product due to the steric stabilizing properties by the presence of a thick secondary layer on the outer side of protein (Abughoush, Al-Mahasneh, Samhouri, Al-Holy, & Herald, 2008).

In rheological properties of beverages with all CMC concentrations, full-fat samples exhibits higher consistency, pseudoplasticity, and thixotropy than low-fat samples; this trend also increases with CMC concentration (Arancibia et al., 2011; Bayarri, Chuliá, & Costell, 2010). CMC can stabilize smoothies at moderate concentrations such as 1.00 – 1.50% and further contribute tohigh polyphenols extractability and flow due to the interactions among stabilizer molecules, between fiber network and stabilizer molecules and between water and soluble or insoluble polymers(Sun-Waterhouse, Bekkour, Wadhwa, & Waterhouse, 2013).

Mouthfeel and texture was improved by adding CMC in juice and reduced calorie in beverages. In a triangular test it was found that the samples with CMC were perceived more sweet and flavorful. Clarity was measured and beverages with CMC remained clear while formulations using pectin had a noticeable haze (Ashland, 2014). CMC is added to the juice formulations as a thickener agent, which coats the tongue allowing for a longer lasting flavor impact. CMC also buffers the acidity of the juice to balance out the flavor and improve the sugar: acid ratio. The pH values of the formulations with CMC were all slightly higher than those of the formulations with pectin; this demonstrates the buffering effect of the CMC. The higher pH allows the sweet fruity notes to linger longer on the tongue (ASHLAND, 2014).

Other Uses of CMC

CMC is compatible with a wide range of other food ingredients, particularly proteins, sugars and other hydrocolloids where synergistic interaction can occur (Arancibia et al., 2011). It is clear that CMC modifies behaviors such as odorless and tasteless; thickens; controls sugar crystal size; improves texture; inhibits syneresis (Arancibia et al., 2011; van Ruth,., de Witte,., & Uriarte, 2004).

It is well known that rheological properties of CMC solutions depend strongly on concentration. Higher concentrations of CMC produce a higher resistance to flow and a stronger time dependence (Arancibia et al., 2011; Sara Bayarri & Costell, 2011; Rehman, S., & Ahmed, 2002). Nowadays, CMC is being used as an alternative thickener to starch in semisolid dairy products (Sara Bayarri & Costell, 2011; Engelen et al., 2005; Jellema, Janssen, Terpstra, De Wijk, & Smilde, 2005).

Few researchers have studied the flow behavior of complex systems like dairy desserts where CMC may interact with carbohydrates and milk proteins, CMC had a significant effect on both emulsion stability and rheological properties of high-fat coconut milk. CMC causes an increase in viscosity of continuous phase and, therefore, retards the gravitational separation of the droplets and CMC had a significant effect on both emulsion stability and rheological properties of high-fat coconut milk. CMC concentration clearly affects the viscoelastic behavior of samples, which range from liquid-like to weak gel (Bayarri et al., 2010; Jirapeangtong, Siriwatanayothin, & Chiewchan, 2008).

Barry and Costell, (Bayarri et al., 2010) in his work observed that with the increase in the concentration of CMC up to 1.25% it is possible to produce a more structured system that partially broke down with increasing shear rates. At low CMC concentrations (0.75 and 1.0%), the system is relatively deformable and the stress–rate curve indeed does not yet show the overshoot.

In the formulation of low-fat CMC-based dairy desserts fat replacers could be used in the formulation According to the rheological results to obtain products with similar rheological behavior to that obtained with full-fat milk. For different CMC concentrations, full-fat samples were perceived as thicker than low-fat samples while the CMC concentration is of 1.3%, the low-fat sample was perceived as less thick and significantly smoother than the control sample (Bayarri et al., 2010). In sweetened coconut milk the results showed that CMC concentrations had marked effect on the emulsion stability and rheological properties (Jirapeangtong et al., 2008).

Syrups

CMC bring some behaviors in syrups like clear, thickens, imparts favorable mouthfeel and body (Donovan, 1988; William 1983), syrup with a softened water containing CMC was used to reduce the brix and increase the viscosity of the filtered syrup.

In syrups, mouthfeel and texture was improved by adding CMC (Bennett, C. J., & Malits, 1988). Carboxymethyl cellulose gums were superior to use in syrup formulation when compared to other gums such as carrageenan, guar and xanthan. In taste tests, off-flavors, off-odor and cloud formation were observed when other were used instead of the CMC (William 1983).

Also, in aqueous table syrup, CMC was used to reduce the sugar and caloric content which is comparable in quality to conventional table syrups with higher content of these values (Bennett, C. J., & Malits, 1988).

Dressings and Sauces

CMC improve thickener, is a suspending aid, used as stabilizers, consistency providers and imparts mouthfeel (Poulin et al., 1997; Sikora et al., 2008). CMC was applied as thickeners and stabilizers of cocoa syrups and sweet and sour sauces. In the production of sauces and salad dressing the trends have been also shown towards reduced sodium content, low calorie, low-fat sauces and salad dressings with functional ingredients and CMC could be a raw material aiming to do it (Sikora et al., 2008).

METHYL CELLULOSE

Definition

Methyl cellulose is white, neutral and not hydrolyzed by acids or alkali, which dissolves readily in cold water forming stable colloidal solutions. Methyl cellulose is manufactured by forming methyl ethers from purified cellulose with caustic soda and methyl chloride. The viscosity of the solution depends on the degree of degradation of the cellulose molecule, and those products are available commercially having viscosities ranging between 15 -4000 centipoises in 2% solution (Tainter, 1943). Methyl cellulose is soluble in water and in ethanol at both high and low temperatures. Methyl cellulose is compatible with poly ethylene glycol and is a good emulsion stabilizer (Debeaufort et al., 1994).

Production

During their elaboration, cellulose fibers are heat-treated with caustic solution and with methyl chloride, yielding the methyl ether of cellulose. The chemical structure of methyl cellulose is shown in Figure 8. The fibrous reaction product is cleared and ground to a fine, uniform powder. The methylcellulose ether dust could possibly cause irritation to the skin and eyes under extreme conditions and is considered a nuisance dust if inhaled, this product is considered to possess no significant health hazard in handling (Tainter, 1943).

Most of the commercially produced cellulose ether production is accounted for methyl cellulose (MC) (approximately 83%) and is useful for applications in agriculture and in the food industry (Park et al., 1994; Psomiadou et al., 1996). The reaction is:

$$\text{Cell} - \text{OH} + \text{NaOH} + CH_3Cl \ \text{Cell} - OCH_{3\,(MC)} + \text{NaCl} + H_2O$$

Figure 8. The chemical structure of methyl cellulose

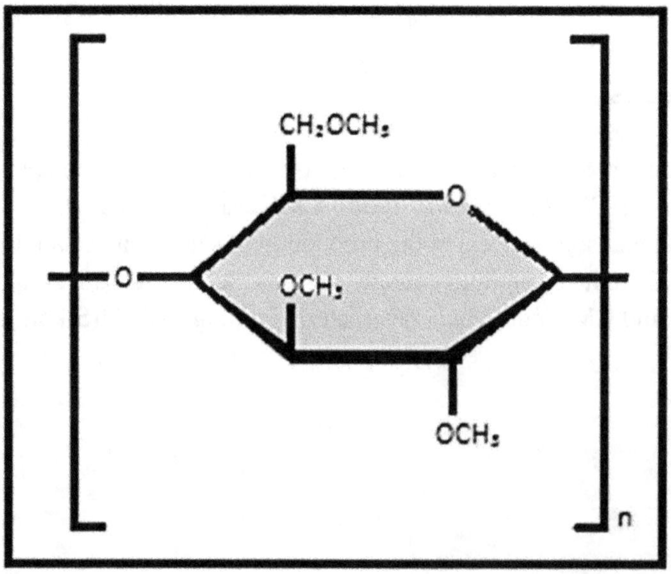

Side Reactions

$CH_3Cl + NaOH\ CH_3OH + NaCl$

$CH_3OH + NaOH + CH_3Cl\ CH_3OCH3 + NaCl + H_2O$

Methylcellulose is regulated by EEC directive 74/329/EEC and its amendments of emulsifiers, stabilizers, thickeners, and gelling agents. Methylcellulose is identified by E461.

Raw Materials for Food

One important behavior of methyl cellulose is the unusual property of forming gels by heating and reverting to the solution state by cooling, with many practical applications in industry (Haque & Morris, 1993). Other important characteristic in food industry is the hygroscopicity; five grams of methyl cellulose swell in a 500 cc graduate of distilled water to about 200 cc, the fineness of the powder determine the final volume, thoroughness of wetting, etc. If the wetting is completed, the compound disperses completely giving a thick solution or gel, the viscosity of which depends on the concentration and grade of the MC (Tainter, 1943).

Several investigations have been focused on the use of surface active cellulose ethers as sole emulsifiers, fine emulsions stabilized by MC were achieved using a high pressure homogenizer for food applications, those solutions shows weaker thickening and stabilizing properties compared to the one of the native MC (Arboleya & Wilde, 2005; Floury et al., 2003; Floury, Desrumaux, Axelos, & Legrand, 2002; Li et al., 2001).

Different applications in food have been registered in scientific papers showing how methylcellulose improving features in different kinds of food products as:

Active Packaging

It is well know that cellulose and its derivative produce high quality films due to the linear structure of the polymer backbone. However, MC film is easily dissolved in water and it is compatible with many water-soluble polymers such as alginate, starch, chitosan and polyvinyl alcohol, with which it is possible to yield complex films with better water resistance properties (Chen, Yeh, & Chiang, 1996).

A film was formed mixing methylcellulose with chitosan and 4% of sodium benzoate or potassium sorbate. Antifungal properties were demonstrated when antimycotic activity of the film on Penicillium-notatum and Rhodotomlawas investigated. At 25 °C, approximately 43-45% of the preservatives were released from the film to a glycerol/water mixture in 30 min. The maximum amount of preservative that could be released from the film at 25 °C was approximately 57-65% in 24 hours. At 4°C, 38-39% of preservatives were released from the film in 30 min, and reached a maximum amount of 49% in approximately 6 h (Chen et al., 1996).

Edible Films

Psomiadou (1996) in his paper shown that aqueous blends of starches and microcrystalline cellulose or methyl cellulose, could be used for the preparation of edible films with improved properties for food packaging applications. The thermal and mechanical behavior, the water and gas permeability of these films were studied in order to find that the addition of methyl cellulose has a beneficial effect upon the aqueous starch/polyol blend, since it acts as a coplasticizer and improve the compatibility of polyol with starch. The incorporation of MC in starch matrix increases the tensile strength and percentage elongation (Psomiadou et al., 1996). The use of methyl cellulose results in a greater increase of water vapour transmission rate because of the higher flexibility of the mix polymer chains and the swelling capability of methyl cellulose compared to microcrystalline cellulose (Debeaufort et al., 1994; Psomiadou et al., 1996).Anedible film made with MC was studied for its water vapor barrier properties.

The initial water content and the water vapor pressure gradient increases the steady-state water vapor transmission rate (WVTR) before the permeation, thus moisture content increases with a decrease in water diffusion. It was due to a clustering phenomenon of water molecules within the film. The WVTR is related to the film water content. For a constant water vapor pressure differential, the steady-state WVTR and water vapor permeability of MC films increases with water content determined after the stationary flow had been established. The permeability and diffusivity of water vapor in MC films were strongly dependent on concentration of MC (Debeaufort et al., 1994).

Low-Fat Products

Aiming to reduce oil uptake during deep-fat frying, Mallikarjunan et al experimented with mashed potato balls of 47 mm diameter which were used as model food system. These samples were coated whit MC film-forming solutions. A reduction of 83.6% in fat uptake by the product was observed for samples coated with methyl cellulose films, in the same way a reduction of 31.1% in moisture loss from the product was observed for the sample coated with MC films. Based on the overall performance in reducing moisture loss and fat uptake, MC exhibits the most significant barrier properties (Mallikarjunan et al., 1997).

Colloid Laxative

This application was evaluated in rats by Tainter showing that methyl cellulose is a synthetic material having physical and chemical properties which would make it suitable for use as a colloid laxative. In doses of 10 g a day of methyl cellulose, the volume of the stools was doubled and frequency was also increased in normal human subjects (Tainter, 1943).

HYDROXYPROPYL METHYLCELLULOSE

Definition

Commonly known as hypromellose, hydroxypropyl methylcellulose (HPMC) is a most commonly used cellulose ether and is used in the food industry as an emulsifier, film former, protective colloid, stabilizer, suspending agent, or thickener (Moura, Aouada, Monteiro, & Azeredo, 2012). HPMC is approved for food uses by the FDA and the EU (The Commission of the European Unions, 1995); Even, its safety in food use has been affirmed by the JECFA (Organization, 2006; Tarantino, 2006). HPMCis a methylcellulose modified with a small number of 2-hydroxypropyl groups attached through ether links to anhydroglucose units of the cellulose(Organization, 2006). It is formulated as a powder and has the following typical composition: HPMC 85-99%, Water 1-10%, sodium chloride 0.5-5% (Tarantino, 2006).

Production

In HPMC production, propylene oxide is used in addition to methyl chloride to obtain hydroxypropyl substitution on the anhydroglucose units. The chemical structure of HPMC is shown in Figure 9. The substituent group ($-OCH_2CH(OH)-CH_3$) contains a secondary hydroxyl on the number two carbon and may also be considered to form a propylene glycol ether of cellulose (Company, 2015) These products possess varying ratios of hydroxypropyl and methyl substitution, a factor which influences organic solubility and the thermal gelation temperature of aqueous solutions(JECFA, 2007; Sarkar, 1979).

The raw materials used for the manufacture of HPMC are as follows: pulp (cellulose), methyl chloride, propylene oxide, caustic soda solution, acid, buffering agent, and water. During the manufacture of HPMC, cellulose fibers are heated with a caustic solution, then they are treated with methyl chloride and propylene oxide (Moura et al., 2012; Tarantino, 2006). The fibrous reaction product is purified and ground to a fine and uniform powder. Main characteristics of HPMC is known to be as a stable substance and to pick up moisture from the air in storage (Tarantino, 2006).

HPMC is regulated by EEC directive 74/329/EEC and its amendments of emulsifiers, stabilizers, thickeners, and gelling agents. It is identified by E464.

Figure 9. The chemical structure of Hydroxy propyl methyl cellulose

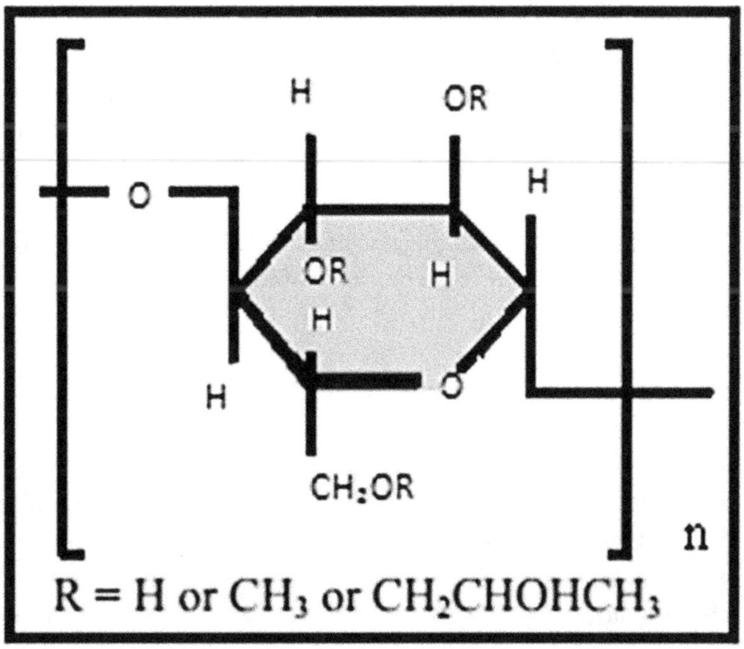

Raw Materials for Food

Generally, the amount of HPMC used in food should not exceed 20 grams/day, this is the limit value recognized as safe. Above this value, one can generate effects such as: undesirable mouth feel, undesirable texture, and off flavor. Low viscosity products are defined as being less than 1500 centipoise and high viscosity products are defined as being greater than or equal to 1500 centipoise. In this way, an advantage of this is that low contents (24%) of HPMC generate high viscosity products having some potential uses such as: peanut butter, frozen dinners and canned pastas, candy bars, cakes, fruit juices, breakfast cereals, veggie burgers and meat substitutes, fish sticks, omelets and egg white substitutes, white breads, sugar substitutes, biscuits, pasta, granola bars, cookies, tortillas and fruit roll-up type snacks (Tarantino, 2006).

Different applications in food have been defined in scientific papers showing how HPMC improves features in different kinds of food products as well as:

Use in Gels

Aqueous solutions of HPMC polymers exhibit a phenomenon of thermoreversible gelation. They gel upon heating and redissolve upon subsequent cooling. When the solutions are heated, the water structure breaks down with an increase in entropy, thus allowing the hydrophobic groups to interact and causing gelation and precipitation (Bodvik et al., 2010; Ford, 1999; Mitchell et al., 1990; Sarkar, 1979; Nitis Sarkar, 1995). Additionally, itis explained that the rates of gelation and gel strengths of MC and HPMC solutions follow a power law relationship with concentration, temperature, substitution grade and molecular weight (Nitis Sarkar, 1995). Ford, in 1999 explained the importance of gel formation in delivery of drugs and how HPMC can form gels with good delivery behavior. The speed of the initial water uptake and the external characteristics of layer gel are important criteria in the control and mechanisms of drug release from matrices (Ford, 1999).

Also sol-gel transformation of aqueous solution of HPMC explaining that thermal gelation is caused by hydrophobic interactions between molecules containing methoxyl group. At low temperatures, the cellulose molecules are hydrated and there is a little polymer–polymer interaction in the entanglement. As the temperature rises the solutions become less viscous and their viscosity decreases (Ford, 1999). Before dehydration is completed, polymer–polymer association occurs leading to an increase of viscosity (Sarkar, 1979). The temperature at which this occurs is called the thermal gelation temperature and is affected by the type of cellulose ether and the presence of ionic materials in solution which may lead to salting out of the polymer (Ford, 1999).

Edible Composite Films

The use of edible coatings films is an alternative method to preserve the postharvest quality and shelf life of fruits and vegetables (Debeaufort, Quezada-Gallo, & Voilley, 1998).A semipermeable barrier to exchange oxygen, carbon dioxide, and water vapor is formed by edible coatings and films, producing weight loss reduction, respiratory rate modification, and senescence delay. Also, it is possible to improve the visual quality of treated product because of the gloss provided by edible coatings and film (Valencia-Chamorro,

S. A., Palou, L., Del Rio, M. A., & Pérez-Gago, 2008). HPMC lipid edible composite films containing low-toxicity chemicals with antifungal properties was studied by Valencia (Valencia-Chamorro, 2008).

HPMCcould be used as edible composite coatings against enzymatic browning (Perez-Gago, 2005; Villalobos-Carvajal, 2009). Villalobos-Carvajal applied coatings made from HPMC containing surfactant mixtures of sorbitanmonostearate and sucrose palmitate. This was made in hydro alcoholic and aqueous media aiming to analyze the hydrophilic–lipophilic balance of the surfactant mixtures, the effect of the hydrocolloid surfactant ratio, and the solvent type used as dispersing media on the water vapour barrier properties of the films and their optical effect when applied to carrot slices. This analysis was carried out taking into account the physical characteristics of the film-forming dispersions (viscosity and surface tension and stability) and their influence on the film structure (Villalobos-Carvajal et al., 2009).

The results shows that the surface tension, the viscosity and stability (particle size and flocculation rate) of the film-forming dispersions were affected by the solvent type, hydrocolloid/ surfactant ratio and the hydrophilic–lipophilic balance. All those aspects affects the water vapour barrier properties of the film and the best water vapour barrier properties were found in the coating made from hydroalcoholic dispersions with intermediate values of surfactant's ratio, the hydrophilic–lipophilic balance (Villalobos-Carvajal et al., 2009).

HPMC Films

Recently Moura et al., employed the microfluidics technology to decrease the size of cellulose microfibers and to study the effect of addition of such fibers on the properties of HPMC films to be applied in food packaging. Application of HPMC improve mechanical and barrier properties of films significantly as well as percentage of elongation of the films, particle size and oxygen permeability (Moura et al., 2012). Moura showed that the oxygen permeability of the films containing fibers was markedly low, this feature being probably associated with the more compact structure of these materials determined by the dispersion of fibers in the HPMC matrix. Transparent, flexible, homogeneous, surface smooth films without pores and cracks were obtained after drying the film-forming solutions containing fibers and HPMC (Moura et al., 2012).

Antihyperglycemic and Antioxidative Effects

The effect of dietary feeding of HEMC and it is on the glucose metabolism and antioxidative status in mice under high fat diet conditions was investigated.

Diet supplementation HPMC was found to counteract the high fat-induced hyperglycemia and oxidative stress via regulation of antioxidant and hepatic glucose-regulating enzyme activities. These findings illustrate that HPMC and HPMC were similarly effective in improving the glucose metabolism and antioxidant defense system in high fat-fed mice and they may be beneficial as functional biomaterials in the development of therapeutic agents against high fat diet- induced hyperglycemia and oxidative stress (Ban, Rico, Um, & Kang, 2012).

The scientific study reveals that diet supplementation of HPMC could lower the serum cholesterol content in hypercholesterolemic human subjects and normalize the metabolic abnormalities in obese mice. It was also reported that consumption of HPMC significantly reduced postprandial glucose and insulin responses in overweight and obese men and women (Ban et al., 2012). Diet supplementation with HPMC reduces the blood glucose concentrations in obese mice hamsters, and human subjects.

This glucose-lowering effect of HPMC has been attributed to its high viscosity, a common property of soluble fibers. When ingested, the HPMC forms a viscous solution in the gastrointestinal tract and acts as a barrier to the gastrointestinal absorption of glucose, and thereby reduces the blood glucose concentrations (Ban et al., 2012).

HPMC Capsules

HPMC polymer is now used as shell material for capsules. HPMC and gelatin capsules the effect of humidity on the mechanical properties was examined and found that gelatin capsules exhibit higher stiffness and hardness values compared to the HPMC capsules. This was due to the fact that gelatin capsules have relatively high moisture content (13-16%) when compared to 4-6% of HPMC capsules. Accordingly, it is possible to retain much of the powder adhering to the inner surfaces of capsules by an interaction of the powdered materials with the gelatin capsule resulting in much of the dose failing to leave the device (Al-tabakha, 2010).

The capsules of HPMC are avowed to enteric and colonic delivery by using coating materials of different pH solubility. The coating of HPMC capsules and gelatin capsules were compared, which showed that the coating of HPMC was straightforward, while gelatin capsules were not suitable for direct coating. Equally, shell surface of HPMC and gelatin capsules were quite different, HPMC capsules were rougher than that of gelatin capsules. Both were examined by scanning electron microscope, which concluded that this may provide good adhesion to the coating (Al-tabakha, 2010).

ETHYL CELLULOSE

Definition

Ethyl cellulose is an ethyl ether of cellulose organo soluble, thermoplastic and versatile used principally as a tablet binder. Such as cellulose, the backbone of the molecule of ethyl cellulose is based on repeating anhydroglucose units (Majewicz, Erazo-Majewicz,. and Podlas, 2002).

Industrial applications are approved for using ethyl cellulose in regulated markets such as food and pharmaceuticals, personal care, food, feed, printing ink. In pharmaceuticals, ethyl cellulose improve the appearance and strength of capsules and tablets, mask the taste of bitter actives and enable controlled release formulations. In food ethyl cellulose can be used as flavor fixatives, film forming and binding, helping to make delicious flavors last longer (AQUALON, 2002; The Dow Chemical company, 2013).

Production

Ethyl cellulose is a cellulose ether obtained by the reaction of ethyl chloride with alkali cellulose (AQUALON, 2002), as expressed by the reaction:

$RONa + C_2H_5CI\ ROC_2H_5 + NaCI,$

Where, R represents the cellulose radical. The chemical structure of ethylcellulose is shown in Figure 10.

Figure 10. The chemical structure of Ethylcellulose

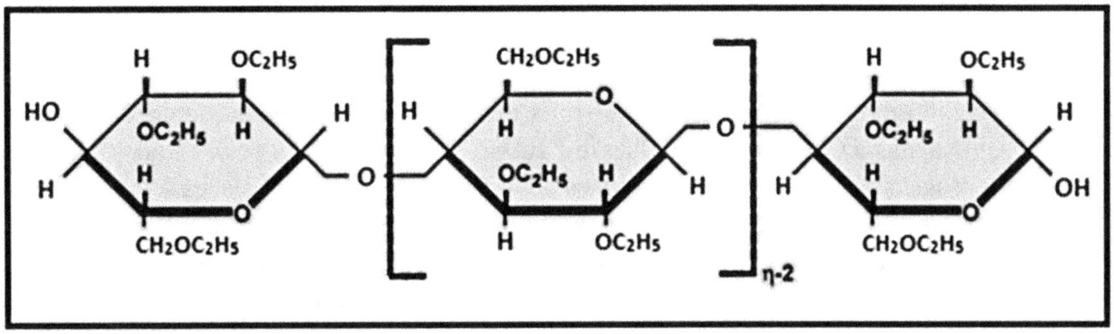

The percentage of ethoxy groups (- OC_2H_5) in the molecule varies between 44% and 51% on a dry weight basis (equivalent to not more than 2.6 ethoxyl groups on average per anhydroglucose unit) (Larsen, Nørby, Reffstrup, & Beltoft, 2004). Ethyl cellulose is prepared from wood pulp or cotton by treatment with alkali and ethylation of the alkali treated cellulose with ethyl chloride. The resulting product is then further steamed and dried (Larsen, Nørby, Reffstrup, & Beltoft, 2004).

Raw Materials for Food

Ethyl cellulose offers different functions in various food categories. Some of the important properties of ethyl cellulose are viscosity modifier, thickener, film-forming, stabilizer, filler and provide thermal gelation for processed foods. Ethyl cellulose exerts functional changes in food products without undergoing or initiating chemical changes that would alter the nutritional value of the food products. Ethyl cellulose is currently used as a food additive (The Dow Chemical company, 2013). Ethyl cellulose is approved as a multipurpose additive for its use as a binder and filler, as a component of protective coating for vitamin and mineral tablets, and as a fixative in flavoring compounds (The Dow Chemical company, 2013).

Ethyl Cellulose Is Widely Used

As a pharmaceutical excipient: Ethyl cellulose is a non-toxic, stable, compressible, passive, hydrophobic polymer that has been commonly utilized to prepare pharmaceutical dosage forms, even though it is not a biodegradable polymer, because of its non-toxicity, good stability, ease of processing and low cost.

The properties of ethyl cellulose sustained release products, including film-coated tablets, microspheres, microcapsules and matrix tablet for both soluble and poorly soluble drugs have been reported. Ethyl cellulose is used for microencapsulation of various pharmaceuticals to stabilize them against unwanted interactions. It has been used as a polymeric coating, for controlled release in oral preparations, in transdermal films and patches, pharmaceutically useful aqueous latex dispersions, and as a particle emulsion stabilizer (Gunduz, Ahmad, Stride, & Edirisinghe, 2012).

Ethyl cellulose in combination with water-soluble polymers, e.g. polyethylene glycols, hydroxypropyl cellulose and HPMC, is widely used in the preparation of sustained release films (Rowe, 1986). Different proportions of ethyl cellulose were evaluated in various tablet formulations as excipient regarding characteristics such as: physical appearance, hardness, friability, weight variation and drug content uniformity

(Gaunø et al., 2013; Gunduz et al., 2012; Phadke, Manjeshwar, & Aminabhavi, 2014; Rajabi-Siahboomi, 2013; Raymond. Rowe, Paul Sheskey, 2006; Rowe, 1986; Sajeev, Vinay, Archna, & Saha, 2002).

Water insoluble films could be produced by dissolving ethyl cellulose in an organic solvent or solvent mixture. More durable and strong films are produced using ethyl cellulose with higher viscosity grades. Ethyl cellulose films may be modified to alter their solubility by the addition of hypromellose or a plasticizer (Raymond C. Rowe & Paul Sheskey, 2006). The effectivity of ethyl cellulose microcapsule to release drugs is a function of the microcapsule wall thickness and surface area. It is possible to produce hard tablets with poor dissolution and low friability by using ethyl cellulose. Additionally in tablet formulations, ethyl cellulose may be employed as a binder, the ethyl cellulose being blended dry or wet-granulated with a solvent such as ethanol (95%). (Rowe, R. C., Sheskey, P. J., & Weller, 2006).

The use of ethyl cellulose in injection molding technology has also been studied, where the core contained drug as hypromellose, and glycerides and was surrounded by a hot-melt extruded ethyl cellulose membrane. The length of the melt extruded device had a greater effect on the drug release rate. Ethyl cellulose melt extruded layer provides a barrier membrane for water ingress and drug release (Upadhye, S. B., & Rajabi-Siahboomi, 2013).

As a Thickener or Binder in Hydrophobic Matrices

Epoxidized soybean oil, which is a biomass-derived resource, was first used as a novel plasticizer for ethyl cellulose film preparation. Surface morphologies, mechanical performances, thermal properties, oxygen and water vapor permeability of plasticized ethyl cellulose films were studied to evaluate the plasticizing effect of epoxidized soybean oil and explore the plasticization mechanisms. Results showed that epoxidized soybean oil was an effective plasticizer that outstripped conventional plasticizers, i.e. dibutyl phthalate and triethyl citrate in producing high-quality films (Yang et al. 2014; Enayatifard et al. 2009; Tiwari et al. 2003 and Kuksal et al. 2006).

The characteristic of ethyl cellulose to prolong the release rate of the drug could be due to the decreased penetration of the solvent molecules in the presence of hydrophobic polymer leading to decreased diffusion of the drug from the matrix. Use of ethyl cellulose in wax based system also resulted in retardation of drug release (Tiwari, Murthy, Pai, Mehta, & Chowdary, 2003).

It is possible to enlarge (50% and 90%) the time to release drugs (3 and 12 hours) by using ethyl cellulose. For example, ethyl cellulose brings rigidity when it is mixed with eudragits helping to retain the drug in matrix and did not allow rapid diffusion of soluble drug from the matrix. The hydrophobic nature of ethyl cellulose seems to have contributed towards the reduction in the penetration of the solvent molecules into the matrix (Kuksal, Tiwary, Jain, & Jain, 2006).

As an Emulsion Stabilizer in (Salad) Dressings and Microencapsulation Agent

Ethyl cellulose was demonstrated to represent a new type of particulate polymeric emulsifier. Measurements of the o/w (water-in-oil) interfacial tensions in the presence of ethyl cellulose indicate the ability of this substance to act as an emulsifier (Das & Rao, 2006; Kar & Choudhury, 2007; Melzer, Kreuter, & Daniels, 2003).

As emulsifier, the ethyl cellulose concentration ranges from 0.2 to 2%.When the concentration of EC is increased the interfacial tension with water is reduced, and also with lipids. It was observed that it exhibits a positive adsorption of the polymer at the oil/water interface.

In the case of colloidal EC particles their affinity to water and lipid obviously depends on the temperature (Melzer et al., 2003).

The affinity of the precipitated colloidal particles for the oil phase becomes better soluble in the lipid with temperature. The affinity for the lipids can be reduced with temperatures below 20°C and oil/water systems are formed. The high oil volume fraction of the system does not influence negatively the emulsion stability. This indicates that stabilization of emulsions with solid particles partitioning between oil and water phases is highly effective and provides an interesting alternative to conventional emulsion stabilization (Melzer et al., 2003). In Rao (Das & Rao, 2006).

It is possible to see Zidovudine loaded EC microspheres by water in oil in oil (w/o/o) double emulsion solvent diffusion method with high entrapment capacity and sustained release. Spherical free flowing microspheres were obtained. The drug-loaded microspheres showed 32 - 55% entrapment capacity.

If formulation parameters and processing are changed, it is likely to alter significantly the in vitro release profile aiming to give sustained release of drug from the microspheres(Das & Rao, 2006).

As a Barrier Layer Controlling Diffusion of Ingredients

To produce these extended release tablet dosage forms, active ingredient is conventionally compounded with cellulose ethers like ethyl cellulose, methylcellulose or HPMC with or without excipients and the resulting mixture is pressed into tablets. Matrix type drug delivery systems as carriers for the active ingredients are interesting and promising option in developing an oral controlled release system. Hydrophilic cellulose polymers are commonly used as excipient base in tableted matrix systems (Maity, Nandy*, Mittal, 2014).

The effectiveness of these hydrophilic matrix systems is based on the successive process of:

- Hydration of the cellulosic polymer.
- Gel formation on the polymer's surface.
- Tablet erosion.
- Subsequent and continuous release of drug. (Maity, Nandy*, Mittal, 2014)

As a Taste Masking Agent in Some Dietary Food Supplements

EC and other cellulose polymers were used to coat vitamins and minerals improving taste and palatability, providing an orally administrable nutritional supplement, such as a chewable prenatal vitamin/mineral supplement (Deepak, Dinesh, Mankaran, & Gurmeet, 2012; Robert Yang, Shri Sharma, Shan-Shan Sheu, 1989; Valley, 2002). The tablet is particularly suitable for administration of vitamins and minerals to women during pregnancy (Deepak et al., 2012; Valley, 2002).

Sharma invented a liquid composition containing "coacervate of cholestyramine with a cellulose hydrocolloid" derivative. The term "coacervate" means the coagulation of two hydrophilic substances of opposite charge (Robert K. Yang, Sharma, Shan-Shan Sheu, 1989).

As a micro-encapsulation agent: Specifically used in cases when water-soluble products cannot be used because of processing problems or water sensitivity of the active ingredient (Uddin,. Hawlader, H., 2001; Prasertmanakit, Praphairaksit, Chiangthong, & Muangsin, 2009; Zheng, Ding, Zhang, & Sun, 2011). Polyphenolic extracts have been microencapsulated using ethyl cellulose as a coating material. The antioxidant activity of polyphenols could be protected by microencapsulation; it was shown by antioxidant capacity assays. The storage stability of bayberry polyphenols against adverse environment was also remarkably improved by this microencapsulation (Zheng et al., 2011).

Features like hydrophobicity, safety, compact film forming, and stability allow the use of ethyl cellulose to protect C vitamin. Microencapsulated ascorbic acid could prevent changes in some characteristics of the ascorbic acid as well as colour, even retard its core release rate, and generally mask its acid taste. Results show that the microencapsulated product size decreases with an increase of molecular weight of ethyl cellulose, the most probable reason is that the aggregation of microcapsules decreases with the increase of molecular weight of ethyl cellulose. (Uddin, Hawlader, 2001).

Ethyl cellulose microcapsules were developed for use as a drug-delivery device for protecting folic acid from release and degradation in the environmental conditions of the stomach, whilst allowing its release in the intestinal tract to make it available for absorption (Prasertmanakit et al., 2009).

HYDROXYETHYL METHYL CELLULOSE

Definition

Hydroxyethyl methylcellulose (HEMC) is also a methylcellulose derivative, but with ethyl group substitution. Hydroxyethyl methyl cellulose is a nonionic cellulose derivative that dissolves readily in cold or hot water. From an industrial viewpoint, HEMC can be produced more cheaply than HEMC though functionally, they are similar. Hydroxyethyl methylcellulose provides many of the same benefits as other methylcellulose derivatives, such as the ability to efficiently thicken and provide water retention. The viscosity of an aqueous HEMC solution is defined by the frictional resistance of a fluid to shear forces or to compressive stress, hence to flow (Ban et al., 2012).

Production

HEMC is a viscous soluble fiber produced from cellulose through a series of chemical modifications (Ban et al., 2012). The cellulose alkalisation carried out in the presence of liquid methyl chloride and the alkali cellulose being reacted with ethylene oxide.

In the production of HEMC at least 4 parts of methyl chloride are used to 1 part of air dry cellulose during the action of propylene oxide on sodium cellulose; i.e. at least 12.8 moles of methyl chloride are required per mole of air-dry cellulose. Alkalization is carried out before the reaction with propylene oxide by immersing the cellulose sheets in 30 to 50% NaOH and subsequently removing the sodium hydroxide under pressure, ranging between 2.5 to 2.8% by weight (Klaus). The chemical structure of HEMC is shown in Figure 11.

Raw Materials for Food

The main application in food is described by Ban (2012) in which HEMC and they are used to suppress high fat-induced hyperlipidemia in mice. The hypoglycemic effect of HEMC, indicates that it may also be beneficial in the treatment and management of high fat diet-induced hyperglycemia (Ban et al., 2012). Supplementation of HEMC in the diet significantly reduces the insulin level in mice relative to those fed with a high fat diet alone. Moreover, the protective effect of HEMC against high fat induced-lipid

Figure 11. The chemical structure of Hydroxyethylmethyl cellulose

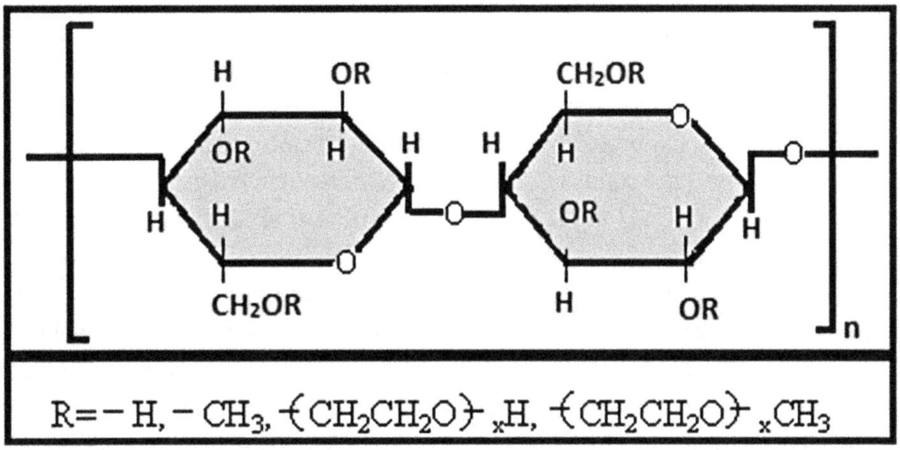

peroxidation and oxidative stress may partly be due to a mechanism involving the elevation of the activities of antioxidant enzymes (Ban et al., 2012).

Cellulose Without Modifications Used in Food

The unmodified cellulose types most commonly used in foods are: bacterial nanocellulose, microcrystalline cellulose and nano-fibrillated cellulose. Cellulose nanofibrils can be used either in the form of low aspect ratio microcrystalline cellulose (MCC) or as high aspect ratio micro-fibrillated cellulose (MFC) (Henriksson, 2008). For nano-fibrillated cellulose, it is necessary to treat the cell wall pulp using homogenization to facilitate the fibrillation. To obtain nanocrystals the pulp need to be treated with a combination of beating and enzymatic treatment prior to homogenization (Sieg, 2009).

BACTERIAL CELLULOSE

Definition

Bacterial cellulose (BC) is an environmentally friendly polymeric material, which is receiving increased attention. Several bacteria of Gluconacetobacter genus produce bacterial cellulose as fine ribbons.

Bacterial cellulose can form hydrogels and its rheological properties have led to uses in food industry such as a bacterial cellulose gel, called Nata, which is eaten as a dessert, multifunctional food ingredients to control the properties of food or beverage as a thickener, stabilizer and texture modifier (Okiyama et al., 1992, 1993; Shi et al., 2014). Bacterial cellulose can also be used as a food packing material because of its edibility and biodegradability (Fan, Wu, Hu, Wu, & Liao, 2011).

Production

Bacterial cellulose is produced by microbial fermentations (Chung & Shyu, 1999; Hong & Qiu, 2008; Jonas & Farah, 1998; Klemm et al., 2005; Nugroho & Aji, 2015) or fermentation of immobilized strain of some genus such as: Glucon*acetobacter, Achromobacter, Aerobacter, Agrobacterium, Alcaligenes, Pseudomonas, Rhizobium, Sarcina and Zooglea.* The most used strain to produce bacterial cellulose is *Gluconacetobacterxylinum* or other stains i.e. *Gluconacetobacterswingsiisp* and *gluconacetobacterme-dellensis* (Castro et al., 2011, 2012; Hong & Qiu, 2008; Martins et al., 2009; Nugroho & Aji, 2015). It is possible to produce BC in different regions and climatic conditions and it is possible too to control the micro-organisms producing bacterial cellulose in the requisite quantities and time span. The optimum pH for bacterial cellulose production range between a neutral to slightly acidic pH (pH between 3 and 7) and temperatures around 28°C. This kind of cellulose contains no lignin or other contaminants so that the energy utilized in the purification of plant cellulose can be avoided. Agro-industrial wastes can be used as a growth medium due to its low cost (Castro et al., 2011; Klemm et al., 2005). Bacterial cellulose could be produce from various fruit juices as well as orange, pineapple, apple, Japanese pear and grape (Kurosumi, Sasaki, Yamashita, & Nakamura, 2009).

In the forming process of BC, polymerization and crystallization occur together. The formed nanoribbons have cross-sectional dimensions in the nm range, which can then aggregate to form microfibrils with a dimension of 50-80 nm in width, and a thickness of 3-8 nm (Tabuchi, 2007). Bacterial cellulose possesses unique physical and mechanical properties, such as its high water holding capacity, crystallinity, tensile strength and Young modulus (Pecoraro, 2008). In bacterial cellulose, it is believed that cellulose molecules synthesized in the interior of bacterial cell are spun out of cellulose export components' or nozzles to form a protofibril(Castro et al., 2012). Schematic illustration of BC biogenesis and fibril formation is shown in Figure 12. Commonly it is assumed that a bacterial produces a certain number

Figure 12. Schematic ilustration of bacterial cellulose biogenesis and fibril formation

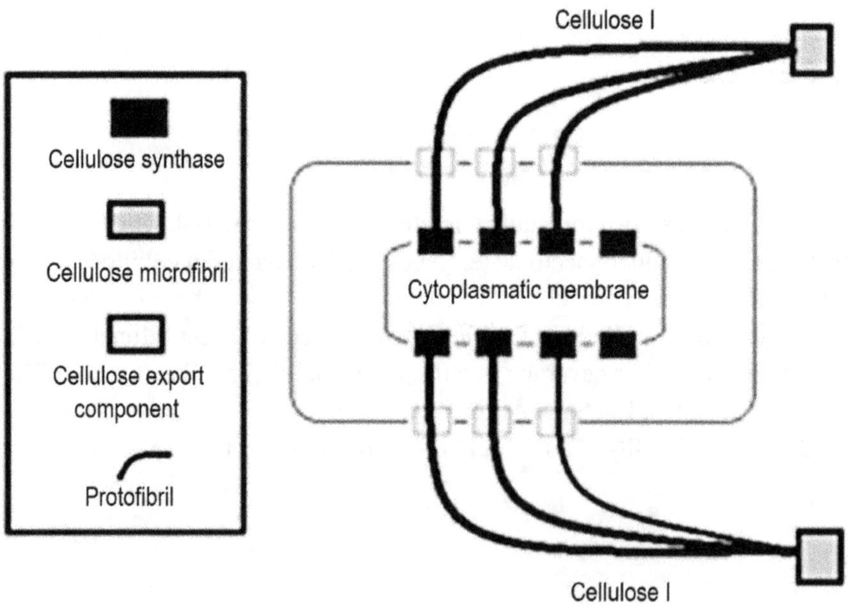

of chain initiators during its generation time to which monomer units are added to form cellulose and the population of bacteria obeys the law of bacterial growth (Iguchi, Yamanaka, & Budhiono, 2000).

Raw Materials for Food

Bacterial cellulose in food was known in the beginning as nata, originated from Philippines during the 1990s and spread to nearby countries, is a traditional dessert in Southeast Asia (Phisalaphong, M., & Chiaoprakobkij, 2012). Due to the smooth mouth feel and simple manufacturing process, Nata de coco has become very popular and is now quickly spreading worldwide. There are many kinds of Nata, such as *Nata de coco and Nata de piña*, with their flavor being controlled by the culture medium resource: Nata de coco uses coconut while Nata de piña uses pineapple flavor (Okiyama et al., 1992).

Bacterial cellulose as an ingredient in food formulations acts to avoid flavor interaction, to enhance the stability of food in a wide range of pH, temperature and freeze-thaw conditions. Potential applications are for thickening, gelling, stabilizing, and water-binding (Okiyama, Motoki, & Yamanaka, 1993; Okiyama et al., 1992, 1993). Since there is no enough literature about the applications of bacterial cellulose in food, Okiyama, in 1992 explained that applications of bacterial cellulose in the food industry are limited to those indicated in the report of Yamanaka *et al.* and *nata* (gelatinous cellulose in the Philippines). This viewpoint is not really different now, when the application for bacterial cellulose in food is limited, especially to beverages products.

Some functions of bacterial cellulose in modifying the rheology of food are:

Thickening Agent

Bacterial cellulose was compared with xanthan gum in terms of thickener, showing extremely high water-holding capacity. After adding BC into pasty condiment, the stickiness of the condiment was markedly improved so that it is easier to spoon quantitatively (Okiyama et al., 1992, 1993; Shi et al., 2014).

Stabilizing Agent

From the viewpoint of fluid property, bacterial cellulose seems feasible as a stabilizer for beverages, because of its network structure of nanoribbons. Product containing bacterial cellulose can retain their humidity for at least one month storage. The shape of ice cream containing BC is retained for at least 60 min after removing from freezer while control ice cream would melt away completely over the same period (Okiyama et al., 1992, 1993; Shi et al., 2014).

Gelling Agent

Bacterial cellulose was tasted in tofu and kamamboko. The gel strength of Tofu increased significantly with 0.2-0.3% bacterial cellulose, which provided better texture and firmness. Bacterial cellulose endows to Kamaboko produces better hardness and fracturability and the springiness almost was eliminated. This modified Kamaboko can better withstand the aging process and still meet the demand of sensory evaluation (Okiyama et al., 1992, 1993; Shi et al., 2014).

Suspending Agent

Bacterial cellulose added to chocolate drink prevents precipitation of the cocoa because the bacterial cellulose mesh can retain the cocoa particles. In addition, bacterial cellulose shows a great heating stability after heat sterilization so that the viscosity remains unchanged after the heat treatment (Shi et al., 2014).

Resistance to Washing, Heat, Freezing, Acidification, and Alkalization of Color

Sheuet al.,2000 (Sheu, Wang, & Shyu, 2000) used bacterial cellulose-nata combined with *monascus-purpureus* as a potential new foodstuff as vegetarian meat or seafood replacement and shows that the complex has good resistance to washing decoloration. The color and texture of the complex were like liver, lean meat, or tuna meat (sashimi). It also provides high fiber content, limited calories, and healthful nutrients. Moreover, the waste broth of fermentation could be further used as a source of water-soluble pigments (Sheu et al., 2000).

MICROCRYSTALLINE CELLULOSE

Definition

Microcrystalline cellulose (MCC) fibers had attracted more interest in the past, recently there has been an upsurge in using lower levels of longer fibers for specialized uses and, mainly, for enhancing the functional properties of foods (Psomiadou et al., 1996).

Microcrystalline cellulose is a purified, partially depolymerized cellulose that occurs as a white, odorless, tasteless, crystalline powder composed of porous particles. It is commercially available in different particle sizes and moisture grades that have different properties and applications (Raymond Rowe, Paul Sheskey, 2006). The degree of polymerization is typically less than 400. Not more than 10% of the particles have a diameter below 5 μm (JECFA, 2000).Micrography of microcrystalline cellulose is shown in Figure 13.

Production

Microcrystalline cellulose (MCC) is produced by treating natural cellulose with hydrochloric acid to dissolve the amorphous regions of the polysaccharide, leaving behind the less reactive crystalline regions as fine crystals. The viscosity of dispersions of this product is both pH and heat invariant. A number of varieties are available — powdered, bulk-dried colloidal, spray-dried colloidal (with CMC), and spray-dried with sweet whey. Microcrystalline cellulose dispersions have been shown to exhibit both thixotropic and pseudoplastic behavior (Stephen, 2006).

Microcrystalline cellulose can be assayed by employing the ferrous ammonium sulfate titration technique. The assay specifies not less than 97% cellulose. Purified cellulose powders for use as dietary fiber of average fiber length varying from 25mm to 120mm (Stephen, 2006).

Microcrystalline cellulose is regulated by EEC directive 74/329/EEC and its amendments of covering emulsifiers, stabilizers, thickeners, and gelling agents. Microcrystalline cellulose is identified by E460.

Figure 13. Micrography of microcrystalline cellulose

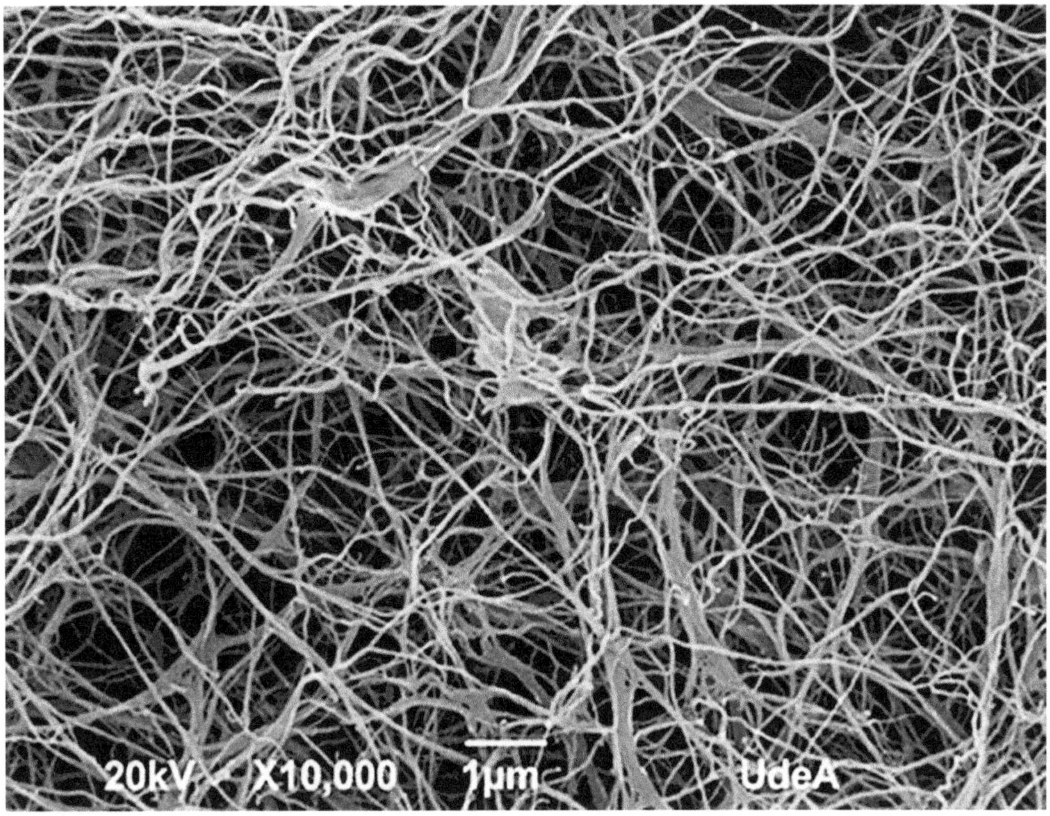

Raw Materials for Food

Many important industrial food applications have been listed and considered such as raw material for food by Food and agriculture organization (FAO) and registered in codex alimentarius.

Functional Classes

Anticaking agent, bulking agent carrier, emulsifier, foaming agent, glazing agent, stabilizer, thickener (JECFA, 2000).

Microcrystalline cellulose is approving to use in products as well as:

- Coffee, coffee substitutes, tea, herbal infusions, and other hot cereal and grain beverages, excluding cocoa (For use in ready-to-drink products and pre-mixes for ready-to-drink products only). In this products microcrystalline cellulose is used as stabilizer (JECFA, 2000; Villagran, F. V., Dria, G. J., Butterbaugh, J. L., & Small, 2004).
- Dried pastas and noodles. In this products microcrystalline cellulose is used as thickener (For use in noodles, gluten-free pasta and pasta intended for hypoproteic diets only) (Battista, 1962; Huang & Merolla, 1994; JECFA, 2000).

- Dried whey and whey products, excluding whey cheeses. In this products microcrystalline cellulose is used as stabilizer (Battista, 1962; Gamay, 1996; JECFA, 2000; Koide, K., Yoneda, Y., & Musashi, 1983).
- Fermented milks (plain), heat-treated after fermentation. In this products microcrystalline cellulose is used as stabilizer or thickener (Gamay, 1996; JECFA, 2000).
- Fermented milks (plain) not heat-treated after fermentation (For use as a stabilizer or thickener only and for use in reconstituted and recombined products only). In this products microcrystalline cellulose is used as stabilizer (JECFA, 2000; Rhodes, 1978).
- Fresh meat, poultry, and game, comminuted (For use in fresh minced meat which contains other ingredients apart from comminuted meat only). In this products MCC is used as stabilizer or thickener (Battista, 1962; JECFA, 2000; Kamada, E., & Mochihara, 2001).
- Fresh meat, poultry, and game, whole pieces or cuts (For use in glaze, coatings or decorations for fruit, vegetables, meat or fish only). In this products microcrystalline cellulose is used as stabilizer or thickener (Battista, 1962; JECFA, 2000; Kamada, E., & Mochihara, 2001).
- Fresh pastas and noodles and like products (For use in noodles only). In this products microcrystalline cellulose is used as stabilizer or thickener (Battista, 1962; JECFA, 2000; Lometillo, J. E., & Wolcott, 1983; Robbins, M. D., & Rodriguez, 1984; Wren, M., & Mullen, 1974).
- Frozen egg products. In this products microcrystalline cellulose is used as thickener (Battista, 1962; JECFA, 2000; Lometillo, J. E., & Wolcott, 1983).
- Liquid egg products. In this products microcrystalline cellulose is used as stabilizer (Battista, 1962; JECFA, 2000).
- Other sugars and syrups (e.g., xylose, maple syrup, sugar toppings). Excluding maple syrup. In this products microcrystalline cellulose is used as low fat (Baccus Jr, 1991; Battista, 1962).
- Pasteurized cream (plain), excluding products conforming to the standard for cream and prepared creams (reconstituted cream, recombined cream, prepackaged liquid cream). In this products microcrystalline cellulose is used as stabilizer (CODEX STAN 288-1976)(Singer, N. S., Chang, H. H., Tang, P., & Dunn, 1990; Tiemstra, 1971).
- Renneted milk (plain). In this products microcrystalline cellulose is used as stabilizer (JECFA, 2000; Tejayadi, 2004; Tiemstra, 1971).
- Sterilized and UHT creams, whipping and whipped creams, and reduced fat creams (plain). In this products microcrystalline cellulose is used as stabilizer (Hogan, D. T., Bertrand, D. T., Tuazon, M. T., & Tuason, 1998; JECFA, 2000; Koide, K., Yoneda, Y., & Musashi, 1983).

NANOFIBRILLATED CELLULOSE

Definition

Nanofibrillated cellulose (NFC) is also referred as microfibrillated cellulose (MFC) or just nanocelullose. The material can be obtained by homogenization of cellulose using a high-pressure homogenizer. The microfibrils have a diameter in the nano scale and a small length the micro size range. It is obtained as a highly viscous gel with a solid content of a few percent (Strom, G., Ohgren, C., & Ankerfors, 2013).

Production

The preparation of this material has been disclosed in a patent from the ITT Corporation. Essentially, slurry of cellulose is passed through a small orifice under conditions of high shear and at a great pressure differential. This treatment disrupts the cellulose into microfibrillar fragments. The assay for nanofibrillated cellulose (NFC) is described in the Food Chemicals Codex (FCC) and involves titration by ferrous ammonium sulfate of a potassium dichromate and sulfuric acid-treated solution of the cellulose. (Stephen, 2006).

Nanofibrillated cellulose has considerably water-retention capability and is less prone to precipitation in the cupriethylenediamine residue test (cuene test). The change in DP (Degree of polymerization) with respect to the pulp feed stock is kept at a minimum during microfibrillation. Suspensions of nanofibrillated cellulose are shear thinning (pseudoplastic), exhibit slight thixotropic behavior, and apparently do not suffer a viscosity drop on heating (Stephen, 2006).

Raw Material for Food

A food product containing nanofibrillated cellulose is prepared by mixing together an edible liquid which swells cellulose, a food additive and fibrous cellulose to form a liquid suspension. The suspension is repeatedly passed through a small diameter orifice in which the mixture is subjected to a pressure drop of at least 3000 psig and a high velocity shearing action followed by a high velocity decelerating impact. The process converts the cellulose into nanofibrillated cellulose and forms a stable homogeneous suspension (Strom, G., Ohgren, C., & Ankerfors, 2013).

The principal function, at the moment of nanofibrillated cellulose in food is in packaging or films as a barrier against air, oxygen and water vapour permeability (Hult,Iotti, & Lenes, 2010; Lavoine, Desloges, & Bras, 2014; Liu, Lin, Chen,Huang, Cao & Wang, 2013). Nonetheless, it has been used for producing fillings, crushes, soups, gravies, puddings, dips, toppings, low calorie products and other food products acting as a thickener, flavor carrier and suspension stabilizer (Domingo C. Tuason, Jr., Thomas A. Ruszkay, 1961; Ling Chen, Rongsheng Roger Ruan, Jack Edgar Johnson, Paul Bradley Addis, 1998; Paul Bradley Addis, Rongsheng Roger Ruan, Joseph Keenan, 2014a, 2014b)

CONCLUSION

This chapter embodies the most important applications of cellulose in food from chemical modified cellulose to physic modified cellulose. Different applications are discussed in this article including environmentally friendly cellulose fiber technologies for food packaging, bacterial cellulose in foodstuff, and other applications as well as thickening agent (Shi et al., 2014), stabilizing agent, gelling agent, suspending agent were highlighted with future aims, strategies, and perspectives of cellulose research and its applications.

In summary, it is feasible to combine cellulose and cellulose derivatives with other additives as well as xanthan gum, guar gum, locus bean gum and other gums aiming to improve the texture of dressings. Due to the use of cellulosic and cellulose derivatives in sauces and soup it is possible to know that instant sauces and soup rapidly thicken while cooling. As coat material, cellulose and cellulose derivatives can be used for coating of products sensitive to humidity or the water absorption properties. Some problems with exuded fluids from food products can be solved with cellulose and cellulose derivatives.

Cellulose and cellulose derivatives have important due to the current trend for healthier foods. Normally the calories reduction imply modification of product formulations because this process is accompanied by a reduction of the sugar. Actually, it is possible to find many information about cellulose and cellulose derivatives in low calories products, even about the reduction of oil absorption and the formation of edible barriers in composite foods.

In bakery products cellulose or cellulose derivatives are often coated (i.e. extruded products) with a batter and breadcrumbs improving adhesion of batter to the stick, giving stability and reducing the oil absorption.

Bacterial cellulose is the less studied type compared to other cellulose or cellulose derivatives in food applications. This brings a big possibility to develop scientific studies in different fields when bacterial cellulose could act as thickening agent (Shi et al., 2014), stabilizing agent, gelling agent and suspending agent.

REFERENCES

Abughoush, M., Al-Mahasneh, M., Samhouri, M., Al-Holy, M., & Herald, T. (2008). Formulation and Fuzzy Modeling of Viscosity of an Orange-Flavored Carboxymethylcellulose-Whey Protein Isolate Beverage. *International Journal of Food Engineering*, *4*(7), 12. doi:10.2202/1556-3758.1496

Addis, P. B., Ruan, R. R., Keenan, J. M., & Geleva, D. (2011). Medical and nutritional applications of highly refined cellulose. *United States Patent No 8,026,226, 2*(12).

Addis, P. B., Ruan, R. R., Keenan, J. M., & Geleva, D. (2014). Medical and nutritional applications of highly refined cellulose. *United States Patent No 8,623,841, 2*(12).

Adeyeye, M. C., Jain, A. C., Ghorab, M. K. M., & Reilly, W. J. (2002). Viscoelastic Evaluation of Topical Creams Containing Microcrystalline Cellulose/Sodium Carboxymethyl Cellulose as Stabilizer. *AAPS PharmSciTech*, *3*(2), 16–25. doi:10.1208/pt030208 PMID:12916945

Al-tabakha, M. M. (2010). HPMC Capsules: Current Status and Future Prospects. *Journal of Pharmacy & Pharmaceutical Sciences*, *13*(3), 428–442. doi:10.18433/J3K881 PMID:21092714

AQUALON. (1999). *Sodium Carboxymetylcellulose Physical and Chemical Properties*. Hercules.

AQUALON. (2002). *Ethylcellulose (EC) Physical and Chemicals Properties*. Retrieved from http://www.ashland.com/industries/pharmaceutical/oral-solid-dose/aqualon-ethylcellulose

Arancibia, C., Costell, E., & Bayarri, S. (2011). Fat replacers in low-fat carboxymethyl cellulose dairy beverages: Color, rheology, and consumer perception. *Journal of Dairy Science*, *94*(5), 2245–2258. doi:10.3168/jds.2010-3989 PMID:21524514

Arbuckle, W. S. (1977). Ice cream. Westport, CT: The AVI. Publishing Inc.

ASHLAND. (2014). *Improved Mouthfeel in Low-juice and Reduced-calorie Beverages.* Food Technology Report. Retrieved from http://www.ashland.com/Ashland/Static/Documents/ASI/Food/FTR_020_Low_Juice_Reduced_Calorie_Beverages.pdf

Baccus, J. B., Jr. (1991). Process for makin a low calorie beverage. *United States Patent No 4,986,994.*

Ban, S. J., Rico, C. W., Um, I. C., & Kang, M. Y. (2012). Antihyperglycemic and antioxidative effects of hydroxyethyl methylcellulose (HEMC) and hydroxypropyl methylcellulose (HPMC) in mice fed with a high fat diet. *International Journal of Molecular Sciences, 13*(3), 3738–3750. doi:10.3390/ijms13033738 PMID:22489179

Battista, O. A. (1962). Food Composition Incorporation Cellulose Crystallite Aggregates. *United States Patent Office No 3,023,104.*

Bayarri, S., Chuliá, I., & Costell, E. (2010). Comparing λ-carrageenan and an inulin blend as fat replacers in carboxymethyl cellulose dairy desserts. Rheological and sensory aspects. *Food Hydrocolloids, 24*(6–7), 578–587. doi:10.1016/j.foodhyd.2010.02.004

Bayarri, S., & Costell, E. (2011). Flow Properties of Carboxymethylcellulose Dairy Systems With Different Fat Content. *Journal of Food Process Engineering, 34*(6), 1903–1914. doi:10.1111/j.1745-4530.2009.00383.x

Bennett, C. J., & Malits, S. J. (1988). Aqueous table syrup with reduced sugar content. *United States Patent No 4,786,521*, 3–7.

Bodvik, R., Dedinaite, A., Karlson, L., Bergström, M., Bäverbäck, P., Pedersen, J. S., & Claesson, P. M. et al. (2010). Aggregation and network formation of aqueous methylcellulose and hydroxypropylmethylcellulose solutions. *Colloids and Surfaces. A, Physicochemical and Engineering Aspects, 354*(1–3), 162–171. doi:10.1016/j.colsurfa.2009.09.040

Brown, C. J., & Houghton, A. A. (1941). The chemical and physical properties of carboxymethylcellulose and its salts. *Journal of the Society of Chemical Industry, 60*(10), 254–258. doi:10.1002/jctb.5000601004

Castro, C., Zuluaga, R., Álvarez, C., Putaux, J.-L. L., Caro, G., Rojas, O. J., & Gañán, P. et al. (2012). Bacterial cellulose produced by a new acid-resistant strain of Gluconacetobacter genus. *Carbohydrate Polymers, 89*(4), 1033–1037. doi:10.1016/j.carbpol.2012.03.045 PMID:24750910

Castro, C., Zuluaga, R., Putaux, J. L., Caro, G., Mondragon, I., & Gañán, P. (2011). Structural characterization of bacterial cellulose produced by Gluconacetobacter swingsii sp. from Colombian agroindustrial wastes. *Carbohydrate Polymers, 84*(1), 96–102. doi:10.1016/j.carbpol.2010.10.072

Chen, L., Ruan, R. R., Johnson, J. E., Addis, P. B., Xu, L., & Yi, L. (1998). Cellulose fiber based compositions and film and the process for their manufacture. *United States Patent No 5,817,381.*

Chung, Y., & Shyu, Y. (1999). The effects of pH, salt, heating and freezing on the physical properties of bacterial cellulose-nata. *International Journal of Food Science & Technology, 34*(1), 23–26. doi:10.1046/j.1365-2621.1999.00231.x

Commission of the European Unions. (1995). Food Additives Other Than Colours and Sweeteners. *European Parlament, 1*(1), 1–56.

Company, T. D. C. (2015). *Hydroxypropyl Methylcellulose*. Retrieved from http://www.dow.com/dow-wolff/en/industrial_solutions/polymers/hydroxypropyl-methylcellulose/

Cybulska, J., Szymańska-Chargot, M., Zdunek, A., Psonka-Antonczyk, K. M., & Stokke, B. T. (2011). Crystallinity and nanostructure of cellulose from different sources. *11th International Congress on Engineering and Food*, 1–4.

Das, M. K., & Rao, K. R. (2006). Evaluation of zidovudine encapsulated ethylcellulose microspheres prepared by water-in-oil-in-oil (w/o/o) double emulsion solvent diffusion technique. *Acta Poloniae Pharmaceutica - Drug Research*, *63*(2), 141–148.

De Moura, M. R., Aouada, F. A., Monteiro, H., & De Azeredo, C. (2012). Microfluidizer Technique for Improving Microfiber Properties Incorporated Into Edible and Biodegradable Films. In Advances in Microfluidics (pp. 219–240). doi:10.5772/38922

Debeaufort, F., Quezada-Gallo, J. A., & Voilley, A. (1998). Edible Films and Coatings: Tomorrow ' s Packagings: A Review. *Critical Reviews in Food Science and Nutrition*, *38*(4), 299–313. doi:10.1080/10408699891274219 PMID:9626488

Debeaufort, F., Voilley, A., & Meares, P. (1994). Water vapor permeability and diffusivity through methylcellulose edible films. *Journal of Membrane Science*, *91*(1–2), 125–133. doi:10.1016/0376-7388(94)00024-7

Deepak, S., Dinesh, K., Mankaran, S., & Gurmeet, S. (2012). Taste Masking Technologies: A Novel Approach for the Improvement of Organoleptic Property of Pharmaceutical Active Substance. *International Research Journal of Pharmcy*, *3*(4), 108–116. Retrieved from http://scholar.google.com/scholar?hl=en&btnG=Search&q=intitle:TASTE+MASKING+TECHNOLOGIES+:+A+NOVEL+APPROACH+FOR+THE+IMPROVEMENT+OF+ORGANOLEPTIC+PROPERTY+OF+PHARMACEUTICAL+ACTIVE+SUBSTANCE#0

Dickinson, E. (1998). Stability and rheological implications of electrostatic milk protein - Polysaccharide interactions. *Trends in Food Science & Technology*, *9*(10), 347–354. doi:10.1016/S0924-2244(98)00057-0

Diftis, N., & Kiosseoglou, V. (2003). Improvement of emulsifying properties of soybean protein isolate by conjugation with carboxymethyl cellulose. *Food Chemistry*, *81*(1), 1–6. doi:10.1016/S0308-8146(02)00236-4

Dow Chemical Company. (2013). *Notification of the Determination of Ethylcellulose As Being Generally Recognized As Safe*. Author.

Elliott, J. H., & Ganz, A. J. (1971). Characterization, I Rheological Property, Organoleptic. *Journal of Texture Studies*, *2*(1526), 220–229. doi:10.1111/j.1745-4603.1971.tb00583.x PMID:28371977

Engelen, L., De Wijk, R., Van Der Bilt, A., Prinz, J. F., Janssen, A. M., & Bosman, F. (2005). Relating particles and texture perception. *Physiology & Behavior*, *86*(1–2), 111–117. doi:10.1016/j.physbeh.2005.06.022 PMID:16112152

Fan, H., Wu, Y., Hu, X., Wu, J., & Liao, X. (2011). Characteristics of thin-layer drying and rehydration of nata de coco. *International Journal of Food Science & Technology*, *46*(7), 1438–1444. doi:10.1111/j.1365-2621.2011.02642.x

Floury, J., Desrumaux, A., Axelos, M., & Legrand, J. (2003). Effect of high pressure homogenisation on methylcellulose as food emulsifier. *Journal of Food Engineering*, *58*(3), 227–238. doi:10.1016/S0260-8774(02)00372-2

Ford, J. L. (1999). Thermal analysis of hydroxypropylmethylcellulose and methylcellulose: Powders, gels and matrix tablets. *International Journal of Pharmaceutics*, *179*(2), 209–228. doi:10.1016/S0378-5173(98)00339-1 PMID:10053214

Gamay, A. (1996). Process for preparing low fat cheese products. *United States Patent No 5,549,916*, 1–8.

Gaunø, M. H., Vilhelmsen, T., Larsen, C. C., Boetker, J. P., Wittendorff, J., Rantanen, J., & Østergaard, J. (2013). Real-time in vitro dissolution of 5-aminosalicylic acid from single ethyl cellulose coated extrudates studied by UV imaging. *Journal of Pharmaceutical and Biomedical Analysis*, *83*, 49–56. doi:10.1016/j.jpba.2013.04.028 PMID:23708430

Gunduz, O., Ahmad, Z., Stride, E., & Edirisinghe, M. (2012). Continuous Generation of Ethyl Cellulose Drug Delivery Nanocarriers from Microbubbles. *Pharmaceutical Research*, 1–13. doi:10.1007/s11095-012-0865-7 PMID:22956171

Haque, A., & Morris, E. R. (1993). Thermogelation of methylcellulose. Part I: Molecular structures and processes. *Carbohydrate Polymers*, *22*(3), 161–173. doi:10.1016/0144-8617(93)90137-S

Hayashi, T. (2006). *The Science and Lore of the Plant Cell Wall: Biosynthesis, Structure and Function*. Universal-Publishers.

Heinze, T., & Koschella, A. (2005). Carboxymethyl ethers of cellulose and starch - A review. *Macromolecular Symposia*, *223*(1), 13–39. doi:10.1002/masy.200550502

Henriksson, M. (2008). *Cellulose Nanofibril Networks and Composites: Preparation, Structure and Properties* (Doctoral dissertation). KTH.

Hogan, D. T., Bertrand, D. T., Tuazon, M. T., & Tuason, D. C. (1998). Coprocessed microcrystalline cellulose stabilizer compositions, foods containing the same and stabilization method. *United States Patent No 5,789,004*.

Hong, F., & Qiu, K. (2008). An alternative carbon source from konjac powder for enhancing production of bacterial cellulose in static cultures by a model strain Acetobacter aceti subsp. xylinus ATCC 23770. *Carbohydrate Polymers*, *72*(3), 545–549. doi:10.1016/j.carbpol.2007.09.015

Huang, D. P., & Merolla, T. (1994). Method of making pre-cooked filled pasta products by co-extrusion. *United States Patent No 5,296,247*.

Hult, E. L., Iotti, M., & Lenes, M. (2010). Efficient approach to high barrier packaging using microfibrillar cellulose and shellac. *Cellulose (London, England)*, *17*(3), 575–586. doi:10.1007/s10570-010-9408-8

Iguchi, M., Yamanaka, S., & Budhiono, A. (2000). Bacterial cellulose — a masterpiece of nature ' s arts. *Journal of Materials Science*, *35*(2), 261–270. doi:10.1023/A:1004775229149

JECFA. (2000). *Microcrystalline cellulose. FNP52* (Vol. 4). JECFA.

Jellema, R. H., Janssen, A. M., Terpstra, M. E. J., De Wijk, R., & Smilde, A. K. (2005). Relating the sensory sensation creamy mouthfeel in custards to rheological measurements. *Journal of Chemometrics*, *19*(3), 191–200. doi:10.1002/cem.922

Jirapeangtong, K., Siriwatanayothin, S., & Chiewchan, N. (2008). Effects of coconut sugar and stabilizing agents on stability and apparent viscosity of high-fat coconut milk. *Journal of Food Engineering*, *87*(3), 422–427. doi:10.1016/j.jfoodeng.2008.01.001

Jonas, R., & Farah, L. F. (1998). Production and application of microbial cellulose. *Polymer Degradation & Stability*, *59*(1–3), 101–106. doi:10.1016/S0141-3910(97)00197-3

Josephson, D. V., & Dahle, C. D. (1949). A new cellulose gum stablizer for ice cream. *Ice Cream Rev.*, *28*, 11–32.

Josephson, D. V., Dahle, C. D., & Patton, S. (1943). A comparison of some ice cream stabilizers. *Southern Dairy Prod. J.*, *33*(4), 34.

Kalia, S., Kaith, B. S., & Kaur, I. (Eds.). (2011). *Cellulose Fibers: Bio- and Nano-Polymer Composites*. Springer Science & Business Media. doi:10.1007/978-3-642-17370-7

Kamada, E., & Mochihara, N. (2001). Stabilizer for meat products and meat product composition. *United States Patent No 6,270,830, 1*(12), 1–8.

Kamper, S. L., & Fennema, O. (1985). Use of an Edible Film to Maintain Water Vapor Gradients in Foods. *Journal of Food Science*, *50*(2), 382–384. doi:10.1111/j.1365-2621.1985.tb13408.x

Kar, M., & Choudhury, P. K. (2007). Formulation and evaluation of ethyl cellulose microspheres prepared by the multiple emulsion technique. *Die Pharmazie*, *62*(2), 122–125. doi:10.1691/ph.2007.2.6062 PMID:17341032

Keyser, W. L., & Kinney, D. S. (1983). Low Caliorie Table Syrup Product. *United States Patent No 4,394,399*.

Klemm, D., Heublein, B., Fink, H.-P., & Bohn, A. (2005). Cellulose: Fascinating biopolymer and sustainable raw material. *Angewandte Chemie International Edition*, *44*(22), 3358–3393. doi:10.1002/anie.200460587 PMID:15861454

Klose, R. E., & Glicksman, M. (1972). *Handbook of food additives*. CRC Press.

Kohajdová, Z., & Karovicová, J. (2008). Influence of hydrocolloids on quality of baked goods. *Acta Sci Pol Technol Aliment*, *7*(1), 42–49.

Koide, K., Yoneda, Y., & Musashi, K. (1983). Process for producing a cream cheese-like food. *United States Patent No 4,397,878*, 3–8.

Krässig, H. A. (1993). *Cellulose, structure, accessibility and reactivity*. Philadelphia: Gordon and Breach Publishers.

Kuksal, A., Tiwary, A. K., Jain, N. K., & Jain, S. (2006). Formulation and in vitro, in vivo evaluation of extended- release matrix tablet of zidovudine: Influence of combination of hydrophilic and hydrophobic matrix formers. *AAPS PharmSciTech*, *7*(1), E1–E9. doi:10.1208/pt070101 PMID:16584139

Kurosumi, A., Sasaki, C., Yamashita, Y., & Nakamura, Y. (2009). Utilization of various fruit juices as carbon source for production of bacterial cellulose by Acetobacter xylinum NBRC 13693. *Carbohydrate Polymers*, *76*(2), 333–335. doi:10.1016/j.carbpol.2008.11.009

Larsen, J. C., Nørby, K. K., Reffstrup, T. K., & Beltoft, V. M. (2004). Opinion of the Scientific Panel on Food Additives, Flavourings, Processing Aids and Materials in Contact with Food (AFC) on a request from the Commission related to Ethyl Cellulose as a food additive Question number EFSA-Q-2003-116. *The EFSA Journal*, *35*, 1–6.

Lavoine, N., Desloges, I., & Bras, J. (2014). Microfibrillated cellulose coatings as new release systems for active packaging. *Carbohydrate Polymers*, *103*, 528–537. doi:10.1016/j.carbpol.2013.12.035 PMID:24528763

Liu, K., Lin, X., Chen, L., Huang, L., Cao, S., & Wang, H. (2013). Preparation of microfibrillated cellulose/chitosan–benzalkonium chloride biocomposite for enhancing antibacterium and strength of sodium alginate films. *Journal of Agricultural and Food Chemistry*, *61*(26), 6562–6567. doi:10.1021/jf4010065 PMID:23750871

Lometillo, J. E., & Wolcott, J. M. (1983). Process for producing pasta products. *United States Patent No 4,394,397*.

Maity, , Nandy, & Mittal. (2014). A brief review on recent advances of extended release technology employed to design the oral dosage forms. *International Journal of Medical and Applied Sciences Research*, *1*(1).

Mallikarjunan, P., Chinnan, M., Balasubramaniam, V., & Phillips, R. (1997). Edible Coatings for Deep-fat Frying of Starchy Products1. *Lebensmittel-Wissenschaft Und-Technologie*, *30*(7), 709–714. doi:10.1006/fstl.1997.0263

Martins, I. M. G., Magina, S. P., Oliveira, L., Freire, C. S. R., Silvestre, A. J. D., Neto, C. P., & Gandini, A. (2009). New biocomposites based on thermoplastic starch and bacterial cellulose. *Composites Science and Technology*, *69*(13), 2163–2168. doi:10.1016/j.compscitech.2009.05.012

Melzer, E., Kreuter, J., & Daniels, R. (2003). Ethylcellulose: A new type of emulsion stabilizer. *European Journal of Pharmaceutics and Biopharmaceutics*, *56*(1), 23–27. doi:10.1016/S0939-6411(03)00025-0 PMID:12837477

Mitchell, K., Ford, J. L., Armstrong, D. J., Elliott, P. N. C., Rostron, C., & Hogan, J. E. (1990). The influence of additives on the cloud point, disintegration and dissolution of hydroxypropylmethylcellulose gels and matrix tables. *International Journal of Pharmaceutics*, *66*(1–3), 233–242. doi:10.1016/0378-5173(90)90404-R

Moore, L. J., & Shoemaker, C. F. (1981). Sensory textural properties of stabilized ice-cream. *Journal of Food Science*, *46*(2), 399–402, 409. doi:10.1111/j.1365-2621.1981.tb04870.x

Nugroho, D. A., & Aji, P. (2015). Characterization of Nata de Coco Produced by Fermentation of Immobilized Acetobacter xylinum. *Agriculture and Agricultural Science Procedia*, *3*, 278–282. doi:10.1016/j.aaspro.2015.01.053

O'Sullivan, A. (1997). Cellulose: the structure slowly unravels. *Cellulose*, *4*(3), 173–207.

Okiyama, A., Motoki, M., & Yamanaka, S. (1992). Bacterial cellulose II. Processing of the gelatinous cellulose for food materials. *Food Hydrocolloids*, *6*(5), 479–487. doi:10.1016/S0268-005X(09)80033-7

Okiyama, A., Motoki, M., & Yamanaka, S. (1993). Bacterial cellulose IV. Application to processed foods. *Food Hydrocolloids*, *6*(6), 503–511. doi:10.1016/S0268-005X(09)80074-X

Okiyama, A., Motoki, M., & Yamanaka, S. (1993). Bacterial Cellulose .3. Development of a New Form of Cellulose. *Food Hydrocolloids*, *6*(6), 493–501. 10.1016/S0268-005X(09)80073-8

Organization, W. H. (2006). Compendium of Food Additive Specifications. *FAO JECFA Monographs*, (63).

Panda, H. (2010). *The Complete Book on Gums and Stabilizers for Food Industry*. Asia Pacific Business Press Inc.

Phadke, K. V., Manjeshwar, L. S., & Aminabhavi, T. M. (2014). Microspheres of Gelatin and Poly(ethylene glycol) Coated with Ethyl Cellulose for Controlled Release of Metronidazole. *ACS Publications*, *53*, 6575–6584.

Phisalaphong, M., & Chiaoprakobkij, N. (2012). Applications and Products—Nata de Coco. In Bacterial NanoCellulose: A Sophisticated Multifunctional Material (pp. 143–156). Academic Press.

Prasertmanakit, S., Praphairaksit, N., Chiangthong, W., & Muangsin, N. (2009). Ethyl cellulose microcapsules for protecting and controlled release of folic acid. *AAPS PharmSciTech*, *10*(4), 1104–1112. doi:10.1208/s12249-009-9305-3 PMID:19763838

Psomiadou, E., Arvanitoyannis, I., & Yamamoto, N. (1996). Edible films made from natural resources; microcrystalline cellulose (MCC), methylcellulose (MC) and corn starch and polyols - Part 2. *Carbohydrate Polymers*, *31*(4), 193–204. doi:10.1016/S0144-8617(96)00077-X

Quintavalla, S., & Vicini, L. (2002). Antimicrobial food packaging in meat industry. *Meat Science*, *62*(3), 373–380. doi:10.1016/S0309-1740(02)00121-3 PMID:22061613

Rao, M., Walter, R. H., & Cooley, H. J. (1981). Effect of Heat Treatment on the Flow Properties of Aqueous Guar Gum and Sodium Carboxymethylcellulose (CMC) Solutions. *Journal of Food Science*, *46*(3), 896–899. doi:10.1111/j.1365-2621.1981.tb15374.x

Regand, A., & Goff, H. D. (2003). Structure and ice recrystallization in frozen stabilized ice cream model systems. *Food Hydrocolloids*, *17*(1), 95–102. doi:10.1016/S0268-005X(02)00042-5

Rehman, S., & Ahmed, M. (2002). Effect of carboxy methyl cellulose and carrageenan gum on the shelf life of bread. *Pakistan Journal of Food Science*, *13*(1), 49–52.

Rhodes, K. H. (1978). Preparation of liquid and frozen yogurt products. *United States Patent No 4,110,476.*

Robbins, M. D., & Rodriguez, S. S. (1984). Low calorie baked products. *United States Patent No 4,461,782*, 1–5.

Rowe, R. C. (1986). The effect of the molecular weight of ethyl cellulose on the drug release properties of mixed films of ethyl cellulose and hydroxypropylmethylcellulose. *International Journal of Pharmaceutics*, *29*(1), 37–41. doi:10.1016/0378-5173(86)90197-3

Rowe, R. C., Sheskey, P. J., & Weller, P. J. (Eds.). (2006). Handbook of Pharmaceutical Excipients. APhA Publications. doi:10.1097/PRS.0b013e31822adba3

Saha, D., & Bhattacharya, S. (2010). Hydrocolloids as thickening and gelling agents in food: A critical review. *Journal of Food Science and Technology*, *47*(6), 587–597. doi:10.1007/s13197-010-0162-6 PMID:23572691

Sajeev, C., Vinay, G., Archna, R., & Saha, R. N. (2002). Oral controlled release formulation of diclofenac sodium by microencapsulation with ethyl cellulose. *Journal of Microencapsulation*, *19*(6), 753–760. doi:10.1080/02652040210000022734 PMID:12569023

Sarkar, N. (1979). Thermal Gelation Properties of Methyl and Hydroxypropyl Methylcellulose. *Journal of Applied Polymer Science*, *24*(4), 1073–1087. doi:10.1002/app.1979.070240420

Sarkar, N. (1995). Kinetics of thermal gelation of methylcellulose and hydroxypropylmethylcellulose in aqueous solutions. *Carbohydrate Polymers*, *26*(3), 195–203. doi:10.1016/0144-8617(94)00107-5

Sheu, F., Wang, C. L., & Shyu, Y. T. (2000). Fermentation of Monascus purpureus on bacterial cellulose-nata and the color stability of Monascus-nata complex. *Journal of Food Science, 65*(2), 342–345.

Shi, Z., Zhang, Y., Phillips, G. O., & Yang, G. (2014). Utilization of bacterial cellulose in food. *Food Hydrocolloids*, *35*, 539–545. doi:10.1016/j.foodhyd.2013.07.012

Shirkhani, A. K., & Madadlou, A. (n.d.). *On stabilization of acidified milk drinks – a review*. Academic Press.

Sieg, J. M. B., & J. (2009). Fiber. In *Ingredients in Meat Products: Properties, Functionality and Applications* (pp. 83–109). <ALIGNMENT.qj></ALIGNMENT>10.1007/978-0-387-71327-4

Singer, N. S., Chang, H. H., Tang, P., & Dunn, J. M. (1990). Carbohydrate cream substitute. *United States Patent No 4,911,946*, 2–4.

Stephen, A. M. (2006). *Food Polysaccharides and Their Applications.* CRC Press. http://doi.org/doi:10.1201/9781420015164.ch6

Strom, G., Ohgren, C., & Ankerfors, M. (2013). Nanocellulose as an additive in foodstuff. *Innventia Report*, (403), 1–25.

Su, J. F., Huang, Z., Yuan, X. Y., Wang, X. Y., & Li, M. (2010). Structure and properties of carboxymethyl cellulose/soy protein isolate blend edible films crosslinked by Maillard reactions. *Carbohydrate Polymers*, *79*(1), 145–153. doi:10.1016/j.carbpol.2009.07.035

Sun-Waterhouse, D., Bekkour, K., Wadhwa, S. S., & Waterhouse, G. I. N. (2013). Rheological and Chemical Characterization of Smoothie Beverages Containing High Concentrations of Fibre and Polyphenols from Apple. *Food and Bioprocess Technology*, *7*(2), 409–423. doi:10.1007/s11947-013-1091-y

Tabuchi, M. (2007). Nanobiotech versus synthetic nanotech? *Nature Biotechnology*, *25*(4), 389–390. doi:10.1038/nbt0407-389 PMID:17420738

Tainter, M. L. (1943). Methyl Cellulose as a Colloid Laxative. *Experimental Biology and Medicine*, *54*(1), 77–79. doi:10.3181/00379727-54-14309

Tarantino, L. M. (2006). *Notification of the Determination of Hydroxypropyl Methylcellulose As Being Generally Recognized As Safe*. Keller and Heckman LLP.

Tejayadi, S. (2004). Flavored milk manufacturing processes and compositions. *United States Patent Application No 10/437,317, 1*(19).

Thomas, G. M., Paquita, E. M., & Thomas, J. P. (2002). *Cellulose Ethers*. Encyclopedia Of Polymer Science and Technology.

Tiemstra, P. J. (1971). Microcrystalline cellulose compositions co-dried with hydrocolloids. *United States Patent No 3,573,058*, 1–3.

Tiwari, S. B., Murthy, T. K., Pai, M. R., Mehta, P. R., & Chowdary, P. B. (2003). Controlled release formulation of tramadol hydrochloride using hydrophilic and hydrophobic matrix system. *AAPS PharmSciTech*, *4*(3), 1–6. doi:10.1208/pt040331 PMID:14621963

Tomé, L. C., Freire, M. G., Rebelo, L. P. N., Silvestre, A. J. D., Neto, C. P., Marrucho, I. M., & Freire, C. S. R. (2011). Surface hydrophobization of bacterial and vegetable cellulose fibers using ionic liquids as solvent media and catalysts. *Green Chemistry*, *13*(9), 2464. doi:10.1039/c1gc15432j

Tuason, D. C., Jr., Ruszkay, T. A., & Heese, S. (2004). Fat-like agents for low calorie food compositions. *United States Patent No 6,689,405, 1*(12), 4–9.

Uddin, M. S., Hawlader, M. N. A., & Zhu, H. J. (2001). Microencapsulation of ascorbic acid: Effect of process variables on product characteristics. *Journal of Microencapsulation*, *18*(2), 199–209. doi:10.1080/02652040010000352 PMID:11253937

Upadhye, S. B., & Rajabi-Siahboomi, A. R. (2013). Properties andApplications of Polyethylene Oxide and Ethylcellulose for Tamper Resistance and Controlled Drug Delivery. In *Advances in the Pharmaceutical Sciences* (Vol. 9, pp. 145–158). Springer New York. doi:10.1007/978-1-4614-8432-5

Valencia-Chamorro, S. A., Palou, L., Del Rio, M. A., & Pérez-Gago, M. B. (2008). Inhibition of Penicillium digitatum and Penicillium italicum by Hydroxypropyl Methylcellulose-Lipid Edible Composite Films Containing Food Additives with Antifungal Properties. *Journal of Agricultural and Food Chemistry*, *56*(23), 11270–11278. doi:10.1021/jf802384m PMID:19012404

Valley, L. (2002). *United States Patent 6495177*. Washington, DC: US Patent Office.

van Ruth, S. M., de Witte, L., & Uriarte, A. R. (2004). Volatile Flavor Analysis and Sensory Evaluation of Custard Desserts Varying in Type and Concentration of Carboxymethyl Cellulose. *Journal of Agricultural and Food Chemistry*, *52*(26), 8105–8110. doi:10.1021/jf048647g PMID:15612803

Villagran, F. V., Dria, G. J., Butterbaugh, J. L., & Small, L. E. (2004). Creamy mouthfeel agent for foods and beverages. *United States Patent No 6,673,384*.

Villalobos-Carvajal, R., Hernández-Muñoz, P., Albors, A., & Chiralt, A. (2009). Barrier and optical properties of edible hydroxypropyl methylcellulose coatings containing surfactants applied to fresh cut carrot slices. *Food Hydrocolloids*, *23*(2), 526–535. doi:10.1016/j.foodhyd.2008.02.008

Vojdani, F., & Torres, J. A. (1990). Potassium Sorbate Permeability of Methylcellulose and Hydroxypropyl Methylcellulose Coatings: Effect of Fatty Acids. *Journal of Food Science*, *55*(3), 841–846. doi:10.1111/j.1365-2621.1990.tb05244.x

Wanga, S. T., Barringer, S. A., & Hansenbpt, P. M. T. (1998). Effects of carboxymethylcellulose and guar gum on ice crystal propagation in a sucrose-lactose solution. *Food Hydrocolloids*, *12*(2), 211–215. doi:10.1016/S0268-005X(98)00032-0

Wren, M., & Mullen, J. (1974). Process for producing low calorie pasta. *United States Patent No 3,843,818*.

Yang, Sharma, & Sheu. (1989). *United States Patent 4882152*. Washington, DC: US Patent Office.

Zecher, D., & Van Coilie, R. (1992). Cellulose derivatives. In Thickening and gelling agents for food (pp. 40–66). http://doi.org/ doi:10.1007/978-1-4615-3552-2_3

Zheng, L., Ding, Z., Zhang, M., & Sun, J. (2011). Microencapsulation of bayberry polyphenols by ethyl cellulose: Preparation and characterization. *Journal of Food Engineering*, *104*(1), 89–95. doi:10.1016/j.jfoodeng.2010.11.031

Zorba, M., & Ova, G. (1999). An improved method for the quantitative determination of carboxymethyl cellulose in food products. *Food Hydrocolloids*, *13*(1), 73–76. doi:10.1016/S0268-005X(98)00062-9

Chapter 3

Comparative Study of Mechanical and Tribological Behaviour of Thermoplast-Based Composites:

With Inorganic and Natural Particulate Fillers

Kaushik Kumar
Birla Institute of Technology, India

J. Paulo Davim
University of Aveiro, Portugal

ABSTRACT

The usage of composite material has been increasing day by day over the years due to the excellent properties being offered by them such as lower volume-to-weight ratio, improved toughness, recyclable, environment friendly and also due to their short cycle productivity, easiness in fabrication, long lasting life span etc. This work presents study of mechanical and tribological behaviour of ABS polymer matrix filled with micron-sized inorganic and Natural / Green fillers. The main purpose behind this work is to compare the performance of composites with different class of fillers. The experimental results unveil that the composite with inorganic fillers perform better than with natural based fillers.

INTRODUCTION

Matrix and reinforcement are two main components of a composite. The reinforcement materials are surrounded by matrix material and support them by maintaining their relative positions. The reinforcements impart their special mechanical and physical properties to increase or alter the matrix properties. A synergism produces material properties impossible from the individual constituent materials, whereas

DOI: 10.4018/978-1-5225-3023-7.ch003

the large choice of matrix and associated reinforcement permits the designer of the merchandise or structure to settle on an optimum combination. The reinforcing material will be of the shape of particle (usually known as filler) or fibre. In a particle reinforced composites, the reinforcement can be of macro to nano size. It is generally not the physical dimensions of the particles by that the materials area unit classified; rather it's the mechanism of reinforcement. In an exceedingly little particle strengthened material the mechanism is on a molecular level whereas just in case of huge particle it is at macro level. The particles is also spread into or precipitated from the matrix; thus, the properties will be varied with the quantity and kind of dispersion.

In the recent years, rising concern towards environmental issues and the requirement for multifarious polymer-based materials has focused utilization of polymer composites reinforced with green reinforcement, i.e., materials derived from natural resources like trees they would like for a lot of varied compound-based materials has junction rectifier to increasing interest concerning polymer composites strengthened with inexperienced reinforcement, i.e., materials derived from natural resources like trees etc. The bio degradable composites, or green composites, have shown an enhancement of interest owing to their recyclability, biodegradability and torrential availableness (Thakur et al., 2012).

Globalization is making us more aware of energy conservation, nature preservation and natural solutions to everyday needs. Golden fiber additionally referred as jute fiber is the most vital and versatile natural fiber of choice after cotton with endless prospects (Basu and Roy, 2008). Jute fiber is a natural composite of cellulose, hemicellulose and polymer (Xia et al., 2010). It is fully perishable eco- friendly, renewable and inexpensive (Sanjay and Yogesha, 2016). This natural fiber has drawn the attention of the consumers because a recent study has revealed that the greenhouse gas emissions by the jute are negative. To make more use of biodegradable jute products due to the natural advantages market has started producing rigid packing reinforced plastics made up of jute fiber. Jute is used in making different types of product in different sizes and shapes ranging from shopping to fashion. For instance, carpets, bags, atmospheric friendly coffins, etc. Jute has high hygroscopic property, strength; they are long lasting, flexible and drainage properties. Jute fibers are good for soil and the jute stalks left after the fiber has been extracted may help to meet the worlds need for wood pulp, as well as being a renewable source of cooking fuel and prevents from soil erosion (Datta. 2009). Jute fibers have the ability to get mixed with both the natural as well as synthetic fibers to make various useful products. Some alterations with acids and anhydrides enhance inherent drawbacks such as breaking strength, thermal stability, color fastness, etc) of jute fibers (Mondal et al., 2014).

Acrylonitrile butadiene styrene (ABS) is a thermoplastic polymer found in large quantity. The process of producing ABS involves three monomers which are acrylonitrile, 1-3 butadiene and styrene and involves 3 chemical compounds such as carbon, nitrogen and hydrogen. Acrylonitrile and butadiene are colourless in nature whereas styrene changes its colour from colourless to yellow. All the three are soluble in nature, acrylonitrile soluble in almost all solvents, butadiene soluble in organic solvents only and styrene is slightly soluble in water. ABS polymers can be produced by injection moulding or compression moulding in press (Dong et al., 1993). They have glass transition temperature of around 1500 centigrade. Mechanical properties vary with temperature between -20^0 centigrade to 80^0 centigrade. It does not risk human health but individual should take care while working with ABS because of initiation of fumes while experimentation. ABS has gain importance due to its mechanical, electrical and thermal properties. The most vital mechanical property of ABS is its impact strength, toughness the most important mechanical properties of ABS are impact resistance and toughness. Generally ABS polymers are used because it provides glossy and shiny surface to plastics, durable, light weight, recyclable and cost effective (Rutkowski and Levint. 1986). Fibers can be mixed with ABS to make the final product. ABS polymers are resistant to aqueous acids, alkalis, alcohols and animal, vegetable and mineral oils. It is being used for number of applications and in manufacturing products like toys, automotive, household appliance, computer equipments (computer, keyboards, monitors etc.) (Brennan et al., 2002).

The dry dusty remains of sugarcane after being crushed are bagasse (Bahurudeen et al., 2015). For production of ethanol sugarcane bagasse is being fermented and hydrolyzed (Thakur et al., 2015). Bagasse waste are mainly organic fibers that have high water holding capacity and contains small amount of calcium, potassium, nitrogen (Hughes and Girdlestone., 2001). It is less reactive and has low specific gravity. It is mainly used in generation of heat and electricity in boilers rather than been wasted.

Many researchers have studied the effect of various fillers on mechanical properties of polymer matrix (Ali and Elleithy, 2011, Chang et al., 2011, Chen et al., 2012, Hassan et al., 2012, Liu et al., 2012, Ma et al., 2012, Wang et al., 2012, Zeng et al., 2012, Asyadi et al., 2013, Kaseem et al., 2013, Krishnan et al., 2013, Ahmed and El-Sabbagh, 2014, Balakrishna et al., 2014). Most of the researchers have varied filler content from 0 to 30 wt% to enhance mechanical properties of polymer composites. General trend noticed from the survey is that the tensile modulus and flexural modulus increase with increase in filler content but tensile strength and flexural strength increase up to a certain filler content but beyond that strength start to decrease. High volume of filler content also leads to agglomeration thus reduction in tensile properties.

Polymeric composite have emerged as one of the most significant materials in weight sensitive applications like automotive parts and aerospace industries. With this, tribological behaviours like wear resistance and friction coefficients are come into serious attention. Wear resistance is one of the important properties as the parts made up with composite material may experience sliding motion. Wear resistant affects the coefficient of friction (Axen et al., 1996). Schön (1996) has also developed a model for predicting the coefficient of friction by applied load at different types of contact. Gamelas, et al. (2014) carried Inverse gas chromatography and found that, Lewis basic properties of the material surface increases with reduction of the dispersive component of surface energy. Addition of Ultra-high-molecular-weight polyethylene (UHMWPE) particles reduced wear rate of composites as well as coefficient of friction. Addition of UHMWPE particles improves abrasive behaviour as well as sliding wear resistance of epoxy resin (Chand et al., 2007). Roselman and Tavor (1976) showed variation of coefficient of friction over normal loads between crossed carbon fibers. Eiss and Czichos (1986) carried out experiments and measured the coefficient of friction for rubber reinforced epoxy composite sliding against glass and steel surfaces and obtained a value between 0.9 and 1.15 where coefficient of friction obtained consistently lower on glass surface than on steel surface. Wan et al. (2006) performed friction and wear behaviour of three-dimensional braided carbon fiber/ epoxy composites and found that, higher volume fraction of composite presented lower specific wear rate at steady wear state as well as in running period and the effect of volume fraction on the coefficient of friction was less significant and more complicated throughout wear process. Modifications of tribological behaviour of polymer by addition of filler have been reported to be quite encouraging (Ariffin et al., 2008, Aurrekoetxea et al., 2008, Kocserha and Gömze, 2010, Rasheva et al., 2010, Song et al., 2010, Xie et al., 2010, Ben Difallah et al., 2012, Esthappan et al., 2012, Li and Chen, 2012, Chang et al., 2013, Praveen et al., 2013, Akinci et al., 2014, Chang et al., 2014, Selvam et al., 2014, You et al., 2014). Researchers have used filler content in a range of 0-30 wt% on polymer composites with different normal load and sliding velocity which slides against metallic counter faces. In addition to tribo-performance obtained due to the addition of fillers in polymer composites, there is also cost reduction of components due to less consumption of matrix material.

Figure 1. Structural formula of ABS

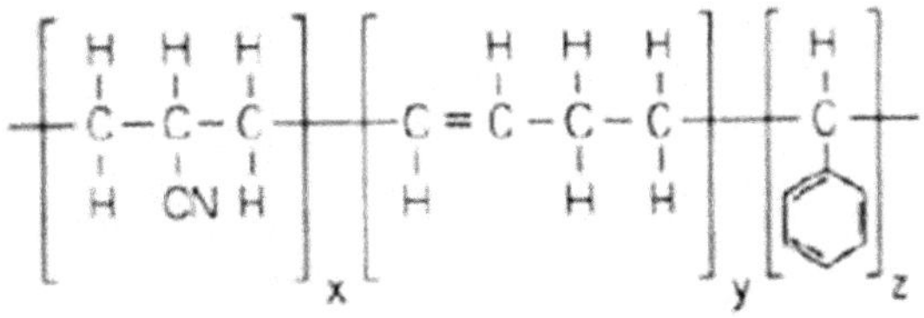

The review of work done by various researchers have revealed that:

1. Most of the researchers have used thermoset matrix (usually epoxy) for the composite preparation due to easy and room temperature processing.
2. The researchers have used one or two types of filler but of same nature i.e., natural, inorganic etc. not of both.
3. People have varied the filler content and studied the effect of the same on various properties.

In the present work, authors have used both natural fillers (jute dust and bagasse dust) and inorganic (Titanium di-oxide and clay) fillers in a thermoplast matrix (ABS) to prepare the composite and study its mechanical behaviour e.g. Tensile, Flexural, Density etc. These samples were also tested for abrasive wear and friction coefficient measurement. Tests were undertaken considering a specific filler content (10%). The microstructures of the composites were studied to analyse the distribution of the filler in ABS matrix using scanning electron microscope.

MATERIALS USED

ABS

In the present work, acrylonitrile-butadiene-styrene (ABS) is used as matrix material. It is a common low cost engineering thermoplastic material derived from the polymerization of Acrylonitrile (a synthetic monomer produced from propylene and ammonia), Butadiene (a petroleum hydrocarbon made from C4 fraction of steam cracking) and Styrene (obtained by dehydrogenation of ethyl benzene) monomers. The ABS material is called "terpolymers" due to the combination of three different monomers to form a single material and draws properties of all three monomers and its molecular formula is (C_8H_8-C_4H_6-C_3H_3N)n (Figure 1). In a typical ABS proportions of three monomers usually lies from 15 to 35% acrylonitrile, 5 to 30% butadiene and 40-60% styrene.

Jute Dust

Jute in the form of particulate or dust (about 70μ) was taken as filler. Golden filler additionally referred as jute filler is the most vital and versatile natural filler of choice as it is fully perishable eco- friendly, renewable and inexpensive (Sanjay and Yogesha, 2016). This natural filler has drawn the attention of the consumers because a recent study has revealed that the greenhouse gas emissions by the jute are negative.

Bagasse Dust

The dry dusty remains of sugarcane after being crushed are bagasse (Bahurudeen et al., 2015). In this work bagasse in the form of particulate or dust (about 70μ) was taken as filler. Bagaase waste are mainly organic fibers that have high water holding capacity and contains small amount of calcium, potassium, nitrogen (Hughes and Girdlestone., 2001). It is less reactive and has low specific gravity. It is mainly used in generation of heat and electricity in boilers rather than been wasted.

Titanium Di-Oxide

Titanium di-oxide is a white solid inorganic substance with chemical structure $TiO_2$and density of 4.23 g/cc which is thermally stable, non-flammable and poorly soluble and non-hazardous. It has been used for many years in a wide range of industrial and consumer applications such as paints, coatings, adhesives, paper, plastics and rubber, coated fabrics and textiles, pharmaceuticals, food colorants, automotive products etc. In this work TiO_2, purchased from Dupont, India of Dupont-R902 grade in the form of powder with a mean particle size of 70μ and density of 4.197 g / cc was used.

Clay

Clay is a fine grained soil of white in colour and brightness which combines one or more clay minerals with traces of metal oxides and organic matter. It has good plasticity, does not smell in water & stands high temperatures and also it is non-conductor of electricity and heat. It is used in manufacturing of porcelain and chinaware, electrical insulators, paint industry, pharmaceuticals and cosmetic industry, paper and textile industry and used as filler in plastics and rubber industries to improve the physical, mechanical and electrical properties. Here Clay filler, supplied by Shree Ram minerals Ltd, India in the form of powder with a mean particle size of 70μ and bulk density of 0.33 g / cc has been used as a filler.

Conditioning of Fillers

Natural fillers are hygroscopic in nature and hence require effective drying. Drying of these is performed in dehumidifier chamber (Figure 2) at 50^0 C for three Hours.

Composite Preparation

For preparation of composite ABS, as the matrix material and all the four fillers as 10% by weight were taken. Initially matrix and fillers (individually) are mixed homogenously in a using a Haake single screw extruder (Rheocord – 9000) shown in Figure 3. The extruder was fitted with a rod die and screw speed of 60 rpm is employed for preparing polymer composites. The temperature profile of the extruder is shown in Table 1

The granules, from the extruder, were put in mould (Figure 4) of 99 X 88 X 4.5 mm. For the formation of the samples the mould was placed in a Carver Press (Figure 5) at a pressure of 5T. By varying the vacuum pressure throughout the heating stage of five metric tons air bubbles and wetness were removed. After hardening of specimens, samples were taken out of the press and mould for testing.

Figure 2. Drying of natural fillers using a dehumidifier
Source: Roy et al., 2016

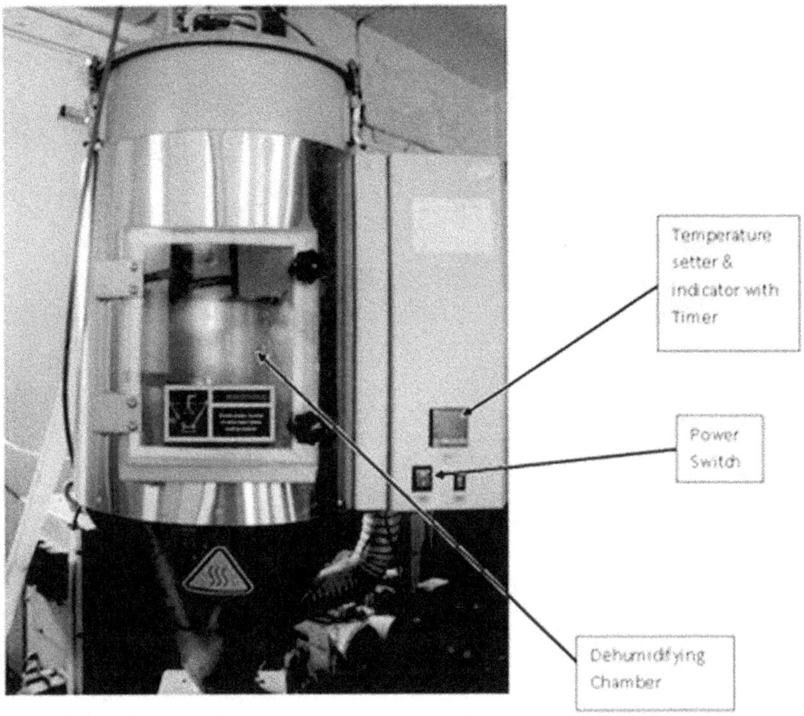

Table 1. Temperature profile along the extruder barrel

Feed Zone	Compression Zone	Metering Zone	Die Zone
210°C	220°C	230°C	240°C

Figure 3. Haake single screw extruder

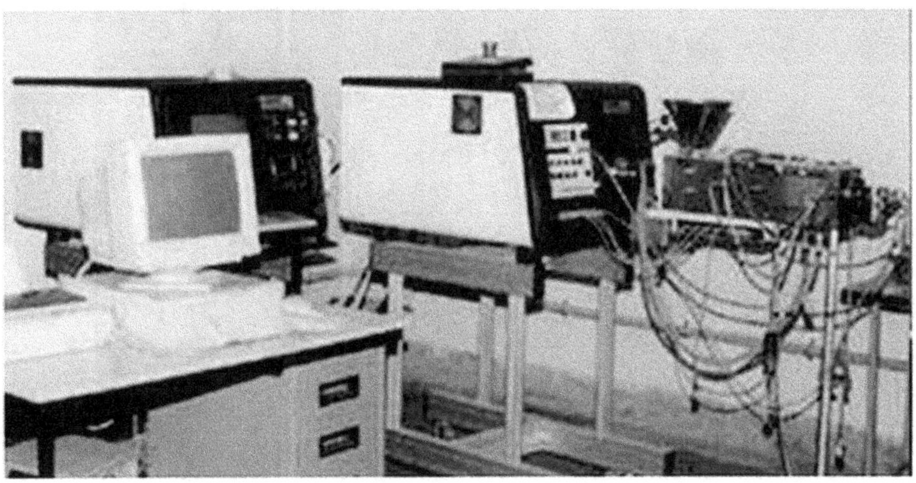

Figure 4. Rectangular mold

Figure 5. Carver press

TESTING

Density and Void Content

The density of polymer composites basically depends upon the composition of the composite. It is basically defined as the ratio of mass to the unit volume of a substance. The theoretical densities of composite material would be calculated by using weight additive principle equation and the actual density of composite would be determined experimentally by using simple water immersion method with the help of Mettler Toledo electronic balance (Figure 6). The difference would indicate the void. Void content, in the fabricated composites, is an important physical property, which can significantly affect the physical and mechanical properties of composite material. Hence estimation of void content is important to evaluate the quality of composites (ASTM D2734 – 16: Standard Test Methods for Void Content of Reinforced Plastics

Figure 6. Mettler toledo electronic balance

Tensile and Flexural Tests

Universal testing machine (UTM) is used to test the tensile and flexural behaviour (tensile strength, tensile modulus, flexural strength and flexural modulus) of composite materials. The mechanical tests (Tensile & Flexural) are carried out using Instron universal testing machine (Instron Ltd, UK) (Figure 7). The tensile tests are conducted according to ASTM D 638. The specimens for mechanical tests are of the shape of a rectangular bar of size 100 X 12 X 8 mm. The flexural or 3 point bending tests are undertaken as per ASTM D 790 under the same condition as that of the tensile tests.

The tensile test was undertaken in accordance to ASTM D 638 where the specimens are placed in the grips and are pulled until failure. The tensile tests are performed with the maximum load capacity of 1 kN with 50 mm gauge length and cross head speed of 2.5 mm / min. The ambient conditions are maintained at 23^{0}C and 55% RH. The data of tensile modulus and tensile strength are recorded on a computer attached to the machine setup. The flexural or 3 point bending tests was performed as per ASTM D 790 where the test was initiated by applying the load on the specimen at the specified rate until rupture of specimen. The tests are carried out with a span length of 50 mm and cross head speed of 2.5 mm / min under the same ambient conditions as in case of flexural tests. Flexural modulus and flexural strength are recorded from the tests. The mechanical tests are carried out for minimum of three samples for each composition and the average results are reported.

Hardness Test

Hardness is the resistance of a material to deformation, indentation or scratching. It is required to ascertain the usage of the composite as a building material for the calipers. It and also reveals the dimensional stability of a composite. The indentation value has high importance for technical applications reflecting the resistance to permanent deformation, which not only decides modulus, strength, elasticity, plasticity but also provides dimensional stability of a material.

Figure 7. Instron universal testing machine a) tensile; b) flexural

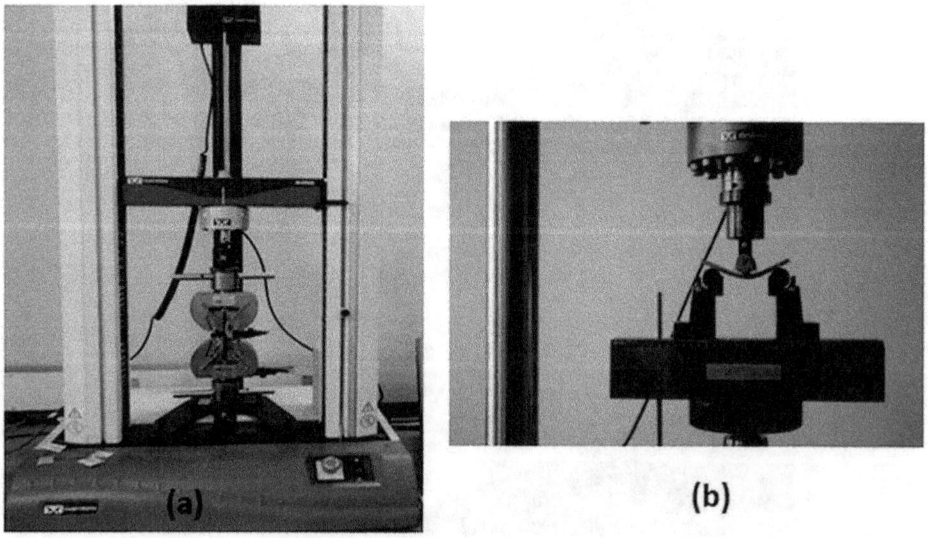

In this study, Vicker's hardness test setup is used to find out the micro-hardness value of polymer composites. The instrument used is a UHL micro hardness tester (Model - VMHT MOT, Sl. No. 1002001, Technische Mikroskopie) (Figure 8) with a Vicker's diamond indenter. The dwell time and speed of indentation are kept constant and a constant load is applied (ASTM E 384-89: Standard test method for microhardness of materials).

Tribological Test

Most of the engineering systems such as machines, instruments, vehicles etc. depend on the processes of motion. The physical definition of motion denotes the change of the object's position with time, and the forces that act on the object to do that change. In the processes of motion of engineering systems moving relatively to each other, there will be some resistance to motion i.e., the occurrence of friction between the objects. Further in engineering system the friction is mostly accompanied by wear, if the moving component consists of a solid body. The tribological characteristics of friction induced energy losses and wear induced material losses have to be analysed carefully. Tribological tests of composites was performed on a block-on-roller multi-tribotester TR25 (Ducom, India) (Figure 9) under dry condition (ambient temperature is about 28°C with a relative humidity at 85%) with the constant time. The composite specimens of size 20 X 20 X 8 mm^3 was held stationary with the help of attachment provided in the test rig and made to slide against the steel roller (diameter 50 mm, thickness 50 mm and material EN8 steel of hardness 55 HRc) which acts as a counter face. A loading lever was used to apply a normal

Figure 8. Vicker's micro hardness tester
Source: Roy et al., 2016

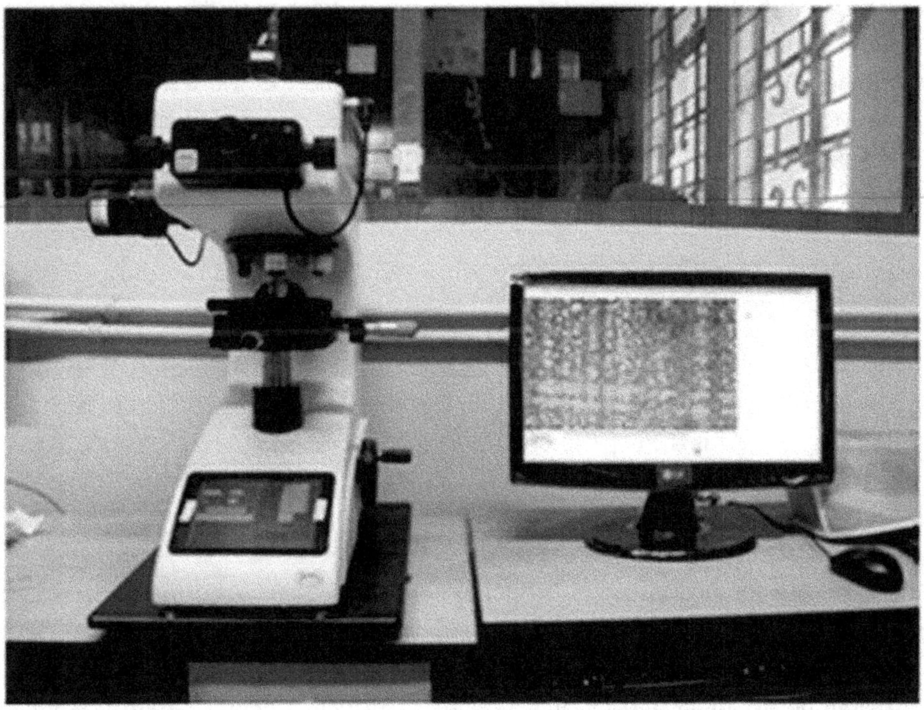

Figure 9. Experimental setup of multi-tribo tester

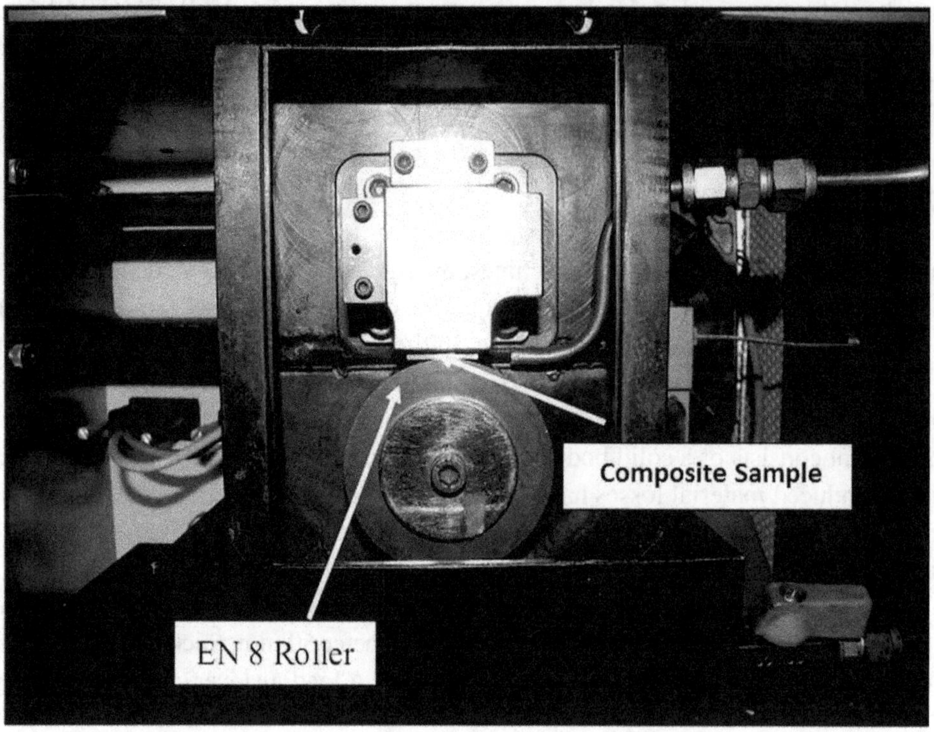

load by placing dead weights on the loading pan on the top of the specimen. The experimental data of Coefficient of Friction (COF) were recorded on a computer attached to the testing apparatus. Weight loss of the composite are used to calculate the specific wear rate. Samples were weighed before and after the experiments in a Mettle Toddler electronic balance. The specific wear rate (Ws) is calculated by the Equation 1 (Wang et al. 2012). The experimental runs are repeated twice and the average values are considered.

$$W_s = \frac{W_1 - W_2}{\rho * P * \upsilon * t} \quad (1)$$

where W_s is the specific wear rate in $mm^3/N.m$, W_1 is the weight before the testing, W_2 is the weight after the test in g, ρ is the density of composites in g / cc, P is the applied normal load in N, υ is the relative sliding velocity in m / s and t is the experimental time in sec.

Morphological Study

It is important to study a composite material which is newly developed to ensure the structure; chemical composition, distribution etc. are exactly similar as desired. The microstructure characterization of a developed material helps one to quickly understand the reasons for variation of its properties, physical behaviour, distribution of fillers into the matrix and performance under certain investigation, hence these techniques are in demand to study and characterize materials. Microstructure characterization of a material is mostly studied by using Scanning Electron Microscopy (SEM) images and it is very useful in order to observe the microstructure in detail.

Scanning electron microscopy examinations is undertaken for the composite material to observe the filler distribution in the matrix. The surfaces of composite is coated with thin platinum film on the surface by sputtering and the images are observed in a JEOL (model JSM 6390LV, Japan) microscope (Figure 10).

RESULTS AND DISCUSSION

Density and Void Content

Density can void content is very important physical properties of polymeric based composites materials to estimate the characteristic of composites. Densities of the matrix and fillers are shown in Table 2.

The experiment is being undertaken using Mettler Toledo electronic balance by applying simple water immersion technique and experimental density is found. Theoretical density is found out by weight additive principle. Void content for the polymeric based composites is determined by taking into consideration the experimental and theoretical data. It is seen that the experimental data is lower than the theoretical indicating presence of voids at the surface forming the common boundary between the composites. It can be seen from the Table 3 that the density of composites with inorganic filler (Composite C & Composite D) is higher than same with natural fillers (Composite A & Composite B). This is due to the higher bulk density value of the fillers. But the void content for composites with inorganic filler is lower than that

Figure 10. Scanning electron microscope
Source: Roy et al., 2016

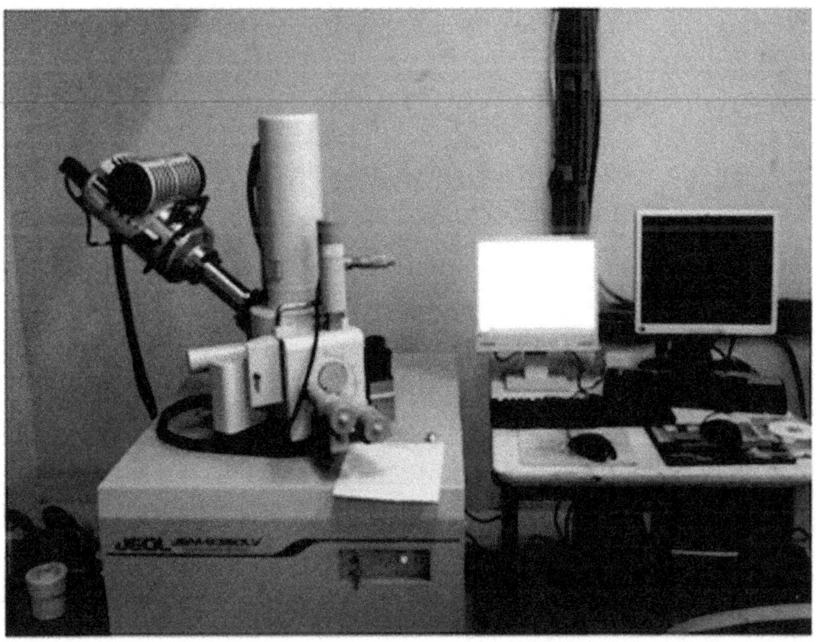

Table 2. Densities of the materials used

Material	Density(g/cc)	Material	Density(g/cc)
ABS	1.04	TiO_2	4.197
Jute Dust	0.037	Clay	0.33
Bagasse Dust	0.12		

Table 3. Density and Void Content of Composites

Material of Construction	Density (g/cc)		Void fraction (%)
	Experimental	Theoretical	
ABS	1.0300	1.0400	0.9615
Composite A (ABS + Jute Dust)	0.9014	0.9397	4.0758
Composite B (ABS + Bagasse Dust)	0.9086	0.9480	4.1561
Composite C (ABS + TiO_2)	0.9327	0.9610	2.9448
Composite D (ABS + Clay)	0.9381	0.9660	2.8882

with natural filler. The variations is due to the hygroscopic nature of the natural filler which tends to create void (resultant of the moisture) when processed at a higher temperature.

But since void content is less than 5% (acceptable level) these can be used for all industrial and household products.

Tensile and Flexural Tests

The results obtained were plotted and are depicted in the various graphs. Figure 11 shows the tensile strength values for each of the materials whereas Figure 12 illustrates variation in tensile modulus for the same. Figure 13 reveals the flexural strength data and Figure 14 indicates the flexural modulus. From all

Figure 11. Tensile strength of fabricated composites samples (with error bars)

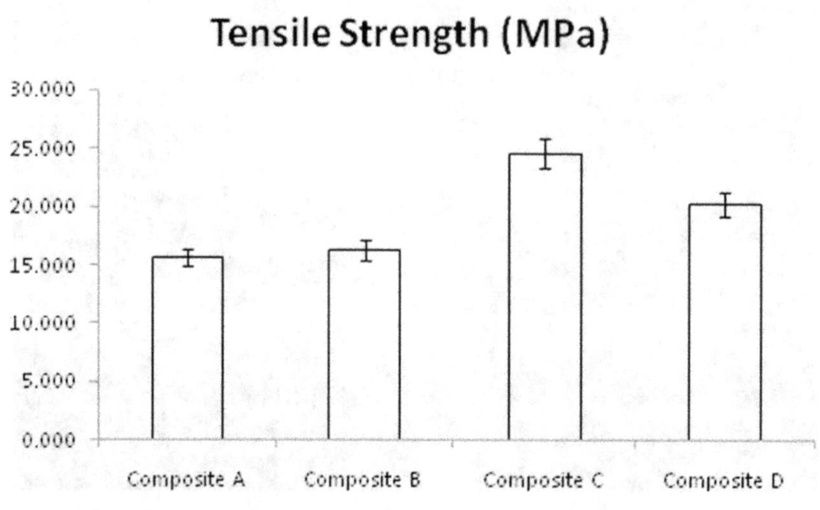

Figure 12. Tensile modulus of fabricated composites samples (with error bars)

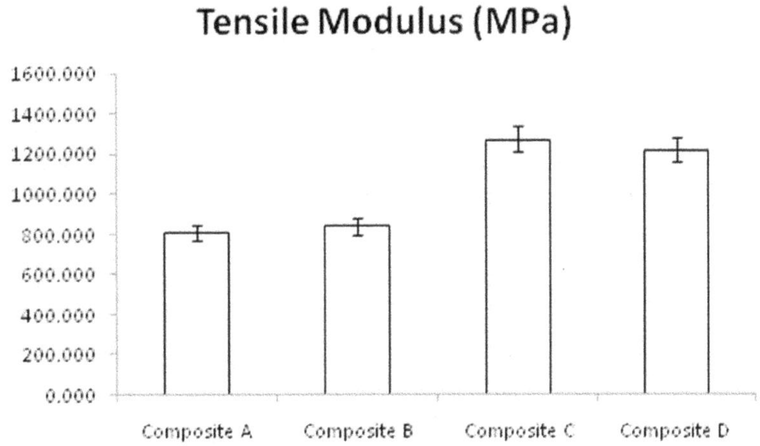

Figure 13. Flexural strength of fabricated composites samples (with error bars)

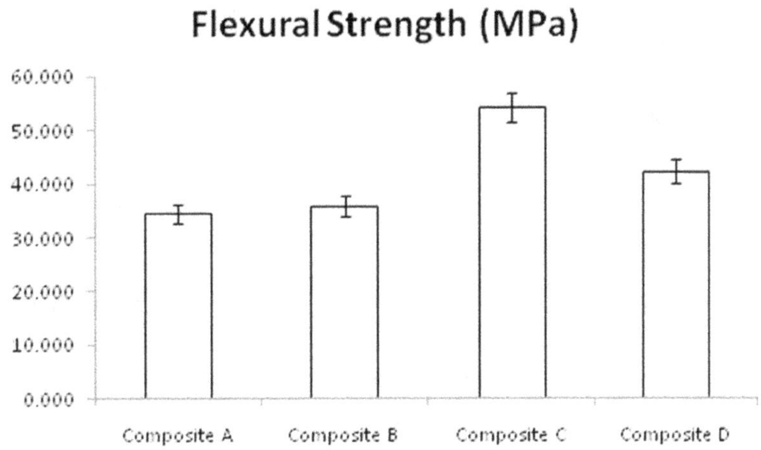

Figure 14. Flexural modulus of fabricated composites samples (with error bars)

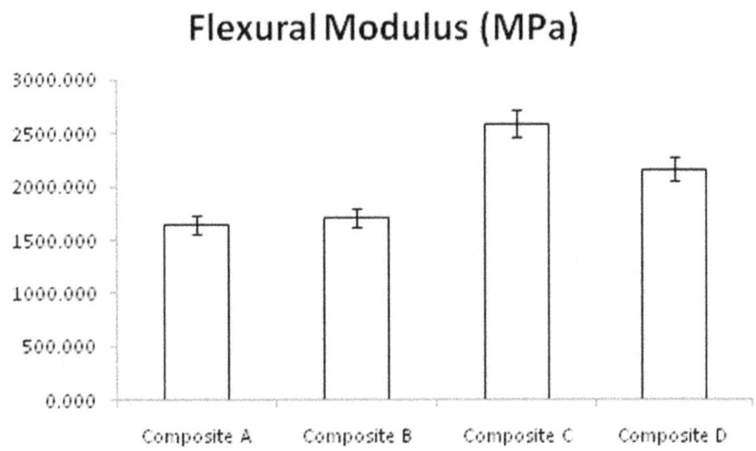

the above graphs it is clear that the mechanical strength of natural filler based composites (Composite A & B) are lower than that of inorganic filler based composites (Composite C & D). This can be attributed to the presence of lignin and cellulose in natural fillers, which tends to lower the mechanical properties. Within the two Composites C & D, Composite D exhibits a bit lower value than Composite C. This may be due to presence of various minerals in clay whereas TiO_2 is a single compound. This indicates that although the best option is Composite C but both Composite C & Composite D can be used to make products required to bear more stress but Composite A & Composite B should be considered if environmental issues are prevalent.

Hardness Test

Hardness is generally classified into three categories with respect to the depth of indentation (d) as follows: (a) Nano-hardness (d <1 μm), (b) Micro-hardness (d = 1-50 μm) and (c) Macro-hardness (d > 50 μm). The test methods commonly used for expressing the relationship between hardness and the size of impression are Brinell, Vicker's and Rockwell hardness tests. In this work, Vicker's hardness test setup is used to find out the micro-hardness value of all three products. The dwell time is kept at 10s while the speed of indentation is set at 50μm/s and indentation load at 100gf. The micro hardness tester is controlled via. a touch screen based system which is part of the tester. The hardness numbers are obtained by processing the indention image. An average of atleast three hardness values for each sample is recorded. The plotted graph of recorded data is shown in Figure 15. The trend replicates the tensile and flexural testing results.

Tribological Test

The experimental procedure and measurement of friction coefficient and specific wear rate are undertaken as discussed earlier. Figures16 and 17 show the variation of friction coefficient and specific wear rate on the various filler content of ABS based composites. From figure, it is seen that the friction coefficient and specific wear rate are lower for Inorganic Based Composites (Composite C & Composite D) than

Figure 15. Vicker's micro hardness number fabricated composites samples (with error bars)

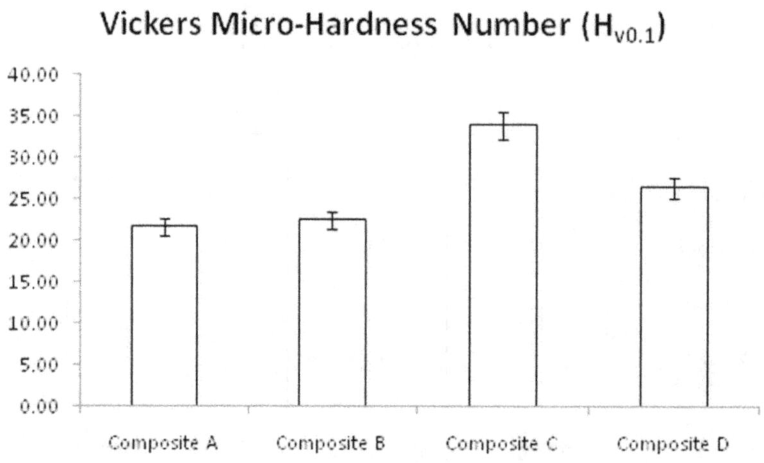

Figure 16. Coefficient of friction of fabricated composites samples (with error bars)

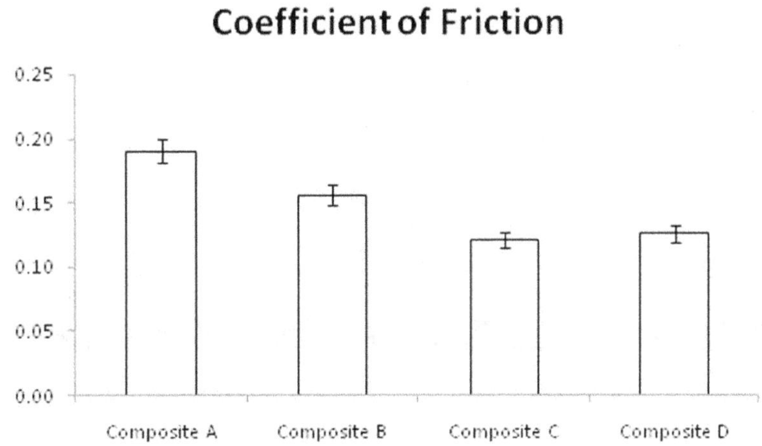

Figure 17. Specific wear rate of fabricated composites samples (with error bars)

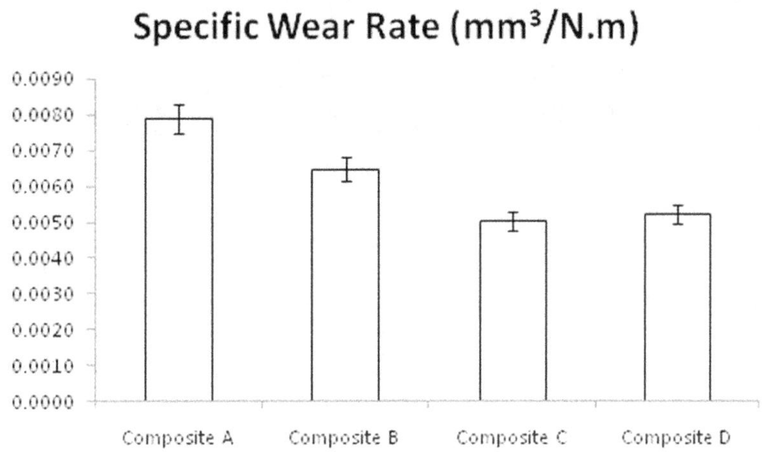

other two. The reason can be attributed towards the low bondage of the natural filler with the matrix. This makes fillers to wear out faster increasing both the friction coefficient (due to surface waviness created by removal of filler contents) and specific wear rate. However, variation in the results is not much as ABS contains Butadiene which in itself has low friction characteristics. But increasing the speed and loading condition may further increase both the outputs due to softening of composite as a result of increase in contact surface temperature due to frictional heating.

Morphological Study

Scanning electron microscopy examinations are carried out for the composite surfaces coated with thin platinum film on the worn out surface by sputtering to get a conducting layer on the surface to observe the filler distribution in the matrix. Figure 18 shows the SEM picture for the composite various fillers.

Figure 18. SEM images of fabricated composites samples (wear track shown with arrow)
((a) composite a; (b) composite b; (c) composite c; and (d) composite d

From figures it can be visualized that the dispersion and the interaction of filler with the matrix is evenly spread for all composites. This indicates that the properties achieved are more or less uniform over the total composite.

CONCLUSION

In this research, an attempt has been made to study and comparison of mechanical and tribological behaviour of natural and inorganic reinforced ABS composite. Void content results supported the fact that processing technique adopted was strong enough to provide better composites. It was novel finding of the work that hardness and strength (tensile as well as flexural) was better for inorganic filler based composites than natural ones. Tribological results although showed same trend but due to nature of the matrix used, variation of results were not much. A different matrix may lead to substantial findings. But it is obvious that normal load and sliding speed would influence the output more for natural fillers than the inorganic filler based composite. Hence, an inference can be drawn that in case of higher load and speed, products should be made of ABS based composites using inorganic filler than natural ones.

REFERENCES

Ahmed, N. M., & El-Sabbagh, S. H. (2014). The influence of kaolin and calcined kaolin on SBR composite properties. *Polymer Composites*, *35*(3), 570–580. doi:10.1002/pc.22697

Akinci, A., Sen, S., & Sen, U. (2014). Friction and wear behavior of zirconium oxide reinforced PMMA composites. *Composites. Part B, Engineering*, *56*, 42–47. doi:10.1016/j.compositesb.2013.08.015

Ali, I., & Elleithy, R. (2011). Toughness of HDPE / CaCO. *Composite Structures*, *48*(1), 49–58.

Ariffin, A., Mansor, A. S., & Jikan, S. S. (2008). Mechanical, morphological, and thermal properties of polypropylene / kaolin composite. Part I. The effects of surface-treated kaolin and processing enhancement. *Journal of Applied Polymer Science*, *108*(6), 3901–3916. doi:10.1002/app.27950

Asyadi, F., Jawaid, M., Hassan, A., & Wahit, M. U. (2013). Mechanical Properties of Mica-Filled Polycarbonate/Poly (Acrylonitrile-Butadiene-Styrene) Composites. *Polymer-Plastics Technology and Engineering*, *52*(7), 727–736. doi:10.1080/03602559.2012.762672

Aurrekoetxea, J., Sarrionandia, M., & Gomez, X. (2008). Effects of microstructure on wear behaviour of wood reinforced polypropylene composite. *Wear*, *265*(5), 606–611. doi:10.1016/j.wear.2007.12.013

Axen, N., Hutchings, I. M., & Jacobson, S. (1996). A model for the friction of multiphase materials in abrasion. *Tribological International*, *29*(6), 467–475. doi:10.1016/0301-679X(95)00104-C

Bahurudeen, A., Wani, K., Basit, M. A., & Santhanam, M. (2015).. . *Assesment of Pozzolanic Performance of Sugarcane Bagasse Ash. American Society of Civil Engineers*, *28*(2), 1–11.

Balakrishna, N. S., Ismail, H., & Othman, N. (2014). Polypropylene / Rattan Powder / Kaolin Hybrid Composites: Processing, Mechanical and Thermal Properties. *Polymer-Plastics Technology and Engineering*, *53*(5), 451–458. doi:10.1080/03602559.2013.844833

Basu, G and Roy, A.N. (2008). Blending of Jute with Different Natural Fibres. *Journal of Natural Fibres*, *4*(4), 13-29.

Ben Difallah, B., Kharrat, M., Dammak, M., & Monteil, G. (2012). Mechanical and tribological response of ABS polymer matrix filled with graphite powder. *Materials & Design*, *34*, 782–787. doi:10.1016/j.matdes.2011.07.001

Bhowmick, M., Mukhopadhyay, S., & Alagirusamy, R. (2012). Mechanical properties of natural fibre reinforced composites. *Textile Progress*, *44*(2), 85–140. doi:10.1080/00405167.2012.676800

Brennan, L. B., Isaac, D. H., & Arnold, J. C. (2002). Recycling of acrylonitrile–butadiene–styrene and high-impact polystyrene from waste computer equipment. *Journal of Applied Polymer Science*, *86*(3), 572–578. doi:10.1002/app.10833

Chand, N., Dwivedi, U. K., & Sharma, M. K. (2007). Development and tribological behaviour of UHMWPE filled epoxy gradient composites. *Wear*, *262*(1-2), 184–190. doi:10.1016/j.wear.2006.04.012

Chang, B. P., Akil, H. M., & Nasir, R. B. M., Bandara, & Rajakpase, S. (2014). The effect of ZnO nanoparticles on the mechanical, tribological and antibacterial properties of ultra-high molecular weight polyethylene. *Journal of Reinforced Plastics and Composites.*

Chang, B. P., Akil, H. M., Nasir, R. M., & Nurdijati, S. (2011). Mechanical and antibacterial properties of treated and untreated zinc oxide filled UHMWPE composites. *Journal of Thermoplastic Composite Materials*, *24*(5), 653–667. doi:10.1177/0892705711399848

Chang, B. P., Md Akil, H., & Bt. Md Nasir, R. (2013). Comparative study of micro-and nano-ZnO reinforced UHMWPE composites under dry sliding wear. *Wear*, *297*(1), 1120–1127. doi:10.1016/j.wear.2012.11.083

Chen, J., Feng, Y., Chen, N., Li, B., Chen, F., & Zhang, X. (2013). Fabrication and Properties of Silver-Doped Acrylonitrile Butadiene Styrene (ABS) Composites. *Polymer-Plastics Technology and Engineering*, *52*(11), 1081–1088. doi:10.1080/03602559.2013.769574

Datta, U. (2009). Application of Jute Geotextiles. *Journal of Natural Fibers*, *4*(3), 87–92.

Dong, L., Greco, R., & Orsello, G. (1993). Polycarbonate/acrylonitrilebutadiene styrene blends: 1. Complementary etching techniques for morphology observations. *Polymers*, *34*(7), 1375–1382. doi:10.1016/0032-3861(93)90848-5

Eiss, N. S. Jr, & Czichos, H. (1986). Tribological studies on rubber-modified epoxies: Influence of material properties and operating conditions. *Wear*, *111*(4), 347–361. doi:10.1016/0043-1648(86)90132-8

Esthappan, S. K., Kuttappan, S. K., & Joseph, R. (2012). Thermal and mechanical properties of polypropylene/titanium dioxide nanocomposite fibers. *Materials & Design*, *37*, 537–542. doi:10.1016/j.matdes.2012.01.038

Gamelas, J. A. F., Ferraz, E., & Rocha, F. (2014). An insight into the surface properties of alcined kaolinitic clays: The grinding effect. Colloids and Surfaces A: *Physicochem. Eng. Aspects*, *455*, 49–57. doi:10.1016/j.colsurfa.2014.04.038

Hassan, A., Akbari, A., Hing, N. K., & Ratnam, C. T. (2012). Mechanical and thermal properties of ABS / PVC composites: Effect of particles size and surface treatment of ground calcium carbonate. *Polymer-Plastics Technology and Engineering*, *51*(5), 473–479. doi:10.1080/03602559.2011.651242

Hughes, J. E., & Girdlestone, S. J. A. (2001). Use of solid waste bagasse from a paper mill as an organic amendment and its effect on some soil properties: A case study. *South African Journal of Plant and Soil*, *18*(3), 128–134. doi:10.1080/02571862.2001.10634416

Kaseem, M., Hamad, K., & Deri, F. (2013). On-line rheological measurements and mechanical properties of acrylonitrile-butadiene-styrene / corn starch composite. *International Journal of Polymeric Materials and Polymeric Biomaterials*, *62*(5), 260–264. doi:10.1080/00914037.2011.641698

Kocserha, I., & Gömze, L. A. (2010). Friction properties of clay compounds. *Applied Clay Science*, *48*(3), 425–430. doi:10.1016/j.clay.2010.01.017

Krishnan, A. K., George, T. S., Anjana, R., & George, K. E. (2013). Effect of modified kaolin clays on the mechanical properties of polypropylene / polystyrene blends. *Journal of Applied Polymer Science*, *127*(2), 1409–1415. doi:10.1002/app.38043

Li, Z. H., & Chen, T. (2012). The mechanical and tribological properties of MAH-grafted HDPE-filled clay / PEI composites. *Polymer-Plastics Technology and Engineering*, *51*(8), 776–779. doi:10.1080/03602559.2012.663848

Liu, H., Dong, L., Xie, H., & Xiong, C. (2012). Novel-modified kaolin for enhancing the mechanical and thermal properties of poly (vinyl chloride). *Polymer Engineering and Science*, *52*(10), 2071–2077. doi:10.1002/pen.23161

Ma, L., He, H., Jiang, C., Zhou, L., Luo, Y., & Jia, D. (2012). Effect of alkali treatment on structure and mechanical properties of acrylonitrile-butadiene-styrene / bamboo fiber composites. *Journal of Macromolecular Science, Part B: Physics*, *51*(11), 2232–2244. doi:10.1080/00222348.2012.669688

Mondal, M. I. H., Hoque, M. A., Latif, M. A., & Islam, M. K. (2014). Improved Characteristics of Acetic Anhydride and Phthalic Anhydride Modified Jute Fibers. *Research Journal of Textile and Apparel*, *18*(1), 64–70. doi:10.1108/RJTA-18-01-2014-B008

Praveen, R. N., Muruli, D., & Nayak, S. S. (2013). Sliding wear behavior of Acrylonitrile Butadiene Styrene (ABS) thermoplastic composites. *International Journal of Mechanical Engineering and Robotics Research*, *2*(4), 361–372.

Rasheva, Z., Zhang, G., & Burkhart, T. (2010). A correlation between the tribological and mechanical properties of short carbon fibers reinforced PEEK materials with different fiber orientations. *Tribology International*, *43*(8), 1430–1437. doi:10.1016/j.triboint.2010.01.020

Roselman, I. C., & Tabor, D. (1976). The friction of carbon fibres. *Journal of Physics. D, Applied Physics*, *9*(17), 2517–2532. doi:10.1088/0022-3727/9/17/012

Roy, S., Davim, J. P., Bhowmik, S., & Kumar, K. (2016). Estimation of mechanical and tribological properties of epoxy based green composites. In D. Verma, S. Jain, X. Zhang, & P. C. Gope (Eds.), *Green Approaches to Biocomposite Materials Science and Engineering* (pp. 96–124). Hershey, PA: Engineering Science Reference. doi:10.4018/978-1-5225-0424-5.ch005

Rutkowski, J. V., & Levint, B. C. (1986). Acrylonitrile-Butadiene-Styrene Copolymers (ABS): Pyrolysis and Combustion Products and their Toxicity-A Review of the Literature. *Fire and Materials*, *10*(3-4), 93–105. doi:10.1002/fam.810100303

Sanjay, M. R., & Yogesha, B. (2016). Studies on Mechanical Properties of Jute/E-Glass Fiber Reinforced Epoxy Hybrid Composites. *Journal of Minerals and Materials Characterization and Engineering*, *4*(01), 15–25. doi:10.4236/jmmce.2016.41002

Schön, J. (1996). Coefficient of friction of composite delamination surfaces. *Wear*, *237*(1), 77–89. doi:10.1016/S0043-1648(99)00315-4

Selvam, B., Marimuthu, P., Narayanasamy, R., Anandakrishnan, V., Tun, K. S., Gupta, M., & Kamaraj, M. (2014). Dry sliding wear behaviour of zinc oxide reinforced magnesium matrix nano-composites. *Materials & Design*, *58*, 475–481. doi:10.1016/j.matdes.2014.02.006

Song, H. J., Zhang, Z. Z., Men, X. H., & Luo, Z. Z. (2010). A study of the tribological behavior of nano-ZnO-filled polyurethane composite coatings. *Wear*, *269*(1), 79–85. doi:10.1016/j.wear.2010.03.011

Thakur, I. S., Ray, A. K., & Singhal, A. (2015). Ethanol from Sugar Cane Bagasse of Pulp and Paper Mill Effluent by Cryptococcus albidus and Saccharomyces cerevisiae. *Energy Sources, Part A: Recovery, Utilization, and Environmental Effects*, *37*(11), 1172–1179. doi:10.1080/15567036.2011.605431

Thakur, V. K., Singha, A. S., & Thakur, M. K. (2012). Green Composites from Natural Fibers: Mechanical and chemical aging properties. *International Journal of Polymer Analysis and. Characterization*, *17*, 401–407.

Wan, Y. Z., Chen, G. C., Raman, S., Xin, J. Y., Li, Q. Y., Huang, H., & Luo, H. L. et al. (2006). Friction and wear behavior of three-dimensional braided carbon fiber/epoxy composites under dry sliding conditions. *Wear*, *260*(9-10), 933–941. doi:10.1016/j.wear.2005.06.007

Wang, L. L., Zhang, L. Q., & Tian, M. (2012). Mechanical and tribological properties of acrylonitrile–butadiene rubber filled with graphite and carbon black. *Materials & Design*, *39*, 450–457. doi:10.1016/j.matdes.2012.02.051

Xia, Z. Y., Yu, J., Cheng, L., Liu, L., & Wang, F. (2010). Study of the Tensile Properties of Jute / Cotton Blended Yarns Using Weibull Distributions. *Research Journal of Textile and Apparel*, *14*(1), 18–25. doi:10.1108/RJTA-14-01-2010-B002

Xie, G. Y., Zhuang, G. S., Sui, G. X., & Yang, R. (2010). Tribological behavior of PEEK / PTFE composites reinforced with potassium titanate whiskers. *Wear*, *268*(3), 424–430. doi:10.1016/j.wear.2009.08.032

You, Y. L., Li, D. X., Si, G. J., Lv, R. Y., Deng, X. (2014). Improvement in the tribological properties of polyamide 6: Talc, glass fiber, graphite, and ultrahigh-molecular-weight polyethylene. *Journal of Thermoplastic Composite Materials.*

Zeng, A., Zheng, Y., Guo, Y., Qiu, S., & Cheng, L. (2012). Effect of tetra-needle-shaped zinc oxide whisker (T-ZnOw) on mechanical properties and crystallization behavior of isotactic polypropylene. *Materials & Design*, *34*, 691–698. doi:10.1016/j.matdes.2011.05.028

Chapter 4
Properties Enhancement of Polymer/Ceramic Nanocomposites:
Polymer Nanocomposites

Noureddine Ramdani
Harbin Engineering University, China

Mehdi Azibi
Independent Researcher, Algeria

ABSTRACT

This chapter reviews the various properties enhancement of polymer/ceramic nanocomposites. Ceramics nanofillers have attracted both academic and industrial interest as they can produce a significant improvement in the properties of polymers even at lower filler loadings. Recently, numerous kinds of polymeric matrices reinforced with ceramic nanoparticles have been reported. The surface-modification of ceramic nanoparticles was reported to provide extra-improvements in the thermal and mechanical properties of these materials. In addition, the type of the used ceramic nanofillers agent determines the final properties of the nanocomposites. Herein, the various effects of adding ceramic nanoparticles on the thermal, electrical, optical, and mechanical properties of polymer/ceramic nanocomposites as well as the reinforcing mechanism are discussed in general along with detailed examples drawn from the scientific literature.

INTRODUCTION

Polymer nanocomposites have been developed to mitigate the thermal and mechanical drawbacks of conventional polymers to make them suitable for diverse applications (Anatolii, 2000). The most used inorganic nanofillers are ceramics based nanofillers which are incorporated into different polymer matrices to improve their inferior mechanical and thermal properties. The fabrication of polymer/ceramic nanocomposites was mainly achieved by adding raw or surface-modified ceramic nanoparticles into

DOI: 10.4018/978-1-5225-3023-7.ch004

polymeric matrices via solution blending or direct mixing methods to improve the lower thermal and mechanical properties of the pure thermoplastic and thermoset matrices. These kinds of nanomaterials are widely used in various industries such as coating, microelectronics, automobiles, and aeronautics due to their good mechanical properties, low cost, high temperature applications, lightweight (Ginés et al., 2015; Kumar, 2003).

Ceramics are categorized into four main different groups, which are oxides, nitrides, carbides, and borides based ceramics. In the last few decades, various kinds of ceramic nanoparticles have produced and used in different industries such as the automotive, electronics, aerospace, medical and telecommunications due to their excellent inherent characteristics, including, a high strength, a good corrosion resistance, improved wear properties, low dielectric constant, superior hardness, higher toughness, and higher operating temperatures (Paul et al., 2012). Therefore, ceramic nanofillers will continue to play a critical role in the development of smart and intelligent materials in future.

The surface-treatment of ceramic nanoparticle is one of the key parameters which control the final properties of polymer/ceramic nanocomposites due to its ability to improve the dispersity and processability with polymeric matrices. For instance, extensive researches have been made on the surface-modification using silane coupling agents as effective routes for the functionalization of ceramic nanofillers (Rahman & Padavettan, 2012). In addition, surface-modification of ceramic nanoparticles has attracted a wide attention because it resulted in excellent integration and better contact in the nano-ceramics/polymers interface. Polymer reinforced with modified ceramic nanoparticles combine both the functionalities of unfilled polymeric matrices, including low weight, ease of processing and the unique features of the ceramic nanoparticles such as high mechanical and thermal properties. Thus, such nanocomposites can exhibit dramatic improvement in several aspects, such as mechanical, electrical, optical, magnetic, and inflammability properties. A detailed study of literature on the surface-modification of ceramic nanofillers was compiled herein (Kango et al., 2013).

This review is focused on the properties enhancements of ceramic nanofillers reinforced-polymeric matrices composites and the mechanisms involved. The effects of the different type of ceramic, ceramic loadings, and surface–treatment on the performance of these also discussed and the applications of the prepared nanocomposites were presented.

BACKGROUND

Ceramic Nanoparticles

Several physical and chemical methods have been used to elaborate ceramic nanoparticles having different shape, dimension, and properties. The physical approach is based on vapor deposition and related to the principle of sub-dividing bulk precursors into finer nanoparticles, while the chemical methods involves reduction of metal ions into metal atoms in the presence of a stabilizer, succeeded by the controlled aggregation of atoms (Chen & Liu, 2011). The synthesis of ceramic nanofillers by chemical methods is preferred over the physical ones. The synthesis methods, properties, and applications of some ceramic nanoparticles are collected in Table 1.

Table 1. Synthesis, properties and applications of some ceramic nanoparticles

Name	Synthesis Methods	Properties	Applications
SiO_2	• Sol–gel. • Water-in-oil microemulsion processes. • Flame synthesis.	• Good optical properties. • High thermal stability. • High mechanical properties.	•Drug delivery. • Carrier for antimicrobial applications. • Biosensor applications.
TiO_2	• Hydrothermal. • Flame spray pyrolysis. • Solvothermal. • Reverse micelles.	• Good anticorrosion. • Good optical properties. • High mechanical properties.	• Photocatalysis. • Dye-sensitized. • Solar cells. • Gas sensor. • Waste water treatment.
Al_2O_3	• Sol-Gel. • Precipitation and freeze drying. • Flame spray pyrolysis. • Reverse microemulsion.	• Good fracture properties. • High mechanical properties. • Good transport properties.	• Waste water and soil treatment. • Ceramic ultrafilters. • Membranes for microorganisms. • Gas separation. • Catalysis and absorption.
B_4C	• Carbothermic reduction of boron oxide. • Metallothermic reduction of boron oxide with Mg/C. • Gas-phase interaction.	• Good hardness. • Good elastic properties. • High melting point. • Specific nuclear and electric characteristics.	• Thermoelectric, • semiconducting, • biological and radiation-resistant • Armour
AlN	• Arbothermal reduction. • Ball milling. • Physical and chemical vapor deposition. • Wet chemistry.	• High thermal conductivity. • High breakdown voltage. • High electrical Resistivity. • Good mechanical properties, • High piezoelectric properties	• Piezoelectric materials. • Electronics industry.
Si_3N_4	• Direct nitridation of silicon. • Diimide synthesis. • Carbothermal nitridation. • Plasmachemicalsynthesis. • Pyrolysis of silicon organic compounds.	• Low thermalexpansion coefficient. • High-temperature strength. • High hardness. • Excellentwear-resistance. • Good shock resistance.	• Gas turbines • Turbocharger • heat exchanger and engines • shell-type components • friction couples

PROCESSING METHODS OF POLYMER/CERAMIC NANOCOMPOSITES

Sol–Gel Method

The ceramic nanofillers can be processed using a sol–gel technique to reinforce the various polymeric matrices, where these nanofillers can be dissolved in either aqueous or organic media at the presence of their matrices, forming interpenetrated networks between inorganic and organic group moieties at mild temperatures; this structure improves the compatibility between hybrid constituents and constructs strong interfacial adhesion between them. This technique has been extensively used to elaborate many kinds of thermosetting nanocomposites reinforced by ceramic fillers such as silica, alumina, calcium oxide, and Titania (Kango et al., 2013). Different sol–gel process routes have been used to prepare the thermosets/ceramic nanocomposites. Anew series of polystyrene/SiO_2 nanocomposites via a sol–gel technique, where a good miscibility between the matrix and nanofillers were obtained by the formation of covalent bands (Hsiue, Kuo, & Jeng, 2000). In addition, sol–gel hydrolysis and condensation of a precursor, such as tetraethyloxysilane (TEOS), tetrabutyl titanate, or aluminium iso-propoxide can be carried out starting from a preformed functional organic polymer, such as epoxy (Liu, Hsu, & Wei,

2003), polybenzoxazine (Agag & Takeichi, 2011), polyetherimide (Nunes et al., 1999), polyvinyl alcohol (Suzuki et al., 1990), polyamides (Sengupta et al., 1990), polyimide (Hsiue, Chen, & Liu, 2000) and several other polymers. Using *in*-situ sol–gel method, (Wu, 2004) elaborated of novel polycaprolactone (PCL)/TiO_2 nanocomposites were prepared using tetra isopropyl ortho-titanate (TTIP) and PCL as the ceramic precursor and the continuous phase, respectively. Nylon-6/SiO_2 nanocomposites were also prepared by a sol–gel method (Garcia et al., 2004). Behniafar & Amirkhalili described a facile method for preparing poly(4,4`-oxydiphenylene-pyromellitimide) (POPI)/$TiO_2$ nanocomposites. 4,4`-Methylene diphenyl diisocyanate (MDI) was used as the surface modifier of TiO_2 nanoparticles as depicted in Figure 1 (Behniafar & Amirkhalili, 2014).

Figure 1. Schematic illustration for surface treatment of nano-TiO_2 by MDI (a), MDI-treated TiO_2 nanopowder incorporated into the macromolecular chains.
Source: Reprinted from Springer. Behniafar & Amirkhalili, 2014

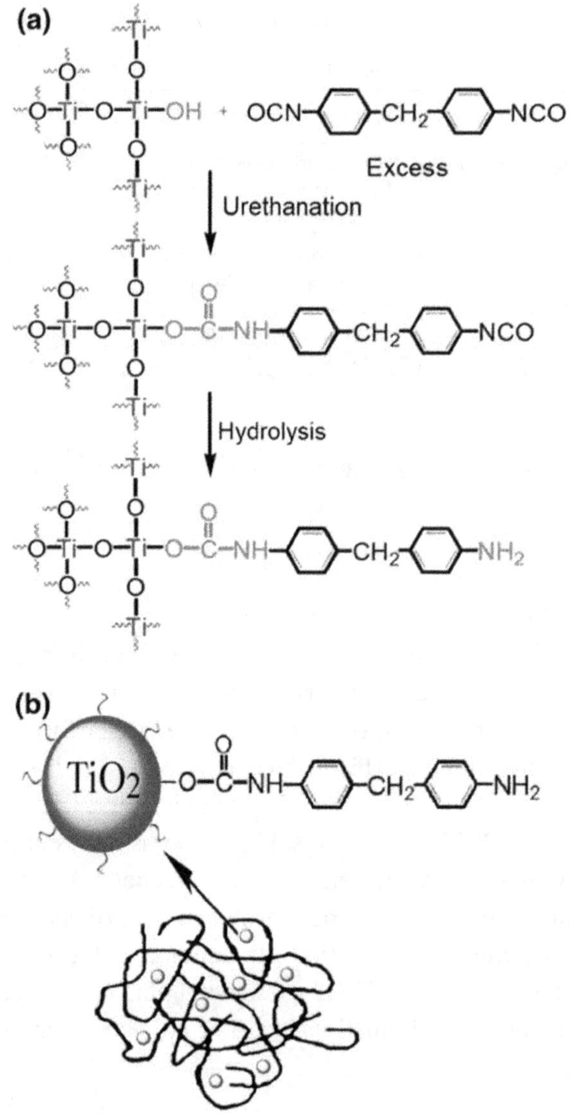

In-Situ Polymerization

In-situ polymerization technique is based on dispersing the ceramic nanoparticles at first in the polymeric monomer, and polymerizing the resulted mixture by a technique similar to bulk polymerization. (Ou, Yang, & Yu, 1997; Ou, Yang, & Yu, 1998) carried out the synthesis of nylon-6/SiO_2 nanocomposites using *in*-situ polymerization method. A series of poly (methyl methacrylate)/TiO_2nanocomposites were developed by an *in*-situ radical polymerization of methyl methacrylate in a toluene solution of surface-treatedTiO_2 nanoparticles (Guan, Lu, & Cheng, 2009; Dzunuzovic et al., 2007). Newkind of polyester/ TiO_2 nanocomposites were synthesized by *in*-situ polymerization. This polymerization method was also applied for the reparation of LLDPE/SiO_2 nanocomposites in the presence of a metallocene catalyst (Evora & Shukla, 2003).

Solution Blending

The most convenient method for producing polymer/ceramic nanocomposites is the direct mixing of nanofillers with matrices, which can be generally carried out by a solution blending method or melt processing. Due to the high tendency of the ceramic nanofillers to form agglomerates, the degree of their dispersion in the various thermosets is low, in this method. The solution blending method is based on mixing the different ceramic fillers and matrices in a liquid state generated from the added solvent or from the melted resin. The good mixing quality, which is generally provided by this technique, increases its use in the processing of such nanomaterials.

Yan et al. used the solution blending method to prepare a series of Bismalimide/ZrO_2 nanocomposites (Yan et al., 2007). They ultrasonically dispersed the silane-treated and untreated nano-$ZrO_2$of 20–40 nm in acetone solvent. Bismalimide prepolymer solution was added and the resulting mixture was heated to 120°C and kept at this temperature with stirring for 30 min to evaporate the solvent. The mixture obtained was then thermally cured at various temperatures followed by a post-cure process at 250°C for 6 hours. Ramdani et al. synthesized polybenzoxazine-based networks reinforced with γ-Aminopropyltrioxysilane-treated Si_3N_4 nanoparticles having a diameter of 50 nm, using a solution blending method in dichloromethane (DCM) (Ramdani et al., 2014). They also fabricated polybenzoxazine/TiC nanocomposites via a solution casting in tetrahydrofuran and using evaporation method (Ramdani et al., 2014). The synthesis of thermally-conductive epoxy nanocomposite was carried out by adding hybrids of Al_2O_3/AlN at different sizes to epoxy matrix by solution blending and using casting method (Choi& Kim, 2013). New kinds of polyimide/modified-alumina (PI/silane-treated Al_2O_3) nanocomposites were recently developed by Ghezelbash et al. A schematic representation of a solution blending method is as shown in Figure 2.

Many polymer nanocomposites filled with ceramic likes SiC, BN, $BaTiO_3$, $CaCO_3$ were also produced by melt processing technique (Lim & Zeng, 2010; Ulus et al., 2014; Nhuapeng, et al., 2008). Recently a wetting process was used to fabricate epoxy/BN nanocomposites by Kim & Kim is as shown in Figure 3.

PROPERTIES ENHANCEMENT OF POLYMER/CERAMIC NANOCOMPOSITES

Recently, extensive researches on polymer/ceramic nanocomposites have been conducted by both academia and industrials. The inclusions of ceramic nanofillers into various polymeric matrices resulted in new materials having substantial thermal, optical, and mechanical performance improvements involving

Figure 2. Preparation of polyimide/Al2O3 nanocomposites
Source: Reprinted with copyright 2011, permission from Indian academy of sciences Ghezelbash, Ashouri, & Mousavian, 2012

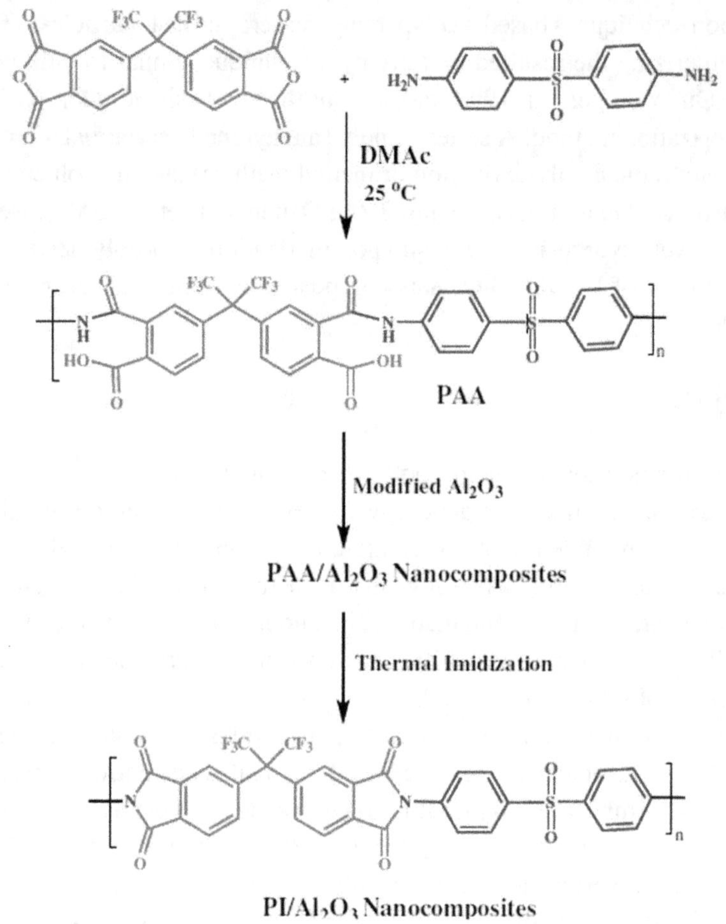

Figure 3. Schematic diagram of the fabrication procedure of the epoxy/BN composite.
Source: Reprinted from Copyright 2014, with permission from Elsevier. Kim & Kim, 2014

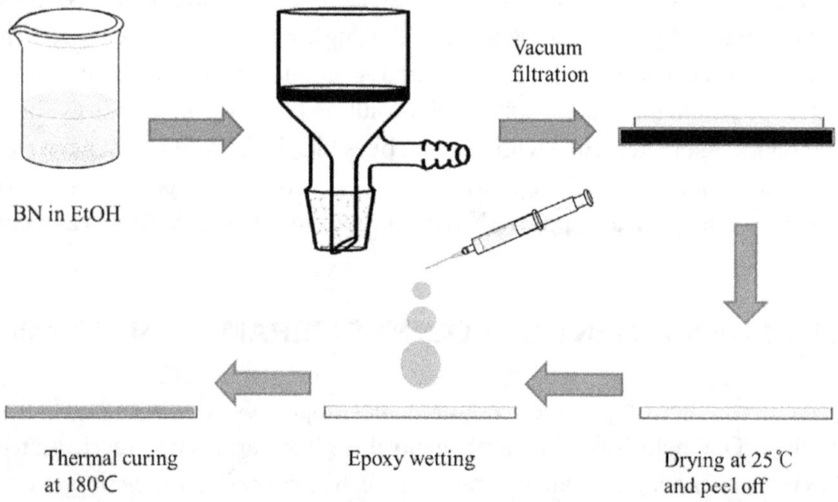

different mechanisms and thus they are suited for many industrial applications. Ceramic nanofillers, includes oxide, boride, and carbide have been successfully used to reinforce thermoplastic or thermoset polymers, such as epoxy, polybenzoxazine, nylon 6, polyamide, phenolic, etc. The choice of ceramic reinforcing agent depends on to the targeted polymer property that needs to be improved. For instance, BN particles are selected for enhancing the high conductivity of thermosets; however theAl_2O_3 and SiC nano-ceramics are used to ameliorate the mechanical and corrosion-resistance properties of some polymers. In this section, the enhanced properties of polymer/ceramic nanocomposites were discussed, compared and explained their improvement mechanisms.

Mechanical Properties

The main objective to fill polymeric matrices with ceramic nanoparticles is to elaborate new materials having improved mechanical properties like tensile strength, flexural strength, hardness, Young's modulus via different mechanisms as explained in theories of polymer nanocomposites (Wang et al., 2005; Zhang & Simon, 2005). Ou et al. investigated the mechanical characteristics of nylon-6 reinforced with SiO_2 nanofillers and found that the addition of 5 wt.% nanoparticles at the size of 50 nm to nylon-6 increases the tensile strength by almost 15%, the strain-to-failure by 150%, the Young's modulus by 23%, and the impact strength by 78% (Ou, Yang, & Yu, 1998). The mechanical properties of acrylonitrile butadiene styrene (ABS) reinforced with various amounts of $CaCO_3$ nanoparticles have been reported by Jiang et al. (Jiang, Lam, & Tam, 2005). The fracture toughness of polyester thermoset resin was significantly improved by incorporating Al_2O_3 nanoparticles (Zhang & Singh, 2004). (Wang, Lim, & Luo, 2006) explored the wear behavior of Al_2O_3/polymer nanocomposite coatings, reported that the inclusion of Al_2O_3 nano-ceramic enhances the scratch and abrasion resistance of the pure polymer. The change in quasi-static fracture toughness of new polyester/TiO_2 nanocomposites as a function of the nanoparticles volume contents. Results shows that the addition of 1 vol.%, 2 vol.% and 3 vol.% TiO_2 nanoparticles led to an increase in the fracture toughness of 57, 42 and 41%, respectively, where compared with that of the unfilled polymer. However, where the volume content is 4 vol.% TiO_2 there was a decrease in the toughness value due to the lack of dispersion in this system caused by weak bonding between the TiO_2 nanoparticles and polyester resin. The addition of SiC nano-filler to phthalonitrile resin significantly increases the flexural strength, modulus, tensile, and the microhardness as reported by Derradjiet al. (Derradji et al., 2015)

Zhang & Yang found that the inferior creep resistance and dimensional stability of Polyamide 6 thermoplastic can be improved by reinforcing it with TiO_2 nanoparticles (Zhang & Yang, 2004). The effect of nano-sized SiC on the mechanical performance of epoxy resin was studied by Chisholm et al. and they reported that the prepared nanocomposites have better mechanical properties compared pure matrix (Chisholm et al., 2005). Ng et al. found that the incorporation of 10 wt.% TiO_2 nanoparticles into epoxy matrix improved both the scratch resistance and the toughness of the resulted nanocomposite as compared to that of virgin thermoset (Ng et al., 1999). Rong et al. investigated the tribological behavior of TiO_2/ epoxy nanocomposites and reported that to be strongly controlled by the homogeneity dispersion of the nanofillers (Rong, Zhang, & Liu, 2001). Schwartz& Bahadur also evaluated the tribological characteristics of Al_2O_3/polyphenylene sulphide nanocomposites showed that the loading of 2 vol.% nanofillers had the lowest wear rate and best bond strength while further increases in filler content increases the wear rate (Schwartz & Bahadur, 2000). These results were attributed to the fact that nano-Al_2O_3fillers have the ability to anchor the transfer film to the counter face. The mechanical properties of silane-modified

Al_2O_3 nanoparticles filled PMMA matrix was studied by Siegel et al. where the inclusion of a 5 wt.% Al_2O_3 nanoparticles content into PMMA engendered an increase in the strain-to-failure of over 28%, which allows a ductile flow in the glassy state (Siegel, Chang, & Ash, 2001). Using micro-tensile tests, Guo et al., evaluated the tensile strength and modulus of the pure resin modified with Al_2O_3 nanofillers which reveals that the filling with surface-modified nano-Al_2O_3 particles increases the tensile modulus, the tensile strength, and the toughness of the pure resin. For instance, the tensile strength and Young's modulus of the 3 vol.% filled nanocomposite was increased by almost 60% and 85%, respectively, compared to that of the unfilled resin (Guo et al., 2006). New series of TiO_2/cyanate ester resin (CE) composites were prepared using a multi-step grafting technique, since up to 4 wt.% TiO_2 loading, the impact and bending strength increased by 75.9% and 78.9%, respectively, (Zhu & Wang, 2015). The impact strength and flexural strengths of the cyanate ester/silane-treated SiO_2nanocomposites were also significantly improved due to the changes of both interface adhesion and dispersibility of filler in matrix (Yan et al., 2014). A novel nano-Si_3N_4/epoxidized silane-cyanate ester was recently developed. The impact and flexural strength values of the nano-system werereached 16 kJ/m^2 and 114MPa, respectively, which represents an increase of about 56% and 21% compared to the unfiled cyanate ester matrix, respectively. Yan et al., studied the influence of the nano-SiC on the wear resistance of the cyanate ester resins/SiC nanocomposites in terms of frictional coefficient and wear rate (Yan et al., 2013). The result shows that the wear rate of the prepared nanocomposites with 1wt.% nano-SiC was the minimum and decreased by almost 51% compared to that of the pure thermoset, while the frictional coefficient was only decreased by 18%. The mild abrasive wear and ploughing were the main wear mechanisms of such improvements (Zhang & Zhang, 2013).

Thermal Properties

Thermal properties of polymer nanocomposites are generally evaluated using various thermal analysis techniques such as dynamic scanning calorimetry (DSC), thermogravimetric analysis (TGA), thermo-mechanical analysis (TMA), and dynamic mechanical analysis (DMA) tests. The filling of polymers with nano-sized ceramic particles lead to an enhanced thermal stability where these nanofillers act as a superior thermal insulator and as a mass transport barrier to the volatile products generated during the thermal decomposition process (Ray & Okamoto, 2003). Ramdani et al., studied the thermal stability of polybenzoxazine/TiC nanocomposites and found that is higher than that of pure resin (Ramdani et al., 2014). Phthalonitrile/silane-treated TiO_2 nanocomposites were prepared by Derradji et al. TGA analysis showed that the $T_{5\%}$ and the char yield at 800°C under nitrogen atmosphere (Y_c) were significantly increased by adding the TiO_2 nanofillers. At 6 wt. % KH550-TiO_2 content, the T_g and the storage modulus were considerably enhanced reaching 364°C and 3.2 GPa, respectively (Derradji et al., 2016). In addition, Phthalonitrile/silane-treated Si_3N_4 nanocomposites were also exhibit an excellent thermal stability, improved modulus and glass transition temperature (T_g) due to the good dispersion and adhesion between the particles and the resin in the presence of KH550 silane (Derradji et al., 2015). The curing reaction of the cyanate ester reinforced with functionalized SiO_2nanoparticleswas facilitated with the addition of F-SiO_2 and the cured nanomaterials exhibit superior thermomechanical and thermal stability (Li et al., 2015).(Wu & Kessler, 2015) prepared h-BN/BECy nanocomposites which exhibited outstanding performance in dimensional stability, dynamic-mechanical properties, and thermal conductivity, along with the controllable dielectric property and preserved thermal stability for high-temperature applications.

Chen et al. observed that the loading of only 5 wt.%nano-Al_2O_3 induced a marked increase in the thermal stability and the $T_g$of polymer whereas higher Al_2O_3 concentration decreases the first decomposition temperatures (Chen et al., 2009). Omrani et al. also observed that the $T_g$was raised up to 5 wt.% loading of Al_2O_3nano-fillers (Omrani et al., 2009). A marked increase in T_g and a significant decrease in the coefficient of thermal expansion (CTE) was also found in epoxy/SiO_2 nanocomposites (Kang et al., 2001). Wu & Ke discovered that the addition of polystyrene-SiO_2 encapsulates increases the crystallization temperature and rate of polyethylene terephthalate (PET) as compared to both untreated SiO_2 and silane-treated nano-SiO_2 due to the nucleation effect of PS-encapsulated SiO_2 nanoparticles in the matrix (Wu & Ke, 2007). The decomposition temperatures of PMMA/SiO_2 and PMMA/ZrO_2 nanocomposites were significantly increased in cases where the maximum weight loss was observed by thermal decomposition of PMMA segments as reported by (Wang et al., 2005). This phenomenon was attributed to the effect of rigid ceramic nanofillers which reduces the mobility of the polymeric chains. Ramdani et al. reported an enhancement in the thermal stability and char yields of different polybenzoxazine/ceramic nanocomposite systemsusing some nano-ceramic agents such as SiC, AlN, and Si_3N_4 (Ramdani et al., 2016). Derradji et al., stated that the thermal stability and the first decomposition temperatures of phthalonitrile/Si_3N_4 nanocomposites increase remarkably with the increase in the nanoparticle loading (Derradji et al., 2015).

Novel SiO_2-ZrO_2/Cyanate ester nanocomposites with good interfacial adhesion and having higher storage moduli and T_g values were successfully fabricated. The resulted improvements were attributed to the double effects of the unique structure of SiO_2-ZrO_2ceramics and the active groups on their surfaces which chemically-bonded them with the matrix, while SiO_2 particles covered the ZrO_2 provide physical interlocking between the matrix and fibers (Qin et al., 2014). The addition of nano-BN significantly improves the thermal conductivity of the cyanate ester resin, while the nanocomposites relatively still retain the good electrical insulation. As the BN content reached the value of 23.6 vol.%, the thermal conductivity of the composite increased to 1.33 W/m.K (Zhao et al., 2014).The silane-treated AlN filled cyanate ester nano-hybrid was found to have the highest synergistic effect, endowing the resin with outstanding thermal conductivityand significantly improved flame retardancy (Mi et al., 2013). The effect of using silane-coupling agent treated-AlN nanoparticles to reinforce cyanate ester resin was studied and the collected results proved that better thermal stability, higher thermal conductivity, and T_g can be obtained compared to the nanocomposites prepared from raw AlN due to better dispersion of treated nano-fillers, improves interfacial adhesion between treated-nanofillers and CE resin, and increases cross-linking density (Zhang et al., 2012). The addition of nano-SiO_2 enhances the thermal decomposition temperature (T_d) and low oxygen index (LOI) value of the phenolic matrix by almost 70 °C and 5, respectively, due to flame retardancy of filler and equally to the nanocomposite integrated structures (Periadurai et al., 2010). Based on their findings the CTE of the polyimide hybrid films decreases gradually with the increase of Al_2O_3 content. The 15 wt. % Al_2O_3/polyimide had a CTE of 41 ppm/°C, in the temperature range of 30–240 °C, which was 24% lower than that of pure polyimide film. This dramatic decrease in CTE could be due to the low-CTE feature of inorganic materials (Al). Besides, the network structure of Al_2O_3 domains could lead to the decrease in the segmental mobility of the polyimide chains. The reduced CTE value of polyimide/Al_2O_3 hybrid film could improve the existed thermal stress between Si_3N_4 barrier film and polyimide /Al hybrid film, and then reduce the separation of Si/Al_2O_3 hybrid film.

Electrical Properties

The majority of polymer–ceramic nanocomposites have been prepared to satisfy the requirements of the advanced electronic and optoelectronic devices which are designed at nano-scale dimension. Using the *in*-situ polymerization (Su & Kuramoto, 2000) developed new polyaniline/TiO_2nanocomposite films those exhibit higher electrical conductivity at the range of 1–10 S/cm.Mo et al. also synthesized polyaniline/TiO_2 nanomaterials from TiO_2 nanoparticles and colloids and reported that the electrical conductivity, the dielectric constant, and dielectric loss of the elaborated nanocomposites gradually increase with increase in the TiO_2 content from 1 to 5 wt.% (Mo, Wang, & Chen, 2008). Ramdani et al., reported that the dielectric properties of polybenzoxazine/AlN nanocomposites synthesized by solution blending method increases as the content of AlN increased (Ramdani et al., 2016). The addition of 60 wt.% of AlN spherical particles caused an increase in the dielectric constant of 50% as compared to that of the pure resin due to the formation of a conductive AlN pathways. Ma et al., detected that using functionalized TiO_2 nanofillers significantly improves the electrical properties of polyethylene/TiO_2 nanocomposites (Ma et al., 2005). Singha & Thomas studied the electrical characteristics of an epoxy thermoset filled with TiO_2 and Al_2O_3 nano-fillers at low nano potential electrical systems (Singha & Thomas, 2008). Using sol-gel approach, Ariraman et al. designed new CE/ZrO_2 nano-hybrid at various nano-ZrO_2 weight fractions. Tests reveal that at 7 wt.% ZrO_2 content the nanocomposites exhibited the highest dielectric constant value of 9.21 due to the homogeneous dispersion of ZrO_2 nanofillers in the matrix and their strong interfacial polarization in the hybrid (Ariraman et al., 2015). Cyanate ester nanocomposites have been developed by polymerizing cyanate ester monomer reinforced with varying weight percentages of glycidyl silane functionalized mesoporous nano-SiO_2. The existence of pores in these nanocomposites was responsible for the reduction of the dielectric constant, which reached its lowest value of 1.98 at 1 MHz for a nanocomposite having 10 wt.% of nanao-SiO_2 (Ariraman et al., 2014). Wang et al. found that the dielectric constant varies with the sweeping frequency and $BaTiO_3$ content (Wang & Cheng, 2009). Figure 4 and Figure 5 show the effects of $BaTiO_3$ content on the dielectric constants and dielectric losses of polyimide/$BaTiO_3$ composite films at the sweep frequencies of 10 kHz, 100 kHz, and 1 MHz, respectively. The dielectric constant of pure polyimide varies between 2.40 and 2.60 and exhibits low frequency dependence, whereas (Li et al., 2007) investigated that the dielectric constant of the nano-Al_2O_3/polyimide hybrids synthesized by in-situ polymerization increases with the increase of the alumina content and exhibits gradually strong frequency dependence. This is because alumina has a higher dielectric constant (8–10) than polyimide (2.5).

As reported by Wu et al., the mechanical properties and thermal stability of the pure polyimide film can be improved by adequate addition of Al_2O_3 (Wu et al., 2005). They found that the polyimide hybrid films with 10 wt.% of Al_2O_3 content exhibited obviously enhanced electrical aging performance with the time to failure of 3.4 times longer than that of pure polyimide film, which is attributed to the nano-sized Al_2O_3 particles highly dispersed in the hybrid film as shown in Figure 6.

Optical Properties

The polymer/ceramic materials have established themselves as functional components in optoelectronic devices such as light-emitting diodes (LEDs), field effect transistors (FETs), and photovoltaic devices due to their advantageous optical properties, such as photoluminescence, light absorption, and extreme refractive index. These optical properties were controlled by the dispersion state of ceramic nanoparticles

Figure 4. The effects of $BaTiO_3$ content on the dielectric constants and dielectric losses of polyimide/ $BaTiO_3$ composites, at the sweep frequencies of 10 kHz, 100 kHz, and 1 MHz
Source: Reprinted with permission from Elsevier (Wang et al., 2009).

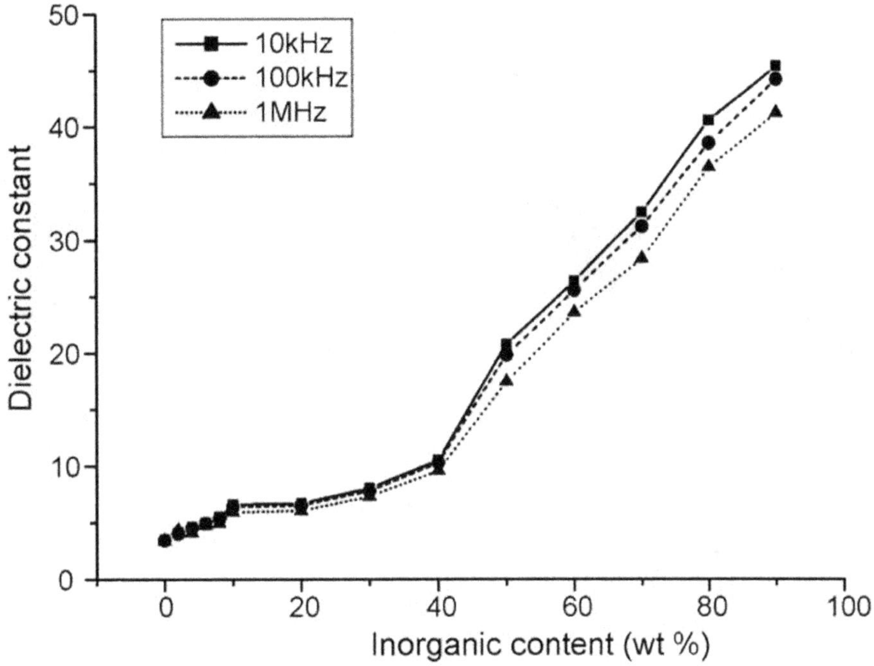

Figure 5. The effects of $BaTiO_3$ content on the dielectric constants and dielectric losses of polyimide/ $BaTiO_3$ composites, at the sweep frequencies of 10 kHz, 100 kHz, and 1 MHz
Source: Reprinted with permission from Elsevier (Wang et al., 2009).

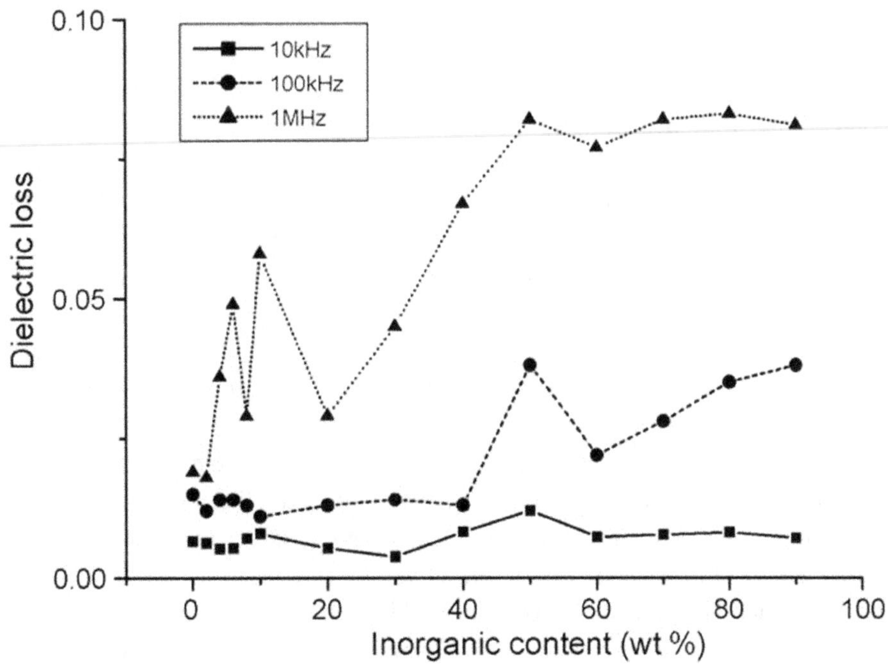

Figure 6. TEM micrographs of nano-sized Al_2O_3 dispersed in DMAc by mechanical stirring (a) and ultrasonic treatment (b).
Source: Reprinted from Elsevier. Wu, Yang, &Gao, 2005

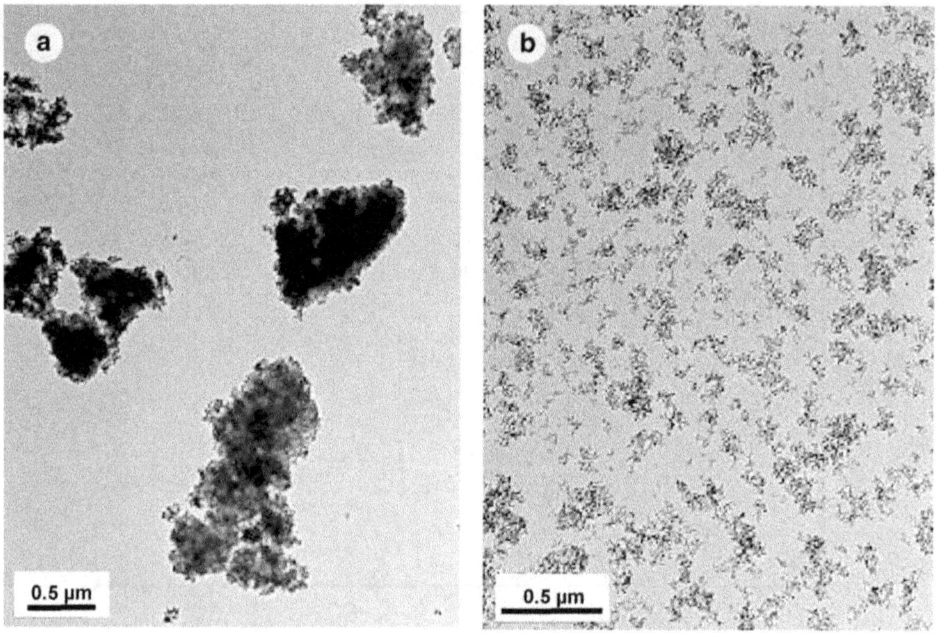

into the polymer matrix. New polymer/TiO_2UV-absorber nanocomposites have been successfully elaborated by the direct inclusion of nano-fillers into the studied polymeric matrices (Yuwono et al., 2004; Khrenov et al., 2005). In addition, the PMMA/TiO_2 nanocomposites show better UV-shielding efficiency than commercially available UV-blocking contact lenses (Li et al., 2010). Ritzhaupt-Kleissl et al. studied the effect of surface-treatment of alumina nanoparticles on the PMMA/Al_2O_3 nanocomposites optical properties. Compared to the untreated Al_2O_3 nanoparticles, silane-treated Al_2O_3 nanoparticles shows good dispersion homogeneity in the PMMA matrix that reduces the nanocomposites refractive index and increased the optical transmittance values in both the visible and infrared domains (Ritzhaupt-Kleissl et al., 2006). The Al_2O_3 and TiO_2 nanoparticles were functionalized using GPTMS and employed to reinforce the polybenzoxazine matrix as reported by Lakshmikandhan et al. UV shielding studies showed that 7 wt.% GPTMS-TiO_2/polybenzoxazine nanocomposite show higher UV shielding value (93%) than those of neat matrix and the other reinforced nanocomposites (Lakshmikandhan et al., 2016). Similar mechanism has also been reported for luminescent oxide/PMMA nanocomposites in other studies (Vollath & Szabo, 2004; Wang, Zu, & Xiang, 2006). The core/shell nanoparticles made of HfO_2, ZrO_2, or Al_2O_3 cores and a PMMA shell also exhibits strong luminescence (Vollath et al., 2004).

It is known that by adding ceramic nanoparticles into the polymer matrix, the refractive index (RI) of their resulting nanocomposites can be tailored. Nanomaterials having high RI can be used in many potential applications like lenses, optical filters, reflectors, optical adhesives, and solar cells (Althues et al., 2007). It was reported that the inclusion of 30 wt.% TiO_2 nanoparticles into the epoxy thermoset formed a nanocomposite coating that exhibits a refractive index of 1.668 (Chau et al., 2008). An optical transparency that exceeded 90% was observed in other kinds of epoxy nanocomposites filled with different amounts of TiO_2.Wang et al. prepared two PMMA nanocomposites reinforced with SiO_2or ZrO_2

nanoparticles via a non-hydrolytic sol–gel technique and their optical properties were investigated in details. The obtained results reveal that the transmittance of the synthesized nanocomposite films in the visible-light region still higher than 95% even at a 20 wt.% nano-ceramics content, and this value was proportionally enhanced as the nanofillers concentration increased (Wang et al., 2005). It was also reported that both the polyamide and its nanocomposite films reinforced with various fractions of SiO_2 nanoparticles were transparent, but the highest transmittance was recorded in a nanocomposite loaded with 5 wt.% nano-SiO_2, which was the optimum fraction for these hybrids (Sarwar et al., 2008). It was found that the addition of nano-Al_2O_3 particles into polycarbonate (PC) resin decrease the overall light transmittance of their nanocomposites (Chandra et al., 2008). The authors also discovered that the poly (styrene-maleic anhydride) copolymer (SMA) modified (PC)/Al_2O_3nanocomposite showed a higher light transmittance than that of the unmodified(PC)/Al_2O_3nanocomposite. Guo et al. also reported the decrease in the light transmittance of the polyvinyl alcohol (PVA) nanocomposites in the presence of TiO_2 nanoparticles (Guo et al., 2006). As thenano-TiO_2 particle loading is increased, a further decreased in the light transmittance was recorded. However, these nanocomposite films exhibited an absorption in the UV region, which make them suitable for a UV-shielding coating. It was also reported that the nanocomposite containing 5 wt.% content of TiO_2 nano-filer was the most transparent even some haziness can still be detected and this transparency was still relatively observed till 10 wt.% of TiO_2 concentration. Al-Bayer et al. stated that both the optical energy gap (E_{opt}) and tail widths (D_E) vary with the dispersion of Al_2O_3 content in the epoxy matrix (Al-Bayer et al., 2013).

Barrier Properties

Generally, barrier properties of polymer/ceramic nanocomposites are very strong due to the high surface area of nano-ceramic that isolates the polymeric matrix and acts as supplementary resistance layers. Polymers ranging from epoxies and good sealants (like siloxanes) to semi-permeable (e.g. polyureas) and highly hydrophilic (e.g. PVA) all are improved up to an order of magnitude by adding low amounts of ceramics nanoparticles (Liang et al., 2008). The dramatic improvement of barrier properties is explained by the concept of tortuous paths. That is, when impermeable nanoparticles are incorporated into a polymer, the permeating molecules are forced to wiggle around them in a random walk, and hence diffuse by a tortuous pathway

Jun et al. developed new the B_4C/epoxy composites having a uniform dispersion of nano-fillers to use them as a neutron shield for spent nuclear fuel cask, by a direct ultrasonic dispersion of the B_4C particles in the hardener (Jun et al., 2011). The strong adhesion between B_4C and epoxy matrix enhances the material properties without any chemical-treatment, avoiding undesired impurities. (Tsekmes et al., 2015) proposed that the dielectric response of the epoxy/BN nanocomposites produced by three different techniques was mainly affected by two possible mechanisms, which were polymer re-organization and water-uptake, and high levels of reproducibility could be achieved for well-dispersed nanoparticles in the polymeric matrix. Ahmad et al., investigated the moisture absorption characteristics of epoxy/ceramic nanocomposites at various temperature and humidity conditions and reported that such nanocomposites exhibited a Fickian-like diffusion only at moderated conditions. Swelling kinetics of the epoxy/SiO_2 nanocomposites showed that for most of the solvents the type of diffusion is non-Fickian and as the SiO_2 content increased deviation from Fickian behavior increases (Ahmad et al., 2011). In addition, diffusivity reduces with the increase in SiO_2 loading but enhances with temperature. Sorption values increases with SiO_2 content and is also with temperature (Kaushik & Kaushik, 2010). The diffusion behaviour for the newly developed polybenzoxazine/B_4C nanocomposites is a Fickian-like behaviour and the adopted mechanism for water-sorption in the tested polymeric nanocomposites is an activated sorption diffusion process started first by dissolving into their surface and then diffusing through the bulk of the material via a series of activated steps depends on temperature, the material structure and its morphology (Ramdani et al., 2016). The addition of impermeable crystalline nano-B_4C can protect the surface of resin and thus inhibiting the diffusion of water.

APPLICATIONS OF CERAMIC/POLYMER NANOCOMPOSTES

The reinforcing with ceramic nanofillers into polymeric matrices remarkably improves the mechanical, optical, electronic and thermal properties for their resulting nanocomposite materials. The enhancement in these material properties increases the opportunity to use these nanocomposites in different industrial fields, including optics, micro-electronics, mechanics, energy storage, environment, biology and medicine. Polymer/ceramic nanomaterials are used in many aspects including membranes devices, smart coatings, solar cells, catalysts, sensors, military articles, automobiles, aerospace, and drug delivery. The improved mechanical properties of polymer/ceramic nanocomposites promote their uses in wide industrial areas that include blades for vacuum cleaners, power tool housings, mower hoods, electronic equipment, such as mobile phones and pagers and glues for the manufacturing of pressure moulds in the ceramic industry. The SiO_2, TiO_2, SiC and BN based polymer nanocomposites are the most investigated materials due to their potential application in multidisciplinary fields as revealed in Table 2.

Table 2. Application of ceramic/polymer nanocomposites

Nanocomposite Type	Applications	References
PMMA/SiO_2	Bioactive bone substitute, Dental application and optical devices.	(Rhee & Choi,2002)
Polyimide/SiO_2	Microelectronics.	(Camargo et al., 2009)
Polycarbonate/SiO_2	Abrasion resistant coating.	(Sanchez et al., 2005)
High-density polyethylene/TiO_2	Bone repair.	(Hashimoto et al., 2006)
Poly(amide-imide)/TiO_2	Composite membranes for gas separation.	(Camargo et al., 2009)
Polystyrene/AlN	Microelectronics packaging.	(Yu et al., 2002)
Polybenzoxazine/AlN	Microelectronics packaging.	(Ramdani et al., 2016)
Polymer/ZrO_2	Wear and erosion	(Liang et al., 2008)
Polyvinyl alcohol–polyethylene glycol–polyvinyl pyrrolidinone/TiO_2	Energy storage	(Agool et al., 2016)
Polymer/Al_2O_3	Gas membranes	(Yu et al., 2002)
Polymer/$BaTiO_3$	High energy capacitors	(Bai et al., 2000)

FUTURE TRENDS

Researches on polymer/ceramic nanocomposites will further expand in the near future due to the great demand in nanotechnologies. From the environmental issues and because of the higher cost of some synthetic ceramic nanofillers, the combination of natural ceramic like layered silica, $CaCO_3$, and clay nanofillers with the various polymeric matrices can result in high-performance nanocomposite materials at a lower cost. In addition, the use of binary or ternary ceramic nanoparticle systems as reinforcing agent for some single polymeric matrices or for a polymer blends would be another direction of research in this filed where the final properties of the nanocomposites can be tailored. For example the addition of SiO_2/BN system into polymer can be resulted in microelectronic packaging nanomaterials which exhibit high thermal conductivity due to the presence of BN and low dielectric constant because of SiO_2. On the other hand, the researches on the surface-treatment of ceramic nanofillers will continuously conducted due the extra improvements that can be obtained.

CONCLUSION

Polymer/ceramic nanocomposite materials exhibit several unique performances which make them suitable for academic as a subject of research and for various industrial applications. These nanomaterials combine the advantageous properties of both polymers and ceramic nano-fillers in one system. These nanocomposites have multifunctional characteristics and they can be potentially used in high-tech applications, including coatings for corrosion-resistance and abrasive-resistance, industrial membranes, gas separation, catalysts, adsorbents of heavy metal ions and devices with optoelectronic and magnetic properties for media purpose. Several methods have been achieved to prepare these nanomaterials such as sol-gel method, solution blending technique, and direct melt processing. The type of ceramic nano-filler and its ratio can determine the final performance of these nanocomposite hybrids. In addition, during the preparation of such nanocomposites, the ceramic nanoparticles tend to construct aggregates; thus, to obtain uniform dispersion and good compatibility of these nanofillers with polymeric matrices, their surfaces generally were treated by grafting polymers or by absorption of silane coupling agents. The process of surface-treatment can enhance the interfacial interactions between the ceramic nanoparticles and organic polymers, which provide some striking properties, such as improved mechanical toughness and good optical properties, better electrical characteristics, efficient gas-barrier and high thermal stability.

ACKNOWLEDGEMENT

The authors are very grateful to many publishers, and especially to InTech – Sage Publications, Elsevier, Springer and Wiley–for permission to reuse in our review fragments of text, figures and schemes taken from their publications.

REFERENCES

Agag, T., & Takeichi, T. (2011). Synthesis and characterization of benzoxazine resin-SiO_2 hybrids by sol-gel process: The role of benzoxazine-functional silane coupling agent. *Polymer*, *52*(13), 2757–2763. doi:10.1016/j.polymer.2011.04.044

Agool, I. R., Kadhim, K. J., & Hashim, A. (2016). Preparation of (polyvinyl alcohol–polyethylene glycol–polyvinyl pyrrolidinone–titanium oxide nanoparticles) nanocomposites, electrical properties for energy storage and release. *International Journal of Plastics Technology*, *20*(1), 121–127. doi:10.1007/s12588-016-9144-5

Ahmad, Z., Ansell, M. P., & Smedley, D. (2011). Moisture Absorption Characteristics of Epoxy Based Adhesive Reinforced with CTBN and Ceramic Particles for Bonded-in Timber Connection, Fickian or Non-Fickian Behaviour. *Materials Science and Engineering*, *17*(1), 012011.

Al-Bayer, R., Zihlif, A., Lahlouh, B., Elimat, Z., & Ragosta, G. (2013). AC electrical and optical characterization of epoxy–Al_2O_3 composites. *Journal of Materials Science Materials in Electronics*, *24*(8), 2866–2872. doi:10.1007/s10854-013-1184-4

Althues, H., Henle, J., & Kaskel, S. (2007). Functional inorganic nanofillers for transparent polymers. *Chemical Society Reviews*, *36*(9), 1454–1465. doi:10.1039/b608177k PMID:17660878

Anatolii, D. P. (2000). Hybrid polymer-inorganic nanocomposites. *Russian Chemical Reviews*, *69*(0), 53–80.

Ariraman, M., Kumar, R. S., & Alagar, M. (2014). Studies on FMCM-41 reinforced cyanate ester nanocomposites for low k applications. *RSC Advances*, *4*(101), 57759–57767. doi:10.1039/C4RA09399B

Ariraman, M., Kumar, R. S., & Alagar, M. (2015). Design of cyanate ester/azomethine/ZrO_2 nanocomposites high-k dielectric materials by single step sol-gel approach. *Journal of Applied Polymer Science*, *131*(24), 41097–41102.

Bai, Y., Cheng, Z. Y., Bharti, V., Xu, H. S., & Zhang, Q. M. (2000). High-dielectric-constant ceramic-powder polymer composites. *Applied Physics Letters*, *76*(25), 3804–3806. doi:10.1063/1.126787

Behniafar, H., & Amirkhalili, S. K. (2014). Poly(4,4`-oxydiphenylene-pyromellitimide)/TiO_2 nanocomposites with surface modified titanium dioxide using methylene diphenyl diisocyante. *Polymer Bulletin*, *71*(4), 775–785. doi:10.1007/s00289-013-1091-0

Camargo, P., Satyanarayana, K., & Wypych, F. (2009). Nanocomposites, synthesis, structure, properties and new application opportunities. *Materials Research*, *12*(1), 1–39. doi:10.1590/S1516-14392009000100002

Chandra, A., Turng, L. S., Gopalan, P., Rowell, R. M., & Gong, S. (2008). Study of utilizing thin polymer surface coating on the nanoparticles for melt compounding of polycarbonate/alumina nanocomposites and their optical properties. *Composites Science and Technology*, *68*(3-4), 768–776. doi:10.1016/j.compscitech.2007.08.027

Chau, J. L. H., Tung, C. T., Lin, Y. M., & Li, A. K. (2008). Preparation and optical properties of titania/epoxy nanocomposite coatings. *Materials Letters*, *62*(19), 3416–3418. doi:10.1016/j.matlet.2008.02.058

Chen, D., Qiao, X., Qiu, X., & Chen, J. (2009). Synthesis and electrical properties of uniform silver nanoparticles for electronic applications. *Journal of Materials Science*, *44*(4), 1076–1081. doi:10.1007/s10853-008-3204-y

Chen, H. M., & Liu, R. S. (2011). Architecture of metallic nanostructures: Synthesis strategy and specific applications. *The Journal of Physical Chemistry C*, *115*(9), 3513–3527. doi:10.1021/jp108403r

Chisholm, N., Mahfuz, H., Rangari, V. K., & Jeelani, S. (2005). Fabrication and mechanical characterization of carbon/SiC-epoxy nanocomposites. *Composite Structures*, *67*(1), 115–124. doi:10.1016/j.compstruct.2004.01.010

Choi, S., & Kim, J. (2013). Thermal conductivity of epoxy composites with binary-particle system aluminium oxide and aluminium nitride filers. *Composites. Part B, Engineering*, *51*, 140–147. doi:10.1016/j.compositesb.2013.03.002

Derradji, M., Ramdani, N., Zhang, T., Wang, J., Gong, L. D., Xu, X. D., & Liu, W. B. (2016). Effect of silane surface modified titania nanoparticles on the thermal, mechanical, and corrosion protective properties of a bisphenol-A based phthalonitrile resin. *Progress in Organic Coatings*, *90*, 34–43. doi:10.1016/j.porgcoat.2015.09.021

Derradji, M., Ramdani, N., Zhang, T., Wang, J., Lin, Z. W., Yang, M., & Liu, W. (2015). High thermal and thermomechanical properties obtained by reinforcing a bisphenol-A based phthalonitrile resin with silicon nitride nanoparticles. *Materials Letters*, *149*, 81–84. doi:10.1016/j.matlet.2015.02.122

Dzunuzovic, E., Jeremic, K., & Nedeljkovic, J. M. (2007). In situ radical polymerization of methyl methacrylate in a solution of surface modified TiO_2 and nanoparticles. *European Polymer Journal*, *43*(9), 3719–3726. doi:10.1016/j.eurpolymj.2007.06.026

Evora, V. M. F., & Shukla, A. (2003). Fabrication, characterization, and dynamic behavior of polyester/TiO_2 nanocomposites. *Materials Science and Engineering A*, *361*(1-2), 358–366. doi:10.1016/S0921-5093(03)00536-7

Garcia, M., van Vliet, G., Cateten, M. G. J., Chavez, F., Norder, B., & Kooi, B. et al.. (2004). Large-scale extrusion processing and characterization of hybrid nylon-6/SiO_2 nanocomposites. *Polymers for Advanced Technologies*, *15*(4), 164–172. doi:10.1002/pat.458

Ghezelbash, Z., Ashouri, D., Mousavian, S., Ghandi, A. H., & Rahnama, Y. (2012). Surface modified Al_2O_3 in fluorinated polyimide/Al_2O_3 nanocomposites: Synthesis and characterization. *Bulletin of Materials Science*, *35*(6), 925–931. doi:10.1007/s12034-012-0385-4

Ginés, R., Libanori, R., Studart, A. R., Bergamini, A., Motavalli, M., & Ermanni, P. (2015). Ceramic–polymer composites with improved dielectric and tribological properties for semi-active damping. *Composites. Part B, Engineering*, *72*, 80–86. doi:10.1016/j.compositesb.2014.11.035

Guan, C., Lu, C. L., Cheng, Y. R., Song, S. Y., & Yang, B. A. (2009). Facile one-pot route to transparent polymer nanocomposites with high ZnS nanophase contents via in situ bulk polymerization. *Journal of Materials Chemistry*, *19*(5), 617–621. doi:10.1039/B818876A

Guo, Z., Pereira, T., Choi, O., Wang, Y., & Hahn, H. T. (2006). Surface functionalized alumina nanoparticle filled polymeric nanocomposites with enhanced mechanical properties. *Journal of Materials Chemistry*, *16*(27), 2800–2808. doi:10.1039/b603020c

Hashimoto, M., Takadama, H., Mizuno, M., & Kokubo, T. (2006). Enhancement of mechanical strength of TiO_2/high-density polyethylene composites for bone repair with silane-coupling treatment. *Materials Research Bulletin*, *41*(3), 515–524. doi:10.1016/j.materresbull.2005.09.014

Hsiue, G. H., Kuo, W. J., & Jeng, R. J. (2000). Microstructural and morphological characteristics of PS-SiO_2 nanocomposites. *Polymer*, *41*(8), 2813–2825. doi:10.1016/S0032-3861(99)00478-4

Jiang, L., Lam, Y. C., Tam, K. C., Chua, T. H., Sim, G. W., & Ang, L. S. (2005). Strengthening acrylonitrile-butadiene-styrene (ABS) with nano-sized and micron-sized calcium carbonate. *Polymer*, *46*(1), 243–252. doi:10.1016/j.polymer.2004.11.001

Jun, J., Kim, J., Bae, Y., & Seo, Y. C. (2011). Enhancement of dispersion and adhesion of B_4C particles in epoxy resin using direct ultrasonic excitation. *Journal of Nuclear Materials*, *416*(3), 293–297. doi:10.1016/j.jnucmat.2011.06.014

Kang, S., Hong, S. I., Choe, C. R., Park, M., Rim, S., & Kim, J. (2001). Preparation and characterization of epoxy composites filled with functionalized nanosilica particles obtained via sol–gel process. *Polymer*, *42*(3), 879–887. doi:10.1016/S0032-3861(00)00392-X

Kango, S., Kalia, S., Celli, A., Njuguna, J., Habibi, Y., & Kumar, R. (2013). Surface modification of inorganic nanoparticles for development of organic–inorganic nanocomposites—A review. *Progress in Polymer Science*, *38*(8), 1232–1261. doi:10.1016/j.progpolymsci.2013.02.003

Kaushik, A., & Kaushik, J. (2010). Solvent absorption characteristics of epoxy-colloidal silica nanocomposites. *Journal of Reinforced Plastics and Composites*, *29*(18), 2821–2833. doi:10.1177/0731684409360995

Khrenov, V., Klapper, M., Koch, M., & Mullen, K. (2005). Surface functionalized ZnO particles designed for the use in transparent nanocomposites. *Macromolecular Chemistry and Physics*, *206*(1), 95–101. doi:10.1002/macp.200400213

Kim, K., & Kim, J. (2014). Fabrication of thermally conductive composite with surface modified boron nitride by epoxy wetting method. *Ceramics International*, *40*(4), 5181–5189. doi:10.1016/j.ceramint.2013.10.076

Kumar, B. (2003). Polymer–ceramic composite protonic conductors. *Journal of Power Sources*, *123*(2), 132–136. doi:10.1016/S0378-7753(03)00530-5

Lakshmikandhan, T., Chandramohan, A., Sethuraman, K., & Alagar, M. (2016). Development and Characterization of Functionalized Al_2O_3 and TiO_2-reinforced polybenzoxazine nanocomposites. *Design of Monomers and Polymers*, *19*(1), 67–76. doi:10.1080/15685551.2015.1092014

Li, H., Liu, G., Liu, B., Chen, W., & Chen, S. (2007). Properties of polyimide/Al_2O_3 hybrids synthesized by in-situ polymerization. *Materials Letters*, *61*(7), 1507–1511. doi:10.1016/j.matlet.2006.07.063

Li, J. P., He, Q. H., Xu, R. F., & Hu, B. X. (2015). Cyanate ester/functionalized silica nanocomposite, synthesis, characterization and properties. *Chinese Journal of Chemistry*, *33*(11), 1259–1268. doi:10.1002/cjoc.201500498

Li, S., Lin, M. M., Toprak, M. S., Kim, D. K., & Muhammed, M. (2010). Nanocomposites of polymer and inorganic nanoparticles for optical and magnetic applications. *Nano Review*, *1*(1), 5214–5219. doi:10.3402/nano.v1i0.5214 PMID:22110855

Liang, X. H., King, D. M., Groner, M. D., Blackson, J. H., Harris, J. D., George, S. M., & Weimer, A. W. (2008). Barrier properties of polymer/alumina nanocomposites membrane fabricated by atomic layer deposition. *Journal of Membrane Science*, *322*(1), 105–112. doi:10.1016/j.memsci.2008.05.037

Lim, S. H., Zeng, K. Y., & He, C. B. (2010). Morphology, tensile and fracture characteristics of epoxy-alumina nanocomposites. *Materials Science and Engineering A*, *527*(21-22), 5670–5676. doi:10.1016/j.msea.2010.05.038

Liu, Y. L., Hsu, C. Y., Wei, W. L., & Jeong, R. J. (2003). Preparation and thermal properties of epoxy-silica nanocomposites from nanoscale colloidal silica. *Polymer*, *44*(18), 5159–5167. doi:10.1016/S0032-3861(03)00519-6

Ma, D., Hugener, T. A., Siegel, R. W., Christerson, A., Martensson, E., Onneby, C., & Schadler, L. S. (2005). Influence of nanoparticle surface modification on the electrical behavior of polyethylene nanocomposites. *Nanotechnology*, *16*(6), 724–731. doi:10.1088/0957-4484/16/6/016

Mi, Y. N., Liang, G. Z., Gu, A. J., Zhao, F. P., & Yuan, L. (2013). Thermally conductive aluminum nitride-multiwalled carbon nanotube/cyanate ester composites with high flame retardancy and low dielectric loss. *Industrial & Engineering Chemistry Research*, *52*(9), 3342–3353. doi:10.1021/ie3029569

Mo, T. C., Wang, H. W., Chen, S. Y., & Yeh, Y. C. (2008). Synthesis and dielectric properties of polyaniline/titanium dioxide nanocomposites. *Ceramics International*, *34*(7), 1767–1771. doi:10.1016/j.ceramint.2007.06.002

Ng, C. B., Schadler, L. S., & Siegel, R. W. (1999). Synthesis and mechanical properties of TiO_2-epoxy nanocomposites. *Nanostructured Materials*, *12*(1-4), 507–510. doi:10.1016/S0965-9773(99)00170-1

Nhuapeng, W., Thamjaree, W., Kumfu, S., Singjai, P., & Tunkasiri, T. (2008). Fabrication and mechanical properties of silicon carbidenanowires/epoxy resin composites. *Current Applied Physics*, *8*(3-4), 295–299. doi:10.1016/j.cap.2007.10.074

Nunes, S. P., Peinemann, K. V., Ohlrogge, K., Alpers, A., Keller, M., & Pires, A. T. N. (1999). Membranes of poly(ether imide) and nanodispersed silica. *Journal of Membrane Science*, *157*(2), 219–226. doi:10.1016/S0376-7388(98)00379-2

Omrani, A., Simon, L. C., & Rostami, A. A. (2009). The effects of alumina nanoparticle on the properties of an epoxy resin system. *Materials Chemistry and Physics*, *114*(1), 145–150. doi:10.1016/j.matchemphys.2008.08.090

Ou, Y., Yang, F., & Chen, J. (1997). Interfacial interaction and mechanical properties of nylon 6 potassium titanate composites prepared by in-situ polymerization. *Journal of Applied Polymer Science*, *64*(12), 2317–2322. doi:10.1002/(SICI)1097-4628(19970620)64:12<2317::AID-APP7>3.0.CO;2-O

Ou, Y., Yang, F., & Yu, Z. (1998). A new conception on the toughness of nylon 6/silica nanocomposite prepared via in situ polymerization. *Journal of Polymer Science. Part B, Polymer Physics*, *36*(5), 789–795. doi:10.1002/(SICI)1099-0488(19980415)36:5<789::AID-POLB6>3.0.CO;2-G

Paul, A. I., Justine, I. N., Alabi, O. O., & Adamu, W. U. (2012). Recent Trends and Developments in Surface Hardening Technology, A Focus on New Ceramics for Surface Modification in Industry. *International Journal of Science & Engineering Research*, *3*(11), 1–22.

Periadurai, T., Vijayakumar, C. T., & Balasubramanian, M. (2010). Thermal decomposition and flame retardant behaviour of SiO_2-phenolic nanocomposite. *Journal of Analytical and Applied Pyrolysis*, *89*(2), 244–249. doi:10.1016/j.jaap.2010.08.010

Qin, D. K., Liang, G. Z., Gu, A. J., & Yuan, L. (2014). Fabrication of SiO_2 covered ZrO_2 electrospun fibres with controllable structure and their novel high-performance composites with outstanding thermal, mechanical and dielectric properties. *Journal of Composites Materials*, *48*(26), 3201–3214. doi:10.1177/0021998313507619

Rahman, I. A., & Padavettan, V. (2012). Synthesis of Silica Nanoparticles by Sol-Gel, Size-Dependent Properties, Surface Modification, and Applications in Silica-Polymer Nanocomposites—A Review. *Journal of Nanomaterials*, *2012*, 1–15. doi:10.1155/2012/132424

Ramdani, N., Derradji, M., Wang, J., Mokhnache, E. O., & Liu, W. (2016, September). Improvements of Thermal, Mechanical, and Water-Resistance Properties of Polybenzoxazine/Boron Carbide Nanocomposites. *Journal of the Minerals Metals & Materials Society*, *68*(9), 2533–2542. doi:10.1007/s11837-016-2040-9

Ramdani, N., Derradji, M., Wang, J., Mokhnache, E.-O., Liu, W., Liu, Y., & Dong, W. (2016). Experimental and modeling of thermal and dielectric properties of aluminum nitride-reinforced polybenzoxazine hybrids. *Journal of Thermal Analysis and Calorimetry*, *126*(2), 561–570. doi:10.1007/s10973-016-5492-9

Ramdani, N., Liu, W., Wang, J., & Derradji, M. (2017). Ceramic Based Polybenzoxazine Micro- and Nanocomposites. In Advanced and Emerging Polybenzoxazine Science and Technology. Amsterdam: Elsevier.

Ramdani, N., Wang, J., & Liu, W. (2014). Preparation and thermal properties of polybenzoxazine/TiC hybrids. *Advanced Materials Research*, *887*(1), 49–52. doi:10.4028/www.scientific.net/AMR.887-888.49

Ramdani, N., Wang, J., Wang, H., Feng, T., Derradji, M., & Liu, W. (2014). Mechanical and thermal properties of silicon nitride reinforced polybenzoxazine nanocomposites. *Composites Science and Technology*, *105*, 73–79. doi:10.1016/j.compscitech.2014.10.006

Ray, S. S., & Okamoto, M. (2003). Polymer/layered silicate nanocomposite, a review from preparation to processing. *Progress in Polymer Science*, *28*(11), 1539–1541. doi:10.1016/j.progpolymsci.2003.08.002

Rhee, S. H., & Choi, J. Y. (2002). Preparation of a bioactive poly(methyl methacrylate)/silica nanocomposite. *Journal of the American Ceramic Society*, *85*(5), 1318–1320. doi:10.1111/j.1151-2916.2002.tb00269.x

Ritzhaupt-Kleissl, E., Bohm, J., Haubelt, J., & Hanemann, T. (2006). Thermoplastic polymer nanocomposites for applications in optical devices. *Materials Science and Engineering C*, *26*(5-7), 1067–1071. doi:10.1016/j.msec.2006.02.004

Rong, M. Z., Zhang, M. Q., Liu, H., Zeng, H. M., Wetzel, B., & Friedrich, K. (2001). Microstructure and tribological behavior of polymeric nanocomposites. *Industrial Lubrication and Tribology*, *53*(2), 72–77. doi:10.1108/00368790110383993

Sanchez, C., Julian, B., Belleville, P., & Popall, M. (2005). Applications of hybrid organic–inorganic nanocomposites. *Journal of Materials Chemistry*, *15*(35-36), 3559–3592. doi:10.1039/b509097k

Sarwar, M. I., Zulfiqar, S., & Ahmad, Z. (2008). Polyamide-silica nanocomposites, mechanical, morphological and thermomechanical investigations. *Polymer International*, *57*(2), 292–296. doi:10.1002/pi.2343

Schwartz, C. J., & Bahadur, S. (2000). Studies on the tribological behavior and transfer film-counterface bond strength for polyphenylene sulphide filled with nanoscale alumina particles. *Wear*, *237*(2), 261–273. doi:10.1016/S0043-1648(99)00345-2

Sengupta, R., Bandyopadhyay, A., Sabharwal, S., Chaki, T. K., & Bhowmick, A. K. (2005). Polyamide 6,6/in situ silica hybrid nanocomposites by sol–gel technique, synthesis, characterization and properties. *Polymer*, *46*(10), 3343–3354. doi:10.1016/j.polymer.2005.02.104

Siegel, R. W., Chang, S. K., Ash, B. J., Stone, J. A. P. M., Doremus, R. W., & Schadler, L. S. (2001). Mechanical behavior of polymer and ceramic matrix nanocomposites. *Scripta Materialia*, *44*(8-9), 2061–2064. doi:10.1016/S1359-6462(01)00892-2

Singha, S., & Thomas, M. J. (2008). Dielectric properties of epoxy nanocomposites. *IEEE Transactions on Dielectrics and Electrical Insulation*, *15*(1), 12–23. doi:10.1109/T-DEI.2008.4446732

Su, S. J., & Kuramoto, N. (2000). Processable polyaniline-titanium dioxide nanocomposites, effect of titanium dioxide on the conductivity. *Synthetic Metals*, *114*(2), 147–153. doi:10.1016/S0379-6779(00)00238-1

Suzuki, F., Onozato, K., & Kurokawa, Y. (1990). A Formation of compatible poly (vinyl alcohol)/alumina gel composite and its properties. *Journal of Applied Polymer Science*, *39*(2), 371–381. doi:10.1002/app.1990.070390214

Tsekmes, I. A., Kochetov, R., Morshuis, P. H. F., & Smit, J. J. (2015). The role of particle distribution in the dielectric response of epoxy–boron nitride nanocomposites. *Journal of Materials Science*, *50*(3), 1175–1186. doi:10.1007/s10853-014-8674-5

Ulus, H., Üstün, T., Eskizeybek, V., & Sinan, Ö. (2014). Boron nitride-MWCNT/epoxy hybrid nanocomposites, Preparation and mechanical properties. *Applied Surface Science*, *318*, 37–42. doi:10.1016/j.apsusc.2013.12.070

Vollath, D., & Szabo, D. V. (2004). Synthesis and properties of nanocomposites. *Advanced Engineering Materials*, *6*(3), 117–127. doi:10.1002/adem.200300568

Vollath, D., Szabo, D. V., & Schlabach, S. (2004). Oxide/polymer nanocomposites as new luminescent materials. *Journal of Nanoparticle Research*, *6*(2), 181–191. doi:10.1023/B:NANO.0000034624.56366.ac

Wang, H., Xu, P., Zhong, W., Shen, L., & Du, Q. (2005). Transparent poly(methyl methacrylate)/silica/zirconia nanocomposites with excellent thermal stabilities. *Polymer Degradation & Stability*, *87*(2), 319–327. doi:10.1016/j.polymdegradstab.2004.08.015

Wang, K., Chen, L., Wu, J., Toh, M. L., He, C., & Yee, A. F. (2005). Epoxy nanocomposites with highly exfoliated clay, mechanical properties and fracture mechanisms. *Macromolecules*, *38*(3), 788–800. doi:10.1021/ma048465n

Wang, S. F., Wang, Y. R., Cheng, K. C., & Hsaio, Y. P. (2009). Characteristics of polyimide/barium titanate composite film. *Ceramics International*, *35*(1), 265–268. doi:10.1016/j.ceramint.2007.10.004

Wang, Y., Lim, S., Luo, J. L., & Xu, Z. H. (2006). Tribological and corrosion behaviors of Al_2O_3/polymer nanocomposite coatings. *Wear*, *260*(9-10), 976–983. doi:10.1016/j.wear.2005.06.013

Wang, Z. G., Zu, X. T., Xiang, X., & Yu, H. J. (2006). Photoluminescence from TiO_2/PMMA nanocomposite prepared by gamma radiation. *Journal of Nanoparticle Research*, *8*(1), 137–139. doi:10.1007/s11051-005-7527-1

Wu, C. S. (2004). *In*-situ polymerization of titanium isopropoxideinpoly caprolactone, properties and characterization of the hybrid nanocomposites. *Journal of Applied Polymer Science*, *92*(3), 1749–1757. doi:10.1002/app.20135

Wu, H. C., & Kessler, M. R. (2015). Multifunctional Cyanate Ester Nanocomposites Reinforced by Hexagonal Boron Nitride after Noncovalent Biomimetic Functionalization. *ACS Applied Materials & Interfaces*, *7*(10), 5915–5926. doi:10.1021/acsami.5b00147 PMID:25726956

Wu, J., Yang, S., Gao, S., Hu, A., Liu, J., & Fan, L. (2005). Preparation, morphology and properties of nano-sized Al_2O_3/polyimide hybrid films. *European Polymer Journal*, *41*(1), 73–81. doi:10.1016/j.eurpolymj.2004.08.014

Wu, T., & Ke, Y. (2007). Melting crystallization and optical behaviors of poly (ethylene terephthalate)-silica/polystyrene nanocomposite films. *Thin Solid Films*, *515*(13), 5220–5226. doi:10.1016/j.tsf.2006.12.029

Yan, H. X., Zhang, M. M., Feng, S. Y., Wang, Q. Q., & Zhang, J. P. (2014). Effects of macromolecular silane coupling agent on the mechanical properties of nano-SiO_2/cyanate ester nanocomposites. *Journal of Reinforced Plastics and Composites*, *33*(2), 127–136. doi:10.1177/0731684413509701

Yan, H. X., Zhang, M. M., Liu, C., & Zhang, J. P. (2013). Nano-Si_3N_4/epoxidized silane/cyanate ester composites for electronic packaging. *Polymer Bulletin*, *70*(11), 2923–2933. doi:10.1007/s00289-013-0997-x

Yan, X. H., Ning, R. C., Liang, G. Z., Huang, Y., & Lu, T. L. (2007). The effect of silane coupling agent on the sliding wear behavior of nanometer ZrO_2/bismaleimide composites. *Journal of Materials Science*, *42*(3), 958–965. doi:10.1007/s10853-006-0008-9

Yu, S., Hing, P., & Hu, X. (2002). Thermal conductivity of polystyrene-Aluminium nitride composites. *Composites. Part A, Applied Science and Manufacturing*, *33*(2), 289–292. doi:10.1016/S1359-835X(01)00107-5

Yuwono, A. H., Liu, B., Xue, J., Wang, J., Elim, H. I., Ji, W., & White, T. J. et al. (2004). Controlling the crystallinity and nonlinear optical properties of transparent TiO_2-PMMA nanohybrids. *Journal of Materials Chemistry*, *14*(20), 2978–2987. doi:10.1039/B403530E

Zhang, M., & Singh, R. (2004). Mechanical reinforcement of unsaturated polyester by Al_2O_3 nanoparticles. *Materials Letters*, *58*(3-4), 408–412. doi:10.1016/S0167-577X(03)00512-3

Zhang, M. C., Gu, A. J., Liang, G. Z., & Yuan, L. (2012). Preparation of high thermal conductive aluminum nitride/cyanate ester nanocomposite using a new macromolecular coupling agent. *Polymers for Advanced Technologies*, *23*(11), 1503–1510. doi:10.1002/pat.2073

Zhang, W. G., & Zhang, X. Y. (2013). Study on the wear resistance of nm-SiC modified cyanate ester resins. *Thermosetting Resin*, *28*, 46–48.

Zhang, X., & Simon, L. C. (2005). In situ polymerization of hybrid polyethylene alumina nanocomposites. *Macromolecular Materials and Engineering*, *290*(6), 573–583. doi:10.1002/mame.200500075

Zhang, Z., Yang, J. L., & Friedrich, K. (2004). Creep resistant polymeric nanocomposites. *Polymer*, *45*(10), 3481–3485. doi:10.1016/j.polymer.2004.03.004

Zhao, C. B., Su, L., Xu, S. C., & Yang, X. J. (2014). Study on KH550 modified boron ntride/cyanate ester resin thermal conductive composites. *New Chemical Materials*, *42*, 119–121.

Zhu, B. L., & Wang, J. L. (2015). Grafting modification onto anatase TiO_2 by multistep and its effect on disoersion properties of CE composite materials the properties test of static mechanical, dynamic mechanical and heat resistance. *Applied Chemical Industry*, *44*, 1661–1667.

Chapter 5
Structure-Property Relations in Polymers and Polymer Blends Using X-Ray Diffraction Technique

Somashekarappa H
University of Mysore, India

Manjunath Alabur
Government Science College Chitradurga, India

ABSTRACT

This chapter discusses about the synthesis and characterization of polymers, polymer blends, polymer composites, and polymer nanocomposites. Electrically conductive blends polyaniline dodecyl benzene sulphonic acid (Pani:DBSA)/Styrene Butadiene Styrene (SBS) block copolymer have been prepared by melt mixing and using in situ polymerization method. The microstructural parameters were computed using wide angle X-ray scattering (WAXS) and Small angle X-ray scattering (SAXS). Bimodal distribution method has been studied for SBS, DCH32 cotton fiber and Pure Mysore multi-voltine silk (PMS) samples using Line profile analysis.

INTRODUCTION

Materials have become a symbol of the progress of human civilization. Human society has experienced the Stone Age, Bronze Age and Iron Age. In the 20th century high performance plastics and polymer composites became the substituent for traditional materials. Polymer composites exhibit superior properties to those of individual component materials. This is obviously the driving force for the development of composites.

The properties of polymer composites like weight reductions, corrosion resistant, high specific mechanical properties, superior stiffness and improved fatigue life leads to the production of many new

DOI: 10.4018/978-1-5225-3023-7.ch005

materials from low cost plastics to experience emerging materials. These tailored materials are useful in transportation, medical field, construction sectors etc.,.

Classification of composites can be polymer-polymer blends, fibre inforced composites, polymer-matrix composites (PMCs), polymer-nanocomposites, polymer electrolytes etc.

Materials are generally classified as insulators, semiconductors, conductors and superconductors based on their electrical properties. A material with conductivity less than 10^{-7} S/cm is regarded as an insulator. Metals have conductivity larger than 10^{3} S/cm whereas the conductivity of a semiconductor varies from 10^{-4} to 10 S/cm depending upon the degree of doping. It was generally believed that plastics (polymers) and electronic conductivity were mutually exclusive and the inability of polymers to carry electricity distinguished them from metals and semiconductors. As such, polymers were traditionally used as inert, insulating and structural materials in packaging, electrical insulations and textiles where their mechanical and electrically insulating properties were paramount. In fact, any electrical conduction in polymers was generally regarded as an undesirable phenomenon.

The evolution of conducting polymers began in 1975 with the discovery of a linear conjugated organic polymer, polyacetylene by Shirakawa (Shirakawa. H, et al., 1977). However, the material showed metallic properties only after its oxidation by iodine, which was reported two years after its discovery (Shirakawa. H, et al., 1977; Skotheim. T J, et al., 1998; Chandrashekar. P, et al., 1999). The 2000 Nobel Prize in Chemistry, recognized the discovery of conducting polymers after over 25 years of progress in this field (MacDiarmid. A G, 1999; N. Hall, 2003). In recent years, there has been growing interest in research on conducting polymer since they combine the advantages of organic conductors with low-dimensional systems and therefore create interesting physicochemical properties and potentially useful applications (Sailor. M J, et al., 1994; Neves. S, et al., 2004; Gangopadhyay. R, et al., 2004; Wallace. G G, et al., 2004; Epstein. A J, et al 2001; Martin. C R, 1995). Traditionally, an advantage of polymeric materials is that they can be synthesized and processed on a large scale at relatively low cost. Many of the applications (sensors, functional coatings, catalysts, etc.) of conducting polymers indeed need bulk quantity materials. Therefore, developing bulk syntheses for conducting polymers would be especially important for practical reasons. It has been found that the nanofibers significantly improve the processability of polyaniline and its performance in many conventional applications involving polymer interactions with its environment. This leads to much faster and more responsive chemical sensors, new inorganic/ polyaniline nanocomposites, and ultra-fast non-volatile memory devices.

Among the family of conjugated polymers, polyaniline is one of the most useful since it is air- and moisture-stable in both its doped, conducting form and in its de-doped, insulating form (Huang. W S, et al., 1986; MacDiarmid. A G, 1997; MacDiarmid. A G, et al., 1985). It has a great variety of potential applications including anticorrosion coatings, batteries, sensors, separation membranes, and antistatic coatings (Skotheim. T J, et al., 1998; Chandrashekar. P, et al., 1998). Conventional polyaniline synthesis is known to produce particulate products with irregular shapes. Therefore, many methods have been developed to make nanostructures of polyaniline (with diameters smaller than 100 nm) by introducing structural directing agents during the chemical polymerizing reaction. A great variety of such agents have been reported in the literature, and these include: surfactant, (Zhang. X, et al., 2004; Yu. L, et al., 2003; Michaelson. J C, et al., 1994; Li. G C, et al., 2004).

CONDUCTING POLYMERS

Conducting polymers can be classified mainly into three types:

1. Intrinsically / inherently conducting polymers (ICPs),
2. Conducting polymer composites,
3. Ionically conducting polymers.

POLYMER BLENDS

Polymer blends can be simply defined as the mixture of two or more polymers or copolymers. Blends of two polymers are usually prepared to achieve desired properties, which cannot be given by the individual polymers. To attain processibility in polyaniline, it can be blended with other highly processable thermoplastics or elastomers. There are many techniques which can be adopted for blending of two polymers like melt blending, powder blending, solution blending, reactive blending, etc. Among these, melt blending is the most feasible and industrially viable processes. The blend of two polymers can be of different kind depending on the compatibility of these polymers. Depending on this, there can be different classifications like miscible blends, immiscible blends, compatible blends and polymer alloys. The main factors affecting the miscibility of the blend components are polarity of the components, specific group attraction, molecular weight, ratio of the components taken and crystallinity (Deanin. R D, et al., 1999). The properties and performance of the blends are highly dependent on the morphology. Morphology of any blend is mainly decided by the miscibility of the components. Depending on this polymer blends can be classified as miscible and immiscible blends.

ADVANTAGES

Controllable Range of Conductivity

The electrical conductivity of polyaniline based compositions can be closely controlled over a wide range. For neat polyaniline compositions conductivity levels as high as 100 S/cm can be achieved. The full range of conductivity levels from less than 10^{-10} to 10^{-1} S/cm (melt processing) and 10 S/cm (solution processing), can be achieved for polymer blends containing polyaniline compositions.

Conductive Blends with Many Commodity Polymers

Electrically conductive polyaniline based blends with commodity polymers can be produced by using common solution and melt processing techniques. Examples of commodity polymers are polyethylene, polypropylene, polystyrene, PVC, phenol-formaldehyde resins, and different types of thermoplastic elastomers. Unlike conventional filled materials, the mechanical properties of these blends are close to those of the insulating matrix polymer.

Coloured and Transparent Electrically Conductive Products

Electrically conductive, coloured and transparent thin films and coatings, which would otherwise be difficult to achieve with conventional filled materials, can be made using polyaniline based compositions.

Applications Using Polyaniline Based Conductive Polymers

Neat Materials, Blends, Compounds and Solutions

Polyaniline based conductive polymers can be used neat, or as blends and compounds with commodity polymers. Several prototype objects and articles have been made of these compounds with polyethylene, polypropylene, polystyrene, soft PVC, poly--(methylmetacrylate), phenol-formaldehyde resins, melamine formaldehyde resins, epoxies, and thermoplastic elastomers. The solution processing properties of polyaniline based conductive polymers make totally new product ideas possible. These include electro-active inks, paints, coatings, and adhesives, electrochromic "smart" windows, electrically conductive transparent films, and conductive high performance fibres. Solution processing can be carried out using the neat material or a blend.

Electrostatic Discharge (ESD) Protection Materials

One target application of these materials is the protection from Electro Static Discharge (ESD). ESD causes problems when handling sensitive electronic components, explosive chemicals or dry powders. The use of, e.g. antistatic materials in specific environments reduce ESD related problems, such as the amount of damaged valuable end products in an assembly line. The controllable conductivity in ESD protection materials is one of the basic benefits of conductive polymer technology.

Examples of Industries and Products

Conductive polymers are attractive for new product concepts and for replacing conventional materials due to their unique combination of processability, stable and controllable conductivity, and optical and mechanical properties. There is a diversity of different industries and products where these compositions can be used:

Packaging Industry

1. Injection moulded antistatic products
2. Antistatic films

Electronics

1. Antistatic packaging of components
2. Printed circuit boards

Textile Industry

1. Conductive fabrics

Automotive Industry

1. Antistatic charge dissipation
2. Paint primers
3. Electrochromic rear vision systems

Construction

1. Antistatic floors
2. Antistatic work surfaces

Mining

1. Conductive pipes for explosives
2. Antistatic packaging

TOMORROW'S APPLICATIONS

Products Tailored to Customer's Needs

The aim of manufacturers is to develop and market a new family of proprietary and high-quality conductive polymer materials that are superior to other materials, and exactly match customer needs. Close co-operation between all those involved is the key to success. Polyaniline products can be tailored to needs. These efforts are focussed on customers' technical specifications, such as required conductivity, mechanical properties, processing conditions and operating conditions. Once the right property spectrum of these novel materials has been defined and specified, totally new applications and product concepts can be designed and developed into value added end products.

Future Possibilities

Most of the current development is focused on further developing polyaniline compounds for the existing applications, i.e. films, fibres and coatings. This will include development to achieve compatibility with new matrixes, binders and solvents. As this technology further develops, there will be new applications and market segments emerge. These may include for example shielding of electronic equipment from Electro-Magnetic Interference (EMI) or conductive and semi-conductive cable jackets. Newly synthesised conducting polymers are studied to understand the structure-property relation in these materials. There are several techniques one among them is X-ray technique.

X-Ray Technique

In this method, there are two ways of characterizing the polymer materials.

1. WAXS (Wide Angle X-ray Scattering), normally used to study the phases and crystallinity in polymers and polymer blends. It is normally recorded in the range of 10-90°
2. SAXS typically are concerned with scattering angles 2 Theta=0-2°. As per Bragg's law, the diffraction information about structures with large d-spacings lies in this region. Therefore the SAXS technique is commonly used for probing large length scale structures such as high molecular weight polymers, biological macromolecules (proteins, nucleic acids, etc.), and self-assembled superstructures (e.g. surfactant templated mesoporous materials). SAXS measurements are technically challenging because of the small angular separation of the direct beam (which is very intense) and the scattered beam. Large specimen-to-detector distances (0.5 m - 10 m) and high quality collimating optics are used to achieve good signal-to-noise ratio in the SAXS measurement.

Analysis of SAXS data from a polymer gives information about long periodicity, amorphous and crystalline length and their distributions.

Analysis

WAXS Data of Polymers and Polymer Blends

For the analysis of WAXS data of polymers and polymer blends, we have developed the one-dimensional model based on paracrystalline statistics. The preliminary results of this approach are already reported by earlier investigators (Somashekar. R, et al., 1993). Using paracrystalline statistics, we can simulate one-dimensional and three-dimensional X-ray diffraction patterns from polymers and polymer blends for various values of crystal size and lattice distortion parameters. This approach has been used to compute microstructural parameters of polymers and polymer blends (Souza Jr. F R, et al., 2006).

Recent investigation reports that the field of polymer has been considered as one of the most promising and rapidly emerging research area due to its vast applications in day today life. In this concept, polymers are complexed or doped with inorganic salts or alkalies which are being probed for their benefit in high performance energy storage devices like batteries, fuel cells and super capacitors for enhancing their performance. In such devices, the electrolyte which carries electrical charges is vital component and generally liquid or solid electrolytes are used. A substance that dissociates into free ions when dissolved in solvent to produce an electrically conductive medium is an electrolyte. Thermal stability, sustainability, longitivity and non-leakage are few properties due to which solid electrolytes are preferable over liquid electrolytes, which shows high conductivity for certain amount of the dopant. Polymer electrolytes are high ion conducting polymers and are solid material which are synthesized by dissolving the salt of alkali metal salts of type MX (M= Na, Li, Ag, NHI, Cu, etc and X= F, I, CI, etc) in polymers like poly (vinyl alcohol) polyethylene oxide, polypropylene oxide, polyurethane (Behera. P S, et al., 2006). Polymer electrolytes are mainly prepared by solution casting method, sol gel method, electro deposition method etc., which is in the form of bulk as well as thin films.

SAMPLE PREPARATION

PANI:DBSA was synthesized by one step route in toluene according to the procedure reported elsewhere (Osterholm. J E, et al., 1994; Barra. GMO, et al., 2001). Polyaniline – Dodecyl Benzene Sulphonic Acid (PANI:DBSA)/Styrene Butadiene Styrene (SBS) block co-polymer have been prepared by two methods namely melt mixing and polymerization of aniline in the presence of SBS using in situ polymerization method

Synthesis of Polyaniline Doped With DBSA (PANI:DBSA)

PANI:DBSA used in mechanical mixing was synthesized by one-step route in toluene according to the procedure reported elsewhere (Somashekar. R, et al., 1993; Souza Jr. F R, et al., 2006). In a typical procedure, 4.7 mL (0.051 mol) of aniline and 16.7g (0.051 mol) of DBSA were dissolved into 250 mL of toluene under constant stirring. The reaction mixture was kept at 0^0C at an aqueous solution containing 11.36g (0.051 mol) of APS in 40 mL of water was slowly added over a period of 20 min. After 6 h, the reaction medium was poured into a non-solvent like methanol, filtered, washed several times with methanol and dried.

Blend Preparation

Melt Mixing Method

Different amounts of SBS copolymer was first introduced into a Haake Rheocod 9000 internal mixer operating with a Cam rotor at 100^0C and 60 rpm. After 2 min, PANI:DBSA powder was added and mixing was continued for 8 min. The blends were then quickly sheeted in a two-roll mill.

In Situ Polymerization Method

A 10(%) solution of SBS in toluene was first prepared. Then, different amount of aniline (Ani) and DBSA were added under continuous stirring. The reaction medium was kept at 0^0C and an aqueous solution of APS was added drop-wise into the emulsion, with stirring. The APS/Ani/DBSA molar ratio was 1:1:1.

THEORETICAL ASPECTS / METHODOLOGY

In WAXS studies, microstructural parameters like crystal size $<N>$ and lattice strain (g in %) of PANI:DBSA are determined by employing Fourier method of Warren and Averbach (Warren. B E, et al., 1950 - Warren. B E, et al., 1969) and the multi-dimensional minimization algorithm of the SIMPLEX method was used (Somsahekar. R, et al 1989).

In SAXS studies, a_1 is the intensity related to the volume function of the independent scatterers and a_2 is the half width of the non-periodic structures with maximum at the origin. The parameters a_4, a_5 and a_6 are respectively the height, position and width of Gaussian peak that represent lamellar reflection. a_4

represents the lamellar intensity, a_5 and a_6 respectively represent the lamellar spacing (L) and coherence length of the lamellar stack (L_{coh}) and are given by;

$$L = 2\pi \,/\, a_5$$

$$L = \frac{4\pi}{[a_6\sqrt{(8\ln 2)}]}$$

using the fitted parameters a_3, a_4, a_5 and a_6 and using the standard expression for the area under Gaussian peaks. We have developed a FORTRAN program using SIMPLEX multidimensional minimization method to determine these six parameters. The observed scattered intensity can be directly fitted to the intensity calculated from suitable six parameter models of Murthy et al. (Murthy. N S, et al., 1998) and one dimensional paracrystalline model (Soares. B G, et al., 2007).

Intensity profile of X-ray reflection from a crystalline material is a function of the distribution of crystal sizes in the sample and of lattice disorder. Line profile analysis (LPA) using single order method employing various bimodal functions for crystal size distribution is used in the study of SBS, DCH32 and PMS samples (Samir. O M, et al., 2008).

CHARACTERIZATION TECHNIQUES

The prepared PANI:DBSA/SBS films were characterized by mechanical, morphological and electrical properties. The microstructural parameters were also computed using wide angle x-ray scattering (WAXS) and also using small angle x-ray scattering (SAXS).

RESULTS AND DISCUSSIONS

The influence of synthetic routes on the microcrystalline behaviour of SBS/PANI blends has been studied by WAXS. The <N>, value increases with increase in crystalline PANI content in the amorphous SBS

Figure 1. WAXS patterns of (a) PANI:DBSA and (b) SBS

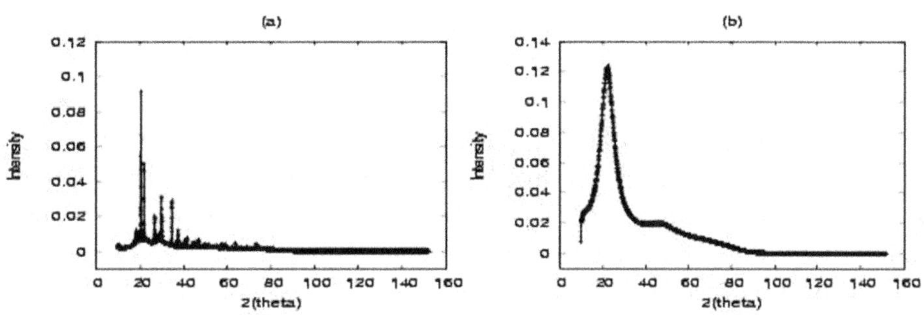

Figure 2. WAXS patterns of SBS/PANI:DBSA blends with (c) 15%, (d) 30% of PANI:DBSA by melt mixing method

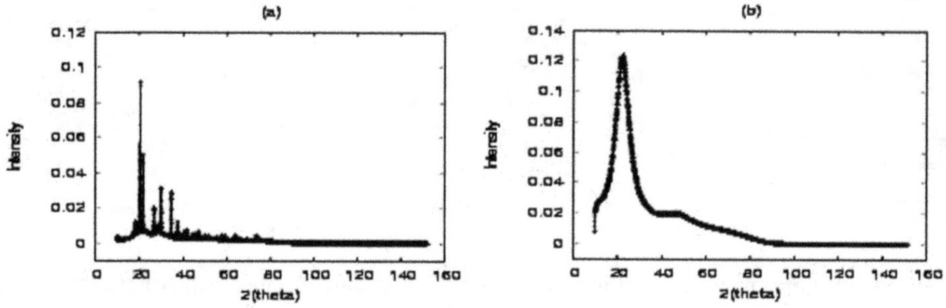

Figure 3. WAXS patterns of SBS/PANI:DBSA blends with (f) 16.2%, (g) 33.3% (h) 43.9% of PANI:DBSA by using insitu polymerization method

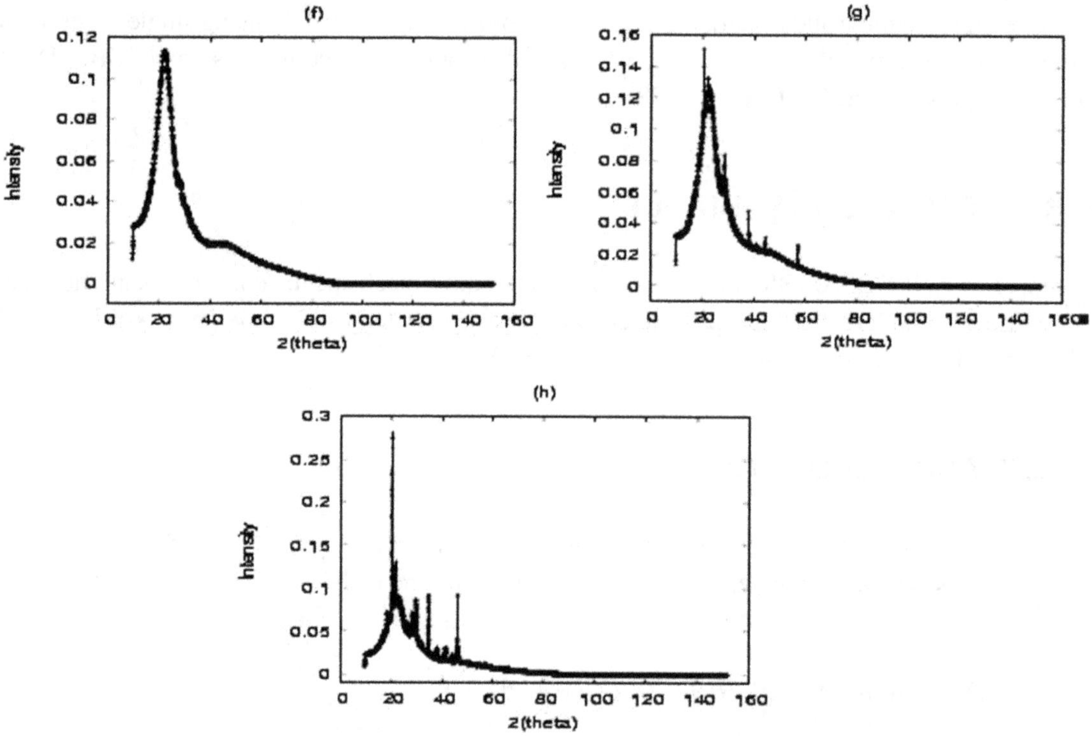

matrix, as expected. The tensile strength and volume resistivity decrease with increase in PANI content, whereas <N>, value increases. This is due to the increase in high crystalline PANI phase in SBS matrix. The changes in physical properties are attributed to the crystal imperfection, wherein parameters like <N>, and D_s increases with increase in PANI contents. Blend prepared by in situ polymerization resulted in lower resistivity because of a favorable morphology characterized by the formation of interconnected micro tubules, which form the conducting pathway with lower amount of PANI. In spite of lower resis-

Figure 4. Experimental and simulated X-ray profiles for PANI and SBS using Exponential distribution function

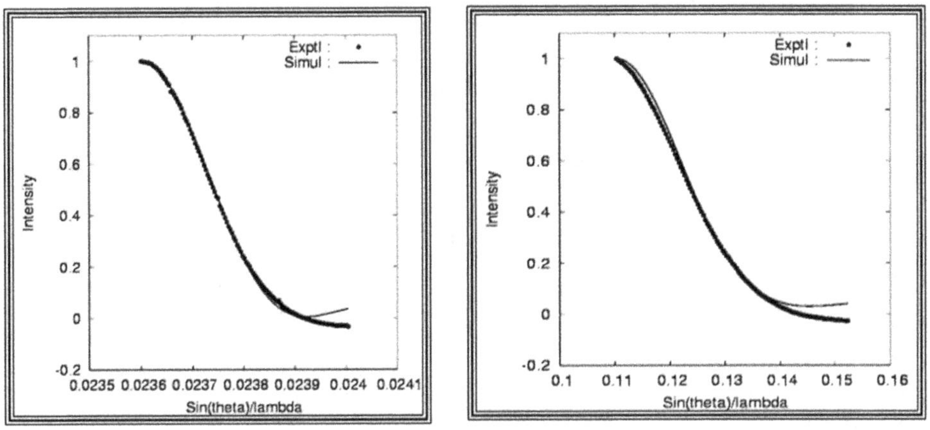

Figure 5. The crystallite ellipsoid shapes of SBS/ PANI:DBSA blends with (a) 15%, (b) 30%, (c) 45% of PANI:DBSA

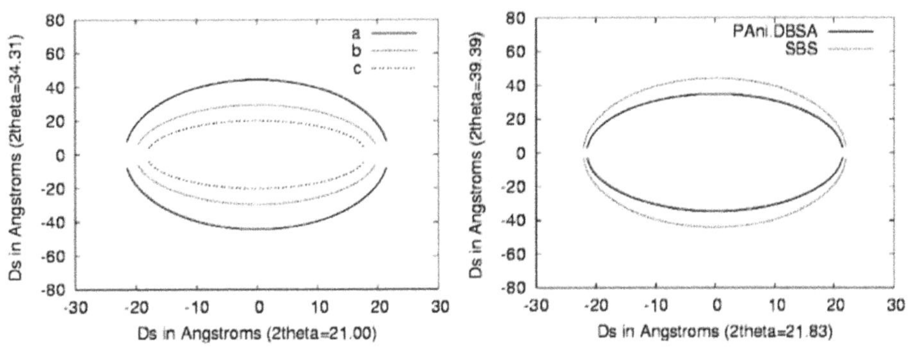

Table 1. Microstructural parameters of SBS, PANI obtained from WAXS data employing Exponential distribution function

Composition of Blends (SBS/PANI),(wt/wt,%)	2theta (deg)	d in (Å)	<N>	g (%)	Ds in (Å)	Δ
0/100	20.21	2.523	128.15 ± 6.4	0.4 ± 0.02	443.12	0.05
100/0	21.59	2.369	3.40 ± 0.17	0.5 ±0..02	15.4	0.05
	26.66	1.943				
	29.84	1.752				
	34.60	1.535				
	22.18	2.309				

tivity, the crystallinity level of these blends is lower than the corresponding mechanical mixing blends. Blends prepared by in situ polymerization method can be considered a physical inter penetrating network. Therefore, large aggregates constituted by PANI chains did not form in large extent because of the

Table 2. Microstructural parameters of SBS/PANI blends obtained from WAXS data employing Exponential distribution function (Melt mixing method

Composition of Blends (SBS/PANI),(wt/wt,%)	2theta (deg)	d in (Å)	<N>	g (%)	Ds in (Å)	δ
85/15	21.85	2.342	3.48 ± 0.17	0.5	16.0	0.05
70/30	20.10	2.536	107.4 ± 6.4	0.5	379.5	0.06
55/45	21.07	2.425	107.6 ± 4.3	0.3	438.26	0.04
	29.39	1.776				
	34.24	1.549				
	18.25	2.783				
	29.56	1.767				
	20.09	2.538				
	34.35	1.545				

Figure 6. The crystallite ellipsoid shapes of SBS/ PANI:DBSA blends with (a) 16.2%, (b) 32.3%, (c) 43.9% of PANI:DBSA

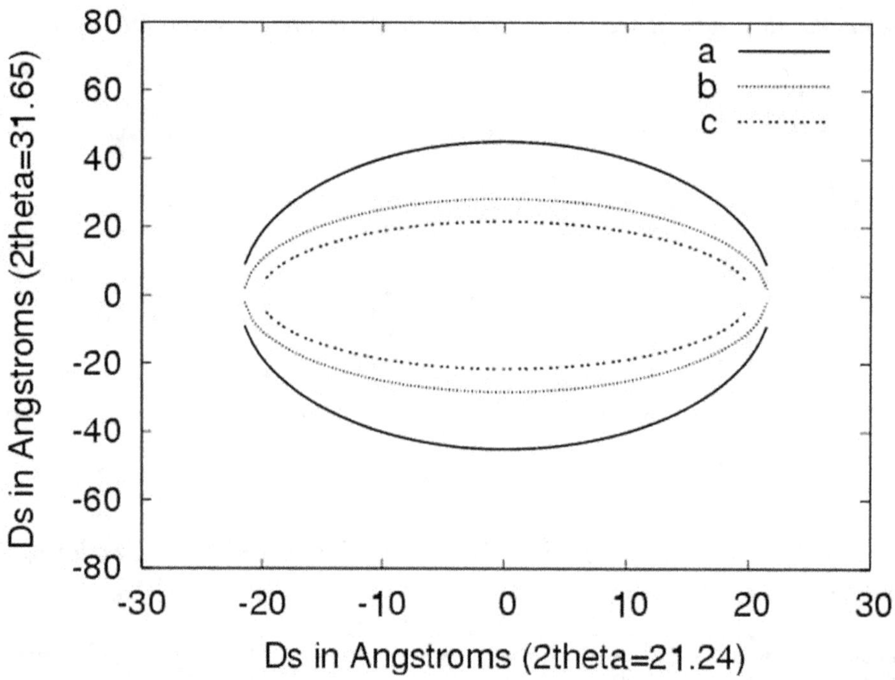

interconnection of SBS chains. The absence of large agglomerates decreases the crystallinity level but promotes the formation of conducting pathway at lower PANI concentration (lower percolation threshold).

From the wide angle X-ray scattering (WAXS) study of electron irradiated polymer films, we have observed that even though there is not much change in the position of the X-ray Bragg reflections, a significant change in the values of microstructural parameters occurs. We have shown that among the three asymmetric crystallite size distributions, Exponential gives a better fit in films. The only justification for the good fit that we observe with Exponential distribution in these polymers can be interpreted

Table 3. Microstructural parameters of SBS/PANI blends obtained from WAXS data employing Exponential distribution function (Chemical method)

Composition of Blends (SBS/PANI),(wt/wt,%)	2theta (deg)	d in (Å)	<N>	g (%)	Ds in (Å)	δ
83.8/16.2	21.95	2.332	3.92 ± 0.11	0.5	17.94	0.03
67.7/32.3	20.43	2.497	13.37 ± 0.66	0.3	60.34	0.05
56.1/43.9	21.57	2.371	41.8 ± 1.67	0.2	60.34	0.04
	28.29	1.839				
	18.33	2.772				
	20.21	2.523				
	21.61	2.367				

Figure 7. (a) Saxs X-ray diffraction pattern of PANI, (b) Saxs X-ray diffraction pattern of SBS PANI: Melt mixing method

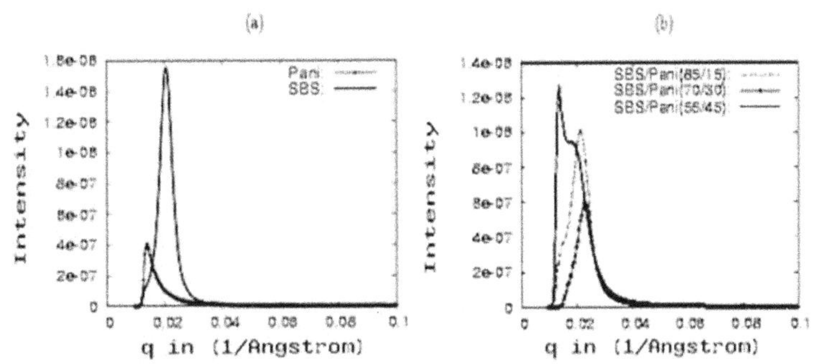

Figure 8. Experimental and simulated SAXS profiles of PANI and SBS using paracrystalline model.

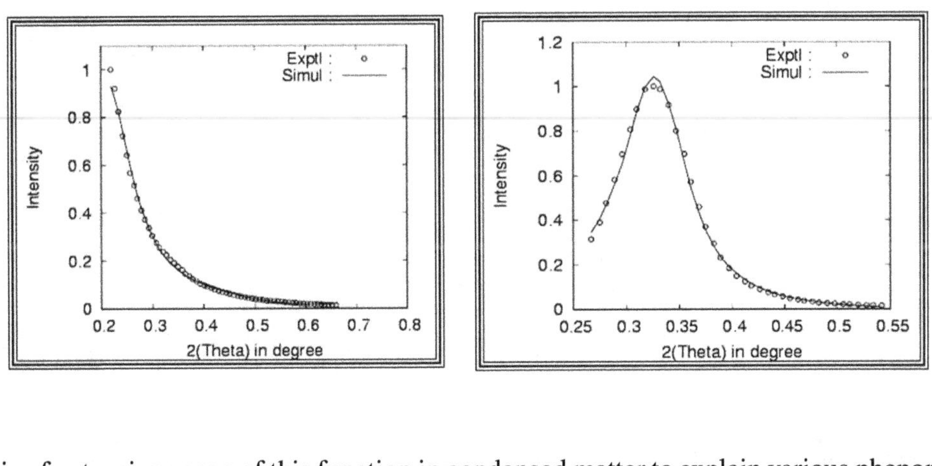

on the basis of extensive usage of this function in condensed matter to explain various phenomenon like dielectric relaxation, luminescence decay law and other physical properties. Single order method that we have used here is capable of estimating both the size and the distortion parameters and could in general measure crystallite size, only upto a certain limit. The changes in polymer network with different dose rates are quantified here in terms of microstructural parameters. Our study indicates that the 40 Mrad

Figure 9. The normalized experimental probability distribution of the lengths of the phases (a) amorphous and (b) Crystalline

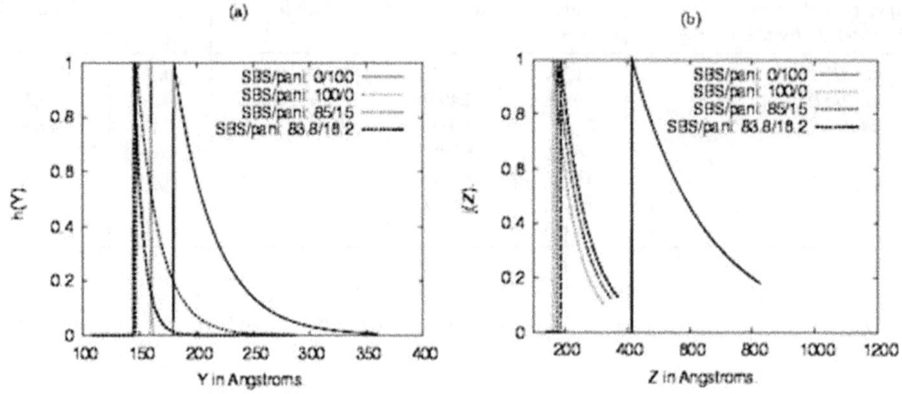

Table 4. Microstructural parameters of SBS, PANI and their blends using SAXS

Composition of Blends (SBS/PANI),(wt/wt,%)	2 Theta (deg)	Periodicity (Å)	Phase ratio	γ_y	γ_z
0/100 100/0	0.22 3.58	594 27.9	0.3 0.92	0.098 0.052	0.289 0.001
(a)Blends(Melt Mixing)					
85/15 70/30 55/45	0.33 0.36 0.29	321 294 393	0.5 0.4 0.4	0.067 0.081 0.030	0.238 0.202 0.302
(a)Blends(Chemical Method)					
83.8/16.2 67.7/32.3 56.1/43.9	0.34 0.33 0.28	329.7 341.9 471.2	0.44 0.39 0.32	0.027 0.025 0.098	0.242 0.231 0.347

Table 5. Microstructural parameters of SBS, PANI and their blends using SAXS

Composition of Blends (SBS/PANI),(wt/wt,%)	2 theta (deg)	<Y> in (Å)	<Z> in (Å)	Crystallinity	Tensile Strength
0/100 100/0	0.22 3.58	180 25.76	414 2.2	0.70 0.92	---- ----
(a) Blends (Melt Mixing)					
85/15 70/30 55/45	0.33 0.36 0.29	147 128 146	174 167 248	0.54 0.57 0.63	7.2 ± 0.3 2.4 ± 0.8 2.0 ± 0.2
(a) Blends (Chemical Method)					
83.8/16.2 67.7/32.3 56.1/43.9	0.34 0.33 0.28	144.3 130.9 149.4	185.3 211 322	0.56 0.94 0.68	4.6 ± 0.9 3.7 ± 0.4 3.0 ± 0.1

Figure 10. Experimental and simulated SAXS profiles of SBS and PANI using Six parameter model of Murthy et al.

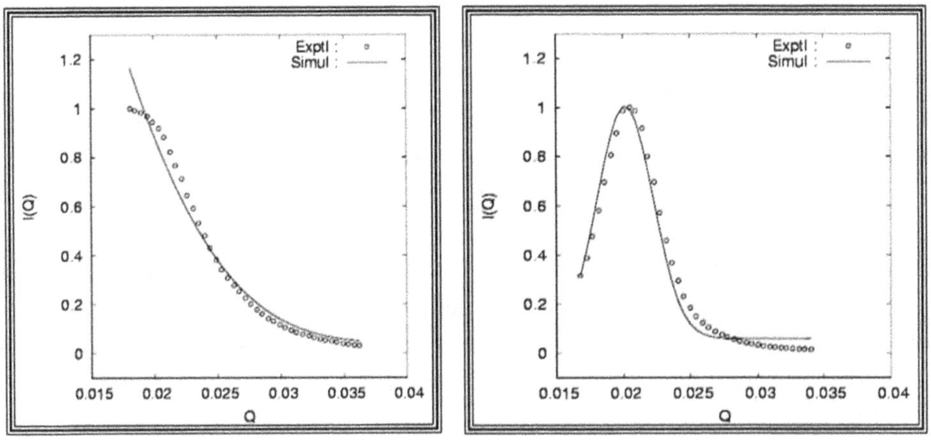

Table 6. Six-parameters used to simulate SAXS profiles for SBS, PANI and their blends

Composition of Blends (SBS/PANI),(wt/wt,%)	a1	a2	a3	a4	a5	a6
100/0	0.002	2.62	64.9	0.02	0.01	0.01
0/100	0.01	3.10	0.43	0.14	0.17	0.03
	0.20	5.13	29.55	0.02	0.02	0.003
(a) Blends (Melt Mixing)						
85/15	0.16	0.004	10.13	0.02	0.02	0.004
70/30	0.26	0.08	15.39	0.02	0.02	0.003
55/45	0.22	0.29	11.34	0.14	0.12	0.004
	0.03	2.80	79.39	0.02	0.02	0.01
	0.17	2.47	4.71	0.12	0.12	0.004
(a) Blends (Chemical Method)						
83.8/16.2	0.004	1.98	74.87	0.02	0.02	0.01
67.7/32.3	0.23	1.85	17.53	0.02	0.02	0.004
56.1/43.9	0.17	2.73	22.13	0.02	0.02	0.004

Table 7. Microstructural parameters obtained by SAXS data of SBS, PANI and their blends

Composition of Blends (SBS/PANI),(wt/wt,%)	Profile No.	Lc in (Å)	L_A in (Å)	LCOR in (Å)	LM in (Å)	(Lc/Lc+LA) Crystallinity
0/100	1	326.2 ±3.3	129.7 ± 1.3	529.5 ± 0.5	455.9	0.71
100/0	1	245.7 ±2.5	59.1 ± 0.6	1847.2 ± 1.9	304.3	0.81
(a) Blends (Melt Mixing)						
85/15	1	265.3 ±2.7	20.3 ± 0.2	1291.8 ± 1.3	285.6	0.93
70/30	1	243.8 ±2.4	30.8 ± 0.3	1565.1 ± 1.6	224.6	0.89
55/45	1	187.3 ±1.9	158.8 ± 1.6	529.2 ± 0.5	346	0.54
(a) Blends (Chemical Method)						
83.8/16.2	1	247.8 ±2.5	44.3 ± 0.4	1308.8 ± 1.3	292.1	0.85
67.7/32.3	1	263.9 ±2.6	35.1 ± 0.3	1418.4 ± 1.4	299	0.88
56.1/43.9	1	135.7 ±1.4	149.7 ± 1.5	534.7 ± 0.5	285.5	0.48

irradiated polymer sample has a higher value of crystallite shape and surprisingly we observe that the intrinsic strains are very small. It is evident from this study that irradiation of polymer films changes the polymer network and hence the physical properties, leading to a better quality copolymer films, depending on the nature of applications.

From SAXS studies in the region of 15-45% of PANI SBS, we observe that the long periodicity increases with increase in the content of PANI in the blends obtained by chemical method. This also favors lowering of the tensile strength in these blends. This is due to formation of microcrystals at the expense of the un-oriented amorphous chains between the lamellae. These microcrystals inhibit the mobility polymer chains in both the amorphous and crystalline domains resulting in decreasing tensile strength. We also observe from these studies that blends formed by chemical method is more suitable than blends formed by melt mixing or mechanical method.

Blends of PANI in SBS prepared by in-situ polymerization show higher conductivity and hence balanced by appropriate of crystallinity, which has been observed by SAXS analysis. The reason for such a behaviour in these blends is due to entanglement of PANI with SBS in in situ polymerization. The computed morphological parameters using a six-parameter model for the observed SAXS data showed a good fit at lower and higher scattering angles. There is a linear variation of the crystalline length with the tensile strength of the blends studied here. Our results suggest that in situ polymerization promotes regular stacking of the lamellar structure, in spite of almost invariant thickness of the crystal-amorphous interfacial regions.

A linear combination of exponential and Reinhold functions, which we refer as bimodal distributions for crystallite sizes in polymers leads to improved fitting of the experimental profiles with reliable micro structural parameters. The present analysis of X-ray data on polymers can be considered as very close to realistic method. By using bimodal distributions, we have reduced the errors in the computation of micro structural parameters arising due to truncation and background estimations.

In the case of PANI:DBSA/SBS films characterized by WAXS, the authors observed that the presence of PANI:DBSA in SBS drastically reduced the ultimate properties (tensile strength and elongation at break). Whatever may be the method used for the blend preparation this behaviour may be attributed to the interactions between the PANI molecules with the SBS matrix. Blends prepared by in situ polymerization also displayed lower values of resistivity. With 15% of PANI:DBSA, the difference in resistivity was around 10^7 of magnitude, indicating the insulator-conductor transition at lower percolation threshold. X-ray studies in SBS/PANI:DBSA blends shows decrease in crystallite size and increase in lattice strain. From the SAXS studies we observed that there is a linear relationship between long periodicity and concentration of blends. These parameters are compared with tensile strength to find the structure property relation.

CONCLUSION

Blends of PANI:DBSA in SBS were prepared by melt mixing method and in situ polymerization methods. The prepared samples were characterised by mechanical, morphological and electrical studies. The microstructural parameters were computed by WAXS and also by SAXS. The blends prepared by in situ polymerisation resulted in lower resistivity. The WAXS study of electron irradiated polymers shows a significant change in the microstructural parameters. Using bimodal distribution method we have reduced the errors in the computation of microstructural parameters arriving due to truncation and background estimation. Conducting polymers exhibits two main groups of applications. The first group utilizes their conductivity as its main property and second group utilizes their electro activity.

REFERENCES

Shirakawa, H., Louis, E. J., MacDiarmid, A. G., Chiang, C. K., & Heeger, A. J. (1977). Synthesis of electrically conducting organic polymers: Halogen derivatives of polyacetylene,(CH) x. *Journal of the Chemical Society. Chemical Communications*, (16), 578–580. doi:10.1039/c39770000578

Skotheim, T. A., & Reynolds, J. (Eds.). (2007). *Handbook of Conducting Polymers* (Vol. 1-2). CRC Press.

Chandrasekhar. (1999). *Conducting Polymers, Fundamentals and Applications: A Practical Approach.* Kluwer Academic.

MacDiarmid, A. G. (2001). Synthetic metals: A novel role for organic polymers (Nobel lecture). *Angewandte Chemie International Edition*, *40*(14), 2581–2590. doi:10.1002/1521-3773(20010716)40:14<2581::AID-ANIE2581>3.0.CO;2-2 PMID:11458347

Hall, N. (2003). Bringing inorganic chemistry to life. *Chemical Communications*, (7): 803–806. PMID:12739626

Sailor, M. J., & Curtis, C. L. (1994). Conductiong polymer connections for molecular devices. *Advanced Materials*, *6*(9), 688–692. doi:10.1002/adma.19940060917

Farchioni, R., & Grosso, G. (Eds.). (2013). *Organic electronic materials: Conjugated polymers and low molecular weight organic solids* (Vol. 41). Springer Science & Business Media.

Martin, C. R. (1995). Template synthesis of electronically conductive polymer nanostructures. *Accounts of Chemical Research*, *28*(2), 61–68. doi:10.1021/ar00050a002

Huang, W. S., Humphrey, B. D., & MacDiarmid, A. G. (1986). Polyaniline, a novel conducting polymer. Morphology and chemistry of its oxidation and reduction in aqueous electrolytes. *Journal of the Chemical Society, Faraday Transactions 1: Physical Chemistry in Condensed Phases*, *82*(8), 2385–2400.

MacDiarmid, A. G. (1997). Polyaniline and polypyrrole: Where are we headed? *Synthetic Metals*, *84*(1-3), 27–34. doi:10.1016/S0379-6779(97)80658-3

MacDiarmid, A. G., Chiang, J. C., Halpern, M., Huang, W. S., Mu, S. L., Nanaxakkara, L. D., & Yaniger, S. I. et al. (1985). Polyaniline: Interconversion of metallic and insulating forms. *Molecular Crystals and Liquid Crystals*, *121*(1-4), 173–180. doi:10.1080/00268948508074857

Zhang, X., & Manohar, S. K. (2004). Polyaniline nanofibers: chemical synthesis using surfactants. *Chemical Communications*, (20), 2360-2361.

Yu, L., Lee, J. I., Shin, K. W., Park, C. E., & Holze, R. (2003). Preparation of aqueous polyaniline dispersions by micellar-aided polymerization. *Journal of Applied Polymer Science*, *88*(6), 1550–1555. doi:10.1002/app.12127

Michaelson, J. C., & McEvoy, A. J. (1994). Interfacial polymerization of aniline. *Journal of the Chemical Society. Chemical Communications*, (1): 79–80. doi:10.1039/c39940000079

Li, G., & Zhang, Z. (2004). Synthesis of dendritic polyaniline nanofibers in a surfactant gel. *Macromolecules*, *37*(8), 2683–2685. doi:10.1021/ma035891k

Deanin, R. D., & Manion, M. A. (1999). Article. In C. B. Shonaike & P. G. Simon (Eds.), Polymer Blends and Alloys (p. 1). New York: Marcel Dekker Inc.

Somashekar, R., & Gopalkrishne Urs, R. (1993). Simulation of X-ray diffraction patterns for silk fibres using paracrystalline statistics. *Pramana*, *40*(5), 335–343. doi:10.1007/BF02847494

Souza, F. G. Jr, Soares, B. G., Mantovani, G. L., Manjunath, A., Somashekarappa, H., Somashekar, R., & Siddaramaiah, . (2006). Blends of styrene butadiene styrene TRI block copolymer/polyaniline-Characterization by WAXS. *Polymer*, *47*(6), 2163–2171. doi:10.1016/j.polymer.2006.01.033

Behera, P. S. (2012). *Dielectric properties of polymer electrolyte* (Doctoral dissertation).

Barra, G. M., Leyva, M. E., Gorelova, M. M., Soares, B. G., & Sens, M. (2001). X-ray photoelectron spectroscopy and electrical conductivity of polyaniline doped with dodecylbenzenesulfonic acid as a function of the synthetic method. *Journal of Applied Polymer Science*, *80*(4), 556–565. doi:10.1002/1097-4628(20010425)80:4<556::AID-APP1130>3.0.CO;2-N

Warren, B. E., & Averbach, B. L. (1952). The separation of cold-work distortion and particle size broadening in X-ray patterns. *Journal of Applied Physics*, *23*(4), 497–497. doi:10.1063/1.1702234

Warren, B. E. (1955). A generalized treatment of cold work in powder patterns. *Acta Crystallographica*, *8*(8), 483–486. doi:10.1107/S0365110X55001503

Somashekar, R., Hall, I. H., & Carr, P. D. (1989). The determination of crystal size and disorder from X-ray diffraction photographs of polymer fibres. 1. The accuracy of determination of Fourier coefficients of the intensity profile of a reflection. *Journal of Applied Crystallography*, *22*(4), 363–371. doi:10.1107/S0021889889004085

Murthy, N. S., Akkapeddi, M. K., & Orts, W. J. (1998). Analysis of lamellar structure in semicrystalline polymers by studying the absorption of water and ethylene glycol in nylons using small-angle neutron scattering. *Macromolecules*, *31*(1), 142–152. doi:10.1021/ma9707603

Soares, B. G., Souza, F. G. Jr, Manjunath, A., Somashekarappa, H., Somashekar, R., & Siddaramaiah, . (2007). Variation of long periodicity in blends of styrene butadiene, styrene copolymer/polyaniline using small angle X-ray scattering data. *Pramana*, *69*(3), 435–443. doi:10.1007/s12043-007-0144-z

Samir, O. M., Manjunath, A., Parameswara, P., Somashekarappa, H., & Somashekar, R. (2008). Microstructural Parameters of Polymer Materials using WAXS data: Bimodals Crystal Size Distribution. *Journal of Polymer Materials*, *25*(4), 529–540.

Chapter 6
Advances in Polymer-Based Piezoelectric Systems

Nithin Kundachira Subramani
The National Institute of Engineering, India

Shilpa K. N.
Sri Jayachamarajendra College of Engineering, India

Sachhidananda Shivanna
Sri Jayachamarajendra College of Engineering, India

Jagajeevan Raj B. M.
The National Institute of Engineering, India

Siddaramaiah Hatna
Sri Jayachamarajendra College of Engineering, India

ABSTRACT

Lately, polymer based piezoelectric materials that harness energy from mechanical vibrations and/or impact are being increasingly investigated as radical alternates to conventional batteries that are hard to service once deployed. Nevertheless, the optimization of energy outputs of piezoelectric energy harvesters is one of the prime challenges faced by the scientific community. This chapter provides an overview of polymer based piezoelectric energy harvesters with special emphasis on current state of research on polymer composites/nanocomposites for vibrational energy harvesting. A detailed summary of piezoelectric phenomenon in polymers is also presented. An in-depth narration detailing the enhancement of piezoelectric behavior of one of the most commonly employed piezoelectric polymer (PVDF) is presented with special emphasis on some of the promising filler materials towards realizing high efficient piezoelectric modules. This chapter is intended to give an insight on the recent advances in the field of polymer based piezoelectric materials.

DOI: 10.4018/978-1-5225-3023-7.ch006

INTRODUCTION

Energy is much obligatory and the ever decreasing size of electronic gadgets coupled with augmented ask for sustainable energy have led the scientific community to design and develop self powered gadgets that do not ask for any external power, either continuous or punctuated (Atzori et al., 2010). These self powered gadgets mostly make use of ambient and harness-able energies. Piezoelectricty is one such approach of energy recovery that essentially involves vibration-to- electricity energy conversion (Henry, 2004) via buttressing of electrical charges under applied stress. The most distinguished characteristic of this kind of mechanical energy harvesting was initially identified for low frequency power generation applications. Therefore, one of the prominent targeted applications of vibration to electricity energy conversion is to power micro-electronic devices.Nevertheless, vibrational-to-electrical energy conversion involves one of the three thriving mechanisms of energy transduction namely; electrostatic, electromagnetic and piezoelectric (Williams & Rob 1996). All of which have been thoroughly investigated over the last few decades. If the vibrational force is converted electrical energy via electric filed, then the transduction is electrostatic, if magnetic field is the transduction medium, then electromagnetic and if strain is involved, then it is termed piezoelectric.

Piezoelectric Energy Harvesting

Piezoelectric energy harvesting is one of the most popular and widely followed vibrational energy harvesting technique, owing to its design simplicity and inherent ability to harness high power densities (Priya et al., 2007; Henry et al., 2004). Literatures till date support piezoelectric materials as future energy materials with power densities comparable to conventional lithium ion batteries and as phonon to electron converting thermo-electric generators (Cook-Chennault et al., 2008). Additionally, piezoelectric systems are capable enough to generate relatively higher potential (voltage) to be used directly for real time operations, in contrast to other energy harvesting techniques that demand further processing and up-gradation to generate usable amounts of voltage. The combination of excellent power density and voltage potential gives piezoelectric materials an edge as boundless replacements for conventional batteries in low-power wireless electronic devices, thus acknowledgingtheir increasing research interest in recent decades.

The word piezoelectricity has its roots linked to the Greek word **'piezein'**, meaning to crush or press. Piezoelectric materials are no-centrosymmetric that enables them to display a net symmetry distortion on applying electrical or mechanical stress. In simpler terms,piezoelectric effect is a relative phenomenon that involves inter-conversion of mechanical and electrical signals categorized as; Direct piezoelectric effect that involves the conversion of mechanical stress into electrical charge and Converse piezoelectric effect, whereinapplied electrical potential is converted into mechanical stress (Ballato et al., 1995; Jaffe et al., 1971; Cady et al., 1946).The mechanism of direct piezoelectric effect is commonly encountered in strain sensors and vibrational energy scavenging devices, while converse effect is widely observed among actuators (Kim et al., 2008; Anton et al., 2007; Yu et al., 2009; Auld 1973). Further, the strain co-efficient(d) and voltage co-efficient(g) are the key piezoelectric factors that characterizes the vibration to electricity conversion efficacies. The strain co-efficient (d) and voltage co-efficient (g) values are expressed as d_{ij} and g_{ij} , where the subscripts i and j represent the relative directions. Generally, $d_{ij} = d_{33}$ and $g_{ij} = g_{33}$ denotes the induced polarization per unit stress and induced electric field per unit stress respectively. The two key factors d_{ij} and g_{ij} are related as;

$$\mathbf{d_{ij} = g_{ij}\mu_0\mu_r} \tag{1}$$

Where, μ_0 and μ_r are the permittivity of air and dielectric constant of the material respectively. As can be seen from the above relation, higher the dielectric constant of the material, higher will be the strain co-efficient $\left(d_{ij}\right)$ and lower will be voltage co-efficient $\left(g_{ij}\right)$.However, competent piezoelectric systems must offer higher values of d_{ij} and g_{ij} with added advantages of design flexibility and ease of processing. While, the coefficients of strain and voltage quantifies the energy generated per unit applied force. The energy conversion efficacy is however characterized by electro-mechanical coupling factor (k) expressed by the empirical relation;

$$\mathbf{k_{ij} = \frac{d_{ij}}{\sqrt{\mu_0\mu_r s_{ij}}}} \tag{2}$$

Where, s_{ij} is the stiffness component. From the above relation, one may expect a stiffer material to be experiencing a larger internal strain and hence reduced energy conversion. Additionally, higher dielectric constant may also lead to larger capacitance and greater force to bring about charge displacement, thereby reducing the energy conversion efficiencies. Although most of the piezoelectric energy harvesting applications make use of quartz or ceramic materials, owing to their high strain co-efficient, yet polymers such as Polyvinylidene fluoride (PVDF) with lower strain and voltage co-efficients are often employed as pressure and acoustic sensors, owing to their increased sensitivities, requiring only slightest of the force to induce piezoelectric response.

Piezoelectric Generators

When a sheet of piezoelectric material is made to experience longitudinal mechanical stress, i.e., stress in direction parallel to polarization, a certain voltage is generated in the system, which seeks to restructure the material to its virgin thickness. Correspondingly, when the direction of applied stress is transversal, i.e., perpendicular to the direction of polarization, the voltage generated will now seek to return the material to its initial length and/or width. In simpler terms, a piezoelectric material bonded to a structural unit, that is stretched or flexed will generate a potential (voltage). There are mainly two basic techniques to induce piezoelectric response in a material namely,

- **Extension Generator:** Generation of piezoelectric response via stretching.
- **Bending Generator:** Generation of piezoelectric response via bending.

The performance of piezoelectric generator is however presented in terms of open-circuit potential $\left(V_{OC}\right)$ and closed-circuit current $\left(I_{CC}\right)$, with I_{CC} specifying the total current built up at maximum possible strain level and operational frequency. Moreover, the charge associated with closed-circuit current is totally free to voyage from one electrode to the other. While, V_{OC} represents the potential developed at maximum applied strain, with the developed charge limited from electrode to electrode itinerant. Ideally, any piezoelectric generator would present an optimal performance, when it is capable of delivering the required voltage at one half of its closed-circuit current.

POLYMER BASED PIEZOELECTRIC SYSTEMS

Although, piezoelectric response is commonly observed in materials such as quartz, ceramics and polymers. Yet, most of the piezoelectric based energy harvesting applications make use of ceramic materials owing to higher d values, closely to few hundred pico coulombs per Newton (Table 1).While, the best performing polymer(poly(vinylidene fluoride) (PVDF)) displays a maximum piezoelectric charge constant of 33 units only. However, PVDF possesses a large g_{33}value due to its relatively low dielectric constant, which makes it highly sensitive and capable enough to exhibit piezoelectric response, even with slightest of applied strain (Priya, 2009).

However, in recent years, various classes of elastically yielding electro-active polymeric materials have established themselves as ideal piezoelectric systems. The first class of such flexible materials is the **bulk polymers** that yield piezoelectric performance through their molecular structure. The second category includes voided charged polymers, in which the polarization of internal dipoles (upon applying external stress) induced by the creation of gas voids in the polymeric structure yields piezoelectric response. While, the final class involves polymer based composite systems that are essentially hybrids of piezoelectric ceramic intercalated polymeric materials, which take advantage of mechanical flexibility of polymers (Figure 1) and high electromechanical coupling abilities of ceramic materials to yield piezoelectricity.A pictorial representation of different types of elastically yielding polymers with piezoelectric characteristic is depicted in Figure 2.

PVDF and Its Co-Polymer Based Piezoelectric Systems

PVDF is one of the most routinely employed electro-active bulk polymer with excellent electro-mechanical properties. The relatively larger piezoelectric coefficient (20-30 pCN^{-1}) of PVDF over its bulk polymeric counterparts renders it highly employable in vibrational sensing applications (Harrison et al., 2002). During polymerization of $H_2C{=}CF_2$ monomeric unit to produce PVDF, a defect is said to be originated, when the terminal H_2C- or -CF_2 groups of the polymer are fused together. These defect levels in the polymeric chain facilitate increased charge separation and hence better piezoelectric response. The semi-crystalline PVDF molecules are found to exhibit five distinct crystalline phases (α, β, γ, δ and

Table 1. Comparitive account on piezoelectric response of most widely used polymer and ceramic

Material type	ε_r	d33 (pC/N)	g_{33} (10^{-3} V m/N)
PVDF	13	-33	-339
PZT-4	1300	289	25

Figure 1. Advantages of polymer based piezoelectric materials

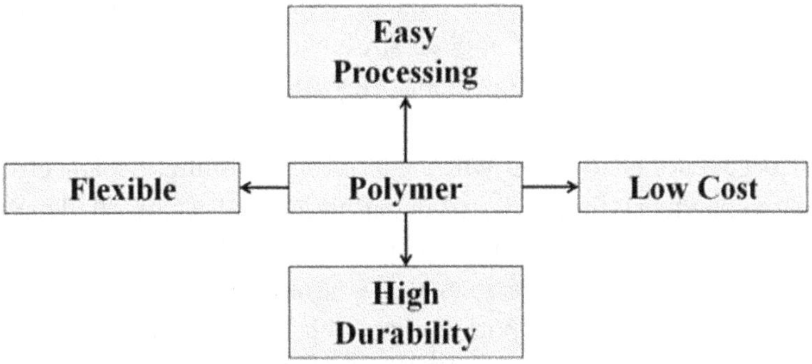

Figure 2. Various classes of elastically yielding electro-active polymers

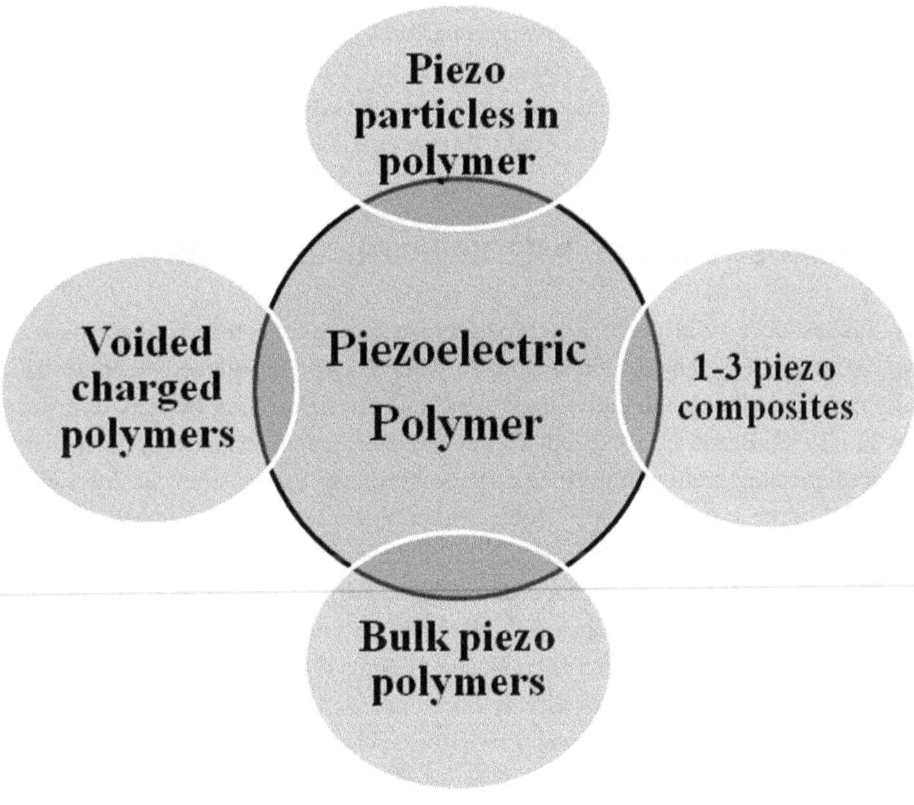

ε), each of which is a characteristic of its distinct chain conformations (Deshmukh et al., 2010). The β-phase of PVDF is the most studied, owing to their intrinsic abilities to show larger charge separation, well beyond other phases and hence increased piezoelectric response (Zhang et al., 2011). In contrast to piezoelectric ceramics that show piezoelectric response upon dipole re-orientation, piezoelectricity in PVDF systems is said to be governed by volume electrostriction, which in turn leads to negative values ofstrain and voltage coefficients in PVDF (Matt et al., 2007). In recent years, consisitent research efforts are focused on diverse strategies to develop electro-active (β) phase of PVDF,either by active flexing of α-phase or by using various filler materials $BaTiO_3$,clay, hydrated ionic salt and so forth.

The stretching of α-phase PVDF under applied stress will lead to an active alignment of polymer chains that allows the electrical dipoles to align themselves in the direction applied stress, thereby leading to increased piezoelectric response. The alternate technique involves the development of PVDF-based copolymers that have been reported to improve the material crystallinities (Zhang et al., 2011; Matt et al., 2007). One such PVDF copolymer is the poly(vinylidenefluoride-co-trifluoroethylene) (PVDF-TrFE), that has been reported with piezoelectric coefficient (38 pC/N) nearly twice that of pristine PVDF systems (Smith et al., 1989). The chemical inertness, increased resistance to hydrolysis and minimal cellular response upon exposure to human tissues support their possible bio-medical applications as active sensing structures (Baek et al., 2012).The limitations of PVDF based systems includes their chemical inertness, lack of ability to form flexible films and low thermal stabilities (Sameoto et al., 2007) in addition to high cost and anisotropic piezoelectric effects (Tiwari et al., 2011; Jean-Mistral et al., 2010; Atkinson et al., 2003; Doring et al., 2012).The intrinsic ability of PVDF based materials to display good combination of both dielectric and piezoelectric properties enabled the realization of PVDF based self-powered wearable electronic devices (Wang et al., 2011; Hillenbrand et al., 2000). It is also effectively involved towards the fabrication of electro-mechanical transducers, actuators and vibrational sensors (Tiwana et al., 2012; Lindahl et al., 2009; Buchberger et al., 2008; Petroni et al., 2012)

POLYMER COMPOSITE BASED PIEZOELECTRIC SYSTEMS

The shortcomings associated with piezoelectric performance of ceramic materials and piezoelectric polymers alone were first addressed by Newnham and co-workers, who introduced piezoelectric composites that essentially combined the distinct advantages of each material arising from synergistic interactions (Newnham et al., 1978; Bernd et al., 2000; Cui et al., 1997; Venkatragavaraj et al., 2001). Further, they observed a relative distinction in piezoelectric properties of composites with particle connectivity patterns, which led to the annotation of piezoelectric composites in "a-b" format, with "a" representing number of dimensions in which the particles are interconnected and "b" representing dimensional inter-connectivity between the matrix. In accordance with "a-b" format, a randomly oriented composite would be designated as 0-3 system, while a pillar like and polymer-ceramic would be addressed as 1-3 and 1-0 respectively. It is this, 1-3 polymer composite systems that are extensively probed in the recent past. A detailed view of some of the extensively investigated 1-3 composite systems and their related properties are presented in the Table below.

Nanocarbon Reinforced Polymer Based Piezoelectric Composites

Carbon nanotubes have been incessantly used to enhance the electrical and piezoelectric performance of polymeric matrices in the recent past. Although, Sreekumar et al., reported considerable enhancement in piezoelectric performance of PVDF via net increase in Young's modulus (Sreekumar et al., 2004; Lan et al., 2005; Salehi-Khojin et al., 2009). More lately, Baur et al. observed a considerable increase (~2X) in piezoelectric coefficient of PVDF composites by using Buckminster fullerenes (C60) and single-walled carbon nanotubes (SWNTs). The figure below (Figure 3 and 4) gives a comparative account on the changes in Young's modulus of PVDF with varying weight fractions of Buckminster fullerenesand

Table 2. Some of the important ceramic/polymer composites of piezoelectric importance

Ceramic Filler	Dopant Volume (%)	Polymeric Host	Piezoelectric Charge Constant (d33)	Reference
calcium modifiedlead titanate	50	Epoxy	22	(Garner, 1989)
lead titanate	70	Acrylic	32	(Hanner, 1989)
lead zirconate titanate	67	PVDF	48	(Takeshi, 1982)
(Pb-Bi)TiO_3	35	Epoxy	41	(Kwonhoon, 1991)

single-walled carbon nanotubes (Cary et al., 2014). As can be seen, the incorporation of 0.05wt% SWNTs enhances the Young's modulus of PVDF by 8.33% from 1.2Gpa to 1.3Gpa and shows a maximum of ~1.35Gpa at 0.1wt%. Any further increase in filler content is however found to have converse effect owing to filler entanglements and non-homogeneous dispersions at higher loadings. The C60 reinforced composites on the other hand permits favorable mixing, thereby facilitating higher loading levels and hence increased moduli, as high as 1.55Gpa.

The TSC curves of PVDF based composite films showed an appreciable increase in amount of current discharged with an increase in C60 and SWNT loadings (Bauer et al., 2003). Further, the TSC curves displayed four prominent peaks (P_1-P_4).The first peak(P_1) corresponds to depolarization of dipoles, while

Figure 3. Variation in Young's Modulus of PVDF-C60 and PVDF-SWNT as a function Offiller content
Source: Cary, 2014

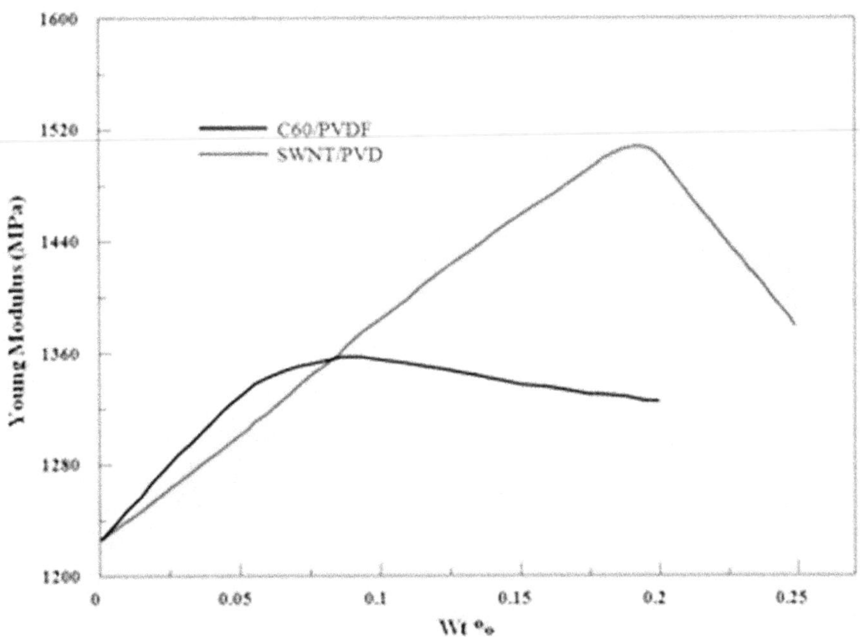

Figure 4. TSC spectrum of PVDF-SWNT and PVDF-C60 system
Source: Cary et al., 2014

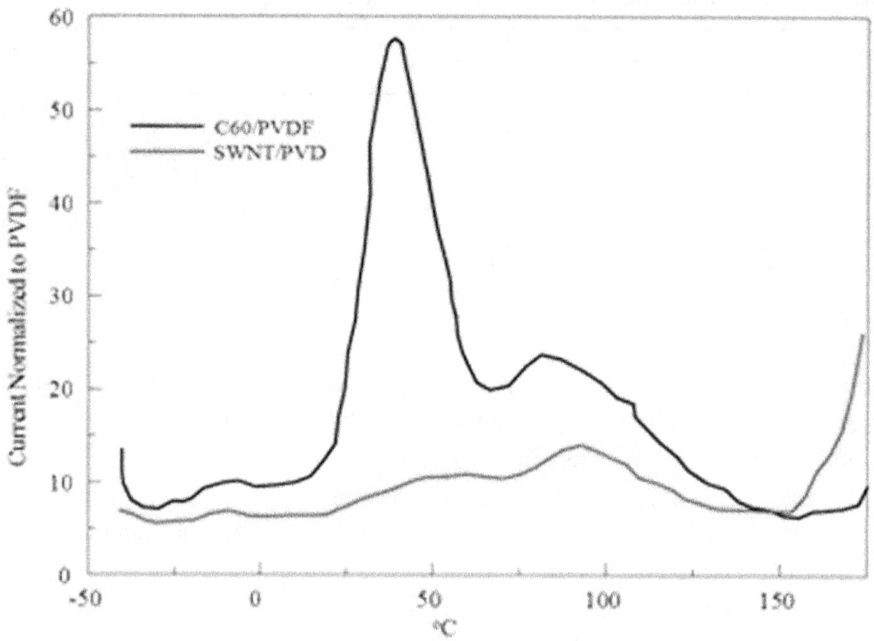

the second(P_2) illustrates the release of subsurface interfacial charge. The C60 molecules reinforced PVDF composites showed a highly intense release of subsurface interfacial charge peak (P_2) centered ~15⁰C, with a magnitude 50 times greater than pristine PVDF, however in case of SWNTs the increase is only marginal (6 times greater).The observed increase may be correlated to symmetric geometry of C60 molecules, that allows the formation of homogeneous composites with near uniformly dispersed interfacial charges that can be easily mobilized through charge hopping. In contrast, the SWNT composites display relatively lower values owing to the increased interaction between the filler materials that lead to low resistance pathways allowing charge indulgence.

EMBLEMATIC DESIGN AND MODELING OF PIEZOELECTRIC ENERGY HARVESTING DEVICES

The relative performance of any piezoelectric energy harnessing device may be predicted via various mathematical modeling and simulation techniques, which assist the further optimization of design parameters and device functionalities to achieve increased performance. In this context, the scientific community is striving to develop novel piezoelectric models to enable piezoelectric response computation.

Figure 5. Pictorial representation of d_{33} bimorph piezoelectric energy harvester

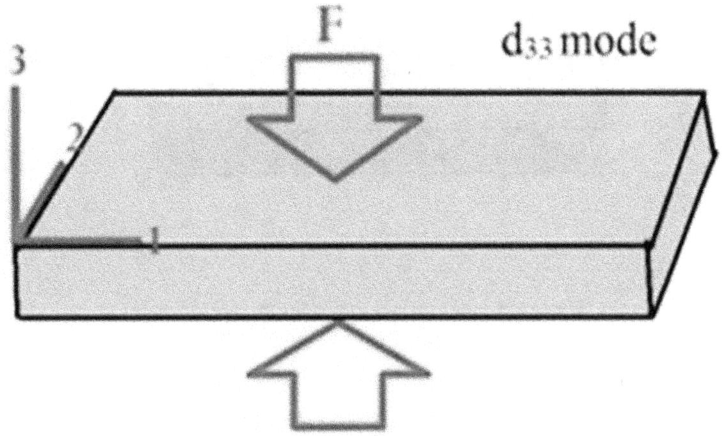

Lumped Parameter Modeling

In the opening stages, lumped parameter modeling that consider only a single mechanical degree of freedom were employed to envisage the coupled systems dynamics of piezoelectric devices (Roundy et al., 2004; Dutoit et al., 2005). It is one of the most convenient technique, which entail only those lumped parameters related to mechanical domain, such that a transformer relation may be built up via coupling of mechanical equilibrium and the electrical loop equations. Although, the lumped parameter modeling makes use of simple expressions, yet it fails to considered quite a few important factors of a coupled system such as dynamic mode shapes, accurate strain distributions and their effects on the electrical response.

Cantilevered Energy Harvesters

In these systems, the excitation occurs by base motion and the lumped parameter harmonic base excitation relation can be directly employed (Meirovitch et al., 2001). In view of the fact that, lumped-parameter models neglected the contribution of distributed mass to the forcing amplitude in the base excitation. An improved model that takes consideration, the vibration mode shapes obtained from Euler–Bernoulli beam theory, along with constitutive equations that offers electric displacement relative to the electrical outputs to vibration mode shape were employed to predict the piezoelectric response (Lu et al., 2003; Chen et al., 2006; Lin et al., 2007). Nevertheless, these approaches also presented several limitations such as missing of resonance phenomenon, misrepresentation of the forces due to the base excitation and so forth (Erturk, 2008). Recent developments in modeling of piezoelectric devices include analytical expressions that take advantage of distributed-parameter electromechanical modeling (Erturk et al., 2008; Erturk et al., 2009).

Vibrational Energy Harvester Design for Polymer Based Composite Systems

The added flexibilities of polymer based composite systems aid their increasing application as piezoelectric energy harvesters involving stretching, compression and vibrational motions. One of the most

Figure 6. Cross sectional representation of piezoelectric energy harvester and forces applied on it (Red arrow = compressive force, blue arrow = tensile force)

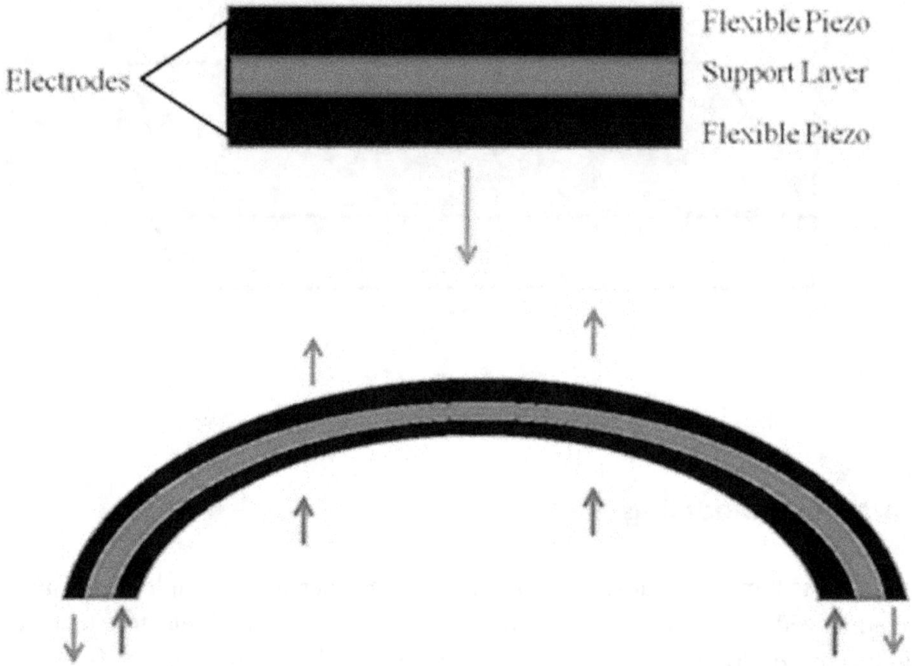

common applications is to harness energy from human activity (Junior et al., 2009), while other applications include microsurgical devices, pacemakers and so forth. The figure below (Figure 5) shows a d_{33} bimorph piezoelectric energy harvester, which essential a three layered structure with the support layer sandwiched between the two layers of flexible piezoelectrics.

One of the major advantages of using polymer based piezoelectric systems is their inherent ability to harness power from both longitudinal (d_{33}) and transverse (d_{31}) modes, in contrast to conventional systems that rely only on transverse vibrations. Further, the added flexibility of polymer based piezoelectrics enable tensile and compressive forces to be applied in all directions, thereby creating additional strain and hence increased charge generation (Figure 6).

APPLICATIONS OF PIEZOELECTRIC POLYMER COMPOSITES

Polymer based composite systems are extensively used in sensing, energy harvesting and actuation applications. The significant advancements in polymer based piezoelectric energy harvesters are tabulated in Table 2. Recent literatures support that platforms with increased surface areas and larger strechability offers maximum power output. Some of the other applications of polymer based piezoelectric systems involve smart clothing, wristwatch, energy generating footwear and so forth. Nevertheless, for efficient energy harvesting, the polymer based systems must overcome the following shortcomings namely;

Table 2. Some of the commercially important polymer based piezoelectric systems and their related applications

Piezoelectric material	Application	Reference
(1–3) PZT/epoxy composite	25 MHz ultrasonic transducer	(Jadidian et al., 2009)
(1–3) PMN–PT/Epoxy composite	2.5 MHz ultrasonic transducer	(Yaoyao et al., 2011)
Cellular PP	Aircoupled ultrasonic transducer	(Doring et al., 2010)
PVDF-PZT	Automobile tyre	(Van den Ende et al., 2011)
PVDF	Backpack strap	(Granstrom et al.,2007)

1. Handling related issues such as torsion and rough handling; and
2. Temperature related problems such as decrease in piezoelectric response with temperature.

The future energy generation technologies using polymer based systems may involve a shift from mobile environments to stationary structures, which could produce considerable amounts of power upon integration to door closers; door push bars, window blinds and so forth. The devices constituted by such materials may conveniently be employed as power switches for lighting and vibrational energy sensing applications.

Piezoelectric Tactile Sensors

Tactile sensors are those analytical devices that measures physical phenomenon such as contact and/or tough (Tiwana et al., 2012) via changes in temperature, shape or pressure. Such type of electro-active materials is of great interest in robotics and medicine (Lindahl et al., 2009). As tactile sensors present a combinatorial effect of sensing via changes in pressure and temperature, polymer based piezoelectric systems combined with ferroelectric polymers are of significant interest in tactile sensors. The combination of piezoelectric and pyroelectric phenomenon in a single system may often lead to complexity in reliable measurement of indivual phenomenon (Ingrid, 2009). This complexity may however be addressed through selective poling as in case of ceramic/PVDF systems. Piezoelectric tactile sensors always prefer polymers over inorganic systems owing to their ease of processing and added mechanical flexibilities. Some of the commonly employed.

Further, when inorganic materials are used in tactile sensors, they are either used as composites with polymers (Ingrid et al., 2009) or as thin films deposited over flexible substrates (Petroni et al., 2012; Akiyama et al., 2007). One of the most common and widely employed material as tactile sensor is the PVDF–TrFE (trifluoroethylene) system. Most of the tactile sensors are two plate capacitor units with piezoelectric polymer as the dielectric layer. Some of the other polymer based tactile sensors include nanograss PVDF with sensitivities 2.8 times higher than conventional flat surfaces (Chen et al., 2012), additionally Cellular Polypropylene films are also employed in touchpad large area sensors.

Vibration Energy Harvesters

Vibrational energy harvesting is one of effective means of realizing self powered electronic systems. An ideal vibrational energy harvester is one that presents highest power harvesting abilities with smallest area at optimal vibrational frequency. Such energy harvesting devices demand materials with high coupling co-efficient such as the ceramic materials. Nevertheless, the advantages of using polymer based systems over ceramic materials in their added mechanical flexibility, which enhances device life time by preventing fatigue (Kim et al., 2011). Energy harvesting generators are attractive as boundless replacements for batteries in low-power wireless electronic devices and have acknowledged increasing research interest in recent years. Ambient motion is one of the main sources of energy for harvesting, and a wide range of motion- powered energy harvesters have been anticipated or demonstrated, particularly at the micro scale. Human power harnessing is the next great entity. The human body is one of the most resourceful stockrooms of energy, and it allots the parts of the stockpiled energy during daily movement. Some researchers have predicted that the average power disbursed throughout various daily activities, shown in Table 3 (80 to 1630 watts), as well as the extent of energy stockpiled in an ordinary human being, depicted in Figure 7.The development of self powered wearable devices by harvesting ambient vibrations from human motion was achieved using $BaTiO_3$ NP/MWCNT/PDMS based composite sheets that generated an open circuit voltage of nearly 3.2 V and short circuit current of 250–350 nA under periodical low frequency vibrational tapping (Kwi et al., 2012).The figure (Figure 8) below depicts piezoelectric energy generation under mechanical tapping during walking. The polymer composite based piezoelectric systems extremely favorable for low frequency vibration energy harvesting.

A perfect vibrational energy harvester demands a synergistic balance between high coupling coefficient and mechanical flexibility, as in case of polymer composites. One such vibration energy harvester PZT/Epoxy composite developed by Patel et al. (Patel et al., 2010) was reported to have generated voltage, relatively higher than the conventional PVDF and cellular PP, while cantilevers developed with ZnO reinforced SU8 nanocomposite systems were reported to generate a power of 0.025μW, when connected to $100\times10^3\Omega$ with an operating frequency of 4KHz. Furthermore, to enhance the energy harvesting abilities, novel design approaches were proposed, that involves power generation via broad band frequencies in place of a single resonant frequency. One such systems involving PVDF based material was demonstrated by Jo et al., which displayed a tunable resonant frequency broadening.

Table 3. Energy associated with human activities

Nature of Human Activity	Associated Energy (Kilocal/hr)	Associated Power(Watts)
Snoozing	60	70
Hill-walking 3-5 mph.	300	400
Swimming	400	600
mountaineering	500	700
Far-reaching run	800	1050
Sprinting	1350	1650

Figure 7. Pictorial representation of energy associated with human activities

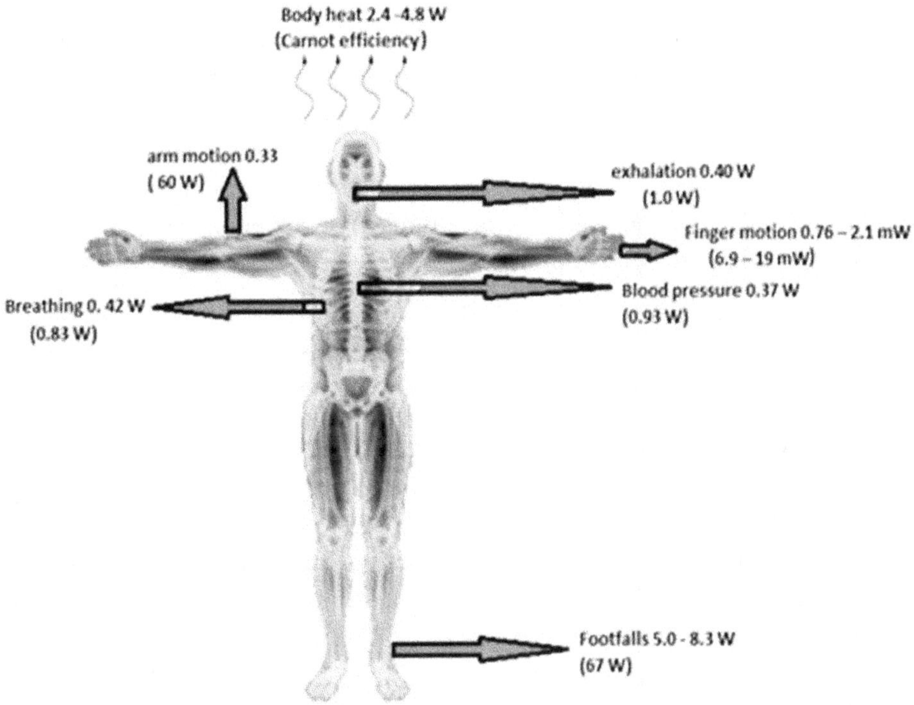

Figure 8. Pictorial depiction of piezoelectric energy harvesting using human footsteps

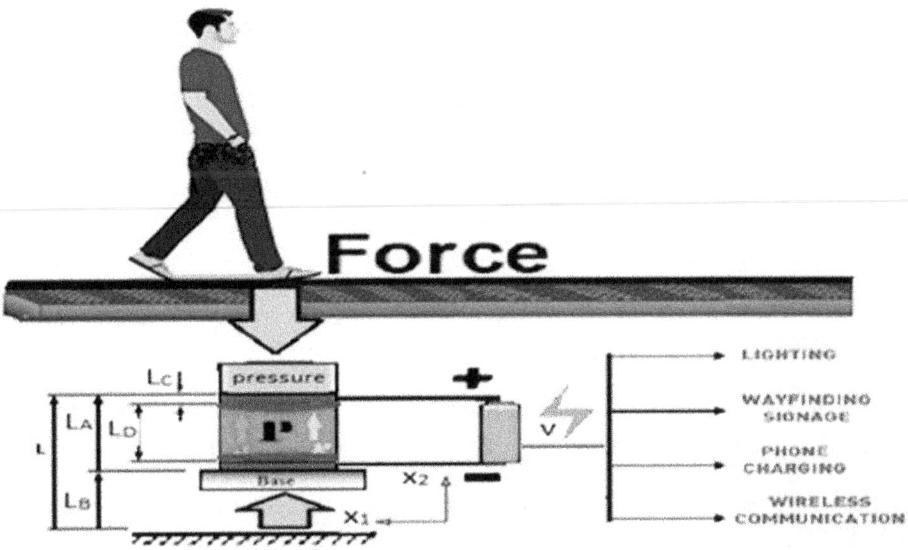

Acoustic Transducers

Acoustic transducers are those electro-mechanical devices that maneuver in the acoustic spectral range that involves audio, ultrasonic and infrasonic frequencies (Rossing et al., 2007). Acoustic emission sensors (AES) are those analytical devices that sense the acoustic signals via structural changes in the system employed. Acoustic emission sensors that sense acoustic signal passively are used in those applications that ask for continuous monitoring. While, ultrasonic transducers that sense structural changes through reflected signals are commonly employed in medical imaging and automated machinery (Zhou et al., 2011) applications.

CONCLUSION

With ever increasing growth in wireless technology and device connectivity, there exists significant demand for novel hybrid systems as self-powered wearable devices with increased sensing and energy harvesting abilities. Although piezoelectric ceramic materials offer potential power outputs, yet they suffer from shortcomings such as increased brittleness and decreased vibrational sensitivity. Piezoelectric polymers on the other hand suffered from low piezoelectric response and thermal stabilities. The advent of composite material technology has enabled the scientific community to realize polymer based piezoelectric composites that essentially offers combinatorial properties that essentially perform beyond their individual components. Nevertheless, in recent years the fabrication of mechanically stable large area and high power output nanocomposites on flexible plastic substrates is gaining increasing technological significance with several research groups fabricating lead-free piezoelectric harvesters using titanates and niobates. Further, the nanocomposite based piezoelectric generators are expected to improvise by further espousing nanotechnology enabled controlled synthesis of lead free piezoelectric ceramic systems with desired morphologies in addition to employing nanostructured metal reinforced conducting polymers alongside novel printing techniques may open up the technological gateway towards the development of environmentally benign self-powered electronic gadgets.

REFERENCES

Atzori, L., Iera, A., & Morabito, G. (2010). The internet of things: A survey. *Computer Networks*, *54*(15), 2787–2805. doi:10.1016/j.comnet.2010.05.010

Henry, A. S., Daniel, J. I., & Park, G. (2004). A review of power harvesting from vibration using piezoelectric materials. *Shock and Vibration Digest*, *36*(3), 197–206. doi:10.1177/0583102404043275

Williams, C. B., & Rob, B. Y. (1996). Analysis of a micro-electric generator for microsystems. *Sensors and Actuators. A, Physical*, *52*(1-3), 8–11. doi:10.1016/0924-4247(96)80118-X

Priya, S. (2007). Advances in energy harvesting using low profile piezoelectric transducers. *Journal of Electroceramics*, *19*(1), 167–184. doi:10.1007/s10832-007-9043-4

Henry, A. S., Inman, D. J., & Park, G. (2004). A review of power harvesting from vibration using piezoelectric materials. *Shock and Vibration Digest*, *36*(3), 197–206. doi:10.1177/0583102404043275

Cook-Chennault, K. A., Thambi, N., & Sastry, A. M. (2008). Powering MEMS portable devices a review of non-regenerative and regenerative power supply systems with special emphasis on piezoelectric energy harvesting systems. *Smart Materials and Structures*, *17*(4), 043001. doi:10.1088/0964-1726/17/4/043001

Ballato, A. (1995). Piezoelectricity: Old effect, new thrusts. *IEEE Transactions on Ultrasonics, Ferroelectrics, and Frequency Control*, *42*(5), 916–926. doi:10.1109/58.464826

Jaffe, B., Cook, W. R., & Jaffe, H. (1971). *Piezoelectric ceramics*. Academic.

Cady, W. G. (1946). *Piezoelectricity: An introduction to the theory and application of electromechanical phenomena in crystals*. McGraw-Hill.

Junhee, Loh, & L. (2008). Piezoelectric polymeric thin films tuned by carbon nanotube fillers. In *The 15th International Symposium on: Smart Structures and Materials &Nondestructive Evaluation and Health Monitoring*. International Society for Optics and Photonics.

Anton, Steven, & Henry. (2007). A review of power harvesting using piezoelectric materials (2003–2006). *Smart Materials and Structures*, *16*(3), R1.

Yu, Y., Zhang, X. N., & Xie, S. L. (2009). Optimal shape control of a beam using piezoelectric actuators with low control voltage. *Smart Materials and Structures*, *18*(9), 095006. doi:10.1088/0964-1726/18/9/095006

Auld, B. A. (1973). *Acoustic fields and waves in solids*. Рипол Классик.

Priya, S., & Inman, D. J. (Eds.). (2009). Energy harvesting technologies. New York: Springer.

Harrison, J. S., & Ounaies, Z. (2002). *Polymers, piezoelectric*. John Wiley & Sons, Inc.

Deshmukh, S., & Ounaies, Z. (2010). Active single walled carbon nanotube -polymer composites. *IUTAM Symposium on Multi-Functional Material Structures and System*, 103-110. doi:10.1007/978-90-481-3771-8_11

Zhang, Y., Wang, S., Liu, D., Zhang, Q., Wang, W., Ren, B., & Luo, H. et al. (2011). Fabrication of angle beam two-element ultrasonic transducers with PMN–PT single crystal and PMN–PT/epoxy 1–3 composite for NDE applications. *Sensors and Actuators. A, Physical*, *168*(1), 223–228. doi:10.1016/j.sna.2011.04.004

Matt, H. M., & Scalea, F. L. (2007). Macro-fiber composite piezoelectric rosettes for acoustic source location in complex structures. *Smart Materials and Structures*, *16*(4), 1489–1499. doi:10.1088/0964-1726/16/4/064

Smith, & Wallace, A. (1989). The role of piezocomposites in ultrasonic transducers. *Ultrasonics Symposium, 1989. Proceedings, IEEE*, 755-766.

Baek, S., Rzchowski, M. S., & Aksyuk, V. A. (2012). Giant piezoelectricity in PMN-PT thin films: Beyond PZT. *MRS Bulletin*, *37*(11), 1022–1029. doi:10.1557/mrs.2012.266

Sameoto, D., See-Ho, T., & Parameswaran, M. (2007). Polymer MEMS processing for multi-user applications. *Sensors and Actuators A: Physical, 134*(2), 457-464.

Tiwari, R., & Garcia, E. (2011). The state of understanding of ionic polymer metal composite architecture: A review. *Smart Materials and Structures*, *20*(8), 083001. doi:10.1088/0964-1726/20/8/083001

Jean-Mistral, C., Basrour, S., & Chaillout, J. J. (2010). Comparison of electroactive polymers for energy scavenging applications. *Smart Materials and Structures*, *19*(8), 085012. doi:10.1088/0964-1726/19/8/085012

Atkinson, G. M., Pearson, R. E., Ounaies, Z., Park, C., Harrison, J. S., & Midkiff, J. A. (2003). Piezoelectric polyimide tactile sensors. *Proc. 15th Biennial University/Government/ Industry Microelectronics Symp*, 308-311. doi:10.1109/UGIM.2003.1225750

Döring, J., Bovtun, V., Gaal, M., Bartusch, J., Erhard, A., Kreutzbruck, M., & Yakymenko, Y. (2012). Piezoelectric and electrostrictive effects in ferroelectret ultrasonic transducers. *Journal of Applied Physics*, *112*(8), 084505. doi:10.1063/1.4759052

Wang, J.-J., Hsu, T.-H., Yeh, C.-N., Tsai, J.-W., & Su, Y.-C. (2011). Piezoelectric polydimethylsiloxane films for MEMS transducers. *Journal of Micromechanics and Microengineering*, *22*(1), 015013. doi:10.1088/0960-1317/22/1/015013

Hillenbrand, J., & Sessler, G. M. (2000). Piezoelectricity in cellular electret films. *IEEE Transactions on Dielectrics and Electrical Insulation*, *7*(4), 537–542. doi:10.1109/94.868074

Tiwana, M., Redmond, S. J., & Lovell, N. H. (2012). A review of tactile sensing technologies with applications in biomedical engineering. *Sensors and Actuators. A, Physical*, *179*, 17–31. doi:10.1016/j.sna.2012.02.051

Lindahl, O. A., Constantinou, C. E., Anders, E., Yoshinobu, M., & Hallberg, P. (2009). Tactile resonance sensors in medicine. *Journal of Medical Engineering & Technology*, *33*(4), 263–273. doi:10.1080/03091900802491188 PMID:19384701

Buchberger, G., Schwödiauer, R., & Bauer, S. (2008). Flexible large area ferroelectret sensors for location sensitive touchpads. *Applied Physics Letters*, *92*(12), 123511. doi:10.1063/1.2903711

Petroni, S., Guido, F., Torre, B., Falqui, A., Todaro, M., Cingolani, R., &DeVittorio, M. (2012). Tactile multisensing on flexible aluminum nitride. *Royal Society of Chemistry, 137*(22), 5260-5264.

Newnham, R. E., Skinner, D. P., & Cross, L. E. (1978). Connectivity and piezoelectric-pyroelectric composites. *Materials Research Bulletin*, *13*(5), 525–536. doi:10.1016/0025-5408(78)90161-7

Bernd, P., Beatrix, P., Shin, F. G., Chan, H. L. W., & Choy, C. L. (2000). Pyroelectric activity of ferroelectric PT/PVDF-TRFE. *IEEE Transactions on Dielectrics and Electrical Insulation*, *7*(4), 517–522. doi:10.1109/94.868071

Cui, C., Baughman, R. H., Iqbal, Z. T., Kazmar, R., & Dahlstrom, D. K. (1997). Improved piezoelectric ceramic/polymer composites for hydrophone applications. *Synthetic Metals*, *85*(1), 1391–1392. doi:10.1016/S0379-6779(97)80287-1

Venkatragavaraj, E., Satish, B., Vinod, P. R., & Vijaya, M. S. (2001). Piezoelectric properties of ferroelectric PZT-polymer composites. *Journal of Physics. D, Applied Physics*, *34*(4), 487–492. doi:10.1088/0022-3727/34/4/308

Garner, G. M., Shorrocks, N. M., Whatmore, R. W., Goosey, M. T., Seth, P., & Ainger, F. W. (1989). 0–3 piezoelectric composites for large area hydrophones. *Ferroelectrics*, *93*(1), 169–176. doi:10.1080/00150198908017341

Hanner, K. A., Safari, A., Newnham, R. E., & Runt, J. (1989). Thin film 0–3 polymer/piezoelectric ceramic composites: Piezoelectric paints. *Ferroelectrics*, *100*(1), 255–260. doi:10.1080/00150198908007920

Takeshi, Y., Toshinobu, U., & Toyoki, K. (1982). Piezoelectricity of a high-content lead zirconatetitanate/polymer composite. *Journal of Applied Physics*, *53*(6), 4328–4332. doi:10.1063/1.331211

Kwonhoon, H., Ahmad, S., & Richard, E. R. (1991). Colloidal Processing for Improved Piezoelectric Properties of Flexible 0–3 Ceramic–Polymer Composites. *Journal of the American Ceramic Society*, *74*(7), 1699–1702. doi:10.1111/j.1151-2916.1991.tb07165.x

Sreekumar, T. V., Tao, L. M., Guo, B. G., Kumar, H., Hauge, S., & Smalley, R. H. (2004). R.E. *Advanced Materials*, *16*(58).

Lan, W., & Zhi-Min, D. (2005). Carbon nanotube composites with high dielectric constant at low percolation threshold. *Applied Physics Letters*, *87*(4), 042903. doi:10.1063/1.1996842

Salehi-Khojin, A., Hosseini, M. R., & Jalili, N. (2009). Underlying mechanics of active nanocomposites with tunable properties. *Composites Science and Technology*, *69*(3), 545–552. doi:10.1016/j.compscitech.2008.11.033

Cary, B., Daniel, J. A., Deepam, M., Shashank, P., & Walter, V. (2014). Advances in Piezoelectric Polymer Composites for Vibrational Energy Harvesting. In Polymer Composites for Energy Harvesting, Conversion, and Storage. American Chemical Society.

Bauer, S., Bauer-Gogonea, S., Dansachmuller, M., Graz, I., Leonhartsberger, H., Salhofer, H., & Schwodiauer, R. (2003). Modern electrets. *Ultrasonics, IEEE Symposium on, IEEE,* (1), 370-376.

Roundy, S., & Wright, P. K. (2004). A piezoelectric vibration based generator for wireless electronics. *Smart Materials and Structures*, *13*(5), 1131–1142. doi:10.1088/0964-1726/13/5/018

Dutoit, N. E., Wardle, B. L., & Kim, S. G. (2005). Design considerations for MEMS-scale piezoelectric mechanical vibration energy harvesters. *Integrated Ferroelectrics*, *71*(1), 121–160. doi:10.1080/10584580590964574

Meirovitch, L., & Parker, R. G. (2001). Fundamentals of vibrations. *Applied Mechanics Reviews*, *54*(6), B100–B101. doi:10.1115/1.1421112

Lu, F., Lee, H. P., & Lim, S. P. (2003). Modeling and analysis of micro piezoelectric power generators for micro-electromechanical-systems applications. *Smart Materials and Structures*, *13*(1), 57–63. doi:10.1088/0964-1726/13/1/007

Chen, S. N., Wang, G. J., & Chien, M. C. (2006). Analytical modeling of piezoelectric vibration-induced micro power generator. *Mechatronics*, *16*(7), 379–387. doi:10.1016/j.mechatronics.2006.03.003

Lin, J. H., Wu, X. M., Ren, T. L., & Liu, L. T. (2007). Modeling and simulation of piezoelectric MEMS energy harvesting device. *Integrated Ferroelectrics, Taylor & Francis*, *95*(1), 128–141. doi:10.1080/10584580701756649

Erturk, A., & Inman, D. J. (2008). Issues in mathematical modeling of piezoelectric energy harvesters. *Smart Materials and Structures*, *17*(6), 065016. doi:10.1088/0964-1726/17/6/065016

Erturk, A., & Inman, D. J. (2008). A distributed parameter electromechanical model for cantilevered piezoelectric energy harvesters. *Journal of Vibration and Acoustics*, *130*(4), 041002. doi:10.1115/1.2890402

Erturk, A., & Inman, D. J. (2009). An experimentally validated bimorph cantilever model for piezoelectric energy harvesting from base excitations. *Smart Materials and Structures*, *18*(2), 025009. doi:10.1088/0964-1726/18/2/025009

Junior, C. D. M., Erturk, A., & Inman, D. J. (2009). An electromechanical finite element model for piezoelectric energy harvester plates. *Journal of Sound and Vibration*, *327*(1), 9–25. doi:10.1016/j.jsv.2009.05.015

Jadidian, B., Hagh, N. M., Winder, A. A., & Ahmad, S. (2009). 25 MHz ultrasonic transducers with lead-free piezoceramic, 13 PZT fiber-epoxy composite, and PVDF polymer active elements. *IEEE Transactions on Ultrasonics, Ferroelectrics, and Frequency Control*, *56*(2), 368–378. doi:10.1109/TUFFC.2009.1046 PMID:19251524

Yaoyao, Z., & Sheng, W. (2011). Fabrication of angle beam two-element ultrasonic transducers with PMN–PT single crystal and PMN–PT/epoxy 1–3 composite for NDE applications. *Sensors and Actuators. A, Physical*, *168*(1), 223–228. doi:10.1016/j.sna.2011.04.004

Doring, J., Bovtun, V., Bartusch, J., Erhard, A., Kreutzbruck, M., & Yakimenko, Y. (2010). Nonlinear electromechanical response of the ferroelectret ultrasonic transducers. *Applied Physics. A, Materials Science & Processing*, *100*(2), 479–485. doi:10.1007/s00339-010-5752-7

Van den Ende, D. A., Van de Wiel, H. J., Groen, W. A., & Van der Zwaag, S. (2011). Direct strain energy harvesting in automobile tires using piezoelectric PZT–polymer composites. *Smart Materials and Structures*, *21*(1), 015011. doi:10.1088/0964-1726/21/1/015011

Granstrom, J., Feenstra, J., Sodano, H. A., & Farinholt, K. (2007). Energy harvesting from a backpack instrumented with piezoelectric shoulder straps. *Smart Materials and Structures*, *16*(5), 1810–1820. doi:10.1088/0964-1726/16/5/036

Ingrid, G., Markus, K., Simona, B. G., Siegfried, B., Stephanie, P. L., Bernd, P., & Sigurd, W. (2009). Flexible active-matrix cells with selectively poled bifunctional polymer-ceramic nanocomposite for pressure and temperature sensing skin. *Journal of Applied Physics*, *106*(3), 034503. doi:10.1063/1.3191677

Petroni, S., Guido, F., Torre, B., Falqui, A., Todaro, M., Cingolani, R., &DeVittorio, M. (2012). Tactile multisensing on flexible aluminum nitride. *Royal Society of Chemistry, 137*(22), 5260-5264.

Akiyama, M., Yukari, M., Toshihiro, K., Keiko, N., Yasunobu, O., Masayoshi, T., & Naohiro, U. (2007). Preparation of oriented aluminum nitride thin films on polyimide films and piezoelectric response with high thermal stability and flexibility. *Advanced Functional Materials*, *17*(3), 458–462. doi:10.1002/adfm.200600098

Chen, A., Kai-Lun, L., Chien-Chong, H., Tong, M. L., Jiann, S., & Szu, H. C. (2012). Flexible tactile sensors based on nanoimprinted sub-20 NM piezoelectric copolymer nanograssfilms. In *Sensors* (pp. 1–4). IEEE.

Kim, H. S., Kim, J. H., & Jaehwan, K. (2011). A review of piezoelectric energy harvesting based on vibration. *International Journal of Precision Engineering and Manufacturing, 12*(6), 1129-1141.

Kwi-Il, P., Minbaek, L., Ying, L., San, M., Geon, T. H., Guang, Z., & Keon, J. L. (2012). Flexible nanocomposite generator made of BaTiO3 nanoparticles and graphitic carbons. *Advanced Materials*, *24*(22), 2999–3004. doi:10.1002/adma.201200105 PMID:22549998

Patel, I., Siores, E., & Shah, T. (2010). Utilisation of smart polymers and ceramic based piezoelectric materials for scavenging wasted energy. *Sensors and Actuators. A, Physical*, *159*(2), 213–218. doi:10.1016/j.sna.2010.03.022

Jo, S. E., Kim, M. S., & Kim, Y. J. (2011). A resonant frequency switching scheme of a cantilever based on polyvinylidene fluoride for vibration energy harvesting. *Smart Materials and Structures*, *21*(1), 015007. doi:10.1088/0964-1726/21/1/015007

Rossing, T. D. (2007). *Springer Handbook of Acoustics*. Berlin: Springer. doi:10.1007/978-0-387-30425-0

Zhou, Q., Lau, S., Wu, D., & Shung, K. K. (2011). Piezoelectric films for high frequency ultrasonic transducers in biomedical applications. *Progress in Materials Science*, *56*(2), 139–174. doi:10.1016/j.pmatsci.2010.09.001 PMID:21720451

Chapter 7
Effect of Cenosphere on the Performance of Polyurethane/Polystyrene Interpenetrating Polymer Network Green Composites

Roopa S.
Sri Jayachamarajendra College of Engineering, India & JSS Science and Technology University, India

Siddaramaiah
Sri Jayachamarajendra College of Engineering, India

ABSTRACT

The effect of cenosphere content on the performances of polyurethane/polystyrene (PU/PS, 90/10) interpenetrating polymer network (IPN) based green composites have been studied. The PU/PS IPNs have been prepared using castor oil, toluene diisocyanate and styrene. IPN/cenosphere composites have been prepared with different weight fractions viz., 0, 5, 10, 20 and 30 wt % of cenosphere. The prepared IPN composites have been characterized by physico – mechanical, chemical and thermal behavior. The tensile strength of unfilled IPN was 1.79 MPa and a significant improvement in tensile strength (34%) was noticed for 10% cenosphere loaded IPN composite. The swelling behavior of the composites has been studied in different organic solvents. Thermal characteristics of the composites have been measured using differential scanning calorimeter, thermogravimetric analysis and dynamic mechanical analysis (DMA). A slight improvement in thermal stability was noticed for filler loaded specimens. Morphological features of cryo-fractured IPN/cenosphere green composites have been analyzed using SEM.

DOI: 10.4018/978-1-5225-3023-7.ch007

INTRODUCTION

An interpenetrating polymer network (IPN), can be defined as a combination of two polymers in network form, at least one of which is synthesized and/or crosslinked in the immediate presence of the other. An IPN can be distinguished from simple polymer blends, blocks and grafts in two ways: (a) an IPN swells, but does not dissolve in solvents and (b) creep and flow are suppressed (Sperling, 1981). They have been among the fastest growing areas in the field of blends during the past twenty years.

During the formation of IPNs, the polymer networks are interlocked permanently, which shows the better thermodynamic compatibility of the two polymers that results in fine disperse microphase-separated morphologies. The majority of the IPNs except homo-IPNs, undergo phase separation, which is essential for the development of newer materials with good properties. Due to excellent compatibility of the two polymers, IPNs exhibit enhanced mechanical strength, broad frequency range of damping, high conductivity and stimuli-responsive behavior as compared to polymer blends, alloys, graft and block copolymers. In addition, IPNs can keep the separate phases together when they are subjected to stress (Sperling, 1981; Kim, 2007; Wu 2007; James, 2016).

Over the past decade polyurethane (PUs) the 6th most used classes of heteropolymer has gained tremendous attention in several applications like coatings, adhesives, furniture, foams electrical/electronic potting and encapsulation, construction, water proofing membranes, asphalt extended membranes, highway sealants, sound and vibration damping, automotive and rubber parts, etc., due to their unique properties i.e., it offers the elasticity of rubber combined with the toughness and durability of metals, owing to their exceptional mechanical properties and impeccable chemical resistance and scratch resistance (Krol, 2007; Zhang, 2012; Houton, 2015; Li, 2015; Rokicki, 2015; Das, 2017). Despite the substantial benefits of PU, it even presents some drawbacks, including poor degradability and toxicity due to the use of isocyanates, which have evoked researchers to get more environmentally friendly starting materials. Accordingly, there have been recent extensive research interests in producing bio-based polyols and PUs from renewable resources (Mariano, 2014). The physical properties of PUs are derived from their molecular structure, as well as the supramolecular structure caused by interaction between the polymer chains. The segmental flexibility, the chain entanglement, the interchain forces and the crosslinking influence on the properties and they determine the use of the end products. Mainly two routes can adjust the properties of the PUs. The first route is based on the chemistry of PUs, formulating the PU based on different isocyanate/polyol ratio and using different amounts of chain extender. The second route is altering the properties of the PUs with different fillers and reinforcements (Zhang, 2003). The addition of reinforcing filler particles into PUs greatly increases their versatility by extending the range of mechanical properties that can be achieved and possibly incorporating further functionalities.

Fillers play an important role in modifying the properties of various polymers. The properties of composites were affected by fillers, their concentration, nature, particle size, shape and interaction with the matrix. Azizah et al (2015) synthesized PU composites filled with the activated carbons by a chemical foaming method to investigate their complex permittivity and the microwave absorption properties for use in electromagnetic interference (EMI) shielding. They have reported that, compared with conventional materials like polyethylene (PE) and polyester filled with metal additives, PU composite is suitable for microwave absorption and is a potential candidate for EMI shielding applications. Chen and coworkers studied the static and dynamic mechanical properties of blocked PU and vinyl ester simultaneous full IPN filled with kaolin (1999). The results revealed that the hardness, increased with increasing filler loading whereas tensile strength and flexural strength of IPNs reached to maximum at 20% filler loading and then decreased. The notched izod impact strength of IPNs reduced with increasing kaoline content. The DMA thermograms showed excellent compatibility of PU/vinyl ester IPNs and with increase in kaoline content, the T_g of the filled IPNs was shifted to a higher temperature. Recently cellulose nano whiskers have been used as reinforcing filler in PU synthesis and improvements of both thermal and mechanical properties have been reported (Saba, 2014; Aomar 2016).

Lipatov et al (1988) studied the effect of fillers on viscoelastic properties and compatibility of components of PU/polyacrylate IPNs by DMA and the vapor sorption of organic solvents. It has been found that, the incorporation of filler during IPN formation will affect on the interfacial adhesion of the components. It was also noticed that, the extent of segregation is dependent on the nature of filler surface.

The effect of silica filler on mechanical, thermal, dynamic mechanical and morphological behaviors of simultaneous IPN of PU/poly(ethyl methacrylate) was studied by Trakulsujaritchok et al (2006). The addition of silica filler reduces the T_g of the IPNs. The tan δ values and nature of thermogram is dependent on filler content. The effect of gaurgum on physico-mechanical, free volume and nanocrystalline behaviors of PU/polyacrylonitrile (PU/PAN) IPNs has been reported by Kumar et al (2007, 2008). It was found that, the tensile strength reached to maximum when the gaur gum content is 20% and then decreased. The effect of filler on free volume fraction was estimated from positron annihilation lifetime spectroscopy (PALS) and the author reported that, the addition of gaurgum upto 20%, the fractional free volume of the IPNs increased. The addition of gaurgum also affect on the nanocrystallite, which intern reduces the rate of biodegradation process.

Electrical conductivity, mechanical and morphological behaviors of conducting polyaniline/PU–silica hybrid IPNs was reported by Jing Luo and coworkers (2007). The prepared hybrid IPNs shows improvement in mechanical strength and conductivity. The surface resistivity of the 10% polyaniline loaded hybrid IPN is 10^8 Ω and this formulation is suitable for antistatic application. Wang et al (2010) reported the damping properties, thermal stability, tensile and impact properties of short carbon fiber (CF) and micro hollow glass bead (HGB) loaded PU/epoxy IPN composites. Tang et al have studied on nanocomposites of modified PU and unsaturated polyester (UP) resin IPNs with $BaTiO_3$ nano powders and fibers (2002a, 2002b, 2002c, 2003a, 2003b, 2004, 2007). The polymerization processes were traced through IR techniques. The parameters such as hard and soft block ratio, crosslink density, physical nature and amount of filler influences on the mechanical properties. The addition of fillers from renewable resources to flexible polyurethane foams (FPFs) shows the improvement in thermal and mechanical properties (Anna, 2016).

Nanocomposites with varying contents of organophilic montmorillonite (o-MMT) were prepared by intercalating o-MMT with IPNs of PU and epoxy resin (PU/EP) (Jia, 2005, 2006a, 2006b, 2006c, 2007, 2008). The PU/epoxy nanocomposites were characterized by FTIR, SEM, TEM, WAXS, water absorption and tensile behaviors. The addition of o-MMT into PU/EP IPN matrix provides two fold benefit to the properties of the IPN nanocomposites. PALS and DSC measurements were used to investigate the effect of o-MMT and PU compositions on free volume and T_g of these nanocomposites. IPNs of PU/EP nanocomposites with varying amounts of nanosized silicon dioxide (SiO_2) particles were prepared and evaluated by DMA, SEM, WAXS and SAXS (Zhang, 2003). The result showed that the addition of nanosize SiO_2 improves the damping and phase structure of IPN matrices. An organophilic palygorskite (o-PGS) was incorporated into PU/EP IPN and a series of PU/EP/clay nanocomposites were obtained by a sequential polymeric technique (Lei, 2008). From XRD and SEM analysis, it was found that, the addition of nanosize o-PGS promote the compatibility between PU and epoxy component in IPN. Due to the special fibrillar structure of palygorskite and the synergistic effect on the mechanical and thermal properties of PU/EP nanocomposites were superior as compared to that of pristine PU/EP IPN. The effect of carbon nano fiber loading and the structure of soy polyols with different hydroxyl value on the thermal and mechanical properties of soy polyol-based PU nanocomposites have been studied by Weijing et al (2015). The morphology, thermal, and mechanical properties of the resulting bio-based-PU/CNF nanocomposites were then investigated by using SEM and TA techniques.

Very recently, Das et al (2017) have investigated on the biodegradability of neat PU and PU/nanosilica green composite obtained from transesterified castor oil and isocyanate derived from palm oil under composting condition for an exposure period of 90 days. The biodegradation of the films have been investigated in terms of change in structural, morphological, weight loss and thermal properties. The results indicate that a faster rate of degradation has been observed in case of PU/nanosilica green composite owing to its surface hydrophobicity, lower crystallinity and the presence of surface -OH groups that facilitates the attack of microorganisms. The IPN technology has been the subject of intense research in both academic and industrial aspects. IPN technology is simple, flexible and inexpensive method to generate newer materials with tailor made properties by changing the composition and morphology of the components. There is an ample scope to continue research on the modifications of IPNs using different reinforcing fillers to meet the demand of newer applications in many sectors.

Cenosphere is a type of hollow ceramic microsphere found in the fly ash by-product created through the burning of coal at power stations (Gupta, 2001). Its cost is primarily associated with cleaning and transportation, it possesses very good mechanical properties on account of its constituents like silica and alumina. Cenosphere particles composed of crystalline compounds of quartz, mullite, hematite and glassy compounds (silica and other metal oxides). The major source of electricity generation in India (about 61.5% of total installed power capacity) is coal based thermal power plants. This result in the production of 184 million tons of coal ash and the utilization of coal ash is only about 55% (CEA, 2015). Disposal and management of flyash are a serious problem as it requires a huge area of precious land and also pollutes air, water and soil. Hence, there is a tremendous scope for utilization of cenosphere as a starting raw material or as a filler to develop newer composite materials.

In their primary application, flyash is mixed with cement to decrease the mixture density. Furthermore, flyash mixed cement has exhibited improved insulation and better vibration damping properties (Daoud, 2007; Yilmaz, 2008). Use of flyash particles as filler in metal- matrix composites is also expanding (Bienias, 2003; Rohatgi, 2006; Daoud, 2007). The yield strength and Young's modulus increased and elongation decreases as an increase in the fly ash volume fraction in polypropylene (PP) matrix (Liang, 2006; 2007). Epoxy/fly ash and vinyl ester/cenosphere composites have also been characterized and similar results are obtained (Kishore, 2002; Kulkarni, 2002; Porfiri, 2009).

The multifunctional role of castor oil and its polymers have been well documented in the literature in the last few years and have left many options to derive useful products out of this agricultural byproduct. Literature is in vogue in utilizing the polyol functionality and long unsaturated hydrocarbon by modifying it with various diisocyanates, chain extenders, vinyl monomers, thermoset resins, etc., polymerizing them to get all kinds of polymer products like elastomers, thermoplastics, thermosets, blends and IPNs of PU. In addition, literature survey reveals some studies reported the influence of reinforcement on the performance of PU based IPN composites. In the present study, PU based IPNs are chosen since it has many excellent characteristic properties such as, good weathering and abrasion resistance. The thermal properties and especially the mechanical properties of these composites have received particular attention from researchers for obvious needs in the areas of automotive, packaging and construction.

Despite the wide spectrum of experimental work that has been carried out with various systems such as, epoxy/cenosphere, vinyl ester/cenosphere, polyolefins/cenosphere etc., to the best of our knowledge, cenosphere filled PU based IPN biocomposites have not been reported yet. The present investigation intends to study the effect of cenosphere filler content on the performance of PU/PS IPN composites. In addition, to generate the wealth of data on IPN composites, this may be useful to widen the utilization of cenosphere in developing newer materials.

EXPERIMENTAL

Castor Oil

The key components to prepare PUs are a long chain oligomeric polyol, a diisocyanate and low molecular weight, chain extender. The polyol may be a polyether or an oligomer. As the oil crisis and environment problem deepen, the use of green and renewable resources to synthesize bio-based chemicals and products have attracted considerable attention (Desroches, 2012). Vegetable oil-based PUs are new emerging bio-based materials, which are catching more and more attention in recent years (Ligadas 2012). Currently in the world, there is an increasing interest in modern technologies, which are based on raw materials obtained from renewable sources. The introduction of plant components to polyurethane systems meets all the ideas of sustainable development and is an important challenge for polymer based industries. Vegetable oils come from various plants (soybean, palm, rapeseed, etc.) and mainly contain triglyceride molecules where the three -OH groups of glycerin are esterified with fatty acids. And the vegetable oils are considered to be one of the cheapest and most abundant biological sources available. Due to the specific structures and abundance, these vegetable oils have become an attractive resource that can convert to bio-based polyols to produce PUs (Zhang, 2015; Wang, 2015; Jin, 2012).

Among the vegetable oils, castor oil (CO) possessing –OH groups appear to be one of the best polyol for the synthesis of biobased PU elastomers (Kuliev, 2006). One of the most naturally occurring vegetable oils is castor oil (Fenigstein, 2001; Oliveira, 2004), which is not only has the advantage of possessing relatively high purity but is also abundantly available in India. The oil is obtained by extracting the seed of a plant which has the botanical name Ricinus communis of the family Eurphorbiacae (Jinbo, 2006). The oil is not only a naturally occurring resource; it is inexpensive, nontoxic and environmentally friendly. Castor oil is viscous, pale yellow, non-volatile and non-drying oil with a bland taste. Amidst, all the non edible natural vegetable oil, castor oil (CO) a trifunctional polyester polyol with the inherent hydroxyl group has gained huge attention over the petro-based polyol for the synthesis of biodegradable PUs (Kunduru, 2015; Yeganeh, 2007). Isocyanate derived from palm oil may be an efficient alternative to the petro-based isocyanate, owing to their comparable reactivity and isocyanate content. Moreover, the properties of PUs derived from vegetable oil based polyol and isocyanate can be tailored to incorporate inorganic nanofillers that can pave a path for designing prospective materials with high end applications. In recent years nanosilica has been extensively used as a reinforcement to enhance the mechanical, chemical, thermal and biodegradable properties of PUs (Das, 2016; Guo, 2015; Han, 2014). Castor oil can possess different physical and chemical properties. However, regardless of its origin or season in which it was grown, its chemical composition remains relatively constant (Monteavaro, 2005). Like other vegetable oils, castor oil is a triacylglycerol composed of various fatty acids and glycerol. The fatty acids consist of up to 90% ricinoleic acid and varying small amounts of saturated and unsaturated fatty acids. The high content of ricinoleic acid is the reason for the high value of castor oil and its versatile application possibilities in the chemical industry.-

The chemistry of castor oil is centered on its high content of ricinoleic acid and three functionality existing in the molecule. They are, (i) the carboxyl group, which can provide a wide range of esterifications; (ii) an unsaturation, which can be altered by hydrogenation or epoxidation or vulcanization, and (iii) the -OH groups, which can be acetylated, or alkoxylated, may be removed by dehydration to increase the unsaturation of the compound to give semi-drying oil.

The trifunctional nature of castor oil contributes toughness to the structure and the long fatty acid chain imparts flexibility. This oil has been known from primitive ages for its medicinal values. Castor oil reacts with polyfunctional isocyanates to form PUs that range from rigid TPUs to elastomers (Desai, 2003; Kim, 2007). The -OH groups of a castor oil can also impart plasticization effect in many polymer blends.

Toluene-2,4-diisocyanate (TDI) is one of the aromatic diisocyanate, its boiling point is 251 ^{0}C and refractive index is 1.567. TDI is a toxic, severe irritant sensitizer and moisture sensitive reagent. Each of the isocyanate functional groups can react with a hydroxyl group of castor oil to form a urethane linkage. Dibutyl tin dilaurate (DBTL) is a colorless liquid with pungent smell. It has a density of 1.05 g/cc at 20 °C. It is used as a catalyst for PU synthesis. The styrene molecular mass is 104.5, boiling point is 146 °C and specific gravity is 0.906. Pure styrene is a colorless liquid that evaporates easily and has a sweet smell. It is soluble in alcohol, ether, acetone and carbon disulphide.

Cenosphere

The cenosphere was obtained from Raichur thermal power plant, Karnataka, India. The specific gravity of the cenosphere is 0.64 and the particle size is 40 μm. Table 1 summarizes the chemical composition of cenosphere used in the present research investigation.

Castor oil was obtained from S.D. Fine Chemicals Ltd., and it was dried overnight at 90 ^{0}C before use. Toluene diisocyanate (TDI), styrene (monomer), benzoyl peroxide (initiator) and dibutyl tin dilaurate (DBTL) (catalyst) were supplied by Fluka and were used as received. Methyl ethyl ketone (MEK) of AR grade was used after distillation.

Table 1. Chemical composition of cenosphere

Constituents	Weight (%)
Silica (SiO_2)	57.77
Alumina (Al_2O_3)	30.33
Iron (III) oxide (Fe_2O_3)	5.33
Titanium (TiO_2)	2.03
Sulfates (SO_3)	1.76
Potassium oxide (K_2O)	1.60
Magnesium oxide (MgO)	0.61
Sodium oxide (Na_2O)	0.58

Synthesis of Cenosphere Filled IPN Composites

A typical procedure to prepare cenosphere filled PU/PS (90/10) IPN composites are as follows; Castor oil (0.001 mole) was initially dissolved in 50 ml of methyl ethyl ketone (MEK) and placed in three-necked reaction flask. The calculated amount of TDI (0.0015 mole) was added, followed by 2-3 drops of DBTL catalyst and the contents of the flask was stirred continuously for about 30 min under oxygen free nitrogen gas purge at 60-70 °C, to get pre PU. The known amount of styrene monomer and 0.5 wt % of benzoyl peroxide (initiator) were added to the pre PU and stirred again for 1 hr at 60-70 °C. Then calculated amount of cenosphere filler was added and stirred to get uniform dispersion. The solution was degassed under vacuum and poured into a cleaned and releasing agent coated glass mold. The mold was kept in preheated circulating hot air oven at 70 °C for 8-10 hrs. The cenosphere filled PU/PS (90/10) sheets, thus formed was left in the oven and allowed to cool gradually to ambient temperature and removed from the mold. The above procedure was repeated to obtain IPN composites with varying amounts viz., 0, 5, 10, 20 and 30 wt % of cenosphere.

Techniques

IR spectra of IPN composites were recorded using Fourier transform infrared spectroscopy (FTIR) in ATR mode using a Perkin Elmer spectrophotometer, in the wave number range 4000-400 cm^{-1}. The PU/PS/cenosphere composites have been characterized for density and surface hardness according to ASTM 792-86 and ASTM D 2240 method respectively. The mechanical properties were performed as per ASTM D 638 method using a 4302 model Houns Field universal testing machine (UTM), UK. The thermal stability of all IPN composites was analyzed using a thermogravimetric analyzer (TGA) Universal V4.3A, TA Instrument, USA under a nitrogen atmosphere at a heating rate of 20 °C/min in the temperature range 30–800 °C. All calorimetric measurements were made by differential scanning calorimetry (DSC) Universal V4.3A (Q2000), TA Instrument, USA. The dynamic mechanical properties of IPN composites were carried out with dynamic mechanical analysis (DMA), RDA-III Rheometrics, TA Instruments, USA. Morphological studies of the cryo-fractured samples were performed by using scanning electron microscopy (SEM) of FEI-Quanta 400 ESEM, USA.

RESULTS AND DISCUSSION

The effect of cenosphere on the mechanical, thermal, microstructural and morphological characteristics of IPN composites have been investigated.

Fourier Transform Infrared Spectroscopic Studies

FTIR spectra of IPN composites are shown in Figure 1 and Table 2 summarizes the expected and the observed IR data for characteristic groups of biocomposites. From the table it can be noticed that, IPN composites showed all the bands corresponding to PU and PS networks (Table 2) and don't show appearance of additional bands thereby ruling out the possibility of any chemical interaction between the cenosphere filler and IPN networks.

Figure 1. FTIR spectra of IPN composites with (a) 5 and (b) 20 wt % cenosphere content

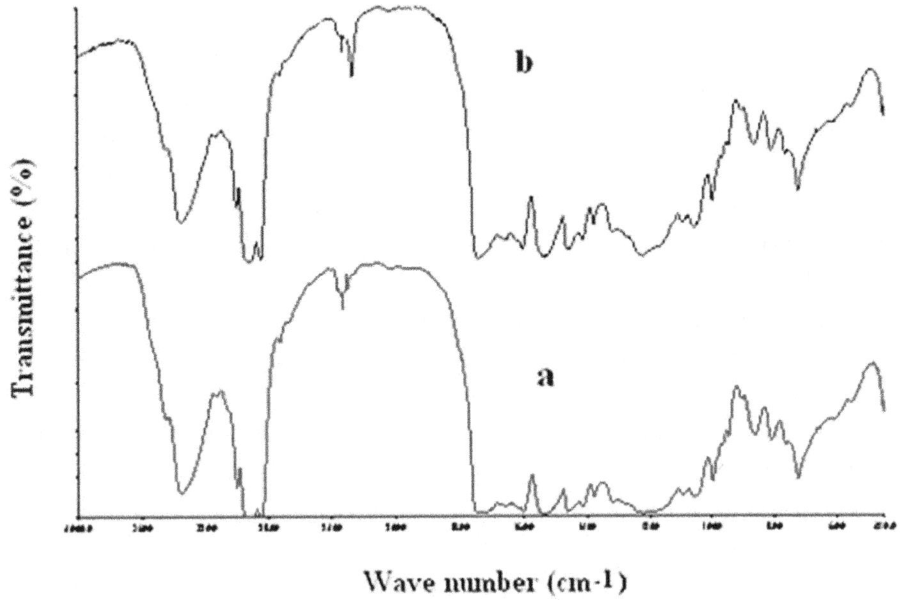

Table 2. Important band assignments of FTIR spectra of IPN composites

Group	Expected Peaks (cm^{-1})	Observed Peaks (cm^{-1}) IPN/Cenosphere (wt %)		
		100/0	95/5	80/20
C=O	1630-1690	1600	1600	1600
N-H stretching with hydrogen bonding	3200-3400	3361	3355	3355
Aromatic C-H stretching	3000-3100	3010	3009	3010
C=C aromatic ring	1450	1456	1456	1455
1,4 – substituted phenyl ring	860, 762	863, 724	865, 724	867, 724
O ǀǀ - C - O (ester)	1750-1700	1742	1734	1739
O ǀǀ -NH - C - NH- (urethane peak)	1528	1537	1533	1538
Aromatic C-H bending	680-860	809	813	813

Physico-Mechanical Properties

The measured physico-mechanical properties such as density, surface hardness, tensile strength, percentage elongation at break and tensile modulus for IPN composites are tabulated in Table 3.

The PU/PS and cenosphere had an average density of 1.0507 and 2.10 g/cc respectively. The measured density values for all PU/PS IPN composites (both theoretical and experimental) is nominated in Table 3. The theoretical density values are calculated according to rule of mixture. From the table it can be noticed that, the density of the cenosphere filled PU/PS IPN composites increases as increase in cenosphere content as expected. This is due to increase in high dense filler in low density PU/PS matrix. The experimental values are slightly lower than the corresponding theoretical values and the difference between theoretical and experimental density values becomes wider at higher concentration of cenosphere filler as compared to that of lower dosage cenosphere loaded composite. This can be attributed to that there is a poor interaction between IPN matrix and cenosphere filler.

Hardness is defined as the resistance of a material to deformation, particularly permanent deformation, indentation or scratching. The experimentally obtained surface hardness values are tabulated in Table 3. The surface hardness of cenosphere filled PU/PS composites lies in the range 62-85 shore A. The surface hardness value for unfilled PU/PS IPN is 62 shore A and for 30% cenosphere filled PU/PS IPN composite is 85 shore A. It was observed that a significant improvement in surface hardness of the IPN composites as an increase in rigid cenosphere content in soft IPN matrix as expected. This is because cenosphere is an inorganic filler (composed of metal oxide) and rigid in nature. The incorporated filler interlocking between IPN networks and reduces the flexibility of the IPNs, which led to improvement in dimensional stability of the composites. This result clearly indicates that the filled IPN composites have good dimensional stability than the unfilled IPN.

Tensile properties, in a broad sense, is a measurement of the ability of a material to withstand forces that tend to pull it apart and to determine to what extent the material stretches before breaking. The basic understanding of stress-strain behavior of composite material is of utmost importance to design engineers. The measured tensile properties of IPN composites are addressed in Table 3. From the table it was noticed that the higher tensile strength for IPN/cenosphere composites as compared to unfilled IPN. Among the filled composites, higher tensile strength (about 44% improvement) was noticed for 10 wt% cenosphere loaded IPN composite. This was not unexpected because, at a lower weight fraction of

Table 3. Physico-mechanical properties of IPN/cenosphere composites

Properties	Cenosphere Content (wt %)				
	0	5	10	20	30
Density (g/cc) Exptl. Theo.	1.0507 -	1.0842 1.1030	1.1234 1.1550	1.1994 1.2600	1.2406 1.3650
Surface hardness (shore A)	62	67	74	78	85
Tensile strength (σ) (MPa)	5.0	7.11	7.20	6.45	5.4
% Improvement in tensile strength	-	42	44	29	8
Tensile modulus (MPa)	0.29	0.43	0.66	0.99	1.55
Elongation at fracture (e) (%)	66	120	125	73	36
Product (σe)	330	853	900	470	194

the filler all cenosphere particles were capable of being completely wet by the matrix material, whereas, at higher weight fractions of the filler the matrix was incapable of wetting the filler. This resulted in heterogeneity in the composites leading to deterioration in tensile strength. Similar kind of observation is reported by Hundiwale et al for cenosphere filled natural rubber (NR) and butadiene rubber (BR) compounds (Hundiwale, 2002; Nabil, 2004).

The percentage elongation at fracture of PU/PS IPN is 66 and after incorporation of cenosphere content up to 10wt% the percentage elongation at fracture increases. The increase in percentage elongation at break is about 89% for composite containing 10wt% cenosphere. This may be due to rearranging of physical interaction between filler and matrix. In case of 20wt% cenosphere loaded system percentage elongation at break is slightly higher than unfilled IPN, but a significant reduction in elongation was noticed in 30wt% cenosphere loaded system. It can be attributed to poor wettability and reduction in network mobility at higher filler loading.

In general, the modulus of composites increases with increase in volume fraction of inorganic filler when the particles are stiffer than the matrix (Jeong, 1994). Significant improvements in tensile modulus of composites were noticed as increase in the filler content in the IPN matrix and it lies in the range 0.29– 1.55 MPa. It has also been observed previously that the use of hollow particles in less than a critical minimum volume fraction result in a reduction in the modulus of the composite (Bardella, 2001). The value of effective modulus depends on the particle wall thickness (Christensen, 2005; Tagliavia, 2009).

Comparing the results it can be noticed that 10% cenosphere filled PU/PS IPN composite exhibited the high tensile strength, elongation at fracture and tensile modulus values as compared to other formulations. Hence, 10% cenosphere loaded composite is referred as the optimized composition. This can be attributed to the fact that, in the absence of cenosphere, the failure would propagate along the loading direction. In the filler filled composites the dispersed cenosphere, which impede the propagation of cracks along the loading direction (Master, 1997). The failure, therefore propagates in a direction as dictated by the dispersed cenosphere particles in the IPN matrix. This means that the failure would propagate easily in those directions where the dispersed concentration is less leading to improvement in tensile strength, elongation at break and tensile modulus.

Lancaster (1972) stated that the product of σe factor (where, σ is the ultimate tensile strength and e is the elongation at fracture) is a very important factor, which controls the abrasive behavior of composites. In the present investigation, at lower filler loading higher will be the tensile strength (σ) and ultimate elongation (e) of neat polymer, and hence their product σe also higher. From the table it was noticed that the σe factor increases with increase in cenosphere content up to 10 wt % and it lies in the range 194 - 900. The higher the values of σe, higher will be the wear resistance of the composites.

Swelling Behavior

The composite specimens were exposed to different organic solvents having different solubility parameters and the effect of the solubility parameter of the solvents on the swelling behavior has been evaluated. From the swelling data it was noticed that the least solvent uptake was occurred for p-xylene of about 98% and higher swelling behavior was observed for dichloromethane of about 232% for unfilled PU/PS IPN. Table 4 reveals that a slightly higher solvent uptake for filled composites. This may be due to the presence of more microvoids in filled composites than unfilled IPN. However, there was no systematic variation in solvent uptake with filler content of the composites because there is a negligible interaction between filler and solvent.

Table 4. Solvent uptake of IPN composites in organic solvents

Chemical Reagents	Solubility Parameter $(cal/cm^3)^{1/2}$	Percentage Solvent Uptake (Cenosphere Content (wt%))				
		0	5	10	20	30
p-Xylene	8.77	98	104	94	114	107
Toluene	8.9	109	121	107	118	112
Ethyl acetate	9.1	108	112	106	100	103
Chlorobenzene	9.6	173	188	184	190	204
Dichloromethane	9.7	232	229	214	226	230

The effect of solubility parameter of solvents on the swelling behaviors of IPN composites is plotted in Figure 2. From the figure, it can be observed that the solvent uptake increases with increase of solubility parameter from 8.77 to 9.7 $(cal/cm^3)^{1/2}$ of the organic solvents. It was also noticed that there was a gradual increase in swelling up to solubility parameters of 9.6 $(cal/cm^3)^{1/2}$. A further increase in the solubility parameter of the solvents from 9.6 to 9.7 $(cal/cm^3)^{1/2}$, a steep increase in solvent uptake values was noticed for all composites. This result clearly indicates that the swelling behavior is strongly depends on the solubility parameter of the solvents. That means the solubility parameter of the IPN composites probably lies in the range 9.6 - 9.7 $(cal/cm^3)^{1/2}$.

Figure 2. Solvent uptake behavior of IPN/cenosphere composites as a function of solubility parameter of the solvents

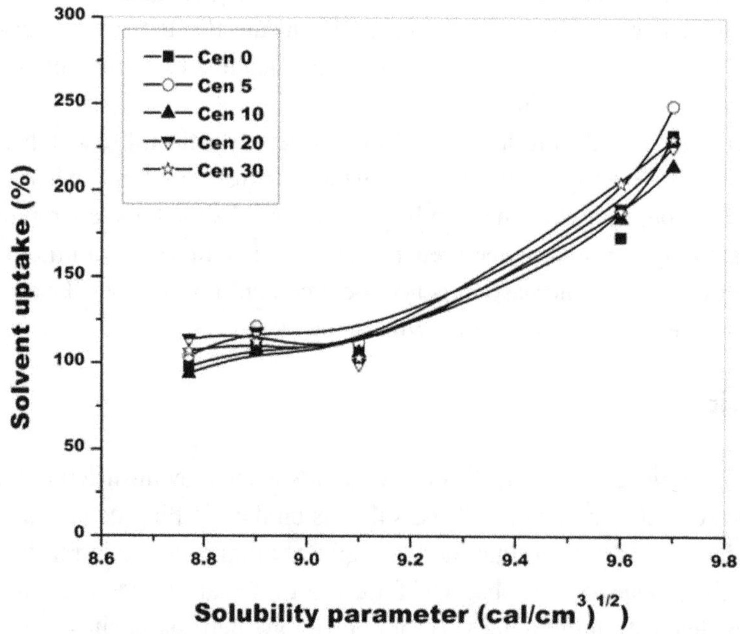

THERMAL BEHAVIOR

Differential Scanning Calorimetry

DSC was used to study the effect of filler loading on the T_g of the composites. The DSC thermograms of the PU/PS/cenosphere biocomposites are presented in Figure 3. The T_g of pristine PU/PS IPN was -5.7 ^{0}C and that of IPN/cenosphere composites varies from -5.8 to -5.0 ^{0}C. The primary relaxation was observed in the temperature range -25 to 30 ^{0}C, is regarded as the α - relaxation and has been designated as the T_g of PU soft segment. From the thermograms it was noticed that there was no significant change in T_g values after incorporation of cenosphere into IPN matrix.

Thermogravimetric Analysis

In order to understand the effect of cenosphere on the relative thermal stability of IPN composites, TGA has been carried out under nitrogen atmosphere. TGA thermograms along with derivative curves of cenosphere and its IPN composites are shown in Figures 4 (a) and (b) respectively. From the thermograms it was observed that the thermal degradation pattern was almost identical for all cenosphere filled PU/PS IPN composites. TGA thermograms reveals that all the samples follow two stage thermal degradation processes. From the figure it was noticed that, in case of cenosphere (Figure 4(a)), the ini-

Figure 3. DSC thermograms of IPN/cenosphere composites

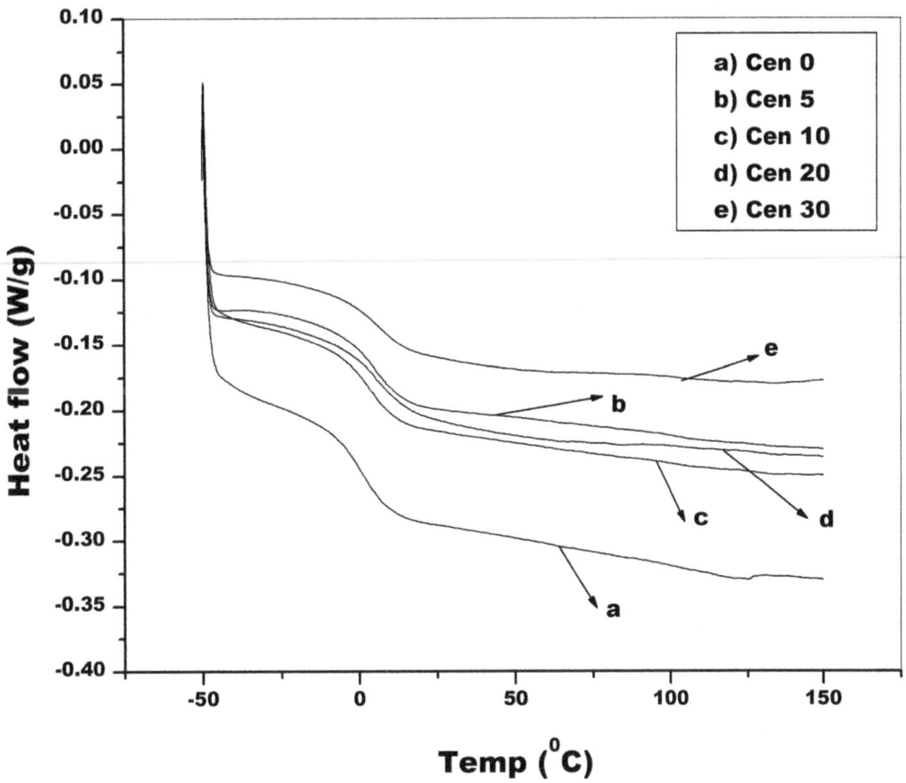

Figure 4. TGA thermograms of, (a) cenosphere and (b) IPN/cenosphere composites

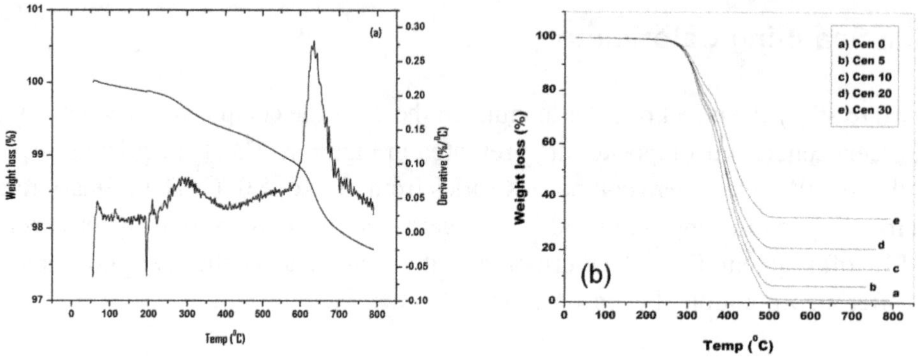

tial weight loss occurred in the temperature range 200–400 ^{0}C with a weight loss of 0.6% which may be due to moisture and volatile impurities present in cenosphere. The second stage degradation process occurred at a temperature range 500–790 ^{0}C with a weight loss of 1.6% and the ash content was found to be 97.8% (Table 5).

The temperature range and percentage weight loss for different thermal degradation steps of all IPN/cenosphere composites are addressed in Table 5. The weight loss occurred in the first step thermal degra-

Table 5. Thermal data obtained from TGA thermograms for IPN composites

Cenosphere Content in IPN (wt %)	Degradation Stages	Temperature (^{0}C) ± 2			Weight Loss (%)
		T_i	T_{max}	T_f	
0	1	255	318	345	27.6
	2	345	380	511	71.2
	Ash	-	-	-	1.2
5	1	254	322	350	26.8
	2	350	399	517	67.2
	Ash	-	-	-	6.0
10	1	254	322	342	24.4
	2	345	384	515	62.8
	Ash	-	-	-	12.8
20	1	250	321	348	22.9
	2	348	383	510	56.5
	Ash	-	-	-	20.6
30	1	252	322	350	19.4
	2	350	399	516	48.6
	Ash	-	-	-	32.0
Cenosphere	1	200	300	400	0.6
	2	500	650	790	1.6
	Ash	-	-	-	97.8

dation process (250-345 ^{0}C) of unfilled and 30% cenosphere filled IPN composite were 27.6 and 19.4% respectively. The weight loss in first step was due to the soft segment of PU and volatile impurities. The main pyrolysis product could be carbon dioxide (Herrera, 2003). In the second thermal degradation step the major weight loss of 71.2% has occurred for unfilled PU/PS IPN and that of IPN/30% cenosphere composite, it was 48.6%, in the temperature range 350-516 ^{0}C. This could be due to de-cross linking of IPN composites, liberation of HCN, nitriles of aromatics, carbon dioxide and ethers (Jeevananda, 2001; Herrera, 2003). The weight loss in second step occurred in the temperature range 345–517 ^{0}C with a weight loss of 48.6 – 71.2%. The weight loss in both degradation steps decreases with increase in cenosphere content.

The effect of cenosphere content on thermal stability of all PU/PS IPN composites is plotted in Figure 5. Figure clearly indicates that the weight loss markedly reduces with increase in cenosphere content and it was also observed that, the onset of decomposition of all IPN composites were almost identical (252-255 ^{0}C). Some of the characteristic degradation transition temperatures such as onset (T_0), 10% (T_{10}), 20% (T_{20}), 30% (T_{30}), 50% (T_{50}) weight loss and complete degradation (T_c) obtained from TGA curves and it is tabulated in Table 6.

For the sake of comparison, the plot of thermal degradation temperatures at different weight losses as a function of filler content is displayed in Figure 8. From the figure it was found that 10% weight loss of the IPN composites with 0, 5, 10, 20 and 30% cenosphere contents were occurred correspondingly at 308, 307, 309, 310 and 316 ^{0}C respectively and that of 50% weight losses of the IPN composites were

Figure 5. Thermal degradation temperatures at different weight losses of IPN/cenosphere composites

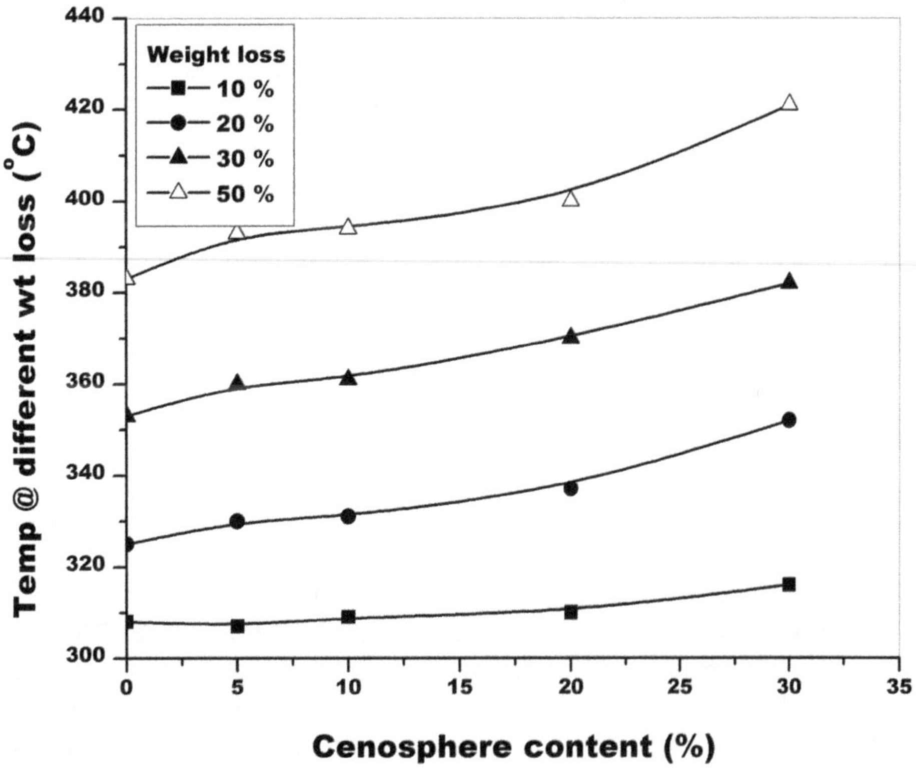

Table 6. Transition temperature data obtained from TGA curves for IPN composites

Cenosphere Content in IPN (wt %)	Temperature at Different Weight Loss (°C) ± 2						OI
	T_o	T_{10}	T_{20}	T_{30}	T_{50}	T_c	
0	255	308	325	353	383	511	0.082
5	254	307	330	360	393	517	0.439
10	254	309	331	361	394	515	0.890
20	250	310	337	370	400	515	1.390
30	252	316	352	382	421	516	2.220

occurred correspondingly at 383, 393, 394, 400 and 421 °C. This result clearly indicates that, at lower filler loading there was no much improvement in thermal stability, but at a higher level of cenosphere loading retards the thermal degradation process of IPN composites and hence, improves the thermal stability.

From TGA measurements, the char yield for unfilled PU/PS IPN was found to be 1.2%. On the other hand, the char yield for all filled PU/PS IPN systems increases with increase in cenosphere content and it lies in the range 6 – 32% (Table 5). From this data oxidation index was calculated and obtained values are tabulated in Table 10. The higher the values of oxidation index, higher will be the thermal stability (Jeevananda, 2001). Annakutty et al (1988) reported that, the char yield is directly correlated to the potency of flame retardation. From the aforesaid investigation, it can be concluded that the flame resistance of IPN/cenosphere composites was markedly increased as the cenosphere content increases.

Kinetic Analysis of Thermal Degradation

Tow mathematical models, namely Coats-Redfern (CR) and Broido (BR) methods are used to evaluate the kinetic parameters of the thermal degradation process. The plots of $\ln[-\ln(1-\alpha)]$ versus 1/T (BR) and $\ln[-\ln(1-\alpha)]/T^2]$ versus 1/T (CR) for first and second stage thermal degradation process of cenosphere filled PU/PS IPN composites are plotted in Figures 6 (a)-(b) and 7 (a)-(b) respectively. Regression analysis carried out for all the plots. The mechanism of R^2 close to unity was chosen. The regression analysis gives the results of slopes, constants and R^2 values corresponding to thermal degradation for the selected temperature range. The R^2 values and calculated activation energy (E_a) for each thermal degradation process and for each method are nominated in Table 7. The E_a results obtained from BR and CR methods are comparable.

To understand the mechanism of thermal degradation, the variation in activation energy as a function of weight percent of the cenosphere content in IPN systems for first and second steps are shown in Figures 8 (a)-(b) respectively. From the figure it can be noticed that, there was a slight reduction in activation energy after incorporation of cenosphere filler for the first step degradation process. The activation energy of the composites for the first step thermal degradation process lies in the range 125–167 kJ/mol and 115–184 kJ/mol for Broido and Coats–Redfern methods respectively. However, there is no systematic variation in the activation energy with respect to the composition of the composites. The lowest E_a values were observed for the first step thermal degradation process as compared to second step degradation for both the methods and for all composites. This is due to the lower energy is required to remove volatile components and low molecular weight (soft component) materials present in IPN composites. Higher

Figure 6. Broido's plots for; (a) first step and (b) second step thermal degradation process of IPN composites

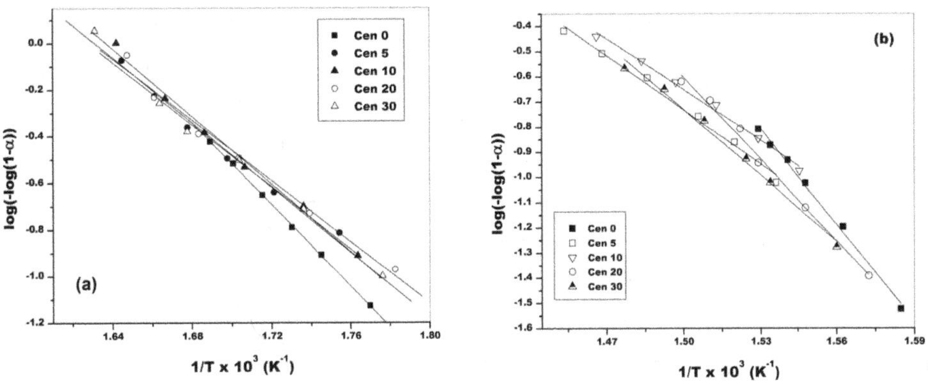

Figure 7. Coats-Redfern plots for (a) first step and (b) second step thermal degradation process of IPN composites

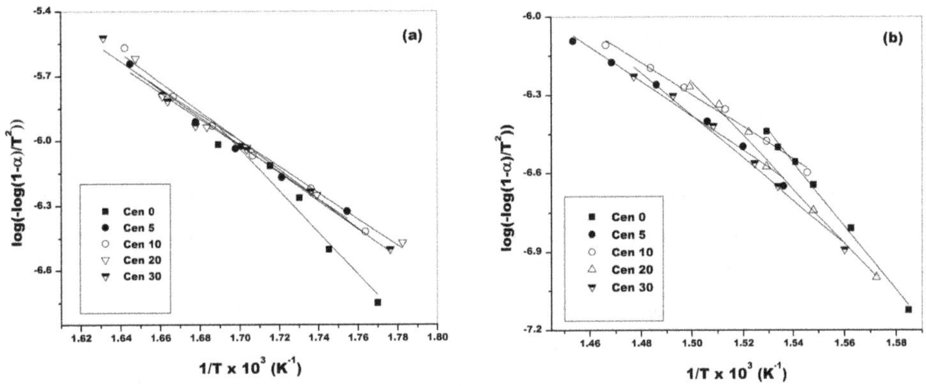

Figure 8. Effect of cenosphere content on activation energy for, (a) first step and (b) second step thermal degradation process of IPN composites

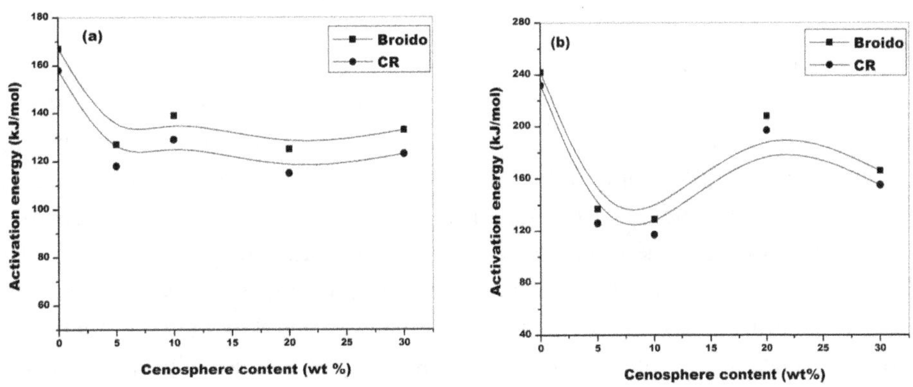

Table 7. Activation energies calculated by CR and BR methods with corresponding concurrency values (R^2) for IPN composites

Cenosphere Content (wt %)	Degradation Stage	Activation energy (E_a) (kJ/mol)			
		Ea (CR)	R^2	Ea (BR)	R^2
0	I	184	0.9995	167	0.9996
	II	232	0.9944	242	0.9949
5	I	118	0.9864	127	0.9983
	II	126	0.9886	137	0.9905
10	I	129	0.9921	139	0.9930
	II	117	0.9923	129	0.9937
20	I	115	0.9846	125	0.9867
	II	197	0.9920	208	0.9937
30	I	123	0.9875	133	0.9891
	II	155	0.9897	166	0.9911

E_a values were observed in the second step degradation process, because higher energies are required for bond scission and unzipping of IPN chains.

Dynamic Mechanical Analysis

Dynamic mechanical measurements are believed to be a good method for gaining insight into the effects exerted by the filler on the polymer matrix. These effects are displayed as a variation in dynamic storage modulus, a shift in peak position and width of the relaxation spectrum because of filler–polymer interaction. Each peak marks a transition, which is interpreted as released of a restricted movement, either in the polymer chain or in lateral groups damping (tan δ) expresses an ability of converting the mechanical energy into heat energy when the material is subjected to external loading. It is defined as;

$$\text{Tan}\ \delta = G'' / G' \qquad (1)$$

where, tan δ is phase angle between stress, G'' is elastic loss modulus and G' is the elastic storage modulus. As a result, tan δ is an important parameter characterizing material's viscoelastic behavior (Liang, 2000).

Figure 9 exhibits the temperature dependence of G', G'' and tan δ for filled PU/PS/5 wt % cenosphere composite. For the sake of clarity, G' is represented on a log scale (Figure 9 (B)). In the temperature range from -50 to 40 ^{0}C, the elastic storage modulus (G') drops sharply with a rise in temperature and reach a platform at around 40 ^{0}C onwards. The curves of the G'' and tan δ both reach maximum around 4 and 17°C respectively, corresponding to the glass transition temperature of the system. Furthermore, the values of G'' and tan δ reaches to zero when the temperature increases over 40 ^{0}C. Similar trend was noticed for other series of filled PU/PS/cenosphere IPN composites (refer Figures 9 and 10).

Figure 11 shows the temperature dependence of tan δ for the IPN/cenosphere composites. The curve reconfirm that all the properties of the composite are greatly affected by the operating temperature and volume fraction of cenosphere. We can see that the tan δ values increases in the initial stage and then

Figure 9. [A] DMA curves of 5 wt % cenosphere filled PU/PS IPN composite and [B] Log G' , G'' and Tan δ curves of 5 wt % cenosphere filled PU/PS IPN composite

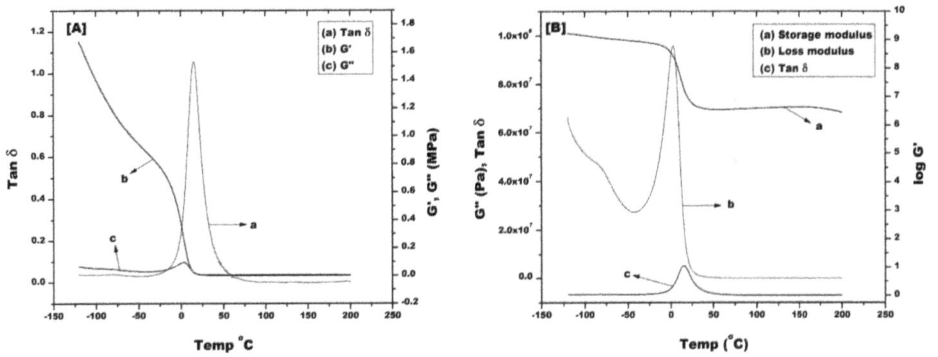

Figure 10. Effect of cenosphere on storage modulus of IPN/cenosphere composites

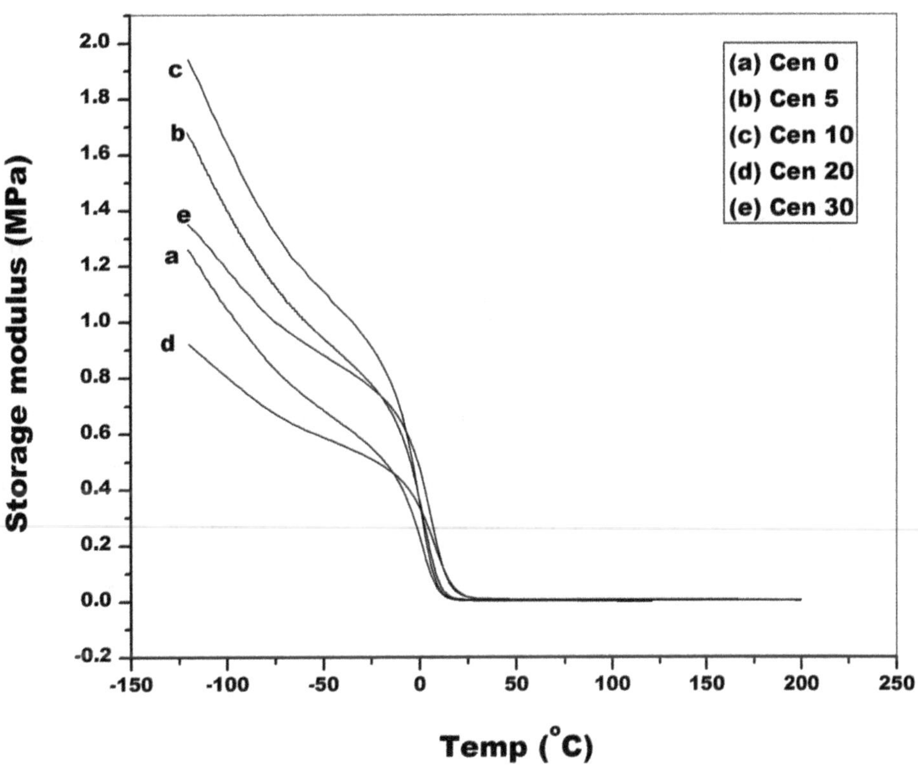

decrease with increasing temperature. The tan δ values of the composites upto 10 wt % cenosphere loading remain unchanged (refer Table 8) when compared with unfilled IPN, which indicates that the addition of cenosphere upto 10 wt% retains the damping property of the PU/PS IPN composites. Furthermore, the tan δ values of these composites reduce with increase in cenosphere content.

Figure 11. Temperature dependence of Tan δ of IPN/cenosphere composites

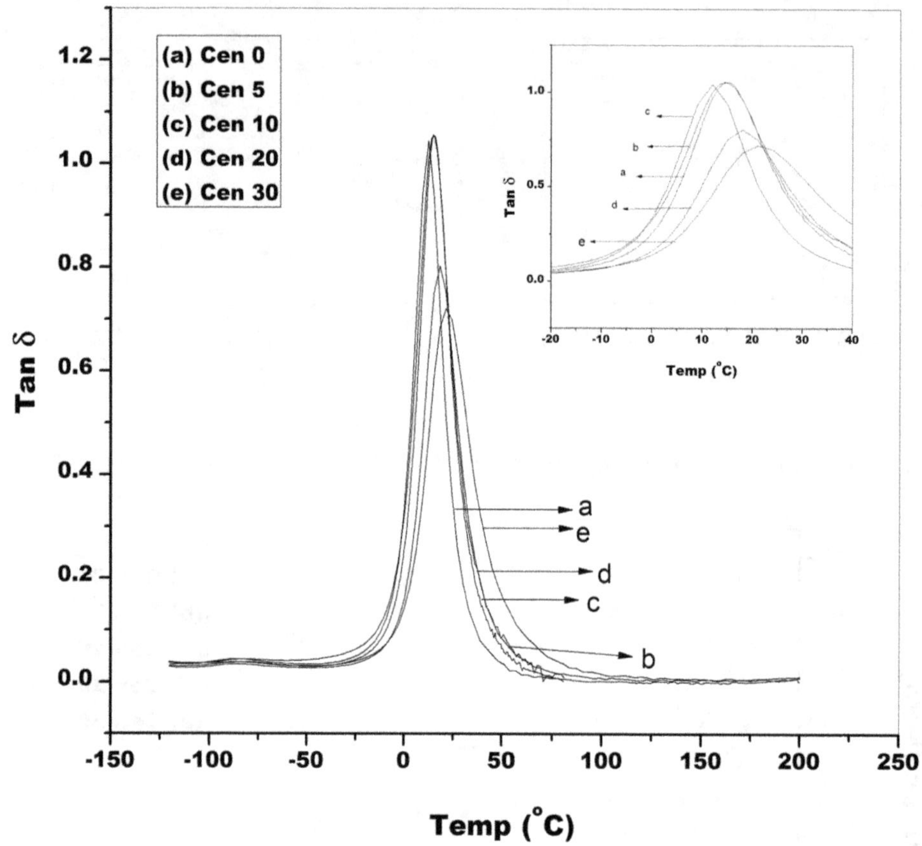

Table 8. Data obtained from DMA thermograms for IPN/ cenosphere composites

Cenosphere Content in IPN (wt %)	Tan Delta (Peak Max)		T_g (°C)
	Experimental	Theoretical	
0	1.06	1.06	15
5	1.06	1.00	14
10	1.05	0.95	12
20	0.84	0.84	21
30	0.74	0.73	18

For the polymer composites, damping capacity is quite sensitive to operating temperature, because their intrinsic damping is the matrix viscoelasticity, which is one typical behavior of polymeric materials. Therefore, the values of tan δ can only reach peak values at the T_g of the composites. For the temperature dependence of the tan δ of such composites, the contribution of the matrix viscoelasticity to composites damping is regnant (Gu, 2007). Therefore, the contribution of the matrix viscoelasticity decreases with an increase in the cenosphere volume fraction and the values of tan δ accordingly reduces. From Table 8 it can be seen that the incorporation of cenosphere leads to small changes in the T_g, however, there is no systematic variation in T_g with cenosphere content. It was also noticed that the, tan delta values determined theoretically are slightly lower than that of experimental values.

Morphological Behavior

Figures 12 (a-d) and Figures 13(a'-d') shows the SEM images of the IPN/cenosphere composites at 250x and 500x magnification respectively. A uniform distribution of cenosphere filler particles is desired in the composite without formation of aggregates to obtain uniform properties. Particle agglomeration can be caused by three adhesion forces, which are mechanical, electrostatic and molecular. As particle size is decreased, the gravitational force decreases proportional to the third power of the diameter, but the natural adhesion force decreases only between the first or second power of the diameter. Therefore, as the particle size decreases, the adhesion forces begin to dominate and the particles agglomerate naturally.

The microstructure presented in Figures 12(b)-(d), did not show an extensive agglomeration of cenospheres, the dispersion seems to be uniform, and it looks as if the beads are floating over the water surface. From SEM photomicrographs, it can also be observed that, most of the cenosphere particles show irregular shapes and rough surface and it contains significant porosity in its walls. The defects

Figure 12. SEM photomicrographs of IPN composites with (a) 0, (b) 5, (c) 20 and (d) 30 wt % of cenosphere content (250x magnifications)

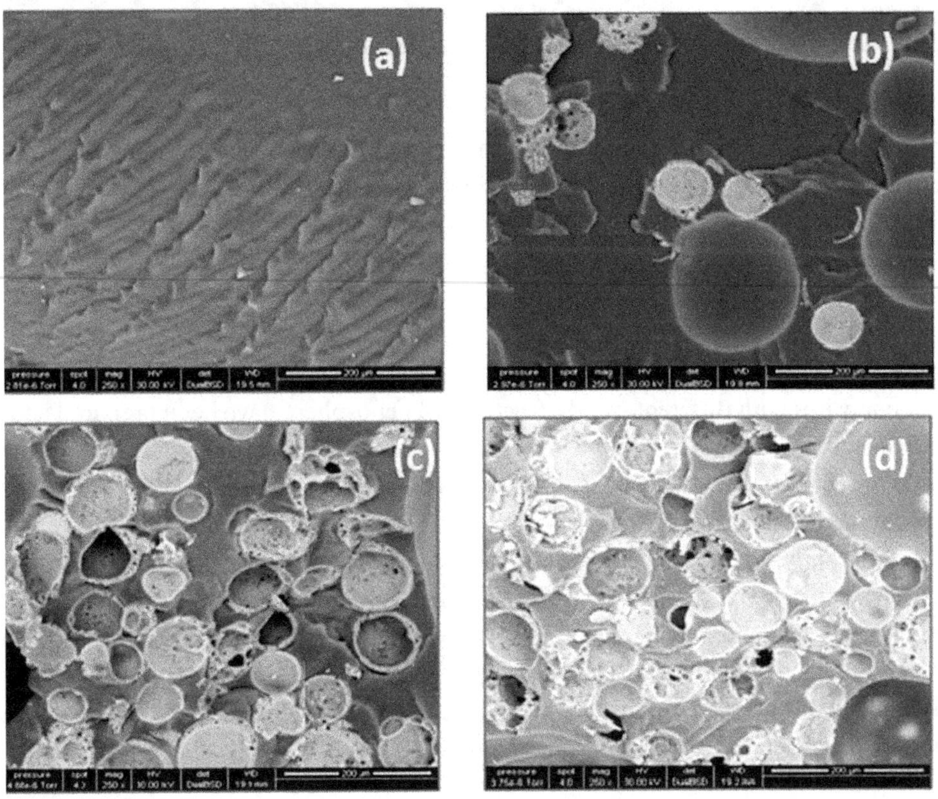

Figure 13. SEM photomicrographs of IPN composites with (a) 0, (b) 5, (c) 20 and (d) 30 wt % of cenosphere content (500x magnifications)

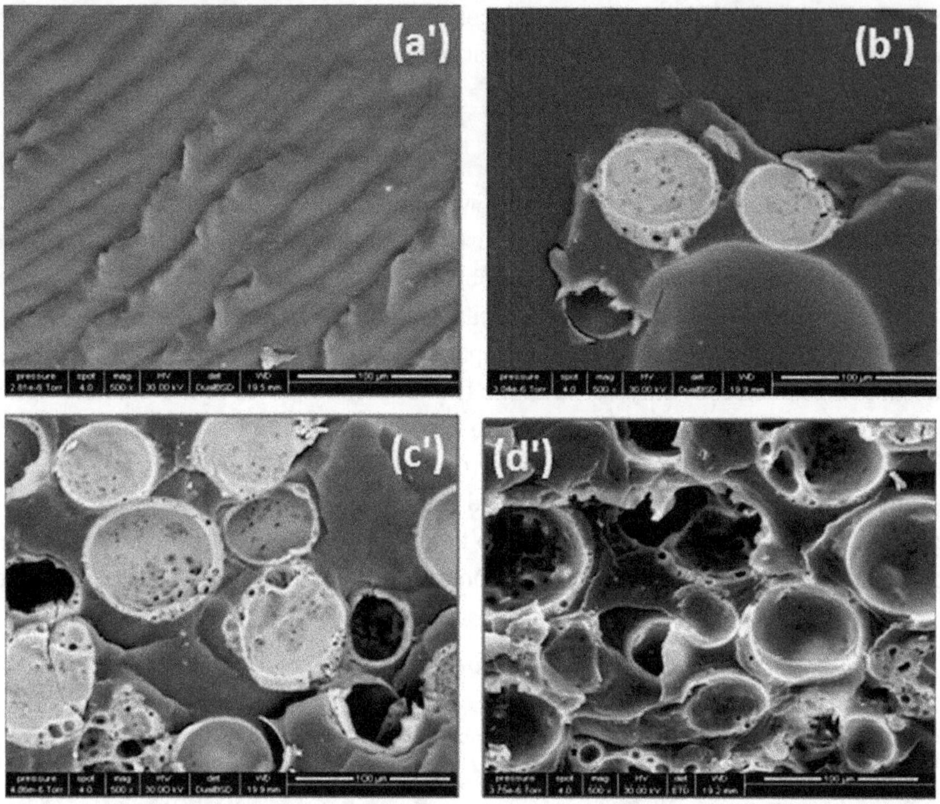

present in the walls play an important role in determining the strength of the systems. The estimation of these defects as a significant challenge, and experimental data can be the best source for such information because of the unique characteristics of each particle (Christensen, 2005).

CONCLUSION

PU/PS IPN composites with different weight fractions of cenosphere have been prepared and characterized for mechanical and thermal behaviors. The tensile strength, percentage elongation at break and σe factor was maximum for 10 wt% cenosphere loaded IPN composite. The dimensional stability and tensile modulus improved prominently with increase in filler content from. The solvent uptake values of the composites significantly depends upon the solubility parameter of the solvents. TGA data reveals that, the improvement in thermal stability of IPN is strongly depends on the filler dosage. DMA analysis reveals that, the tan δ values remain unchanged upto 10 wt% cenosphere content, however, further increase in cenosphere content decreases the damping property of the composites. Morphological characterization indicate a homogeneous dispersion of the cenosphere particles in IPN matrix at lower filler loading and the existence of agglomerates at higher filler content.

In summary, our work clearly indicates that cenosphere can be effectively used as reinforcing filler upto 10 wt% in PU matrix, leading to interesting physico-mechanical properties, chemical resistance and thermal stability. Moreover, the anticipated effects of progress in polymer materials should lead to the improvement of their properties and the reduction of their manufacturing cost. This is one of the reasons to carry on present research in the field of PU composites with the use of low-cost fillers derived from industrial byproduct or waste as raw materials to generate value added product. The current investigation widen the application window of the PU composites for tailor made applications.

REFERENCES

Anna, B., Milena, Z., Katarzyna, P., Piotr, C., & Joanna, R. (2016). Modification of flexible polyurethane foams by the addition of natural origin fillers. *Polymer Degradation & Stability*, *132*, 32–40. doi:10.1016/j.polymdegradstab.2016.05.002

Annakutty, K. S., & Kishore, K. (1988). Flame retardant polyphosphate esters: 1. Condensation polymers of bisphenols with aryl phosphorodichloridates: Synthesis, characterization and thermal studies. *Polymer*, *29*(4), 756–761. doi:10.1016/0032-3861(88)90096-1

Aomar, H., Omar, J., Ahmed, T., Mickael, G., Fouad, M., El Miloud, M., & Lan, T. (2016). Effects of cellulose fiber content on physical properties of polyurethane based composites. *Composite Structures*, *135*, 217–223. doi:10.1016/j.compstruct.2015.09.043

Azizah, S., Sian, M. S., Imran, M. I., & Qumrul, A. (2015). Preparation of rubber wood sawdust-based activated carbon and its use as a filler of polyurethane matrix composites for microwave absorption. *New Carbon Materials*, *30*(2), 167–175. doi:10.1016/S1872-5805(15)60182-2

Bienias, J., Walczak, M., Surowska, B., & Sobczak, J. (2003). Microstructure and corrosion behaviour of aluminium fly ash composites. *Journal of Optoelectronics and Advanced Materials*, *5*(2), 493–502.

Central Electricity Authority (CEA). (2014-15). *Report on fly ash generation at coal/lignite based thermal power stations and its utilization in the country for the year 2014-15*. Retrieved from http://cea.nic.in/reports/others/thermal/tcd/ flyash_final_1415.pdf

Chen, C. H., Chen, W. J., Chen, M. H., & Li, Y. M. (1999). Simultaneous full-inter penetrating polymer networks of blocked polyurethane and vinyl ester. II. Static and dynamic mechanical properties. *Journal of Applied Polymer Science*, *71*(12), 1977–1985. doi:10.1002/(SICI)1097-4628(19990321)71:12<1977::AID-APP7>3.0.CO;2-F

Christensen, R. M. (2005). *Mechanics of composite materials*. New York: Dover Publications.

Daoud, M. T., Abou El-khair, M., Abdelaziz, M., & Rohatgi, P. (2007). Fabrication, microstructure and compressive behavior of ZC63 Mg–microballoon foam composites. *Composites Science and Technology*, *67*(4), 1842–1853. doi:10.1016/j.compscitech.2006.10.023

Das, S., Pandey, P., Mohanty, S., & Nayak, S. K. (2016). Effect of nanosilica on the physicochemical, morphological and curing characteristics of transesterified castor oil based polyurethane coatings. *Progress in Organic Coatings*, *97*, 233–243. doi:10.1016/j.porgcoat.2016.04.012

Das, S., Pandey, P., Mohanty, S., & Nayak, S. K. (2017). Evaluation of biodegradability of green polyurethane/nanosilicacomposite synthesized from transesterified castor oil and palm oil based isocyanate. *International Biodeterioration & Biodegradation*, *117*, 278–288. doi:10.1016/j.ibiod.2017.01.015

Desai, S. D., Emanuel, A. L., & Sinha, V. K. (2003). Polyester polyol-based polyurethane adhesive; Effect of treatment on rubber surface, and biomaterial based polyurethane adhesive for bonding rubber and wood joints. *Journal of Polymer Research*, *10*(3), 141–149. doi:10.1023/A:1026099732272

Desroches, M., Escouvois, M., Auvergne, R., Caillol, S., & Boutevin, B. (2012). From vegetable oils to polyurethanes: Synthetic routes to polyols and main industrial products. *The Polish Review*, *52*(1), 38–79. doi:10.1080/15583724.2011.640443

Fenigstein, M., Eliyahu, S., Gan-Mor, S., & Veierov, D. (2001). Effects of five vegetable oils on the sweet-potato whitefly Bemisia tabaci. *Journal of Phytoparasitica*, *29*(1), 197–206. doi:10.1007/BF02983451

Gu, J., Wu, G., & Zhang, Q. (2007). Preparation and damping properties of fly ash filled epoxy composites. *Materials Science and Engineering A*, *452*, 614–618. doi:10.1016/j.msea.2006.11.006

Guo, J. H., Liu, Y. C., Chai, T., Jing, S. M., Ma, H., Qin, N., & He, W. M. et al. (2015). Synthesis and properties of nano-silica modified environmentally friendly polyurethane adhesive. *RSC Advances*, *5*(56), 44990–44997. doi:10.1039/C5RA01965F

Gupta, N., Brar, B. S., & Woldesenbet, E. (2001). Effect of filler addition on the compressive and of glass fiber reinforced epoxies. *Bulletin of Materials Science*, *24*(2), 219–223. doi:10.1007/BF02710105

Han, Y., Chen, Z., Dong, W., & Zhongyin, X. (2014). Improved water resistance, thermal stability, and mechanical properties of waterborne polyurethane nanohybrids reinforced by fumed silica via in situ polymerization. *High Performance Polymers*, *27*(7), 824–832. doi:10.1177/0954008314563058

Herrera, M., Matuschek, G., & Kettrup, A. (2003). Thermal degradation of thermoplastic polyurethane elastomers (TPU) based on MDI. *Polymer Degradation & Stability*, *78*(2), 323–331. doi:10.1016/S0141-3910(02)00181-7

Houton, K. A., & Wilson, A. J. (2015). Hydrogen-bonded supramolecular polyurethanes. *Polymer International*, *64*(2), 165–173. doi:10.1002/pi.4837

Hundiwale, D. G., Kapadi, U. R., Desai, M. C., & Bidkar, S. H. (2002). Mechanical properties of natural rubber filled with flyash. *Journal of Applied Polymer Science*, *85*(5), 995–1001. doi:10.1002/app.10465

James, J., Thomas, G. V., Akhina, H., & Thomas, S. (2016). Micro-and Nano-Structured Interpenetrating Polymer Networks: State of the Art, New Challenges, and Opportunities. *Micro-and Nano-Structured Interpenetrating Polymer Networks: From Design to Applications*, 1.

Jeevananda, T., & Siddaramaiah, . (2001). Thermal and morphological studies on ethylene-vinyl acetate copolymer–polyaniline blends. *Thermochimica Acta*, *376*(1), 51–56. doi:10.1016/S0040-6031(01)00530-5

Jeong, H., Hsu, D. K., Shannon, R. E., & Liaw, P. K. (1994). Characterization of anisotropic elastic constants of silicon-carbide particulate reinforced aluminum metal matrix composites: Part I. Experiment. *Metallurgical and Materials Transactions. A, Physical Metallurgy and Materials Science*, *25*(4), 799–809. doi:10.1007/BF02665456

Jia, Q. M., Shan, S., Wang, Y., Gu, L., & Li, J. (2008). Tribological performance and thermal behavior of epoxy resin nanocomposites containing polyurethane and organoclay. *Polymers for Advanced Technologies*, *19*(7), 859–864. doi:10.1002/pat.1049

Jia, Q. M., Zheng, M., Chen, H. X., & Shen, R. J. (2005). Synthesis and characterization of polyurethane/epoxy interpenetrating network nanocomposites with organoclays. *Polymer Bulletin*, *54*(1-2), 65–73. doi:10.1007/s00289-005-0372-7

Jia, Q. M., Zheng, M., Chen, H. X., & Shen, R. J. (2006a). Synthesis, characterization and properties of organoclay-modified polyurethane/epoxy interpenetrating polymer network nanocomposites. *Polymer International*, *55*(3), 257–264. doi:10.1002/pi.1931

Jia, Q. M., Zheng, M., Chen, H. X., & Shen, R. J. (2006b). Morphologies and properties of polyurethane/epoxy resin interpenetrating network nanocomposites modified with organoclay. *Materials Letters*, *60*(9-10), 1306–1309. doi:10.1016/j.matlet.2005.11.018

Jia, Q. M., Zheng, M., Chen, H. X., & Shen, R. J. (2006c). The glass transition temperature and microstructure of polyurethane/epoxy resin interpenetrating polymer networks nanocomposites. *Chinese Science Bulletin*, *51*(3), 293–298. doi:10.1007/s11434-006-0293-3

Jia, Q. M., Zheng, M., Zhu, Y. C., Li, J. B., & Xu, C. Z. (2007). Effects of organophilic montmorillonite on hydrogen bonding, free volume and glass transition temperature of epoxy resin/polyurethane interpenetrating polymer networks. *European Polymer Journal*, *43*(1), 35–42. doi:10.1016/j.eurpolymj.2006.10.016

Jin, H., Zhang, Y. G., Wang, C. S., Sun, Y. F., Yuan, Z. R., Pan, Y. Q., & Cheng, R. et al. (2014). Thermal, mechanical and morphological properties of soybean oil-based polyurethane/epoxy resin interpenetrating polymer networks (IPNs). *Journal of Thermal Analysis and Calorimetry*, *117*(2), 773–781. doi:10.1007/s10973-014-3849-5

Kim, D. H., Kwon, O. J., Yang, S. R., Park, J. S., & Chun, B. C. (2007). Structural, thermal, and mechanical properties of polyurethane foams prepared with starch as the main component of polyols. *Journal of Fibers and Polymers*, *8*(2), 155–162. doi:10.1007/BF02875785

Kishore, K., Kulkarni, S. M., Sharathchandra, S., & Sunil, D. (2002). On the use of an instrumented set-up to characterize the impact behaviour of an epoxy system containing varying fly ash content. *Polymer Testing*, *21*(7), 763–771. doi:10.1016/S0142-9418(02)00008-9

Krol, P. (2007). Synthesis methods, chemical structures and phase structures of linear polyurethanes. Properties and applications of linear polyurethanes in polyurethane elastomers, copolymers and ionomers. *Progress in Materials Science*, *52*(6), 915–1015. doi:10.1016/j.pmatsci.2006.11.001

Kuliev, R. S., Kuliev, F. A., Mutalibova, A. A., & Kulieva, S. R. (2006). Improving the antioxidant and anticorrosion properties of vegetable oils. *Chemistry and Technology of Fuels and Oils*, *42*(1), 55–59. doi:10.1007/s10553-006-0029-8

Kulkarni, S. M., & Kishore, . (2002). Studies on fly ash-filled epoxy-cast slabs under compression. *Journal of Applied Polymer Science, 84*(13), 2404–2410. doi:10.1002/app.10507

Kumar, H., Radha, J. C., Ranganathaiah, C., & Siddaramaiah, . (2007). Physico-mechanical and free volume behaviour of guar gum filled polyurethane/polyacrylonitrile biodegradable composites. *European Polymer Journal, 43*(4), 1580–1587. doi:10.1016/j.eurpolymj.2007.01.029

Kumar, H., Siddaramaiah, , Somashekar, R., Mahesh, S. S., & Manjunath, A. (2008). Nanocrystallite regions of guar gum filled PU/PAN composites before and after biodegradation using WAXS. *Journal of Applied Polymer Science, 108*(5), 2762–2771. doi:10.1002/app.27332

Kunduru, K. R., Basu, A., Haim Zada, M., & Domb, A. J. (2015). Castor oil-based biodegradable polyesters. *Biomacromolecules, 16*(9), 2572–2587. doi:10.1021/acs.biomac.5b00923 PMID:26301922

Lancaster, J. K. (1972). *Polymer Science: Material Science Handbook* (A. D. Jenkins, Ed.). Amsterdam: North Holland Publishing Company.

Lei, Z., Yang, Q., Shang, W., & Song, X. (2008). Reinforcement of polyurethane/epoxy interpenetrating network nanocomposites with an organically modified palygorskite. *Journal of Applied Polymer Science, 111*(6), 3150–3162. doi:10.1002/app.29360

Li, J. (2006). High performance epoxy resin nanocomposites containing both organic montmorillonite and castor oil-polyurethane. *Polymer Bulletin, 56*(4-5), 377–384. doi:10.1007/s00289-005-0492-0

Li, M., Qiang, X., Xu, W., & Zhang, H. (2015). Synthesis, characterization and application of AFC-based waterborne polyurethane. *Progress in Organic Coatings, 84*, 35–41. doi:10.1016/j.porgcoat.2015.02.009

Liang, J. Z. (2007). Impact fracture toughness of hollow glass bead-filled polypropylene composites. *Journal of Materials Science, 42*(3), 841–846. doi:10.1007/s10853-006-0074-z

Liang, J. Z., & Li, F. H. (2006). Measurement of thermal conductivity of hollow glass-bead-filled polypropylene composites. *Polymer Testing, 25*(4), 527–531. doi:10.1016/j.polymertesting.2006.02.007

Liang, J. Z., Li, R. K. Y., & Tjong, S. C. (2000). Effects of glass bead size and content on the visco-elasticity of filled polypropylene composites. *Polymer Testing, 19*(2), 213–220. doi:10.1016/S0142-9418(99)00005-7

Lipatov, Y. S., Sergeyeva, L. M., Karabanova, L. V., Rosovitskii, V. F., Skiba, S. I., & Babkina, N. V. (1988). Effect of fillers on viscoelastic properties and compatibility of components in interpenetrating polyurethane-polyacrylate networks. *Polymer Science, USSR., 30*(3), 651–658. doi:10.1016/0032-3950(88)90105-0

Lligadas, G., Ronda, J., Galia, M., & Cadiz, V. (2010). Plant oils as platform chemicals for polyurethane synthesis: Current state-of-the-art. *Biomacromolecules, 11*(11), 2825–2835. doi:10.1021/bm100839x PMID:20939529

Luo, J., Wang, X., Li, J., Zhao, X., & Wang, F. (2007). Conductive hybrid film from polyaniline and polyurethane–silica. *Polymer, 48*(15), 4368–4374. doi:10.1016/j.polymer.2007.05.062

Mariano, M., El Kissi, N., & Dufresne, A. (2014). Cellulose nanocrystals and related nanocomposites: Review of some properties and challenges. *Journal of Polymer Science. Part B, Polymer Physics*, *52*(12), 791–806. doi:10.1002/polb.23490

Master, J. E. (1997). *Basic failure modes of continuous fiber composites, in engineering materials Handbook* (Vol. 1). ASM Intl.

Monteavaro, L. L., DaSilva, E. O., Costa, O. A. P., Samios, D., Gerbase, A. E., & Petzhold, C. L. (2005). Polyurethane networks from formiated soy polyols: Synthesis and mechanical characterization. *Journal of the American Oil Chemists Society*, *82*(5), 365–371. doi:10.1007/s11746-005-1079-0

Nabil, A. N. (2004). Studies on the effect of titanate coupling agent (2.0%) on the mechanical properties of flyash-filled polybutadiene rubber. *Journal of Applied Polymer Science*, *93*(3), 1293–1298. doi:10.1002/app.20548

Oliveira, D. (2004). Optimization of enzymatic production of biodiesel from castor oil in organic solvent medium. *Applied Biochemistry and Biotechnology*, *115*(1-3), 771–780. doi:10.1385/ABAB:115:1-3:0771 PMID:15054231

Porfiri, M., Nguyen, N. Q., & Gupta, N. (2009). Thermal conductivity of multiphase particulate composite materials. *Journal of Materials Science*, *44*(6), 1540–1550. doi:10.1007/s10853-008-3040-0

Rohatgi, P. K., Kim, J. K., Gupta, N., Alaraj, S., & Daoud, A. (2006). Compressive characteristics of A356/fly ash cenosphere composites synthesized by pressure infiltration technique. *Composites. Part A, Applied Science and Manufacturing*, *37*(3), 430–437. doi:10.1016/j.compositesa.2005.05.047

Rokicki, G., Parzuchowski, P. G., & Mazurek, M. (2015). Non-isocyanate polyurethanes: Synthesis, properties, and applications. *Polymers for Advanced Technologies*, *26*(7), 707–761. doi:10.1002/pat.3522

Saba, N., Tahir, P. M., & Jawaid, M. (2014). A review on potentiality of nano filler/natural fiber filled polymer hybrid composites. *Polymers*, *6*(8), 2247–2273. doi:10.3390/polym6082247

Shih, Y. F., Jeng, R. J., & Wei, K. M. (2003). Carbon black containing interpenetrating polymer networks based on unsaturated polyester/epoxy III: Thermal and pyrolysis analysis. *Journal of Analytical and Applied Pyrolysis*, *70*(1), 129–141. doi:10.1016/S0165-2370(02)00102-X

Sperling, L. H. (1981). *Interpenetrating Polymer Networks and Related Materials*. New York: Plenum Press. doi:10.1007/978-1-4684-3830-7

Tagliavia, G., Porfiri, M., & Gupta, N. (2009). Vinyl ester—glass hollow particle composites: Dynamic mechanical properties at high inclusion volume fraction. *Journal of Composite Materials*, *43*(5), 561–582. doi:10.1177/0021998308097683

Tang, D. Y., Cai, W., Zhou, D., Zhao, L., & Gongyi, C. K. Y. (2002a). A morphological study of modified PU/PMMA semi-IPNs and nanocomposites with $BaTiO_3$. *Materials Science and Technology*, *10*, 260–262.

Tang, D. Y., Hong, L., & Weimin, C. (2002b). Synthesis and application studies of castor oil PU/PMMA IPNs with $BaTiO_3$ fiber nanocomposites. *Ferroelectrics*, *265*(1), 259–264. doi:10.1080/00150190208260627

Tang, D. Y., Jin, Z., & Qiang, L. S. (2007). Electric and damping properties of PU based IPNs-$BaTiO_3$ nanocomposites. *Key Engineering Materials*, *118*, 336–338.

Tang, D. Y., Liancheng, Z., Zheng, J., & Weimin, C. (2002c). Preparation, morphology, and thermo-electric property studies of $BaTiO_3$ superfine fiber/castor oil polyurethane-based IPN nanocomposites. *Journal of Applied Polymer Science*, *84*(4), 709–715. doi:10.1002/app.10024

Tang, D. Y., Lili, L., Jusheng, Z., Zheng, J., & Liancheng, Z. (2004). Influence of barium titanate on the damping properties of interpenetrating polymer networks based nanocomposites. *Journal of Aeronautical Materials*, *5*, 40–43.

Tang, D. Y., Qin, C., Cai, W., & Zhao, L. (2003a). Preparation, morphology, and mechanical properties of modified-PU/UPR graft-IPN nanocomposites with $BaTiO_3$ fiber. *Materials Chemistry and Physics*, *82*(1), 73–77. doi:10.1016/S0254-0584(03)00162-7

Tang, D. Y., Xu, Q. Y., Cai, W. M., Zhou, D. R., & Zhao, L. H. (2003b). Preparation and mechanical properties of $BaTiO_3$ nanometer powder/COPU based IPNs nanocomposites. *Materials Science and Technology*, *11*, 181–184.

Trakulsujaritchok, T., & Hourston, D. J. (2006). Damping characteristics and mechanical properties of silica filled PUR/PEMA simultaneous inter-penetrating polymer networks. *European Polymer Journal*, *42*(11), 2968–2976. doi:10.1016/j.eurpolymj.2006.07.028

Wang, C. S., Wang, Y. T., Liu, W. J., Yin, H. Y., Yuan, Z. R., Wang, Q. J., & Cheng, R. et al. (2012). Natural fibrous nanoclay reinforced soy polyol-based polyurethane. *Materials Letters*, *78*, 85–87. doi:10.1016/j.matlet.2012.03.067

Wang, T., Chen, S., Wang, Q., & Pei, X. (2010). Damping analysis of polyurethane/ epoxy graft interpenetrating polymer network composites filled with short carbon fiber and micro hollow glass bead. *Materials & Design*, *31*(8), 3810–3815. doi:10.1016/j.matdes.2010.03.029

Weijing, L., Ke, X., Chengshuang, W., Bin, Q., Yifan, S., Yuge, Z., & Rongshi, C. et al. (2015). Carbon nanofibers reinforced soy polyol-based polyurethane Nanocomposites- Thermal and mechanical characterization. *Journal of Thermal Analysis and Calorimetry*, *123*(3), 2459–2468.

Wu, X., He, G., Gu, S., Hu, Z., & Yao, P. (2007). Novel interpenetrating polymer network sulfonated poly (phthalazinone ether sulfone ketone)/polyacrylic acid proton exchange membranes for fuel cell. *Journal of Membrane Science*, *295*(1-2), 80–87. doi:10.1016/j.memsci.2007.02.039

Yeganeh, H., & Pejman, H. T. (2007). Preparation and properties of novel biodegradable polyurethane networks based on castor oil and poly (ethyleneglycol). *Polymer Degradation & Stability*, *92*(3), 480–489. doi:10.1016/j.polymdegradstab.2006.10.011

Yilmaz, B., & Olgun, A. (2008). Studies on cement and mortar containing low-calcium fly ash, limestone and dolomitic limestone. *Cement and Concrete Composites*, *30*(3), 194–201. doi:10.1016/j.cemconcomp.2007.07.002

Zhang, C. Q., Madbouly, S. A., & Kessler, M. R. (2015). Biobased polyurethanes prepared from different vegetable oils. *ACS Applied Materials & Interfaces*, *7*(2), 1226–1233. doi:10.1021/am5071333 PMID:25541678

Zhang, H., Wang, B., Li, H., Jiang, Y., & Wang, J. (2003). Synthesis and characterization of nanocomposites of silicon dioxide and polyurethane and epoxy resin interpenetrating network. *Polymer International*, *52*(9), 1493–1497. doi:10.1002/pi.1247

Zhang, H., Xu, R., Wu, Z., & Zhou, C. (2003). The synthesis and characterization of polyurethane/clay nanocomposites. *Polymer International*, *52*(5), 790–794. doi:10.1002/pi.1152

Zhang, J., Tu, W., & Dai, Z. (2012). Synthesis and characterization of transparent and high impact resistance polyurethane coatings based on polyester polyols and isocyanate trimers. *Progress in Organic Coatings*, *75*(4), 579–583. doi:10.1016/j.porgcoat.2012.05.005

Chapter 8
Physical Characteristics of CBD Synthesized CdZnS Thin Films:
CdZnS Thin Film Preparation Techniques, Properties, and Characterization

Ayan Mukherjee
The University of Burdwan, India

Partha Mitra
The University of Burdwan, India

ABSTRACT

In recent years, ternary cadmium zinc sulfide (CdZnS) alloy compounds have been paid much attention in the fields of opto-electronics, particularly in photovoltaic devices. CdZnS thin films can be prepared by different techniques among which chemical methods have more advantages. Among different chemical method, Chemical Bath Deposition (CBD) is simple, low cost and widely applicable in industrial applications. In this chapter, we have discussed different methods of preparation of CdZnS thin film and their obtained properties. Also, the films are characterized by XRD, TEM, FESEM, EDAX, UV-Vis spectroscopy, etc. The properties of CdZnS gives insight of the properties of ternary thin film semiconductor and it will help to design semiconductor with tuneable properties for future applications in optoelectronic sector.

INTRODUCTION

In modern age industrial application, semiconductor materials have received great interest due to their novel properties. Among different semiconductor materials, chalcogenides have received a great attention for their ability to easily form binary and ternary compound. The wide-band gap binary II–VI compounds and their alloys have been investigated most extensively for their potential application in different fields such as opto-electronic devices, solar cells, nonlinear optical and luminescence devices because of their quantum size effect and especially due to their tunable optical properties (Rajathi, 2012; Raviprakash, 2010; Baykul, 2010) Modern day's material science needs advanced, low cost smart ternary material

DOI: 10.4018/978-1-5225-3023-7.ch008

for its wide applications. The ternary materials have advantage over binary one as their physical and chemical properties hugely depend on their composition. So the importance of ternary chalcogenides is increasing day by day. Ternary material can be achieved either by alloying three different elements or by alloying two binary materials with same anion. The other way to create ternary material is doping one element on a binary material. Though doping is very popular technique, alloying of semiconductors is also one of the simplest techniques used for tailoring the energy band gap, lattice parameter, electronic and optical properties (Nagamani, 2012).

Those ternary chalcogenides can be fabricated in both particles and thin film form. The main reason for interest in thin films is that they can change both the physical and chemical properties of the surface of the underlying material as well as they can promote desired properties on it. A thin film could improve the hardness and roughness which is beneficial for use as a protective layer in different optoelectronic devices such as solar cells or flexible LED displays (Park, 2011). Many of the device applications requires material in thin film form as it possess specific thickness, well defined phase content, uniform distribution of material over the surface. A thin film also have special benefit of cost effectiveness: instead of manufacturing the total expensive desired material, it is enough to have a cheaper material as a backbone with a thin film with desired properties on top. So, the surface area can be increased using thin film form than particle-form. For example, to fabricate a solid car catalyst we need about 1 kg of platinum in the device. In reality there is only about 3-4 g on a ceramic backbone – lowering the cost by nearly 300 times.

Different chalcogenides such as cadmium sulfide, zinc sulphide, lead sulphide etc. has potential application in different optoelectronics devices. Cadmium sulfide is a II–VI compound semiconductor with a wide direct band gap of 2.42 eV (bulk value) and a small exciton Bohr radius of 2.5 nm. The properties of those semiconductors are greatly influenced by the effects of low dimensionality when the radius of the particle is comparable to the value of Bohr exciton radius in the consequent bulk material. CdS is used as window material for heterojunction solar cells as its wide bandgap helps to avoid the recombination of photogenerated carriers and further improves the solar cells efficiency (Sekhar, 2012). Therefore CdS continues to hold an important position in CdS/ CdTe and CdS/ $CuInSe_2$ solar cell with efficiency nearly 19.2%.

In recent times, transition metal ions-doped II–VI nanoparticles or thin film have engrossed intense interest because transition metal ions incorporated into II–VI nanoparticles can improve significantly the optical properties of nanoparticles. The optical properties of doped nanomaterials greatly differ from the consequent host one. These impurities can strongly improve the luminescence properties by providing impurity centers that can interact with the quantum-confined electron–hole pair. The crystallite luminescence largely depends on the surface characteristics. Usually semiconductor nanoparticles exhibit both excitonic and trap (defect) emission. If the prepared CdS is stoichiometric then it will generate blue light emission. But if it contains excess amount of Cd^{+2} or S^{-2}, several trap states would be generated. CdS containing excess S^{-2} will generate orange or red light emission, while excess Cd^{+2} will generate yellow emission (Sekhar, 2012; Khanna, 2007; Butty, 1996; Chen, 2003).

One can get CdS based ternary material by doping with metal ions such as copper (Cu), zinc (Zn), lead (Pb) etc. The alternative way is alloying the metal sulfides with CdS. Among them $Cd_xZn_{1-x}S$ alloy compounds have attracted much technological interest because their energy gap can be tuned as well as lattice parameters can be varied according to desired requirements (Mukherjee, 2015; Xia, 2011; Gaewdang, 2005). $Cd_xZn_{1-x}S$ ternary compounds can form a continuous series of solid solutions, allowing systematic variation in the band gap of $Cd_xZn_{1-x}S$ from 2.43 eV for CdS to 3.7 eV for ZnS by adjusting the composition (Mukherjee, 2015). As stated earlier, CdS thin films have been demonstrated to be very much effective in solar cell and laser diode applications. But, the absorption of the blue portion of the solar spectrum by CdS window layer results in a decrease in the current density of such solar cells and UV laser diodes. Therefore replacement of CdS with higher band gap ternary materials will help to decrease the window absorption losses with an increase in the short circuit current. $Cd_xZn_{1-x}S$ can play an important role in this regard (Mukherjee, 2015). For the high performance of solar cell device fabrication, it is essential to use a suitable window material. Ternary $CdZnS$ thin films often show better chemical, structural and optical properties which makes them potential material for fabricating window layer of solar cell. Moreover, partial substitution of Zn for Cd increases the optical window of the hetero-junction and the diffusion potential (Song, 2009). $CdZnS$ compound is also useful to fabricate p-n junctions without lattice mismatch using quaternary materials such as $CuInGaSe_2$ and $CuInS_2Se_2$ (Yamaguchi, 1992). In thin film form they found interesting applications in high performance semiconducting LASER diodes, photovoltaic systems and photo conducting devices (Mukherjee, 2015). In nanostructured form, they found important applications in probes for DNA structures and fluorescence probes in peptides etc. (Chen, 2005). The control of the composition $Cd_xZn_{1-x}S$ nanoparticles may lead to the development of ideal materials for short wavelength diode laser applications (Mukherjee, 2015).

Preparation of CdZnS Thin Films

There are several deposition technologies for material formation. A thin film deposition technique involves three steps: (i) Creation of atomic/molecular/ionic species, (ii) transport of these species through a medium, and (iii) condensation of the species on a substrate. The concern with thin film deposition techniques is to form layers in thickness range of a few nanometers to several micrometers. The task of classifying the technologies is made simpler by limiting the number of technologies used. Basically, thin-film deposition technologies are classified mainly in three categories, such as, physical process, chemical process and one hybrid process known as physical-chemical process (Seshan, 2002). In physical methods the film material is moved from a target source with some form of energy to the substrate. Physical Vapor Deposition (PVD), Sputtering, Evaporation etc are physical deposition process. Chemical film fabrication methods involve chemical reactions and the precursors are mostly components undergoing reaction at the substrate surface or in the vicinity of the substrate. The different category of chemical deposition techniques include: Spray pyrolysis, Sol gel, Chemical Bath deposition (CBD), Electroless deposition, Electrodeposition, Anodization, Electrophoresis, Atomic layer deposition and successive ion layer adsorption and reaction (SILAR) etc. A considerable number of processes such as Chemical Vapor Deposition (CVD), Metal Oxide (MOCVD), and Oxidation etc. combines both physical and chemical reactions. These overlapping processes are often categorized as physical-chemical methods or hybrid method (Seshan, 2002).

Conventional physical process (primarily vacuum techniques such as sputtering and evaporation) renders better control over stoichiometry. Further they produces uniform and compact films and generally yields good quality transparent films. They are generally safe (no toxic gas emissions) and achieves high deposition rate at room temperature. On the other hand they require expensive capital instruments. Accordingly they are difficult to expand to large scale. Chemical techniques of thin film deposition involving aqueous solution on the other hand are low cost process and effectively simple compared with physical techniques. Thus they offer the desirable cheapness and possibility of scaling up to large area for industrial application. Accordingly, chemical techniques have come out to be a good alternative process for material preparation during the past few decades.

Among different chemical techniques such as electrodeposition (Al-bassam, 1999), SILAR (Laukaitish, 2000), spray pyrolysis (Ravangave, 2012), CBD is most widely used for depositing $Cd_xZn_{1-x}S$ thin films (Nagamani, 2012; Kumar, 2009; Jia, 2009; Asogwa, 2010; Ashokkumar, 2013) because of its easy processibility and suitability for large area depositions for commercial application (Olivia, 2014). *Chemical bath deposition (CBD)* also known as chemical solution deposition (CSD) is one of the oldest and simplest methods of thin film deposition. This refers to the deposition of films on a solid substrate from a reaction occurring in a solution (almost always aqueous). In this method, deposition of thin films occurs due to substrate maintained in contact with dilute chemical bath containing cationic and anionic solutions (Rajathi, 2012; Chopra, 1983; Jokanovic, 2005; Pathan, 2004). The film formation on substrate takes place when the ionic product (IP) exceeds the solubility product (SP). Nucleation on substrate from liquid phase occurs to form thin film.

According to the solubility product principle, solubility product is the product of the molar concentration of ions of electrolyte (each concentration term being raised to the power of their coefficients or to the power equal to the number of ions of that kind) in the balanced chemical equation in a saturated solution of a weakly soluble compound. This is applicable only to saturated solutions. At a given temperature, its value is constant for a given electrolyte. The following discussion gives the basics of SP and IP.

Sparingly or weakly soluble salt, AB, when placed in water, a saturated solution containing A^+ and B^- ions in contact with undissolved solid AB is obtained and an equilibrium is established between the solid phase and ions in the solution as

$$AB(S) \Leftrightarrow A^+ + B^-$$

Applying the law of mass action,

$$K = \frac{\left[C_A{}^+ C_B{}^-\right]}{C_{AB}}$$

where $C_A{}^+$, $C_B{}^-$ and C_{AB} are the concentrations of A^+, B^- and AB in the solution, respectively (Padam, Malhotra, & Rao, 1988). The concentration of pure solid is a constant number, i.e. $C_{AB}(S) = C_{AB} = K'$ is a constant. Thus

$$K = \frac{\left[C_A{}^+ C_B{}^-\right]}{K'}$$

i.e. $KK' = C_A{}^+ C_B{}^-$

Since K and K' are constants, the product KK' is also constant, say K_S. Thus

$$C_A^{\ +}C_B^{\ -} = K_S$$

The constant K_S is called solubility product (SP) and the product $C_A^{\ +}C_B^{\ -}$ is called the ionic product (IP). Thus ionic product is the product of concentration of ions of the electrolyte, each raised to the power of their coefficients in the balanced chemical equation in solution at any concentration. This is applicable to all types of solutions of any concentration. The value keeps changing with concentration. When the solution is saturated i.e. at equilibrium; the ionic product is equal to the solubility product i.e. $IP = SP$. The ratio of IP and SP is called degree of supersaturation (S). Thus in this case $S = 1$.

When the ionic product exceeds the solubility product, i.e. $IP \rangle SP$ or $\frac{IP}{SP} = S \rangle 1$, the solution is supersaturated. In this situation, precipitation occurs and ions combine on the substrate and in the solution to form nuclei. In other words, at this state, ions in the solution mingle on the surface of the substrates, to form deposition. Temperature, solvent and particle size affect the solubility product. For any formation of thin film, there is some minimum number of ions or molecules, which produce a static phase in contact with solution, called nucleus. Nucleation on the substrate surface starts at local homogeneity. The rate at which nuclei is formed on the surface of the substrate (growth rate) depends on the degree of super saturation. The kinetics of growth of a thin film in this process is determined by the ion-by-ion deposition of the cations and anions on nucleating sites on the immersed surfaces. Initially, the film growth rate is negligible because an incubation period is required for the formation of critical nuclei from a homogeneous system onto a clean surface. Once nucleation occurs, the rate rises rapidly until the rate of deposition equals rate of dissolution i.e. $IP = SP$. Consequently, the film attains a terminal thickness.

If the supersaturation is maintained at low level throughout the precipitation, relatively few nuclei are formed. That will grow to give a small number of large particles with high supersaturation, many more nuclei are formed initially and nucleation may occur through the precipitation process. As a result, there are great numbers of centers upon which growth process can take place, none of the particles grow very large and colloidal suspension is formed. A solution with supersaturation possesses a high Gibbs free energy; the overall energy of the system would be reduced if the solute is set apart from the solution. This lessening in overall Gibbs free energy is the main driving force for both nucleation and growth. The film deposition by CBD depends on different parameter such as pH of the solution, complexing agent, solution temperature, concentration of the solution and also on the nature of the substrate and its temperature. These parameters control the ionic concentrations and hence the reaction rate and deposition rate of a deposition process in supersaturated condition. Figure 1 shows the schematic diagram of CBD process.

CBD technique, on the other hand, is performed normally under the atmospheric pressure at ambient temperatures. The technique is inexpensive and enables the synthesis of thin film materials with complex chemical composition (George, 1992; Niesen, 2002; Lokhande, 1991). The CBD technique can yield stable, uniform, adherent and hard films with good reproducibility. Due to its easy processing, it is suitable for large area deposition making it potential for commercial applications (Olivia, 2014). Thus CBD has become more popular in recent times and is extensively used in thin film fabrication. Also doping with metal ions at low temperatures can be conveniently carried out using this technique. The CBD technique may facilitate materials to be designed on a molecular level as well. The only common problem in CBD is precipitate formation and wastage of material since ions combine to form nuclei in the solution also.

Figure 1. Experimental set-up of chemical bath deposition
Source: Tauc, 1974

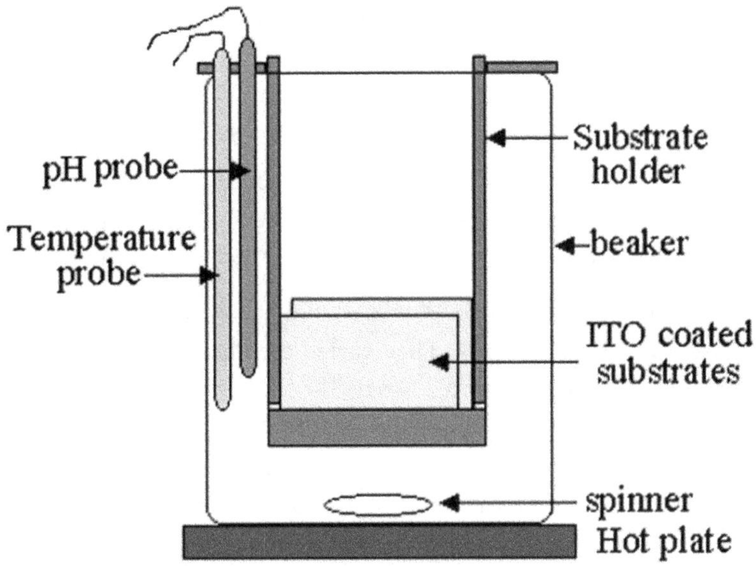

Various researchers have reported synthesis and characterization of CdZnS thin fims by CBD. Kumar et al. (Kumar, 2009) deposited $CdZnS$ thin films on glass substrate by a slightly different route. Microwave assisted chemical bath deposition technique was used by them. They used cadmium chloride, zinc chloride and thiourea as Cd, Zn and S source respectively. This technique required very short duration (about 600 sec.) of deposition compared to the common CBD process. The deposited $CdZnS$ thin films were of hexagonal form with (002), (101), (110) and (112) planes. The band gap increases from the 2.92 eV to 3.26 eV as Zn concentration increased which established the band gap tunability property and thus the aim of formation of ternary material is satisfied. Although transmission was reported to decrease with increasing Zn content in the films, the reason behind such enhancement of absorption was not explained in detail. They observed high intensity of PL spectra with lower Zn content. The crystallinity became poor and the structural and morphological disorder increases with higher Zn content. This causes asymmetry in PL peaks. For each sample, the PL peak energy was always lesser than the corresponding band gap energy. This is obvious that the radiative transition took place due to surface state rather than excitonic transition. Asogwa et al. (Asogwa, 2010) fabricated $CdZnS$ thin film by CBD at room temperature. They deposited the substrate in the bath solution for nearly about 5 hours and then annealed two samples one at 200°C and other at 300°C for 1 hour. The bandgap energy of the $CdZnS$ films lies between 2.10 eV - 2.30 eV which is low compared to other reported data. They also observed that the average reflectance increases with annealing temperature. Nagamani et al. (Nagamani, 2012) also deposited $Zn_xCd_{1-x}S$ thin film by CBD on glass substrate at 70°C. They immersed the substrate on bath solution for 30 minutes to get uniform, adherent film. EDAX micrograph confirmed the presence of Zn, Cd and S along with some oxygen content. Hexagonal structure with (002) preferential plane was obtained from XRD analysis. The grain size was about 45-76 nm and for $Zn = 0.75$ the average transmittance value was nearly 72%. Jia et al. (Jia, 2009) has studied the effect of Zn^{2+} concentration and time of deposition on the properties of $CdZnS$ thin films. They used ammonium nitrate, KOH, cadmium chloride and zinc sulfate to deposit $Zn_xCd_{1-x}S$ thin film on glass substrate. The mixture ratio $x\left(x=\frac{ZnSO_4}{\left[CdCl_2+ZnSO_4\right]}\right)$ was varied from 0.1 and 0.9 in order to investigate the influence of Zn concentration on the growth parameters of the deposited thin films. The competition mechanism of Zn^{2+} and Cd^{2+} with complexing agent ammonia plays a significant role in the formation of $Zn_xCd_{1-x}S$ thin film for low Zn concentration according to their study. The rate of Cd^{2+} integration to $Zn_xCd_{1-x}S$ system is more rapid than the rate of Zn^{2+} insertion. The transmission spectra intensity decreases with increase in time of deposition. The band gap achieved for $Cd_{0.7}Zn_{0.3}S$ was about 3.1 eV which shows it increases with Zn incorporation to CdS. Ashokkumar et al. (Ashokkumar, 2013) deposited $Cd_{0.9}Zn_{0.1}S$ thin film on glass substrate by CBD process. They varied the deposition time from 30 min to 90 min and observed that the grain size increases from 204Å to 347Å gradually with increase in deposition time. The obtained films were homogenous, adherent and are of pure cubic phase. Chynoweth et al. (Chynoweth, 1980) fabricated films with a resistivity of 4.0×10^2 Ω-cm for $x=0.35$ after heat treatment in hydrogen atmosphere at a temperature range 150°C to 300°C. Padam et al. (Padam, 1988) reported hexagonal crystal structure in the range $0\le x\le 0.8$ and cubic crystal structure for $x=1$ for $Zn_xCd_{1-x}S$ films where Chu et al. (Chu, 1991) reported hexagonal $Zn_xCd_{1-x}S$ for the complete range of $0\le x\le 1$. This difference in crystallinity may be due to different growth techniques and different growth conditions. Some researcher prepared $Zn_xCd_{1-x}S$ thin film using cadmium acetate and zinc acetate solutions as Cd and Zn source respectively (Song, 2009). Zinc cadmium sulfide (ZnCdS) thin film layer prepared by CBD (Kumar, 2009) exhibited better blue photon response and very good higher current densities due to its wide bandgap compare to CdS thin layer used in $CuIn_{1-x}Ga_xS_2$ (CIGS2) based solar cells. The CIGS2/ $Zn_xCd_{1-x}S$ heterojunction layer exhibited higher open circuit voltage (V_{oc}) which indicates better junction properties. The *J–V* curves obtained at various temperatures suggest that higher V_{oc} can be obtained by optimizing the conduction band offset. They also discussed the effect of variation of parameters like concentration of compounds, pH of solution and deposition time during CBD process on the properties and composition and crystallinity of obtained film. A better efficiency comparable to CIGS2/CdS devices has been achieved using CIGS2/ $Zn_xCd_{1-x}S$ devices. A group of researcher (Sanap, 2011) deposited $Zn_xCd_{1-x}S$ thin film onto microscopic glass substrate using CBD. The obtained film has both cubic and hexagonal structure. Lattice constant gradually decreased as Zn content increases. The band gap increases with increase in Zn concentration and was in the range of 2.48-3.62 eV. In general, the resistivity of the $Zn_xCd_{1-x}S$ films increased with zinc content. So they doped boron as impurity to decrease the resistivity of the film. The crystal structure was of wurtzite type with preferential orientation along (002) plane and lattice parameters changed from the values of CdS to those for ZnS as the amount of zinc acetate in the solution was increased. $Zn_xCd_{1-x}S$ thin films grown by CBD were also used as the buffer layer between $n-$type ZnO and $p-$type chalcopyrite film for wider applications. By varying the concentration of the Zn and Cd salts in bath, the entire range of overall compositions ranging from primarily cubic ZnS to hexagonal CdS could be deposited easily (Tosun, 2012). Along with band gap, electrical resistivity is also important for a material to be used in solar cells. For application as a window material in solar cell, a wide band gap combined with low resistivity is very much essential.

$Cd_{0.7}Zn_{0.3}S$ was deposited by the chemical bath deposition (CBD) method on glass substrates from an aqueous solution containing cadmium chloride, zinc chloride, and thiourea at 80 ± 5 °C (Al-Jubory, 2012). They studied the influence of Cu doping on the structural and optical properties of $Cd_{0.7}Zn_{0.3}S$ films. The XRD analysis suggests that the obtained film was polycrystalline in nature with presence of hexagonal phase. The energy bandgaps of the $Cd_{0.7}Zn_{0.3}S : Cu$ samples vary between 2.1 eV and 3.8 eV. These films are very much advantageous for different application such as solar cell, optical devices etc. Some researchers have recorded photoconductivity and absorption spectra of $Zn_xCd_{1-x}S$ to study the variation of band gap with change in composition of the mixed system. The growth of photocurrent has been discussed in details. Oxygen absorption criteria were taken into account to explain the rise characteristics (Singh, 1976). Petera et al. (Petera, 2012) have calculated the exciton binding energy as well as the interband emission energy of $Zn_xCd_{1-x}S$ system. They observed that the particles size was in the range of 3-6 nm. A research group (Abbas, 2013) showed variation of energy band with thickness. Band gap decreases from 2.85eV to 2.7 eV with increases in thickness (from 200 nm to 400 nm). Hernandez et al. (Hernandez, 2014) studied the effects of vacuum annealing at different temperatures on the electrical, optical and photocatalytic properties of the ternary semiconductor alloys $Zn_xCd_{1-x}S$, $Cu_xCd_{1-x}S$ and $Cu_xZn_{1-x}S$ obtained by CBD. $Zn_xCd_{1-x}S$ annealed at 250°C was composed of agglomerated sphere (~ 160 nm) consist of small particles sized about 8.5 nm. $Zn_xCd_{1-x}S$ films annealed under vacuum at 300°C shows high transmittance $\geq 85\%$ with band gap energy about 2.5 eV to 2.83 eV for different x value. They also observed extensive photodegradation of methylene blue for the above mentioned films. It was also evident that the $Zn_xCd_{1-x}S$ system is very much economical compared to ZnO and TiO_2 systems in case of application to solar cell devices.It appears that substitution of Zn in CdS on its structural, morphological and optical properties needs to be studied in details.

$Cd_xZn_{1-x}S$ thin films were deposited by chemical bath deposition method from an aqueous solution containing cadmium chloride $\left(CdCl_2\right)$, zinc chloride $\left(ZnCl_2\right)$, thiourea$\left[\left(NH_2\right)_2 CS\right]$, ammonia and triethanolamine (TEA). All the reagent were of analytical grade. Deionised water was used as a solvent. The precursor solution was prepared by mixing appropriate amounts of cadmium chloride and zinc chloride in thiourea solution to get films with different stoichiometry. Requisite amount of zinc chloride and cadmium chloride (0.1 M each) were added to form 100 ml solution. Zn-composition was varied in the range 0 – 0.8 at. % (x = 00, 0.2, 0.5, 0.8) in the solution. 40 ml TEA was added as complexing agent to the solution which was followed by addition of ammonia to make the solution pH in the range 10.5-10.6. The mixture was then heated to 70°C under continuous stirring using a controllable magnetic stirrer. 100 ml thiourea solution (0.1 M) was then added slowly to the heated solution. Ultrasonically precleaned microscope glass substrates were vertically dipped in the deposition bath while stirring of the solution continues. The schematic diagram of CBD setup is shown in Figure 1 (http://www.ece.utep.edu). The deposition of thin film takes place through condensation of the metals and sulfur ions on the initial layer, which acts as a catalytic surface. The deposition of the $CdZnS$ ternary thin film is based on the precipitation followed by condensation. An adherent thin film results upon immersion of the substrate for 60 minutes. After 60 min, the coated substrates are removed and washed with double distilled water to remove loosely bound $CdZnS$ particles and then dried in hot air. The fabricated $CdZnS$ films are bright yellow, uniform, highly adherent with specularly reflective. Next section describes the physical characteristics (structural, morphological and optical) of the as-grown thin layers of $CdZnS$.

Physical Properties of CdZnS Thin Films

X-ray diffraction is a versatile and non-destructive analytical method to uniquely identify the crystalline phases present and to study the structural properties in details. A given substance always generates a characteristic diffraction pattern, whether the substance is present in the pure state or as one component of a mixture of substances. This fact is the basis for the diffraction method of analysis. Qualitative analysis for a particular material is accomplished by identification of the pattern of that substance. Quantitative analysis is also possible, because the intensities of diffraction lines due to one constituent of a mixture depend on the proportion of that constituent in the mixture. The results presented here were taken using a X-ray diffractometer (Bruker, Model D8 Advance-AXS) with Cu-Kα radiation source (λ=1.5406 Å). Figure 1 shows the XRD patterns of $Cd_xZn_{1-x}S$ ternary thin films with different zinc compositions, $x = 1.0$, 0.8, 0.5 and 0.2. The material was scanned in the range 20-80°. The variation of 2θ value was employed with a 0.05 degrees step and a time step of 1 second. Intensity in arbitrary units is plotted against 2θ in Figure 1. The CdS thin film exhibits a strong diffraction peak at $2\theta = 26.75^o$ (Figure 2(a)), which may be associated with the (111) reflection of cubic CdS [CdS (C) (111)] by comparing with standard PDF card [JCPDS file no. 10-0454]. It is very difficult to distinguish between the (111) cubic direction and the (002) hexagonal one for CdS since the corresponding peaks are less than

Figure 2. XRD patterns of the $Cd_xZn_{1-x}S$ films deposited at different x values (a) $x = 1$; (b) $x = 0.8$; (c) $x = 0.5$ and (d) $x = 0.2$

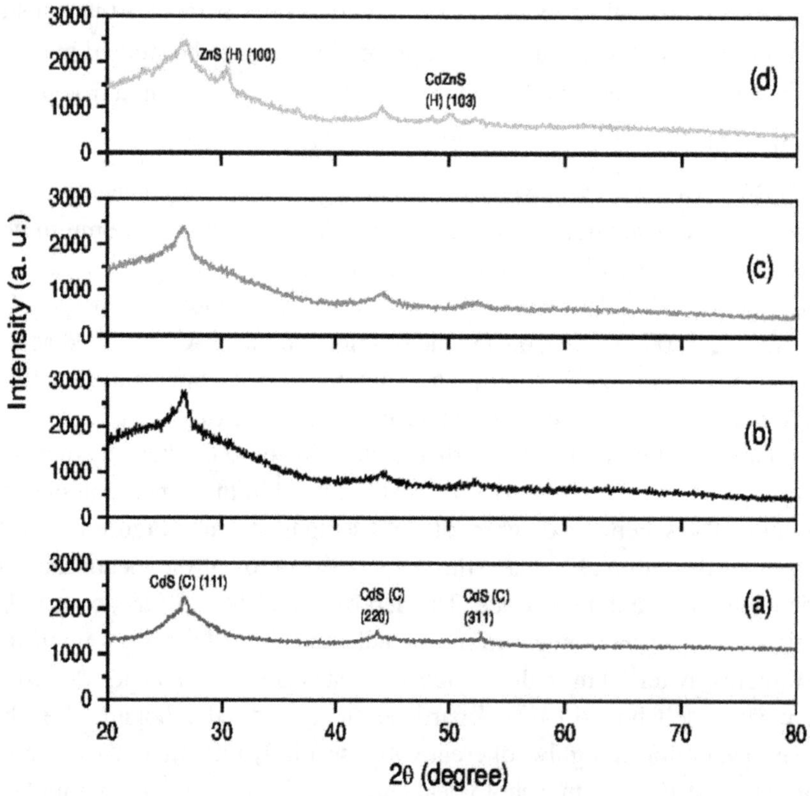

1% apart (Sain, 2012; Sain, 2012). Normally films prepared at low temperatures (up to 100°C) present a high probability of growing with the cubic structure whereas films prepared at high temperatures (300°C or more) show a high probability of growing with hexagonal structure (Olivia, 2003). Since low synthesis temperature is used in our work, cubic structure in our samples seems most probable and we assign the peak to CdS (C) (111). The other two weak peaks at $2\theta = 43.65^o$ and 52.2° may be assigned to CdS (C) (220) and CdS (C) (311) reflections respectively.

An inspection of the peak position for $Cd_{0.8}Zn_{0.2}S$ (with low Zn content) showsno significant change indicating that the ternary peak obtains a nearly binary CdS value for low Zn content. Similar result has been reported for vacuum coevaporated films (Xia, 2011). The position of the major diffraction peak at 26.75° is shifted to higher 2θ value (~26.85°) for maximum Zn -content film $\left(Cd_{0.2}Zn_{0.8}S\right)$ .An increase in diffraction angle means a decrease in the lattice parameter. Evaluation of lattice spacing $\left(d\right)$ for the major diffraction peak at shows a decrease from ~0.3332 nm for pure CdS to ~0.3319 nm for $Cd_{0.2}Zn_{0.8}S$. The corresponding values of parameter $\left(a\right)$ for a cubic system (Deo, 2014) were ~0.5771 nm and ~0.5749 nm respectively. The corresponding values for $Cd_{0.5}Zn_{0.5}S$ are ~0.3323 nm and ~ 0.5754 nm respectively. Thus the shift in the diffraction angle might be due changes took place in the lattice parameters with the enhance of Zn -content. Thus (100) plane of hexagonal ZnS [JCPDS file no. 36-1450] might have superposed on CdS (C) (111). However, while estimating the lattice parameter value for $Cd_{0.2}Zn_{0.8}S$ at 26.85°, we assumed that the peak was actually the shifted major diffraction peak of pure CdS at 26.75° and the shift was mainly due to Zn ion incorporation. This was made on the basis that zinc incorporation is limited in CdS structure (Mukherjee, 2015). Additional peaks appear for $Cd_{0.2}Zn_{0.8}S$ at 30.5° and 50.15°. The peak at 30.5° may be associated with (101) plane of hexagonal ZnS [JCPDS file no. 36-1450] since no peak of cubic CdS or cubic ZnS emerges at this angle. Although a peak of cubic CdS appears at 30.832° [JCPDS file no. 10-0454], it is the fourth highest intensity peak of cubic CdS and it did not appear in case of high Cd content films. The peak at 50.15° might be due to (103) plane of hexagonal $CdZnS$ [JCPDS file no. 35-1468] as no peak of hexagonal or cubic phase of neither CdS or ZnS appears at this angle according to JCPDS data. The shift in the major diffraction peak angle and the appearance of the new peak at 50.15° indicates that there is no phase separation or separated nucleation of ZnS or CdS in the $Cd_xZn_{1-x}S$ phase.

Field emission scanning electron microscopy (FESEM) was used to measure the particle size. The same instrument was used for energy dispersive X-ray (EDX) analyses which was carried out to qualitatively measure the sample stoichiometry. Energy dispersive X-ray (EDX) is an important analytical technique used for the elemental analysis or chemical characterization of a sample. It is one of the variants of X-ray fluorescence spectroscopy which is based on the investigation of a sample through interactions of electromagnetic radiation with matter. Quantitative estimation is made by analyzing the X-rays emitted by the matter when it is bombarded with charged particles. Its characterization capabilities lie on the fundamental principle that each element has a unique atomic structure allowing X-rays that are characteristic of an element's atomic structure to be identified uniquely from one another. Figure 3 (a) to (d) shows the EDX spectra for CdS , $Cd_{0.8}Zn_{0.2}S$, $Cd_{0.5}Zn_{0.5}S$ and $Cd_{02}Zn_{0.8}S$ film respectively. Pure CdS film was found to be nearly stoichiometric with slight excess of cadmium element. Elemental composition was obtained by comparing the relative concentration of the elements; the later was obtained from the normalized atomic percent concentration. Table 1 shows the detail atomic percentage of con-

Figure 3. EDX pattern of films for (a) ; (b) ; (c) and (d)

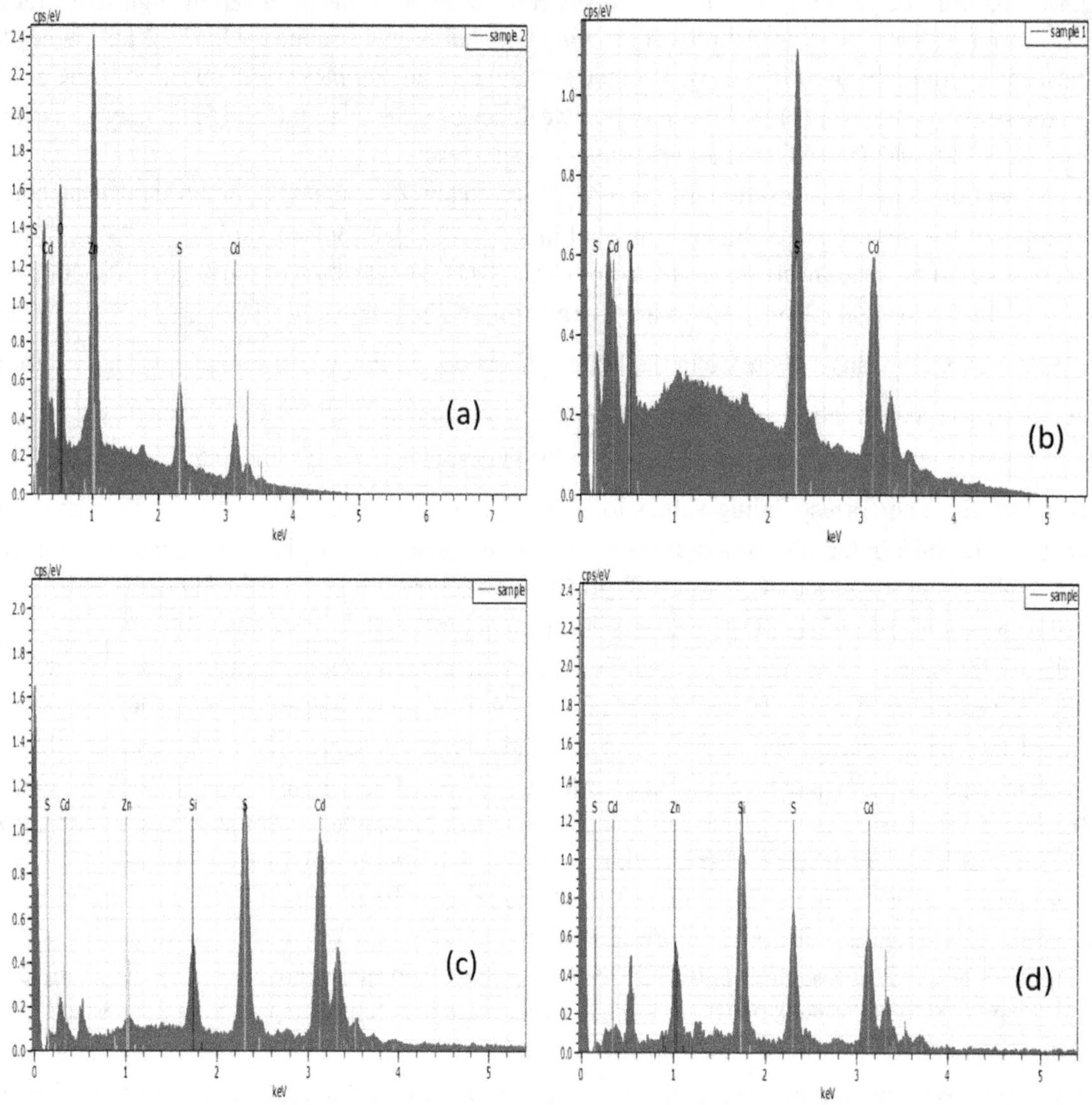

Table 1. Atomic percentage of constituent elements in $Cd_xZn_{1-x}S$ thin film

	$x = 1.0$	$x = 0.8$	$x = 0.5$	$x = 0.2$
Cd	47.29	42.84	25.28	11.73
Zn	--	7.61	22.50	38.88
S	44.87	39.22	41.40	38.08
Si	3.11	4.72	5.01	4.38
O	4.73	5.61	5.86	6.93

stituent elements in $Cd_xZn_{1-x}S$ thin films. The $Cd : Zn$ ratio in $Cd_xZn_{1-x}S$ was observed to be ~85:15, ~53.:47 and ~23:77 for $x = 0.8$, 0.5 and 0.2 respectively. The composition of the layers agrees approximately with that of the elements in the solution with excess of cadmium in all cases. The $Cd : Zn$ ratio of ~53:47 in $Cd_{0.5}Zn_{0.5}S$ indicates moderately good stoichiometry of the deposited films with small excess amount of cadmium. The peaks appeared at 0.52 keV and at 1.72 keV were due to oxygen and silicon coming from the substrate component.

Figure 4 (a) shows the FESEM image of CdS film. Nearly spherical shaped nanoparticles are found to be distributed more or less well covering the substrate with some vacant spaces between the grains. The average size of nano-clusters having minimum dimension was in the range 60-70 nm evaluated us-

Figure 4. FESEM image of (a), (b) film and (c) film

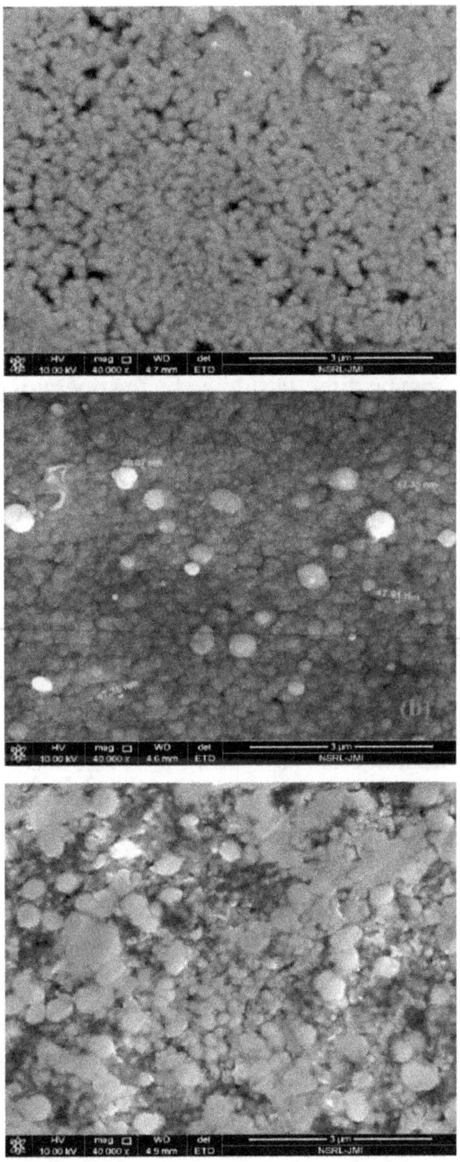

ing IMAGE software. Figure 4 (b) and 4 (c) shows the FESEM images of $Cd_{0.5}Zn_{0.5}S$ and $Cd_{02}Zn_{0.8}S$ films respectively. It is evident from the figures that in many areas nanosized particles form globular aggregates of microdimensions. The average size of nano-clusters having minimum dimension was ~ 45.67 nm for $Cd_{0.5}Zn_{0.5}S$ and it further decreases to ~ 36.6 nm for $Cd_{02}Zn_{0.8}S$ film. Similar observation of reduced mean grain size for Zn-doped CdS film compared to undoped CdS film has been reported for CdS film synthesized by CBD (Olivia, 2014). The surface morphology of $CdZnS$ thin films with lower Zn^{2+} concentration confirms the better surface features than the film at higher Zn^{2+} concentration. It seems that for high Zn-content film, excess doping agent precipitates resulting in zinc coalescence in some specific regions (Teterycz, 2013). This type of excess precipitation has been reported to limit incorporation of zinc ions in CdS structure (Olivia, 2014). Such type of coalescence sometime creates problem by impeding the optical transparency resulting in window absorption loss in solar cells.

Electron microscopes in which electrons are allowed to transmit through the objects are known as transmission electron microscope (TEM). TEM is a very powerful tool for characterization in material science. TEM consists of an electron gun, central column, electromagnetic lenses and a fluorescent screen. The electron gun is the source of electrons which generates a high energy beam of electron. The bean of electron travels through the microscope column which is an evacuated metal tube. The image of a sample is formed by illuminating the sample with an electron beam and detecting the electrons that are transmitted through sample (Spence, 2003). The electromagnetic lenses are used to focus the electrons into a narrow thin beam. Depending on the density of the material present, some of the electrons are scattered and disappear from the beam. At the bottom of the microscope the unscattered electrons hit a fluorescent screen, which gives rise to a "shadow image" of the specimen with its different parts displayed in varied darkness according to the density. The darker area of image represents the denser portion of the sample. The image can be directly studied by the operator or photographed with a camera. TEM is used to study structural properties, lattice parameters, particle size, crystal planes etc. It also gives information about the growth of layers and defects in specimen. The modern age TEM (High resolution TEM or HRTEM) can produce image with resolution upto 1 Å (Chattopadhyay, 2014). According to the desired information from TEM there are four conventional operational modes (Fultz, 2008). These are (i) Bright-Field, (ii) Dark-Field, (iii) Diffraction Contrast and (iv) Selected Area Electron Diffraction (SAD or SAED). Selected area diffraction is a TEM technique to obtain diffraction patterns that came out from the electron beam scattered by the sample lattice. Obeying the Bragg's law, the electrons are scattered elastically by the lattice. So, we can index those diffraction spots in the pattern and identify the phases in the sample, as well as study their structures. Typically, the size of the area of the sample been selected by SAD aperture is in the range of 0.5-1 μm. From a SAED pattern, now a day, the material scientists can acquire structural information of the sample, like crystalline symmetry, unit cell parameter and space group etc. Various models of TEM have been used in our experiments depending on their availability and they have been mentioned in respective places.

Transmission electron microscopy (TEM) investigation using Tecnai F30 G^2, FEI, Hillsboro, Oregon microscope operating at 200 kV are presented here. SAED was used for structural characterization and TEM was used for particle size estimation. The sample was detached from the substrate and dispersed in ethanol using ultrasonic bath, mounted on a carbon coated Cu grid, dried, and used for TEM measurements. The indexed selected area electron diffraction (SAED) pattern of TEM images of $Cd_{0.5}Zn_{0.5}S$ film is shown in Figure 5 (a). The observed d value of ~3.32 Å for the first ring belongs to a single cubic phase (C) (111) in accordance to the consequent XRD pattern. Presence of minor hexagonal ZnS

Figure 5. (a) Selected area electron diffraction (SAED) image of $Cd_{0.5}Zn_{0.5}S$ film shows Debye rings of cubic and hexagonal phases; (b) Presence of bothhexagonal and cubic planes in $Cd_{0.5}Zn_{0.5}S$;(c) Low magnification TEM image

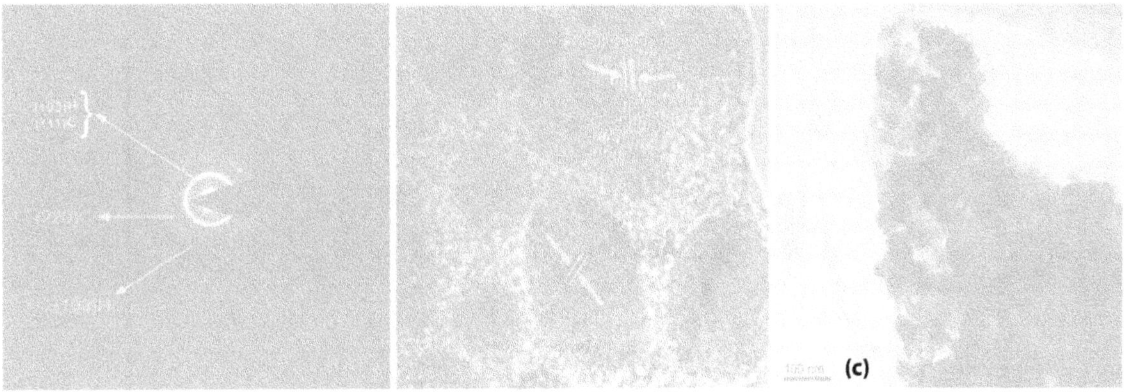

(100) phase cannot be overruled since incorporation of zinc resulted in peak shift. Owing to shorter wavelengthof electron beam, reflections from hexagonal and cubic phases can be entirely overlapped into a single Debye–Scherrer ring in the SAED pattern (Sain, 2012). d value of ~2.208Å for the second ring in the SAED pattern might be due to CdS (C) (220) if the diffraction lines are indexed in agreement to the corresponding XRD pattern. The closest match of d value of ~1.81Å for the third ring is with the (103) reflections of hexagonal $CdZnS$ [1.808Å for (103) plane of $Zn_{0.78}Cd_{0.22}S$; JCPDS file no. 35-1468]. CdS (H) (200) with d value ~1.793 [JCPDS file no. 41-1049] may also be superimposed on this.

The planar view of HRTEM images (Figure 5(b)) confirms the presence of different lattice planes. The observed lattice fringes with d value of ~3.33 Å is from CdS (C) (111). ZnS (H) (100) may be entirely superimposed between these planes. In the same way, observed lattice fringes with d value of ~2.25 Å (measured as ~2.201 Å from SAED ring) is likely close to d_{220} of CdS (C) (220). Figure 5 (c) shows low magnification TEM image which can be used to calculate the average particle size of the CdZnS nanoparticles. The estimated average particle size of the observed sample by TEM analysis was in the range 20 nm. The average value of particle size was slightly lower than the value obtained from FESEM analysis. This may be due to agglomeration in case of FESEM samples. Due to minute particle sizes, peaks are considerably broadened which increases the possibility of overlapping of peaks of both cubic and hexagonal phases and at the same time making it very complex to find individual reflection from minor hexagonal phases in the XRD patterns. Figure 6 shows the XRD pattern of CdS film annealed in air 250°C for 1 hr. The 2θ variation was employed with a 0.05 degrees step and a time step of 5 seconds. The major diffraction peak at 26.75° turns into more intense and sharp indicating improved crystallinity which may be due to heat treatment. The value of full width at half maximum intensity (FWHM) for the major diffraction peak at 26.75° reduces from ~0.46° for as deposited CdS to ~0.34° for heat treated CdS also indicates reduction of peak broadening and improved crystallinity. Also an extra peak appears at ~31.14° which may be attributed to CdS (C) (200) [JCPDS file no. 10-0454]. This helps to compare the changes due to Zn incorporation in CdS.

The optical properties of CdZnS thin films plays an important role in its applications. UV-VIS spectroscopy is used to get details information about optical properties. UV-VIS spectroscopy refers to ab-

Figure 6. XRD patterns of the CdS film air annealed at 250°C

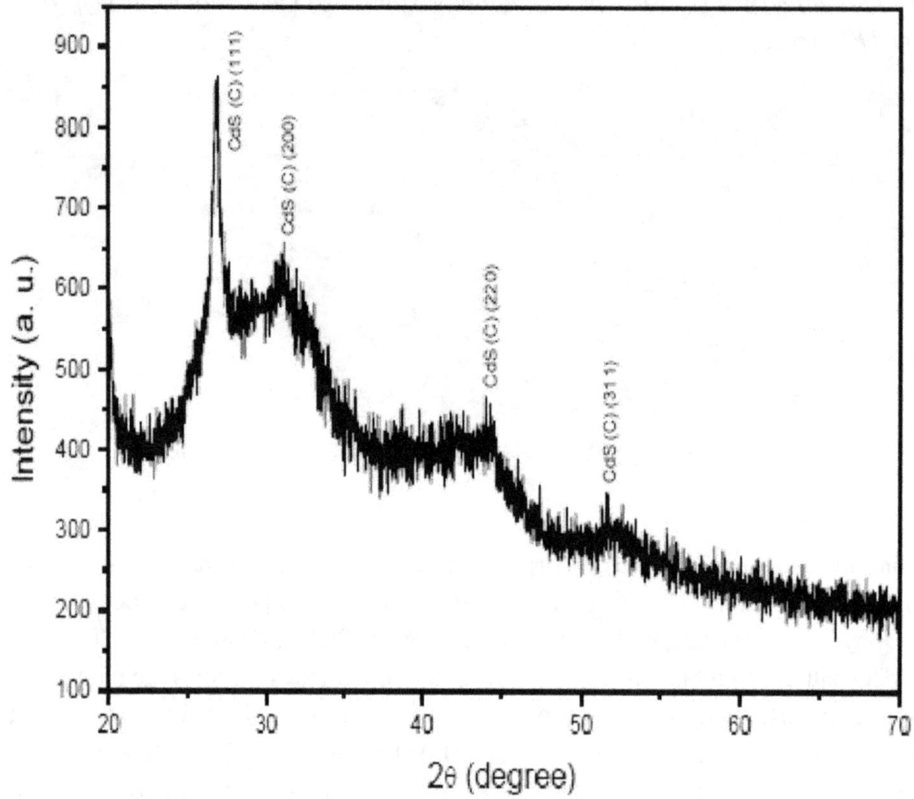

sorption spectroscopy in the UV-visible spectral region. The study of optical absorption explain clearly the band structure of material. There are in general two types of optical transitions, which can occur at the fundamental absorption edge of crystalline as well as non-crystalline semiconducting materials. They are direct and indirect transitions. When a light beam falls on a thin film semiconducting material, a part of the beam will be reflected, another part will be transmitted through the film, and the rest of the beam will be absorbed. Absorption of photons causes transition of the electrons from valance band to conduction band. Figure 6 shows the plot of absorbance $\left(A\right)$ as a function of wavelength $\left(\lambda\right)$ in the wavelength range 400-1000 nm for $Cd_xZn_{1-x}S$ film with lower, intermediate and higher Zn content, respectively. Figure 7 (a, b, c) shows the plots for $Cd_xZn_{1-x}S$ with $x = 0.8, 0.5$ and 0.2 respectively. Increase in absorbance for highest Zn content film indicates lowering of transmittance in zinc excess films. Although excess precipitation of zinc in CdS has been reported by Olivia et al. (Olivia, 2014) and lowering of transmittance in zinc excess films has been reported by Jia et al. (Jia, 2010), no correlation between aggregation of nanocrystals due to zinc enrichment and lowering of transmittance were reported in these works. While higher absorbance for Zn doped CdS compared to Cu doped CdS has been reported by Oliva et al. (Olivia, 2014), no comparison with pure CdS was made. In this work it is shown that such lowering of transmittance with zinc enrichment in the films can be attributed surplus zinc coalescence. It further emerges from Figure 6 that the value of absorbance decreases (increase of transmittance)

Figure 7. Absorbance vs photon wavelength $\left(\lambda\right)$ *of* $Cd_xZn_{1-x}S$ *films for different values of* x *: (a)* $x = 0.8$ *; (b)* $x = 0.5$ *and (c)* $x = 0.2$

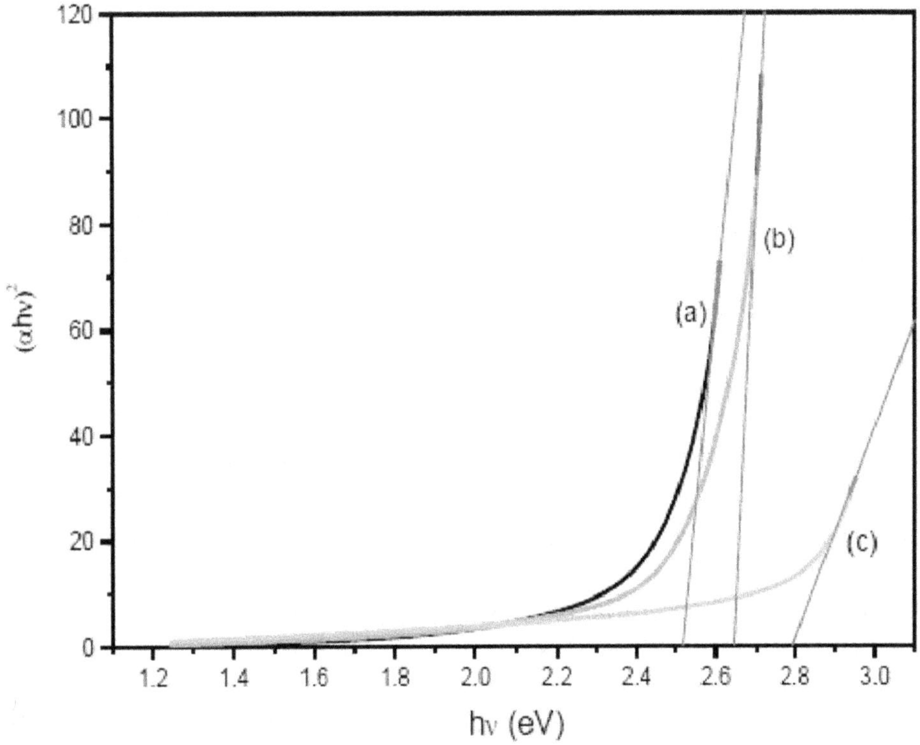

below 600 nm for zinc excess films. This might be related to lowering of film thickness due to zinc enrichment. Although in this work no systematic thickness measurement was carried out, a rough and ready gravimetric technique (in which weight change of the substrate due to film deposition was measured and then it was divided by product of area of deposited film and known density of the material) shows lowering of thickness with increase in zinc content (~800 nm for pure CdS to ~480 nm for highest zinc content films). Unexpected lowering of film thickness with zinc enrichment (even though not monotonically) has also been observed and reported in details by researcher (Jia, 2010). Whereas Mahdi et al. (Mahdi, 2012) observed a monotonic decrease in the film thickness with increasing zinc amount. It is obvious that for lower energy regions (high wavelengths), agglomeration of nanocrystals due to zinc ion coalescence controls the absorption process. However, their contribution to enhanced absorption is neutralized by thickness effect at high energy regions (low wavelengths). This might be due to lessened number of atoms at low thickness leading to decreased number of collisions with incident photons. Thus, it can be concluded from above discussion that for comparable thicknesses, enhancement of zinc ion concentration in CdS structure enhances the absorption. Such lowering of transmittance has adverse effect in window layers designing in solar cells devices.

UV-VIS spectroscopy refers to absorption spectroscopy in the UV-visible spectral region. Studies by optical absorption explain the band structure of material. There are two types of optical transitions, which can occur at the fundamental absorption edge of crystalline as well as non-crystalline semiconducting materials. They are direct and indirect transitions. When a light beam falls on a thin film semiconducting material, a part of the beam will be reflected, another part will be transmitted through the film, and the rest of the beam will be absorbed. Absorption of photons causes transition of the electrons from valance band to conduction band. The absorption ability is measured by its absorption coefficient (α) which is a function of frequency (Pankove, 1975; Klingshirn, 1997;Tauc, 1974) and is defined as:

$$I = I_o e^{-\alpha t}$$

where I is the intensity of the transmitted beam, I_o is the intensity of the incident light and t is the thickness of the film. The nature of transition is determined by using Tauc's relation (Tauc, 1974):

$$\left(\alpha h\nu\right)^n = A\left(h\nu - E_g\right)$$

where $h\nu$ is the photon energy, α is the absorption coefficient, E_g is the band gap energy, A is a constant and it is a function of index of refraction and hole/electron effective masses (Pankove, 1975). The constant n is equal to two (2) for direct transition and equal to one (1) for indirect transition (Pankove, 1975; Klingshirn, 1997; Tauc, 1974).

UV-Visible spectrophotometer instrument requires a source (UV and visible), filter or monochrometer, sample holder and detector to study absorption spectra in UV and visible range. Several models of spectrophotometers of varying degree of sophistication are available. These include single beam, double beam, reflecting and multibeam instruments. Deuterium or hydrogen lamp (for UV light) and tungsten lamp (for visible light) are generally used as the light sources. In dual beam spectrophotometers, the incident light is split into two beams of equal intensity, one of which passes through the reference cuvette and the other through the sample cuvette. The instrument records the change in absorbance in the sample with respect to the reference. The detector records the change in absorbance in the sample. The thickness of the film and the particle size of the specimen thin film have effect on the absorption coefficient. For our experiments, a dual beam UV-VIS spectrophotometer (Shimadzu, Model No. UV-1800; Spectral resolution 1 nm) was used. The microscopic glass slide (on which film deposition was carried out) was used as reference.

The optical absorption measurements performed using a Schimatzu dual beam UV-Vis spectrophotometer (UV-1800) are presented here.The spectra were recorded by using a similar glass as reference and hence the absorption due to the film only was obtained. The band gap of the films was calculated from the absorption edge of the spectrum. The value of band gap energy $\left(E_g\right)$ was derived from the mathematical treatment of the data obtained from the absorbance vs. wavelength spectrum for direct band gap $CdZnS$ thin film (Mukherjee, 2015) using Tauc's relationship. Figure 8 (a, b, c) shows the plot of $\left(\alpha h\nu\right)^2$ as a function of photon energy $h\nu$ for $Cd_xZn_{1-x}S$ film with lower, intermediate and higher Zn content, respectively. Extrapolation of the line to the base line, where the value of $\left(\alpha h\nu\right)^2$ is zero, gives the band gap E_g (Figure 8). The sharp absorption edge corresponding to the band gap confirms the good quality of grown films. The optical band gap vary from ~2.52 eV (for $Cd_{0.8}Zn_{0.2}S$) to

Figure 8. Photon energy (eV) dependence of $(\alpha h\nu)^2$ *of* $Cd_xZn_{1-x}S$ *films for different values of* x *: (a)* $x = 0.8$ *; (b)* $x = 0.5$ *and (c)* $x = 0.2$.

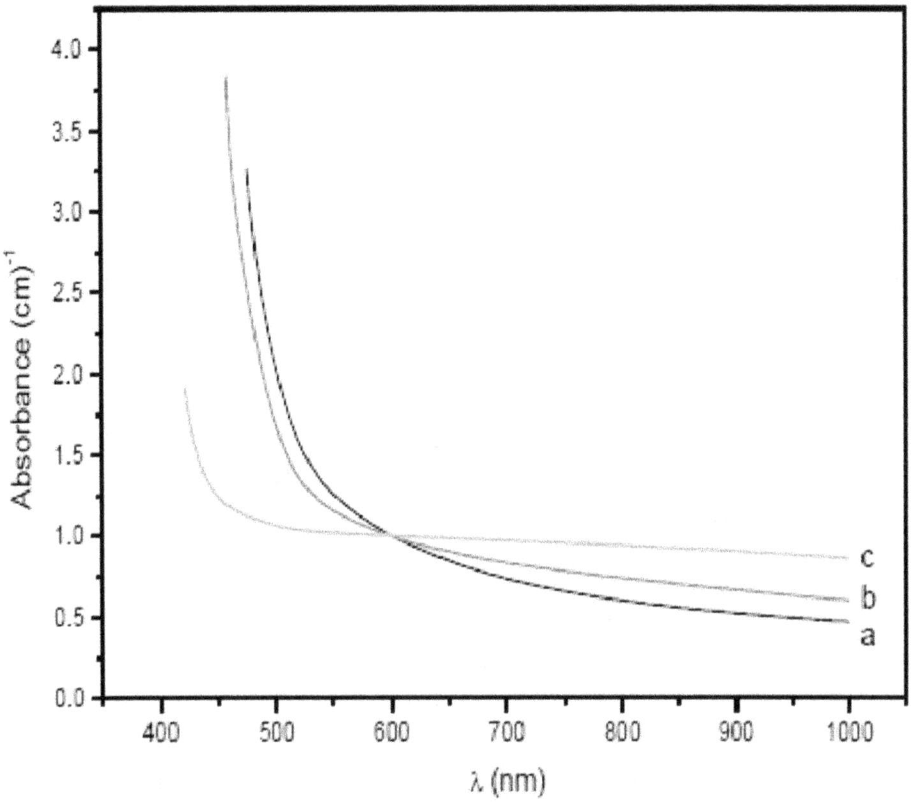

~2.80eV (for $Cd_{0.2}Zn_{0.8}S$). Enhancement in band gap due to zinc incorporation has been reported for CBD films and this indicates solid solution formation (Mukherjee, 2015). The absorption edges of the films were observed to shift towards shorter wavelengths (blue shift) with respect to the zinc enrichment in the film. This is strongly evident that the tuning of the band gap value is mainly due to change in Zn^{2+} concentration in the formation of $CdZnS$ thin films. Figure 9 shows the reflectance spectrum of CdZnS thin films for two different compositions. The reflectance value drops abruptly after 200 nm in both cases.

CONCLUSION

The CdZnS thin film can be prepared by various method. Among them chemical bath deposition (CBD) process is suitable for deposition. The technique of CBD is very simple, cost effective and offers an easily large scalable alternative to industrial level. In this chapter we have discussed in details about the preparation of CdZnS thin film by CBD and their characterization. The crystal structures of the $Cd_xZn_{1-x}S$

Figure 9. Reflectance spectrum of $Cd_xZn_{1-x}S$ films for $x = 0.4$ and $x = 0.6$

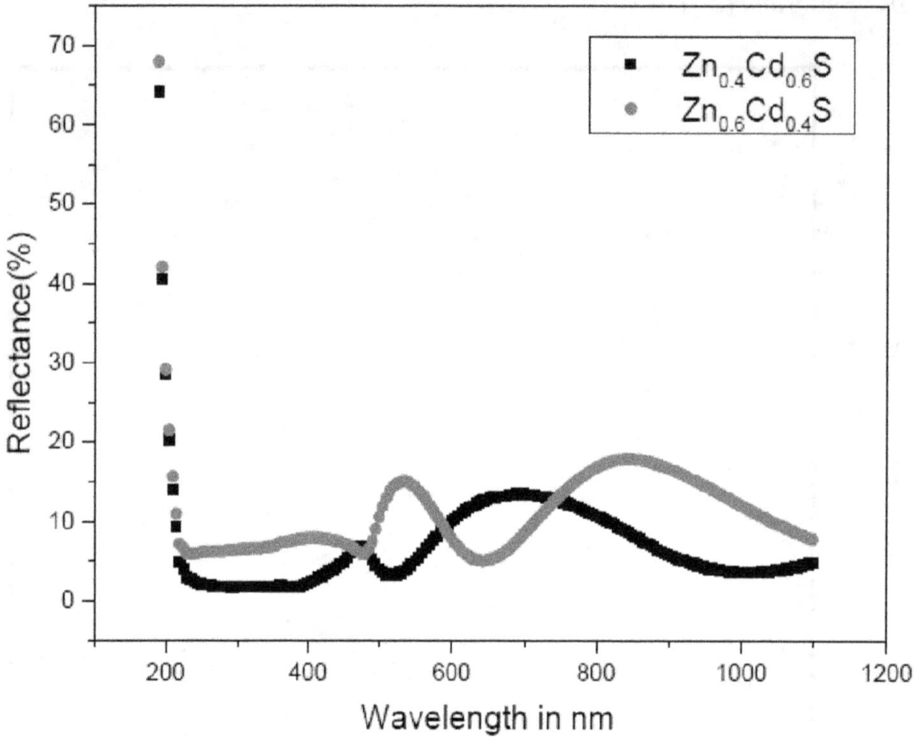

films are primarily cubic with possible coexistence of hexagonal phase component as evidenced from XRD and SAED pattern analysis. EDAX analysis clearly shows nearly stoichiometric films with slight cadmium enrichment. The close packed and less dissociated surface morphology was observed in the $Cd_xZn_{1-x}S$ ternary thin films with lower Zn^{2+} content compared to the higher Zn^{2+} content. TEM shows particle size of ~ 15 nm for as deposited $Cd_{0.5}Zn_{0.5}S$. The band gap of the film increases from ~2.52 eV to ~2.80 eV as the Zn^{2+} compositional parameter increased in the chemical bath. The range of band gap energy for the mixed films may be helpful in designing a suitable window material in fabrication of solar cells and sensors. The band gap can be tuned to develop high quality optical devices. However, for high Zn-content film, excess amount of doping agent precipitates resulting in zinc coalescence in certain regions resulting in enhanced absorption which has adverse effect in window layers in solar cells. It appears from our observations that low value of Zn concentration, although present a band gap value slightly higher than that of CdS bulk, is suitable for application due to its higher transparency and better surface morphology. Future work require more modification of its properties so that its application can be widen more.

REFERENCES

Abbas, N. K., Abbas, L. K., & Muhameed, S. A. (2013). Effect of Thickness on Structural and Optical Properties of $Zn_xCd_{1-x}S$ Thin Films Prepared by Chemical Spray Pyrolysis. *Int. J. Thin Film Sci. Tec.*, *2*, 127–132. doi:10.12785/ijtfst/020208

Al-bassam, A. A. (1999). Photoconductivity and defect levels in $ZnxCd_1\_xSe$ with (x=0.5, 0.55) crystals. *Solar Energy Materials and Solar Cells*, *57*(4), 323–329. doi:10.1016/S0927-0248(98)00154-8

Al-Jubory, A. A. (2012). Study of the effect of copper doping on the structural and optical properties of $Cd_{0.7}Zn_{0.3}S$ nanocrystalline thin films prepared by chemical bath deposition. *International Journal of Science and Technology*, *2*, 707–712.

Ashokkumar, M., & Muthukumaran, S. (2013). Size Dependent Structural and Optical Properties of $Cd_{0.9}Zn_{0.1}S$ Thin Films. *Physics Procedia*, *49*, 137–144. doi:10.1016/j.phpro.2013.10.020

Asogwa, P. U. (2010). Variation of optical properties with post deposition annealing in chemically deposited CdZnS thin films. *Chalcogenide Letters*, *7*, 501–508.

Baykul, M. C., & Orhan, N. (2010). Band alignment of $Cd_{(1-x)}Zn_xS$ produced by spray pyrolysis method. *Thin Solid Films*, *518*(8), 1925–1928. doi:10.1016/j.tsf.2009.07.142

Butty, J., Peyghambarian, N., Kao, Y. H., & Mackenzie, J. D. (1996). Room temperature optical gain in sol-gel derived CdS quantum dots. *Applied Physics Letters*, *69*(21), 3224–3226. doi:10.1063/1.118017

Chattopadhyay, K. K., & Banerjee, A. N. (2014). *Introduction to Nanoscience and Nanotechnology*. Delhi, India: PHI Learning Private Limited.

Chen, X. D., Wang, X. B., Liu, L., Yang, D. C., & Fan, L. (2005). Functionalized semiconductor nanocrystals for ultrasensitive detection of peptides. *Analytica Chimica Acta*, *542*(2), 144–150. doi:10.1016/j.aca.2005.03.071

Chen, Y., Ji, X., Jiang, S., Sun, Q., & Jiang, B. (2003). Synthesis and characterization of CdS nanocrystals in poly(styrene-co-maleic anhydride) copolymer. *Colloid & Polymer Science*, *281*(4), 386–389. doi:10.1007/s00396-002-0750-8

Chopra, K. L., & Das, S. R. (1983). *Thin Film Solar Cells*. Springer Science & Business Media. doi:10.1007/978-1-4899-0418-8

Chu, T. L., Chu, S. S., Britt, J., Ferekides, C., & Wu, C. Q. (1991). Cadmium zinc sulfide films and heterojunctions. *Applied Physics (Berlin)*, *70*(5), 2688–2693. doi:10.1063/1.349384

Chynoweth, T. A., & Bube, R. H. (1980). Electrical Transport in ZnCdS Films Deposited by Spray Pyrolysis. *Journal of Applied Physics*, *51*(3), 1844–1846. doi:10.1063/1.327755

Deo, S. R., Singh, A. K., Deshmukh, L., Paliwal, L. J., Singh, R. S., & Adhikari, R. (2014). Structural, morphological and optical studies on chemically deposited nanocrystalline CdZnSe thin films. *J. Soudi Chem. Soc.*, *18*(4), 327–339. doi:10.1016/j.jscs.2014.01.005

Fultz, B., & Howe, J. (2008). *Transmission Electron Microscopy and Diffractometry of Materials. Springer-Verlag.*

Gaewdang, N., & Gaewdang, T. (2005). Investigations on chemically deposited $Cd_{1-}xZnxS$ thin films with low Zn content. *Materials Letters*, *59*(28), 3577–3584. doi:10.1016/j.matlet.2005.06.031

George, J. (1992). *Preparation of Thin Films*. New York: Marcel Dekker, Inc.

Hernandez, S. A. M., Cruz, D., Flores, F., Garcia, S. A., Jimenez, L., Arrocena, M. C., & Santos-Cruz, J. et al. (2014). Optical, Electrical and Photocatalytic Properties of the Ternary Semiconductors $Zn_{xC}d1_{-xS,}$ $CuxC_{d1}-x_{S\,an}d\,CuxZn1_{-x}S._{Int}$*ernational Journal of Photoenergy, 2014, 158*782. doi:10.1155/2014/158782

Jia, G., Wang, N. A., Gong, L., & Fei, X. (2009). growth characterization of cdzns thin film prepared by chemical bath deposition. Cha*lcogenide Letters, 6, 463–467.*

Jia, G., Wang, N. A., Gong, L., & Fei, X. (2010). optical properties and forming mechanism of cdzns thin film grown by chemical bath deposition. *Chalcogenide Letter*, *7*, 349–355.

Jokanovic, U. (2005). Finely dispersed nanoparticle. CRC Press.

Khanna, P. K., & Singh, N. (2007). Light emitting CdS quantum dots in PMMA: Synthesis and optical studies. *Journal of Luminescence*, *127*(2), 474–482. doi:10.1016/j.jlumin.2007.02.037

Klingshirn, C. F. (1997). *Semiconductor Optics. Springer-Verlag.*

Kumar, B., Vasekar, P., Pethe, S. A., Dhere, N. G., & Koishiyev, G. T. (2009). $ZnxCd_{1-□}xS$ as a heterojunction partner for $CuIn_{1-□}xGa_xS_2$ thin film solar cells. *Thin Solid Films*, *517*(7), 2295–2299. doi:10.1016/j.tsf.2008.10.108

Kumar, T. P., & Sankaranarayanan, K. (2009). tunability of structural, surface texture, compositional and optical properties of cdzns thin films by photo assisted chemical bath deposition technique. *Chalcogenide Letters*, *6*, 617–622.

Laukaitish, G., Lindroos, S., Tamulevicius, S., Leskela, M., & Rackitis, M. (2000). SILAR deposition of $CdxZn_{1-}xS$ thin films. *Applied Surface Science*, *161*(3-4), 396–405. doi:10.1016/S0169-4332(00)00275-0

Lokhande, C. D. (1991). Chemical deposition of metal chalcogenide thin films. *Materials Chemistry and Physics*, *27*(1), 1–43. doi:10.1016/0254-0584(91)90158-Q

Mahdi, M. A., & Al-Ani, S. K. J. (2012). Optical characterization of chemical bath deposition $Cd_{1-x}Zn_xS$ thin films. *Int. J. Nanoelectronics and Materials*, *5*, 11–24.

Mukherjee, A., Fu, M., & Mitra, P. (2015). Influence of Zn incorporation in CdS: Structural and morphological studies. *Journal of Physics and Chemistry of Solids*, *82*, 50–55. doi:10.1016/j.jpcs.2015.03.009

Nagamani, K., Reddy, M. V., Lingappa, Y., Reddy, K. T. R., & Miles, R. W. (2012). Physical Properties of $Zn_xCd_{1-x}S$ Nanocrytalline Layers Synthesized by Solution Growth Method. *Int. J. Optoelectron. Eng*, *2*, 1–4. doi:10.5923/j.ijoe.20120202.01

Niesen, T. P., & De Guire, M. R. (2002). Review: Deposition of ceramic thin films at low temperatures from aqueous solutions. *Solid State Ionics*, *151*(1-4), 61–68. doi:10.1016/S0167-2738(02)00604-5

Olivia, A. I., Corona, J. E., Patino, R., & Aviles, A. I. (2014). Chemical bath deposition of CdS thin films doped with Zn and Cu. *Bulletin of Materials Science*, *37*(2), 247–255. doi:10.1007/s12034-014-0642-9

Olivia, A. I., Rodriguez, R. C., Canto, S., Sosa, V., Quintana, P., & Pena, J. L. (2003). Comparison of properties of CdS thin films grown by two techniques. *Applied Surface Science*, *205*(1-4), 56–64. doi:10.1016/S0169-4332(02)01081-4

Padam, G., Malhotra, G. L., & Rao, S. U. M. (1988). Studies on solution-grown thin films of $Zn_xCd_{1-x}S$. *Journal of Applied Physics*, *63*(3), 770–774. doi:10.1063/1.340069

Pankove, J. I. J. I. (1975). Optical Processes in Semiconductors. Dover Publications Inc.

Park, J. S., Chae, H., Chung, H. K., & Lee, S. I. (2011). Thin film encapsulation for flexible AM-OLED: A review. *Semiconductor Science and Technology*, *26*(3), 034001. doi:10.1088/0268-1242/26/3/034001

Pathan, H. M., & Lokhande, C. D. (2004). Deposition of metal chalcogenide thin films by successive ionic layer adsorption and reaction (SILAR) method. *Bulletin of Materials Science*, *27*(2), 85–111. doi:10.1007/BF02708491

Petera, A. J., Lee, C. W. (2012). Electronic and optical properties of CdS/CdZnS nanocrystals. *Chin. Phys. B*, *21*, 087302 (1-8).

Rajathi, S., Sankarasubramabian, N., Ramsnathan, K., & Senthamizhselvi, M. (2012). Structural and optoelectronic properties of pyralytically sprayed CdZnS thin films. *ChalcogenideLett.*, *9*, 495–500.

Ravangave, L. S., Biradar, U. V., & Misal, S. D. (2012). The Effect of Ionic Composition on Structural and Optical Properties of $Cd_xZn_{1-x}S$ Thin films Grown by Spray pyrolysis. *Int. J. Sci. Res. Publ.*, *2*, 1–5.

Raviprakash, Y., Bangera, K. V., & Shivakumar, G. K. (2010). Growth, structural and optical properties of $Cd_xZn_{1-x}S$ thin films deposited using spray pyrolysis technique. *Current Applied Physics*, *10*(1), 193–198. doi:10.1016/j.cap.2009.05.020

Sain, S., Patra, S., & Pradhan, S. K. (2012). Quickest ever single-step mechanosynthesis of $Cd_{0.5}Zn_{0.5}S$ quantum dots: Nanostructure and optical characterizations. *Materials Research Bulletin*, *47*(4), 1062–1072. doi:10.1016/j.materresbull.2011.12.057

Sain, S., & Pradhan, S. K. (2012). Mechanochemical solid state synthesis of $(Cd_{0.8}Zn_{0.2})S$ quantum dots: Microstructure and optical characterizations. *Journal of Alloys and Compounds*, *509*(10), 4176–4184. doi:10.1016/j.jallcom.2011.01.035

Sanap, V. B., & Pawar, B. H. (2011). study of chemical bath deposited nanocrystalline cdzns thin films. *Journal of Optoelectronics and Biomedical Materials*, *3*, 39–43.

Sekhar, H., & Rao, D. N. (2012). Spectroscopic studies on Zn-doped CdS nanopowders prepared by simple coprecipitation method. *Journal of Materials Science*, *47*(4), 1964–1971. doi:10.1007/s10853-011-5991-9

Seshan, K. (Ed.). (2002). *Handbook of thin film deposition processes and techniques*. Noyes Publication.

Singh, V. P., & Singh, S. (1976). Some physical properties of ZnCdS solid solutions. *Czechoslovak Journal of Physics B*, *26*(10), 1161–1166. doi:10.1007/BF01591089

Song, W. C., & Lee, J. H. (2009). Growth and characterization of $Zn_xCd_{1-x}S$ films prepared by using chemical bath deposition for photovoltaic devices. *Journal of the Korean Physical Society*, *54*(4), 1660–1665. doi:10.3938/jkps.54.1660

Spence, J. C. H. (2003). *High-resolution electron microscopy, Monograph on the Physics and Chemistry of materials* (3rd ed.). New York: Oxford Science Publications.

Tauc, J. (1974). *Amorphous and Liquid Semiconductor*. New York: Plenum Press. doi:10.1007/978-1-4615-8705-7

Teterycz, H., Rac, O., Wozniak, P. S., & Oles, D. (2013). the formation mechanism of colloidal spheres of zno in ethylene glycol. *Digest J. Nanomaterials and Biostructures*, *8*, 1157–1167.

Tosun, B. S., Pettit, C., Campbell, S. A., & Aydil, E. S. (2012). Structure and Composition of $Zn_xCd_{1-x}S$ Films Synthesized through Chemical Bath Deposition. *ACS Applied Materials & Interfaces*, *4*(7), 3676–3684. doi:10.1021/am300771k PMID:22732000

Xia, D., Caijuan, T., Rongzhe, T., Wei, L., Lianghuan, F., Jingquan, Z., & Zhi, L. et al. (2011). Structural and optical properties of $Cd_{0.8}Zn_{0.2}S$ thin films. *J. Semicond.*, *32*(2), 022003. doi:10.1088/1674-4926/32/2/022003

Yamaguchi, T., Matsufusa, J., & Oshida, A. Y. (1992). Optical Transitions in RF Sputtered $CuIn_xGa_{1-x}Se_2$ Thin Films. *Japanese Journal of Applied Physics*, *31*(Part 2, No. 6A), L703–L706. doi:10.1143/JJAP.31.L703

Chapter 9
Studies on Metal Oxide Nanoparticle Doped PVP Polymer Nanocomposites

Madhu B. J.
Government Science College Chitradurga, India

ABSTRACT

Magnesium Oxide (MgO) nanoparticles have been synthesized by solution combustion technique using stoichiometric composition of magnesium nitrate as oxidizer and urea as fuel. Structure of the MgO was studied with the X-ray diffraction (XRD) using Cu-Kα radiation. MgO/polyvinylpyrrolidone (PVP) nanocomposites have been prepared by blending MgO nanoparticles with the polyvinylpyrrolidone. MgO/PVP nanocomposites were characterized by Fourier transform infrared (FTIR) spectroscopy. Frequency dependence of dielectric constant (ε'), dielectric loss tangent (tanδ) and AC conductivity studies have been undertaken on the MgO/PVP nanocomposites in the frequency range 50Hz-5MHz at room temperature. Dielectric properties such as dielectric constant (ε') and dielectric loss tangent (tanδ) are found to decrease with the increase in the frequency. Further, AC conductivity of MgO/PVP nanocomposites was found to increase with an increase in the frequency. Observed variation in the a. c. conductivity with the frequency has been understood on the basis of electron hopping model.

INTRODUCTION

Nanotechnology deals with various structures of matter having dimensions of the order of a nanometer. Nanotechnology is very diverse, ranging from extensions of conventional device physics to completely new approaches based upon molecular self-assembly, from developing new materials with dimensions on the nanoscale to investigating whether we can directly control matter on the atomic-scale. Further, nanotechnology has the potential to create many new materials and devices with a vast range of applications, such as in medicine, electronics and energy production (Franke et al., 2006; Boucle et al., 2007; De et al., 2008; Sarkar, Guibalet al., 2012; Patil et al., 2008; Kulkarni, 2009).

DOI: 10.4018/978-1-5225-3023-7.ch009

In recent decades, nanoparticles have attracted great interest in fundamental science and also in technological applications. Nanoparticles exhibit physical and chemical properties which are quite different from those of their bulk counterparts. Nanocrystalline oxide materials have attracted special attention in the field of electronics and telecommunication industry because of their novel dielectric and electrical properties. Metal oxide nanoparticles, represent great scientific and technological values as a result of the extraordinary physical and chemical features originating from their nanoscale size and increased amount of surface atoms. Due to the fact that their features are dependent on the increased surface area to volume ratio and the quantum confinement effect, their prospective uses span across nearly all discipline (Franke et al., 2006; Boucl é et al., 2007; De, Ghosh & Rotello, 2008; Sarkar et al., 2012; Patil et al., 2008; Kulkarni, 2009). Further, their highly specific surface area combined with their smallness lead to the observation of several other unique behaviors.

A number of physical phenomena become pronounced as the size of the system decreases. These include statistical mechanical effects, as well as quantum mechanical effects, for example the "quantum size effect" where the electronic properties of solids are altered with great reductions in particle size. However, quantum effects become dominant when the nanometer size range is reached, typically at distances of 100 nanometers or less, the so called quantum realm. Additionally, a number of physical properties such as mechanical, thermal, electrical, optical, etc., change when compared to macroscopic systems. One example is the rise in surface area to volume ratio altering mechanical, thermal and catalytic properties of materials (Kulkarni, 2009).

There are large numbers of techniques available to synthesize different types of nanomaterials in the form the colloids, clusters, powders, tubes, rods, wires and thin films etc. Some of the already existing conventional techniques to synthesize different types of materials are optimized to get novel nanomaterials and some new techniques are developed. There are various physical, chemical, biological and hybrid techniques available to synthesize the nanomaterials. The technique to be used depends upon the material of interest and the type of nanostructure. Some of the important techniques are discussed below (Patil et al 2008; Kulkarni, 2009).

High energy ball milling technique is considered as one of the simplest way of making nanoparticles of some metals and alloys in the form of powder. There are many types of mills such as, planetary, vibratory, rod and tumbler etc. Hardened steel or tungsten carbide balls are put in containers along with powder or flakes (<50μm) of a material of interest. When the containers are rotating around the central axis, the material is forced to the walls and is pressed against the walls by controlling the speed of rotation of central axis and container as well as duration of milling. Thus using this method, it is possible to ground the material to fine powder of nanometer range. Some of the materials like Cobalt, Chromium, Tungsten, Ni-Ti, Al-Fe etc. are made nanocrystalline using ball mill technique (Kulkarni, 2009).

Nanoparticles can be synthesized using the chemical method. In the chemical method, some of the wet chemical methods are used to synthesize nanomaterials. There are several advantages of using chemical method. In some cases, nanomaterials are obtained as colloidal particles in solutions, which can be filtered and dried to obtain powder. Colloidal metal nanoparticles are often synthesized by reduction of some metal salt or acid. In a similar manner, metal nanoparticles can be synthesized using appropriate precursors. Further, reaction parameters such as temperature, PH, duration of synthesis etc. are found to affect the particle size; size distribution and shape strongly. Thus reaction parameters can be used to achieve desired results. In addition, it is also possible to synthesize alloy nanoparticles using appropriate precursors. Furthermore, compound semiconductor nanoparticles can be synthesized by wet chemical route using appropriate salts. A wide range of metal oxides and other insulators with wide band gap can be synthesized by chemical precipitation method. Some of the oxides and insulators may be stable and may not agglomerate or coalesce easily depending on the conditions (Kulkarni, 2009).

Sol-gel process involves two types of materials or components, sol and the gel. Sol-gel formation process is usually a low temperature procedure. This means less energy consumption and less pollution too. Although sol-gel process generates highly pure, well controlled ceramics, it competes with other processes like chemical vapor deposition or metallo-organic vapours derived ceramics. Some of the benefits like getting unique materials such as aerogels, zeolites, ordered porous solids by organic-inorganic hybridization are unique to sol-gel process. Further, it is also possible to synthesize nanoparticles, nanorods, nanotubes etc. using sol-gel technique. Sols are solid particles in liquid. They are thus a subclass of colloids. Gels are nothing but a continuous network of particles with pores filled with liquid. A sol-gel process involves formation of 'sols' in a liquid and then connecting the sol particles to form a network. By drying the liquid, it is possible to obtain powders, thin films or even monolithic solid. Sol-gel method is particularly useful to synthesize the ceramics or metal oxides, although sulphides, borides and nitrides also are possible. Synthesis of sol-gel in general involves the hydrolysis of precursors, condensation followed by polycondensation to form particles, gelation and drying processes by various routes (Kulkarni, 2009).

Combustion synthesis or fire synthesis is also called as the self-propagating high temperature synthesis. In this technique, to produce fire, an oxidizer, a fuel and great temperature is needed. Being a solid state reaction it often does not produce homogeneous products and results in coarse powder. In the present work, magnesium oxide (MgO) nanoparticles have been synthesized by solution combustion technique. Details of the combustion technique are discussed in the experimental section.

The growth of organic-inorganic nanomaterials, commonly pursued via the joining of artificial polymers and inorganic particles or by combining altered nanoparticles (NPs) to polymer matrices, is designed to produce materials, which demonstrate enhanced characteristics. The nanomaterials, which consist of both inorganic nanoparticles and organic polymers, can be considered a brand new classification of materials. These new materials demonstrate enhanced features, particularly when placed alongside their micro particle equivalents. Thus, nanomaterials, which incorporate metal oxide nanoparticles, represent great scientific and technological values as a result of the exceptional physical and chemical features deriving from their nanoscale size and increased amount of surface atoms. The existence of a polymer chain boosts the synergy of the particles within the polymer matrix, so that it takes less energy to distribute them (Sarkar et al., 2012).

Nanocrystalline oxide materials and their composites have received remarkable attention due to their potential application in optoelectronic and modern electronic energy storage devices. A polymer matrix is used as the setting for nanoparticle development, as it synergistically brings together the characteristics of the host polymer matrix and the distinct nanoparticles (Patil et al., 2008; Kulkarni, 2009). In particular, there has been increasing research on the investigations of metal oxide nanostructures and their polymer nanostructures. Polymer blends are emerging as advanced polymeric materials and are being explored as matrix resins for advanced composite materials (Rozenberg & Tenne, 2008). The nanoparticles including metal oxides fixed in polymer matrices have attracted a great deal of emphasis and focus, due to the extraordinary features demonstrated by the materials. As a result of the nanometer dimensions of the particles, their physicochemical properties are very different from the properties observed in their corresponding molecular and bulk materials (Kumar et al., 2009). This type of compound materials are found to exhibit interesting catalytic, electrical, optical, mechanical, dielectric and magnetic properties

(Cascone, 1997; Antonietti & Göltner, 1997; Reetz & Helbig, 1994; Gui-Xin Wang et al., 2005; Li Jun et al., 1999).

Recently, great attention has been drawn towards the synthesis and studies of polymer based nanocomposites as they have become a powerful tool for applications in industry, science, and medicine. In particular, the unique magnetic, catalytic and biocompatible properties of polymer nanocomposites make it attractive for use in variety of area such as recording materials, catalysts, pigment, printing ink, magnetic resonance imaging, heating mediators and drug carriers in magnetic drug targeting. The properties of nanoparticles can be altered by controlling their size, which can provide an advantage in formulating new composite materials with optimized properties for various applications. Thus, to control the growth of the nanoparticles, organic stabilizers (polymers), e.g., polyvinyl alcohol (PVA), polyethylene oxide (PEO), polymethacrylic acid (PMAA), and poly(vinyl pyrrolidone) (PVP), are added during the synthesis for capping the surface of the particles (Cascone, 1997; Antonietti & Göltner, 1997; Reetz & Helbig, 1994; Gui-Xin Wang et al., 2005; Li Jun et al.,1999).

Mahmoud Goodarz Naseri et al., (2012) studied the interactions between the PVP capping agent and the metal ions, which shown that the manganese (II) and iron (III) ions are bounded by the strong ionic bonds between the metallic ions and the amide group in a polymeric chain. PVP was found to act as a stabilizer for dissolved metallic salts through steric and electrostatic stabilization of the amide groups of the pyrolidine rings and the methylene groups. Initially, the PVP stabilizer may decompose to as limited extent, thereby producing shorter polymer chains that are capped when they are adsorbed onto the surfaces of metallic ions (Koebel et al., 2008). The metallic ions, which are well dispersed in the cavities and networks, are formed as a result of the shorter polymer chains. These mechanisms continue until they are terminated by the drying step. The influence of PVP is not restricted only to the solution and the drying step; PVP also affects the formation of the nuclei (i.e., nucleation) of the manganese ferrite nanoparticles in the calcination step. In this step, the small nanoparticles with high surface energy levels would become larger via the Ostwald ripening process (Roosen, & Carter, 1998) without the presence of PVP, disrupts steric hindrance, thereby preventing their aggregation. Steric hindrance is a phenomenon that is attributed to large molecular weight (>10,000) and the repulsive forces acting among the polyvinyl groups (Ghosh et al., 2006). These interactions are similar to the stabilization of metallic nanoparticles, i.e., silver and gold (Tsuji et al., 2004)

Santi Maensiri et al. reported the synthesis and characterization of the as-spun $MgFe_2O_4$/ PVP composite nanofibers (Santi Maensiri et al., 2009). Nanostructures of $MgFe_2O_4$ have been successfully fabricated using an electrospinning technique. They found that the crystal structure and morphology of the nanofibers were influenced by the calcination temperature. Room temperature magnetization results showed a ferromagnetic behavior of the calcined $MgFe_2O_4$/PVP composite nanofibers. They concluded that that the electrospun $MgFe_2O_4$ nanostructures could have potential in some new applications as ferromagnetic nanostructures for nanocomposites, separation, anodic material in lithium ion batteries, catalysts, and as electronic material for nano devices and storage devices.

Shen Hong et al., (2005) investigated the nonlinear optical properties of Au/PVP composite thin films fabricated by spin-coating technique. Linear optical absorption measurements of the Au/PVP composite films indicate an absorption peak around 530 nm due to the surface plasmon resonance of gold nano particles. Nonlinear optical properties are studied using Z-scan technique and experimental results show large optical nonlinearities of the Au/PVP composite films. The Au/PVP composite films show high linear transmittance in visible frequency range and large third order optical nonlinearities.

They concluded that these relatively low cost materials make the metal-polymer composite film a valid candidate for applications in optical devices.

Naif Mohammed Al-Hada Elias Saion et al., (2016) investigated the impact of Polyvinylpyrrolidone on the properties of cadmium oxide semiconductor nanoparticles manufactured by heat treatment technique. Cadmium oxide semiconductor nanoparticles were produced using a water based mixture, incorporating cadmium nitrates, polyvinyl pyrrolidone (PVP), and calcination temperature. PVP is found to play important function during the production of CdO nanoparticles. Inclusion of PVP is found to regulate the expansion nucleation of nanoparticles-, which in turn depends on the concentration of PVP. Further, PVP is found to restrict the accretion of the CdO nanoparticles and also found to improve the degree of crystallinity of the nanoparticles. In addition, PVP is found to facilitate the development of nanoparticles with a homogenous dispersal of size and form.

Agarwal et al., (2012) studied the fabrication and characterization of Iron oxide filled polyvinyl pyrrolidone (PVP) nanocomposites. Nanomagnetic polymer composites (MPCs) comprising iron oxide nanoparticles as a filler and PVP as polymer shell were synthesized and analyzed using XRD, FT-IR and SEM. The iron oxide nanoparticles were synthesized under alkaline medium using iron (II) salt. The nanocomposites were synthesized through free radical polymerization under microwave (MW) irradiation. X-ray diffraction reveals the uniform dispersion of iron oxide nanoparticles into polymer matrix. Furthermore, scanning electron microscopy images ensure the development of polymer shell around the nanomagnetic core. The magnetic behaviour of MPCs was studied through vibration sample magnetometry. The thermal durability of MPCs was assessed through simultaneous thermogravimetric-differential thermal analysis. The insertion of iron oxide nanoparticles into PVP matrix has no remarkable effect on thermal stability of nanomagnetic polymer composites. The assessed magnetic properties revealed that, the synthesized nanomagnetic polymer composites may be used for the development of sensor and data storage devices. The proposed microwave radiation assisted route offers an efficient, faster, cleaner and economic alter-native heating source to synthesize the nanocomposites.

Natalia Hoogesteijn von Reitzenstein et al., (2016) studied the morphology, structure, and properties of metal oxide/polymer nanocomposite electrospun mats. Adding nanoparticles into polymer solutions before electrospinning creates unique hierarchical morphologies dispersed throughout small diameter nanoparticle-polymeric fibers. Effects of polymer composition, nanoparticle (NP) type, loading, and electrospinning voltage conditions were studied. Indium, iron, and titanium oxide engineered nanoparticles (NPs) were dispersed into polyvinylpyrrolidone or polystyrene and electrospun. They investigated the effect of NP addition on electrospun polymer fibers and how viscosity, critical voltage, and fiber morphology changed as a result. Viscosity and critical voltage increased with increasing weight percentage of nanoparticles in the polymer solution. The potential for high surface area, low volume, functionalization capability, and ease of synthesis make electrospun fibers good candidates for water treatment applications such as nanofiltration and ion exchange. They concluded that these multifunctional nanomaterial-polymer composite fibers may be employed for wide-ranging applications such as water and air treatment.

Recently polymeric materials have been employed as particle stabilizers in the chemical synthesis of metal colloids and the roles of polymers are generally documented as steric stabilizer or capping agent. The control of reaction kinetics using polymers allows simple and versatile route to the synthesis of metal nanocrystals with well-defined shapes and size. The polymers interact with the small particles, preventing their agglomeration and precipitation. A number of metal-polymer nanocomposites with metals such as Ag, Au, Pd, Pt, Ni, Cu and other noble metal-polymer have been studied. However, the report for the nanocomposites of metal oxides with polymer is limited.

PVP provides numerous potential biomedical and packaging applications as it is reactive, in-expensive, nontoxic, hydrophilic and biocompatible (Galiev et al., 2010). Recently, oxide/polymer composites have been widely used in the area of electromagnetic interference shielding and electronic devices, due to their exceptional collective properties of electrical conduction, corrosion resistance, flexibility and low density which are superior to those of metallic materials (Madhu et al., 2014; Madhu et al., 2016a; Madhu et al., 2016b). The existence of a polymer chain increases the interaction of the particles within the polymer matrix, so that it takes less energy to distribute them. In particular, the nanocrystalline–polyvinylpyrrolidone (PVP) composites are promising candidates for this type of application due to their favorable properties. Polyvinylpyrrolidone (PVP) provides numerous potential biomedical and packaging applications as it is reactive, inexpensive, nontoxic, hydrophilic and biocompatible (Said-Galiev et al., 2010).

In the present study, Magnesium Oxide (MgO) nanoparticles prepared by solution combustion method are doped to polyvinylpyrrolidone (PVP) in order to get MgO/PVP nanocomposites. The structural changes, frequency dependence of dielectric constant, dielectric loss tangent and AC conductivity properties of the MgO/PVP composites have been carried out over a frequency range 50Hz-5MHz in room temperature.

EXPERIMENTAL

The combustion method has been considered to be fast, simple and low-cost, allowing the production of fine, homogeneous crystalline powders without the risk of contamination (Patil et al., 2008; Kulkarni, 2009). In this method, the equivalent ratio of an oxidizer and fuel mixture is expressed in terms of the elemental stoichiometric measurement. The oxidizer/fuel molar ratio required for a stoichiometric mixture is determined by summing the total oxidizing and reducing valences.

To synthesize, given materials are mixed with the urea and dissolved in 15 ml of distilled water. The mixture is taken in a 300 cm^3 Pyrex dish and introduced in a muffle furnace maintained at 500°C. The product formed occupies the volume of whole container. The complete combustion process is executed in about 3 minutes.

Urea is an ideal fuel for the combustion synthesis of high temperature oxides like alumina and alkaline earth aluminates. However, a need to employ alternate fuels to prepare oxides which are unstable above 1000°C such as transition metal aluminates was necessary. In this regard, hydrazine based fuels have low ignition temperature and combustible due to presence of N-N bond that decomposes exothermically to N_2 were found to be suitable (Patil et al., 2008; Kulkarni, 2009).

In the present work, MgO nanoparticles were prepared by solution combustion method using stoichiometric composition of magnesium nitrate as oxidizer and urea as a fuel. The aqueous solution containing redox mixture was taken in a Pyrex dish and heated in a muffle furnace maintained at 500±10°C. The mixture finally yields porous and voluminous powder. Structural analysis of the sample was undertaken using the X-ray diffraction analysis technique. The X-ray diffraction analysis was carried out using the Philips X-ray diffractometer using CuKα radiation ($\lambda = 1.5406$ Å).

MgO/PVP nanocomposites have been prepared by blending MgO nanoparticles with the PVP. 10 g of PVP is dissolved in 100 ml of distilled water to make 10% solution. 7.5 ml of the 10% PVP solution is mixed thoroughly with the prepared MgO powder to make MgO/PVP composite. This mixture is dried and crushed into fine power.

Infrared (IR) spectroscopy is one of the most common spectroscopic techniques used by organic and inorganic chemists. Simply, it is the absorption measurement of different IR frequencies by a sample positioned in the path of an IR beam. The main goal of IR spectroscopic analysis is to determine the chemical functional groups in the sample. Different functional groups absorb characteristic frequencies of IR radiation. Using various sampling accessories, IR spectrometers can accept a wide range of sample types such as gases, liquids, and solids. Thus, IR spectroscopy is an important and popular tool for structural elucidation and compound identification.

Structural analysis of the MgO/PVP nanocomposite was undertaken using the Fourier-transform infrared (FT-IR) spectroscopy. The FT-IR (Infra-red) spectra (500-4000 cm^{-1}) of the MgO/PVP composite was recorded on a Bruker Alpha spectrophotometer in KBr pellets.

The dielectric property of the sample depends mainly on the nature of the sample, synthesis technique, grain size, cation distribution etc. If a time varying field is applied to a dielectric, the polarization of the dielectric material will takes place under this field with frequency. However, due to the presence of damping and frictional forces in the solid, during the interaction of charges with electric field, the polarization will lag behind the electric field E by delta, hence permittivity of a dielectric in an AC field will be a complex quantity ε^*. In the present studies, measurement of dielectric constant (ε'), loss tangent (tan δ) and AC conductivity studies have been undertaken using the impedance analyzer model HIOKI 3532-50 LCR HiTESTER. The measurements were carried out at room temperature in between the 50Hz-5MHz. The dielectric constant can be represented by complex quantities,

$$\varepsilon = \varepsilon' - i\varepsilon'' \tag{1}$$

where ε' represents real or dispersive part of the dielectric constant, while ε'' represents its imaginary or dissipative part and takes in to account the losses. The complex permittivity in general represents the dielectric properties of the materials. The real part represents the electrical energy storage capability and the imaginary part represents the loss of electrical energy. The dielectric loss tangent (tan δ) is defined as the ratio between the imaginary and real part of the losses. That is, the dielectric loss tangent (tan δ) is the ratio between the imaginary and real part of the dielectric constant and is given by,

$$tan\delta = \frac{\varepsilon''}{\varepsilon'} \tag{2}$$

The dielectric constant (ε') is calculated by using the relation,

$$\varepsilon' = \frac{C_P d}{\varepsilon_o A} \tag{3}$$

where, C_P is the parallel capacitance of the sample, d is the thickness of the sample, A is the cross section area of the sample and ε_0 is the permittivity of the free space.

RESULTS AND DISCUSSION

X-ray diffraction studies have been undertaken using Cu-K_α line, on as prepared MgO sample. The X-ray diffraction pattern of the representative annealed nano oxide is displayed in the Figure 1. The average crystallite size of the MgO nano oxide powder sample for the most intense peak plane is determined from the Debye-Scherrer formula.

$$d = \frac{k\lambda}{\beta Cos\theta} \tag{4}$$

where 'd' is the crystallite size, k = 0.9 is a correction factor to account for the particle shapes, β is the full width at half maximum (FWHM) of the most intense diffraction peak (in radians), λ is the wavelength of Cu target = 1.5406 Å and θ is the Bragg angle (in degrees). The average crystallite size of the prepared sample is around 13 nm. Present MgO sample is found to possess cubic system with face centered lattice (PCPDF card No. # 78-0430).

The chemical structure of the MgO/PVP composite was characterized by FT-IR spectroscopy. The FT-IR studies have been carried out on the MgO/PVP composites in the wave number range 500 – 4000 cm^{-1} in KBr medium.

Figure 1. XRD pattern of MgO

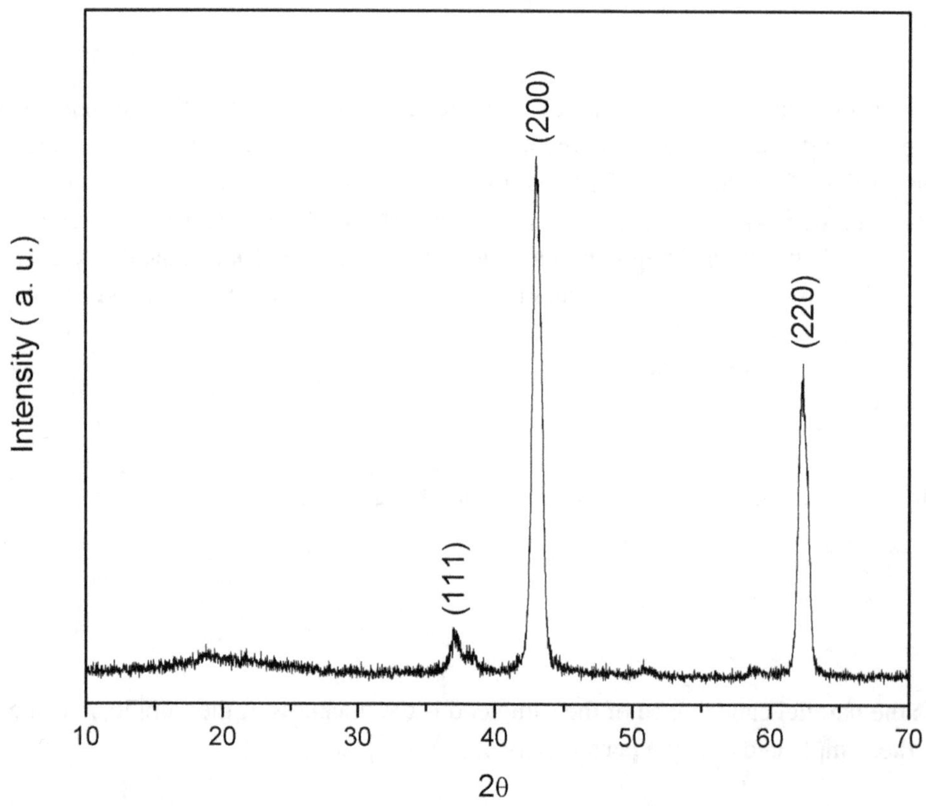

The Fourier Transform Infrared spectrum of the MgO/PVP composites is shown in the Figure 2. The vibrating band at around 500-800 cm^{-1} corresponds to the metal-oxide stretching and nano oxides (Madhu, Kiran, Gurusiddesh, & Shruthi, 2016). The vibrating band at around 1290 cm^{-1} corresponds to the metal-oxide stretching. In the range of 4000–1000 cm^{-1}, vibrations of CO_3^{2-} and moisture were observed. The intensive band at around 1661 cm^{-1} is due to O–H stretching vibration interacting through H bonds. The band at around 2980 cm^{-1} is C–H asymmetric stretching vibration mode due to the $-CH_2-$ groups of the long aliphatic alkyl groups. The υ (C=O) stretching vibration of the carboxylate group (CO_2^{2-}) was observed around 1384 cm^{-1} and the band at 1002 cm^{-1} corresponds to nitrate ion traces (Madhu et al., 2014).

Dielectric materials are basically insulators having the property of storing and dissipating electrical energy when subjected to electromagnetic field. The energy storing property leads to the fabrication of most important constituents of electrical circuit known as capacitors. The dielectric phenomenon arises from the interaction of electric field with different charged particles such as electrons, ions, protons and electron shells, which constitute the dielectric material. All dielectrics can be subdivided into two-polar and non-polar dielectrics. A DC field behaviour, however provides an information about the nature of charge carriers, their mobility, conduction mechanism etc. whereas studies in AC field provides an information into the electrical nature of the molecular or atomic species, which constitutes the dielectric materials (Madhu et al., 2012; Raghavender, & Jadhav, 2009; Sangeeta Thakur & Katyal, 2007; Panwar et al., 2010; Joncher et al., 1977; Joncher, 1981; Nadeem, & Mushtaq, 2009; Younas et al., 2011).

Dielectrics are material with very few electrons to take part in normal electric conductivity. Dielectric materials have interesting electric properties because of the ability of an electric field to polarize the

Figure 2. FT-IR spectrum of MgO/PVP composites

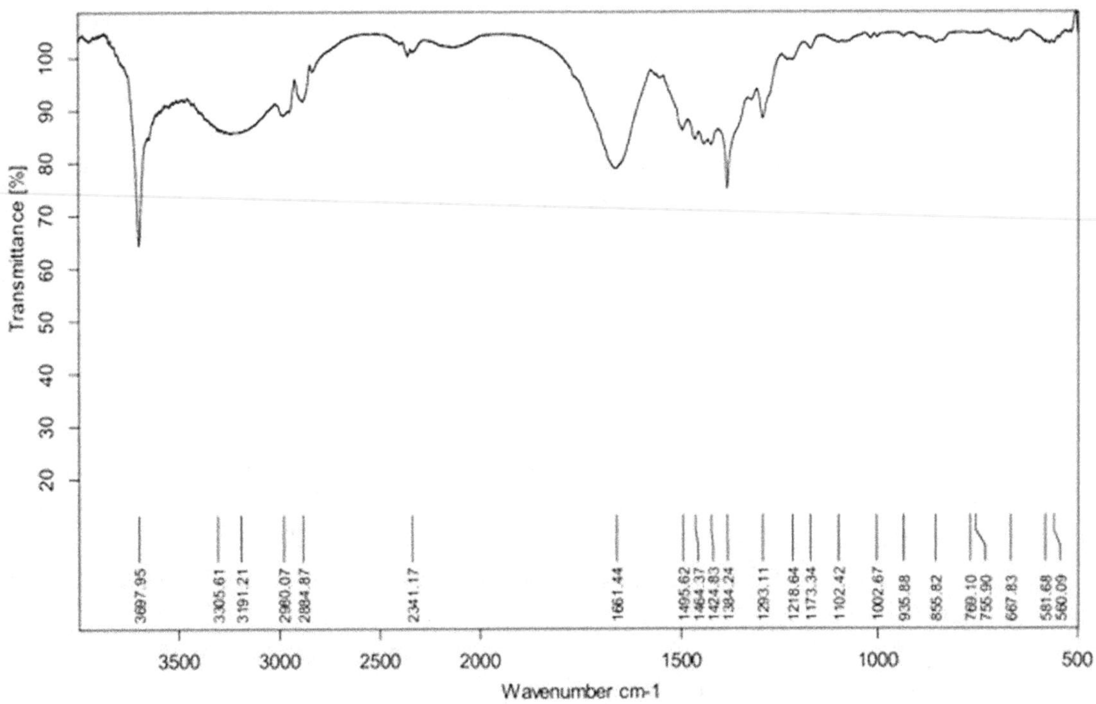

material to create electric dipoles. A dipole is an entity in which equal positive and negative charges are separated by a small distance. Dielectric property of a sample is a characteristic of a short-range electrical conduction of a material under the influence of an applied electric field. When a field is applied across the sample, the applied field displaces the charges within the material and results in the accumulation of charges at the interface, creating dipoles.

The microscopic behavior of a dielectric material under an electric filed can be attributed to the polarization phenomena of the charged particles, which constitutes the dielectrics. The polarization effect will be dependent on the nature of the dipole and frequency of the applied field under an AC field, but the polarization of the dielectric may not follow the field variation. This displacement due to polarization may persist even when the field is stopped. This gives rise to a decay time to attain the equilibrium and the phenomenon is known as relaxation. The decay time is known as relaxation time. The dielectric relaxation, which is due to a number of different polarization mechanisms, is observed under an AC field. The presence of any dielectric relaxation then corresponds to one or more of the possible polarization mechanism that occur on a microscopic scale. Each relaxation process may be characterized by a relaxation time, which describe a decay of polarization with time in a periodic field (Madhu et al., 2012; Raghavender, & Jadhav, 2009; Sangeeta Thakur & Katyal, 2007; Panwar et al., 2010; Joncher, 1977; Joncher, 1981; Nadeem, & Mushtaq, 2009; Younas et al., 2011).

Polymer based nanocomposites with high dielectric constant have attracted much attention due to their important technological applications such as high performance capacitor, electronic packaging and components, electromechanical devices and electric energy storage devices. The dielectric constant is mainly associated with the amount of polarization occurring in the material (Madhu et al., 2012; Raghavender, & Jadhav, 2009; Sangeeta Thakur & Katyal, 2007; Panwar et al., 2010; Joncher, 1977; Joncher, 1981; Nadeem & Mushtaq, 2009; Younas et al., 2011). The real part of permittivity depends mainly on the polarization in the polymer backbone and the interfacial polarizations among the oxide nanoparticles and PVP polymer matrix. In the present studies, the effect of frequency on dielectric constant (ε') was studied in the frequency range 50Hz-5MHz at room temperature. The variation of dielectric constant (ε') with frequency at room temperature is shown in Figure 3. Dielectric constant was found to decrease with an increase in the frequency. This is the normal dielectric behavior and similar behavior has been observed in other nanocomposite materials (Madhu et al., 2014; Madhu et al., 2016a; Madhu et al., 2016b). This is due to the fact that the species contributing to the polarizability are lagging behind the applied field at higher frequency. That is, with increase in the frequency, the dipoles existing in the sample cannot reorient themselves along with the applied electric field, as a consequence of which the dielectric constant decreases. The observed dielectric constant values reach a constant value above 10 KHz without any additional dispersion because of the accumulation of space charge near the electrode due to polarization. Generally, dielectric constant of any material depends on electronic, dipolar, ionic and interfacial polarizations. At lower frequencies, dipolar and interfacial polarizations are known to play most significant role and electronic and ionic polarizations are known to contribute at higher frequencies (Sangeeta Thakur & Katyal, 2007; Panwar et al., 2010; Joncher, 1977; Joncher, 1981; Nadeem, & Mushtaq, 2009; Younas et al., 2011). Thus, at high frequencies, the periodic reversal of the field takes place so rapidly such a way that there is no charge accumulation at the interface, resulting in constant dielectric constant value.

The power of dielectric loss of the absorbing material is also characterized by the dielectric loss tangent tan δ. Good dielectric loss material should have a larger dielectric loss angle tangent independent of the frequency. The effect of frequency on the dielectric loss tangent (tan δ) was studied in the

Figure 3. Variation of dielectric constant (ε′) with frequency

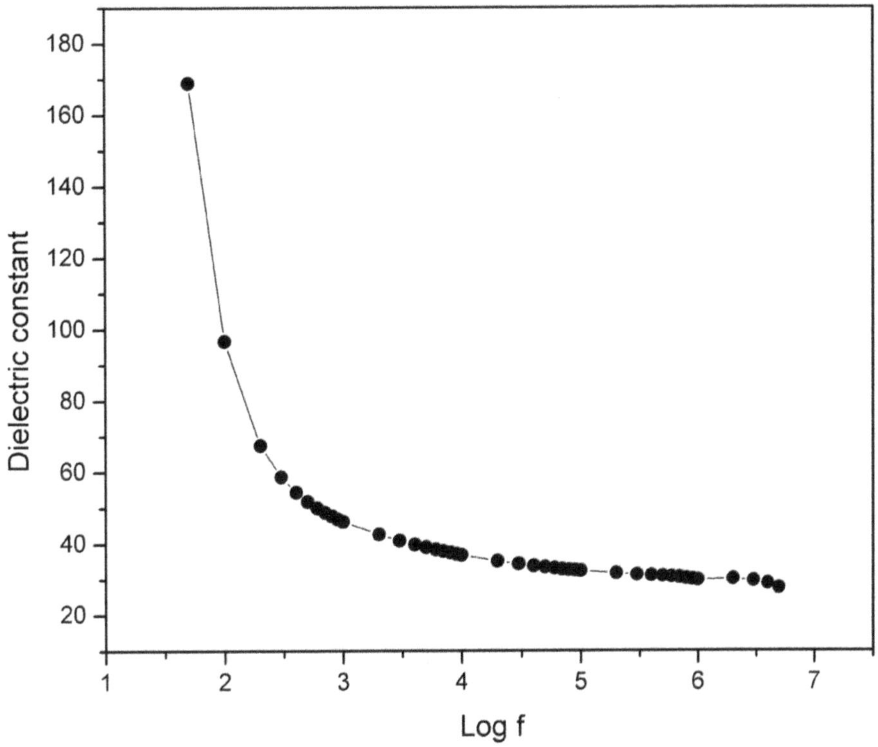

frequency range 50Hz-5MHz at room temperature. The variation of dielectric loss tangent (tan δ) with frequency at room temperature is shown in Figure 4. Dielectric loss tangent was found to decrease with an increase in frequency. Further, it can be observed that the rate of decrease in dielectric loss is found to be much higher at lower frequencies than at higher frequency range. It can be seen that the dielectric loss curves also display the same trend as the dielectric constant curves and can be described on lines similar to those advanced for explaining variation of dielectric constant in the previous section. Present MgO/PVP samples were found to exhibit relatively high values of dielectric loss tangent as compared to the nanoferrite/PVP composites (Madhu et al., 2014). High values of dielectric loss tangent of the present MgO/PVP composites could be utilized for decoupling capacitors (Panwar et al., 2010).

Electrical conductivity is an inherent property of most materials, and ranges from extremely conductive materials like metals to very non-conductive materials like plastics or glass. The actual cause of the resistivity must be sought as a deviation from the periodicity of the potential in which the electrons move. Deviations from the periodicity of the potential causing resistivity may be due to (a) lattice vibration (b) lattice defects such as vacancies, interstitials and dislocations (c) foreign impurity atoms and (d) boundaries. Electrical conductivity is an important experimental tool to probe the structural defects and internal purity of semi-insulating crystalline solids, because unlike metals and semiconductors where electrons are the charge carriers, in semi-insulators the prime factors contributing to their electrical conductivity are the crystal defects (Sangeeta Thakur & Katyal, 2007; Panwar et al., 2010; Joncher, 1977; Joncher, 1981; Nadeem, & Mushtaq, 2009; Younas et al., 2011). Conductivity measures how well a material conducts electricity. The electrical conductivity can be measured in two forms as AC and DC

Figure 4. Variation of dielectric loss tangent with frequency

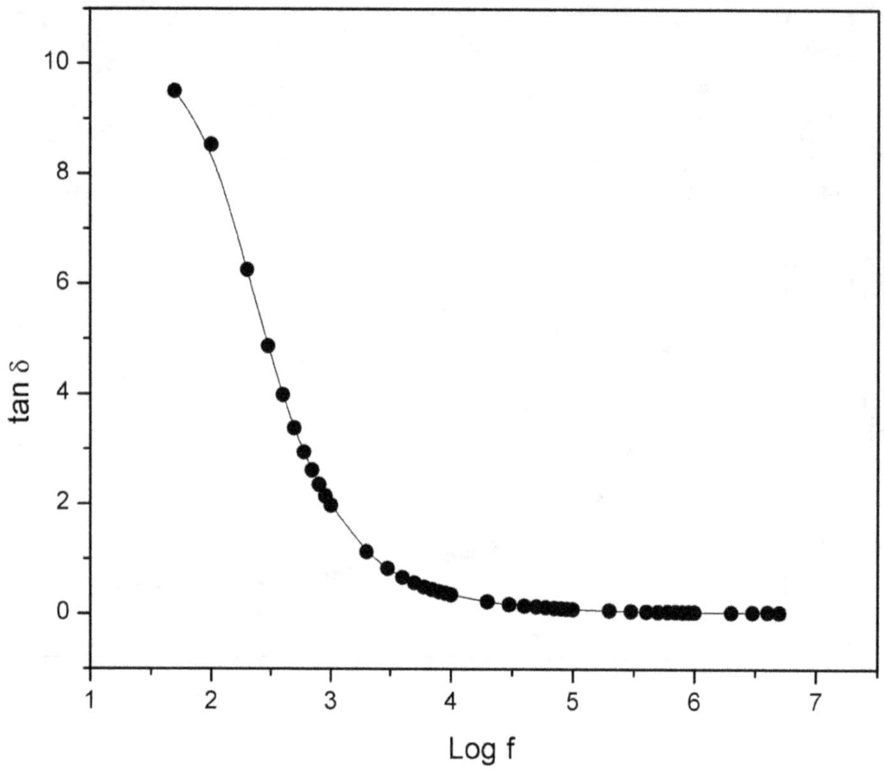

conductivity (σ_{ac} & σ_{dc} respectively). The DC conductivity provides only quantitative measure of the materials, assuming that there is single dynamic response characterizing the sample. Thus, DC conductivity provides less information of the entire dynamic behavior of the sample. Whereas, AC conductivity can shed considerable light on the microscopic nature of the conducting process in the material. Among the available measurement techniques, alternating current (AC) measurements are widely used to overcome certain difficulties in direct current (DC) measurements. Among these are polarization effects in ionic conductors and electrolytes, barriers at internal surfaces and contact resistances. In ac-conductivity measurements a sinusoidal voltage is applied and the current passing through the sample is measured. Electrical conductivity of solid sample as a function of frequency can generally be described as frequency independent, DC conductivity and a strongly frequency dependent components.

Generally, a typical frequency dependence of conductivity spectrum exhibits three separate regions namely a low frequency dispersion region, an intermediate frequency plateau and an extended dispersion region at high frequency. The variation of conductivity in the low frequency region is attributed to the polarization effects at the electrode and electrolyte interface. As the frequency reduces, more and more, charge accumulation occurs at the electrode & electrolyte interface and hence, drop in conductivity. In the intermediate frequency plateau region, conductivity is almost found to be frequency independent and is equal to DC conductivity. In the high frequency region, the conductivity increases with the frequency. The frequency dependence of conductivity or so-called universal dynamic response (UDR] of ionic conductivity is related by a simple expression given by Jonscher's power law (Joncher, 1977; Joncher, 1981).

$$\sigma_{w} = \sigma_{o} + A\omega^{s} \tag{5}$$

where σ_{ω} is the AC conductivity, σ_{o} is the limiting zero frequency conductivity σ_{dc} , A is a pre-exponential constant, $\omega = 2\pi f$ is the angular frequency and 's' is the power law exponent, where $0 < s < 1$. Here, A is a constant for a particular temperature and 's' is a power exponent which represents the degree of interaction between mobile ions and the environment surrounding them.

The AC conductivity study was undertaken over a frequency range (50-5MHz) for MgO/PVP composites at room temperature. In the present composites, it is possible to distinguish two regimes from the conductivity verses frequency plot (Figure 5). The first regime is in the low frequency regime and can be attributed to the frequency independent conductivity corresponding to the DC conductivity. In the high frequency region, frequency dependent conductivity is observed. Figure 5 shows the profile of the AC conductivity of the MgO/PVP composites reported here, revealing an increasing trend in the conductivity with the change of frequency, which is due to an increased density of mobile ions for conduction. The AC conductivity is found to be almost independent at lower frequencies, however, with an increase in frequency, the AC conductivity is found to increase gradually. At low frequency, the AC conductivity of MgO/PVP is found to be weakly frequency dependent due to the nonequilibrium occupancy of the trap charges in the PVP matrix. A further amplification of frequency reduces the occupancy of the trap centers by making them available for conduction (Nadeem, & Mushtaq 2009; Younas et al., 2011). This enables the conductive state to become more active by promoting the hopping of electrons and holes leading to an enhanced conduction at higher frequencies. Thus the electrical conduction mechanism in the present MgO/PVP composites is in accordance with the electron hopping model.

Figure 5. Variation of AC conductivity (σ) with the frequency

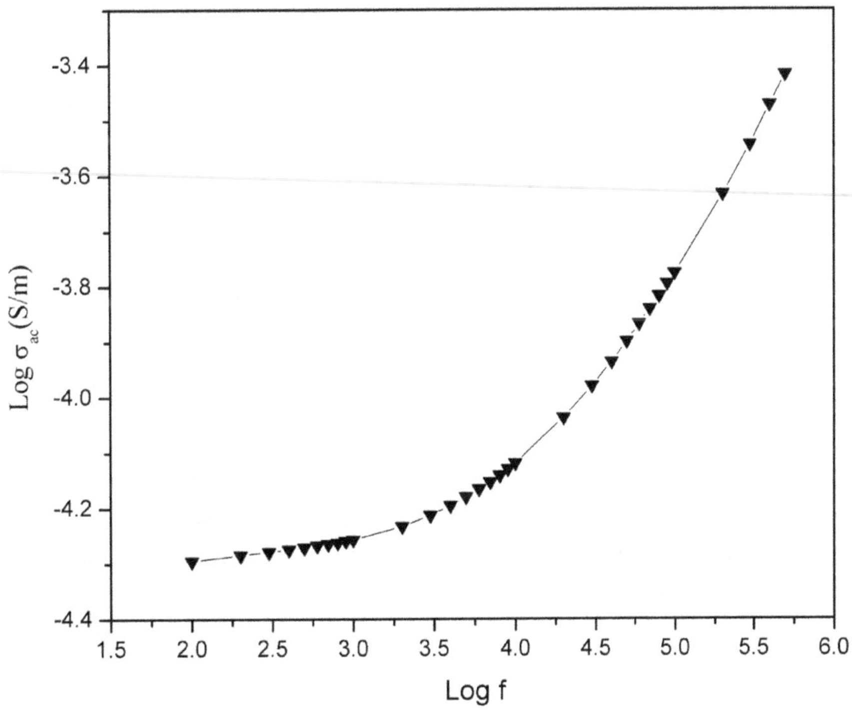

CONCLUSION

MgO nanoparticles were prepared using combustion method. Analysis of the X-ray diffraction patterns revealed the crystalline nature of the as prepared sample. MgO/PVP nanocomposites were prepared by doping MgO nanoparticles to PVP matrix. Frequency dependence of dielectric constant (ε'), dielectric loss tangent ($\tan \delta$) and AC conductivity (σ_{ac}) studies have been undertaken on the MgO/PVP nanocomposites in the frequency range 50Hz-5MHz at room temperature. Dielectric properties such as dielectric constant (ε') and dielectric loss tangent ($\tan \delta$) are found to decrease with the increase in the frequency. Further, AC conductivity (σ_{ac}) of the MgO/PVP nanocomposites was found to increase with an increase in the frequency. Observed variation in the AC conductivity with the frequency has been understood on the basis of electron hopping model.

REFERENCES

Agarwal, T., Gupta, K. A., Alam, S., & Zaidi, M. G. H. (2012). Fabrication and characterization of iron oxide filled polyvinyl pyrrolidone nanocomposites. *International Journal of Composite Materials*, *2*(3), 17–21. doi:10.5923/j.cmaterials.20120203.01

Al-Hada, N. M., Saion, E., Talib, Z. A., & Shaari, A. H. (2016). The Impact of Polyvinylpyrrolidone on Properties of Cadmium Oxide Semiconductor Nanoparticles Manufactured by Heat Treatment Technique. *Polymers*, *8*(4), 113. doi:10.3390/polym8040113

Antonietti, M., & Göltner, C. (1997). Superstructures of functional colloids: Chemistry on the nanometer scale. *Angewandte Chemie International Edition*, *36*(9), 910–928. doi:10.1002/anie.199709101

Bo-Lin, C., Guo-Wei, L., Wei-Tian, W., Dong-Yi, G., Zheng-Hao, C., & Guo-Zhen, Y. (2005). Nonlinear optical properties of Au/PVP composite thin films. *Chinese Physics*, *14*(9).

Bouclé, J., Ravirajan, P., & Nelson, J. (2007). Hybrid polymer–metal oxide thin films for photovoltaic applications. *Journal of Materials Chemistry*, *17*(30), 3141–3153. doi:10.1039/b706547g

Cascone, M. G. (1997). Dynamic–mechanical properties of bioartificial polymeric materials. *Polymer International*, *43*(1), 55–69. doi:10.1002/(SICI)1097-0126(199705)43:1<55::AID-PI762>3.0.CO;2-#

De, M., Ghosh, P. S., & Rotello, V. M. (2008). Applications of nanoparticles in biology. *Advanced Materials*, *20*(22), 4225–4241. doi:10.1002/adma.200703183

Franke, M. E., Koplin, T. J., & Simon, U. (2006). Metal and metal oxide nanoparticles in chemiresistors: Does the nanoscale matter?. *Small*, *2*(1), 36–50. doi:10.1002/smll.200500261 PMID:17193551

Galiev, E. E. S., Stakhanov, A. I., Blagodatskikh, I. V., Kobitskaya, E. M., Khokhlov, A. R., Naumkin, A. V., & Pisarev, S. A. et al. (2010). Synthesis of polyvinylpyrrolidone and its nanosilver_based polymer composites in supercritical carbon dioxide. *Polymer Science, Series B*, *52*(3–4), 165–173. doi:10.1134/S1560090410030061

Ghosh, G., Naskar, M. K., & Patra, A. (2006). Synthesis and characterization of PVP-encapsulated ZnS nanoparticles. *Optical Materials*, *28*(8-9), 1047–1053. doi:10.1016/j.optmat.2005.06.003

Hoogesteijn von Reitzenstein, N., Bi, X., Yang, Y., Hristovski, K., & Westerhoff, P. (2016). Morphology, structure, and properties of metal oxide/polymer nanocomposite electrospun mats. *Journal of Applied Polymer Science*, *133*(33), 33. doi:10.1002/app.43811

Joncher, A. K. (1977). The universal dielectric response. *Nature*, *256*(5613), 673–679. doi:10.1038/267673a0

Jonscher, A. K. (1981). A new understanding of the dielectric relaxation of solids. *Journal of Materials Science*, *16*(8), 2037–2060. doi:10.1007/BF00542364

Jun, L., Rong, L., Jianming, W., & Hang, S. (1999). Effect of cupric oxide addition on the performance of nickel electrode. *Journal of Power Sources*, *79*(1), 86–90. doi:10.1016/S0378-7753(99)00048-8

Koebel, M. M., Jones, L. C., & Somorjai, G. A. (2008). Preparation of size-tunable, highly monodisperse PVP-protected Pt-nanoparticles by seed-mediated growth. *Journal of Nanoparticle Research*, *10*(6), 1063–1069. doi:10.1007/s11051-008-9370-7

Kulkarni, S. K. (2009). *Nanotechnology Principles and Practices*. Capital Publishing Company.

Kumar, A. P., Depan, D., Tomer, N. S., & Singh, R. P. (2009). Nanoscale particles for polymer degradation and stabilization—trends and future perspectives. *Progress in Polymer Science*, *34*(6), 479–515. doi:10.1016/j.progpolymsci.2009.01.002

Madhu, B. J., Ashwini, S. T., Shruthi, B., Divyashree, B. S., Manjunath, A., & Jayanna, H. S. (2014). Structural, dielectric and electromagnetic shielding properties of Ni–Cu nanoferrite/PVP composites. *Material Science and Engineering B. Advanced Functional Solid-State Materials*, *186*, 1–6. doi:10.1016/j.mseb.2014.02.018

Madhu, B. J., Gurusiddesh, M., Kiran, T., Shruthi, B., & Jayanna, H. S. (2016a). Structural, dielectric, AC conductivity and electromagnetic shielding properties of polyaniline/$Ni_{0.5}Zn_{0.5}Fe_2O_4$ composites. *Journal of Materials Science Materials in Electronics*, *27*(8), 7760–7766. doi:10.1007/s10854-016-4764-2

Madhu, B. J., Kavya, M., Razika Banu, S., Shruthi, B., Sowmya, C. P., & Jayanna, H. S. (2012). Frequency Dependence of Dielectric Parameters and AC Conductivity of Mg-Co-Zn Nanoferrites Synthesized Via Combustion Method. *Advanced Materials Research*, *584*, 295–298. doi:10.4028/www.scientific.net/AMR.584.295

Madhu, B. J., Kiran, T., Gurusiddesh, M., & Shruthi, B. (2016b). Synthesis, characterization and dielectric behavior of polyaniline/$Ni_{0.5}Zn_{0.5}Fe_2O_4$ nanocomposites. *Macromolecular Symposia*, *361*(1), 24–29. doi:10.1002/masy.201400265

Maensiri, S., Sangmanee, M., & Wiengmoon, A. (2008). Magnesium ferrite (MgFe 2 O 4) nanostructures fabricated by electrospinning. *Nanoscale Research Letters*, *4*(3), 221–228. doi:10.1007/s11671-008-9229-y PMID:20596441

Nadeem, M., & Mushtaq, A. (2009). Thermal cycling induced reversibility/irreversibility in the ac electrical properties of La 0.50 Ca 0.50 MnO 3+ δ by impedance spectroscopy. *Journal of Applied Physics*, *106*(7), 073713. doi:10.1063/1.3238301

Naseri, M. G., Saion, E. B., & Shaari, A. H. (2012) Role of PVP on the Phase Composition and Morphology of Manganese Ferrite Nanoparticles Prepared by Thermal Treatment Method. *2nd International Conference on Environment Science and Biotechnology IPCBEE*, *48*, 32. Doi:10.7763/IPCBEE

Panwar, V., Park, J. O., Park, S. H., Kumar, S., & Mehra, R. M. (2010). Electrical, dielectric, and electromagnetic shielding properties of polypropylene-graphite composites. *Journal of Applied Polymer Science*, *115*(3), 1306–1314. doi:10.1002/app.29702

Patil, K. C. (2008). *Chemistry of nanocrystalline oxide materials: combustion synthesis, properties and applications*. World Scientific. doi:10.1142/6754

Raghavender, A. T., & Jadhav, K. M. (2009). Dielectric properties of Al-substituted Co ferrite nanoparticles. *Bulletin of Materials Science*, *32*(6), 575–578. doi:10.1007/s12034-009-0087-8

Reetz, M. T., & Helbig, W. (1994). Size-selective synthesis of nanostructured transition metal clusters. *Journal of the American Chemical Society*, *116*(16), 7401–7402. doi:10.1021/ja00095a051

Roosen, A. R., & Carter, W. C. (1998). Simulations of microstructural evolution: anisotropic growth and coarsening. *Physica A: Statistical Mechanics and its Applications*, *261*(1), 232-247.

Rozenberg, B. A., & Tenne, R. (2008). Polymer-assisted fabrication of nanoparticles and nanocomposites. *Progress in Polymer Science*, *33*(1), 40–112. doi:10.1016/j.progpolymsci.2007.07.004

Said-Galiev, E. E., Stakhanov, A. I., Blagodatskikh, I. V., Kobitskaya, E. M., Khokhlov, A. R., Naumkin, A. V., & Pisarev, S. A. (2010). Synthesis of polyvinylpyrrolidone and its nanosilver-based polymer composites in supercritical carbon dioxide. *Polymer Science, Series B*, *52*(3), 165–173. doi:10.1134/S1560090410030061

Sarkar, S., Guibal, E., Quignard, F., & SenGupta, A. K. (2012). Polymer-supported metals and metal oxide nanoparticles: Synthesis, characterization, and applications. *Journal of Nanoparticle Research*, *14*(2), 1–24. doi:10.1007/s11051-011-0715-2 PMID:22448125

Thakur, S., Katyal, S. C., & Singh, M. (2007). Improvement in electric and dielectric properties of nanoferrite synthesized via reverse micelle technique. *Applied Physics Letters*, *91*(26), 262501. doi:10.1063/1.2824454

Tsuji, M., Hashimoto, M., Nishizawa, Y., & Tsuji, T. (2004). Synthesis of gold nanorods and nanowires by a microwave–polyol method. *Materials Letters*, *58*(17), 2326–2330. doi:10.1016/j.matlet.2004.02.020

Wang, G.-X., Zhang, B.-L., Yu, Z.-L., & Qu, M.-Z. (2005). Manganese oxide/MWNTs composite electrodes for supercapacitors. *Solid State Ionics*, *176*(11-12), 1169–1174. doi:10.1016/j.ssi.2005.02.005

Younas, M., Nadeem, M., Atif, M., & Grossinger, R. (2011). Metal-semiconductor transition in NiFe2O4 nanoparticles due to reverse cationic distribution by impedance spectroscopy. *Journal of Applied Physics*, *109*(9), 093704. doi:10.1063/1.3582142

Chapter 10 Preparation and Characterization of Biopolymeric Nanoparticles as Drug Delivery Vehicles: A Comparative Study

Sai S. Sagiri
National Institute of Technology Rourkela, India

Suraj K. Nayak
National Institute of Technology Rourkela, India

S. Lakshmi
Bharatiyar University, India

Kunal Pal
National Institute of Technology Rourkela, India

ABSTRACT

In recent years, the use of biopolymeric nanoparticles as vehicles for drug delivery has increased exponentially. In the present study, chitosan and gelatin nanoparticles were prepared by ionic gelation and desolvation methods, respectively. Salicylic acid was used as the model drug. The nanoparticles were characterized using SEM, XRD analysis and FTIR spectrophotometric studies. In vitro drug release experiments were carried out to understand the mechanism of drug release. SEM micrographs showed the formation of spherical nanoparticles. XRD studies indicated a higher crystalline nature of the chitosan nanoparticles as compared to the gelatin nanoparticles. FTIR studies indicated the presence of salicylic acid within the drug- loaded nanoparticles. Drug release studies indicated that the developed nanoparticles may be used as carriers for various bioactive agents.

DOI: 10.4018/978-1-5225-3023-7.ch010

INTRODUCTION

In the recent years, there has been an exponential increase in the use of nanoparticulate delivery vehicles for controlled drug delivery (Manzoor, et al., 2012; Probst, Zrazhevskiy, Bagalkot, & Gao, 2013). The various nanoparticulate drug delivery vehicles include solid-lipid nanoparticles (Makwana, Jain, Patel, Nivsarkar, & Joshi, 2015), liposomes (Kraft, Freeling, Wang, & Ho, 2014), magnetosomes (Deng, et al., 2013) and biopolymeric nanoparticles (Nitta & Numata, 2013). This increase may be attributed to the advancement of research technologies and the advantages of the nanoparticulate delivery vehicles (Couvreur, 2013; Parveen, Misra, & Sahoo, 2012). The advantages of the nanoparticulate delivery vehicles include their ability to incorporate both hydrophobic and hydrophilic compounds (Aryal, Hu, Fu, & Zhang, 2012), ability to cross blood-brain-barrier (Kreuter, 2013; Wohlfart, Gelperina, & Kreuter, 2012) and increased residence time in the systemic circulation (using stealth technology) (Amoozgar & Yeo, 2012; L. Li, et al., 2013). The above-mentioned properties have helped in designing controlled delivery system using nanoparticulate systems (Coco, et al., 2013; Pathak & Thassu, 2016). Of the various types of nanoparticulate delivery vehicles, the use of biopolymeric nanoparticles is on the rise (Lohcharoenkal, Wang, Chen, & Rojanasakul, 2014; Zhen, et al., 2013). This may be attributed to the easy method of preparation, stability of the vehicle, biodegradable/ bioresorbable and minimum antigenic properties associated with the biopolymers (Elzoghby, Samy, & Elgindy, 2012a; Sundar, Kundu, & Kundu, 2016). The various biopolymers which have been commonly used for devising nanoparticulate delivery vehicles are alginate (Nesamony, Singh, Nada, Shah, & Kolling, 2012; Paul, Shelma, & Sharma, 2013), chitosan (Bernkop-Schnürch & Dünnhaupt, 2012; Garrait, Beyssac, & Subirade, 2014), dextran (Foerster, et al., 2016), hyaluronate (Z. Chen, et al., 2013; Yoon, et al., 2013), cellulose (Abeer, Amin, Iqbal, & Martin, 2014; Rescignano, et al., 2014), albumin (Elsadek & Kratz, 2012; Elzoghby, et al., 2012a), collagen (M.-M. Chen, et al., 2015; Kasoju, Ali, Dubey, & Bora, 2013), heparin (Liang & Kiick, 2014; She, et al., 2013), and gelatin (Elzoghby, Samy, & Elgindy, 2012b; Hu, et al., 2015).

In the present study, attempts were made to develop and characterize chitosan and gelatin based nanoparticles. Chitosan is prepared from chitin, one of the natural and commonly available polymers (Araki, Yamanaka, & Ohkawa, 2012; Zikakis, 2012). Chitosan is a linear polysaccharide with random distribution of deacetylated β-(1-4)-D-glucosamine and N-acetyl-D-glucosamine which has been prepared by the deacetylation of chitin (Zong, Kimura, Takahashi, & Yamane, 2000). Chitin is found in the exoskeleton of crustaceans like crabs and shrimps (Arbia, Arbia, Adour, & Amrane, 2013; Kaya, et al., 2014). Crustacean shell is the major waste from shrimp industry and has the potential to become the foremost source of chitin and chitosan (Anand, et al., 2016; Karthik, Manigandan, Saravanan, Rajesh, & Chandrika, 2016). On the other hand, gelatin is mainly obtained from animal skin and bones (C. Coester, Kreuter, von Briesen, & Langer, 2000). It is one of the most commonly used biopolymer with improved biocompatibility and biodegradation (H. Wang, et al., 2013). Sodium carboxymethyl cellulose (SCMC) was used as a copolymer for the preparation of gelatin nanoparticles.

BACKGROUND

The field of drug delivery technology is an inter-disciplinary field, which requires the expertise of materials science, medical science and cell biology (K. K. Jain, 2014). Research on novel drug delivery systems can help in saving a lot of effort of the pharmaceutical companies, which are involved in synthesizing novel drug molecules (Anselmo & Mitragotri, 2014). Efficient delivery of the drugs to the specific site of action helps in reducing the toxic/ side effects of the drugs, without compromising the therapeutic efficacy of the treatment (A. Jain, Singh, Singh, & Singh, 2013). In the last two decades, there has been an increased research on the development of novel drug delivery systems (Pal, Behera, Roy, Sekhar Ray, & Thakur, 2013). Researchers have developed microparticulate and nanoparticulate delivery systems apart from the macroscale delivery systems (gels). The nanoscale delivery systems have received a special attention of the researchers and the medical practitioners in the last decade due to their beneficial properties (Cheng, Al Zaki, Hui, Muzykantov, & Tsourkas, 2012). The use of nanoparticulate materials for prophylaxis, treatment and diagnosis of the diseases has emerged as a novel field of nanomedicine (Doane & Burda, 2012; Prabhakar, et al., 2013). These systems provide opportunity to manipulate the biological responses by interacting with the biolological components. In other words, nanotechnology allows engineering of the biological systems at a molecular scale (W. Wang, et al., 2014; Zhang, et al., 2012). Hence, a surge in the research on nanoscale delivery systems was observed in the last decade with many products being finding their way into the commercial market (Ragelle, Danhier, Préat, Langer, & Anderson, 2016). The success of nanomedicine is due to its ability to exploit the cellular machinery for delivering the drugs directly to the target site (Pang, et al., 2015). This has been made possible due to the size of the nanomedicine, which is equivalent to the size of the proteins and other macromolecules present within the mammalian cells.

The advancements in the field of polymer chemistry and fabrication technologies relating to the development of nanoparticles have brought revolution in the pharmaceutical industry. The physical and the chemical properties of the nanoparticles are different from that of the bulk properties of the materials (Pfeiffer, et al., 2014). The use of nanoparticles as delivery vehicles not only allows the administration of drugs directly into the systemic circulation, but also to target diseased cells and tissues (Pan, et al., 2012). The afore mentioned advantages have provided many- fold improvement in the treatment efficacies. Polymers have been considered as the choice of materials for designing nanoparticulate drug delivery systems. In general, the drug delivery systems are regarded as nanoparticulate delivery systems if the size of the nanoparticles is less than 500 nm (Couvreur, 2013). But many authors have reported that particles less than 1000 nm can be considered as nanoparticulate drug delivery systems (Bamrungsap, et al., 2012). The smaller size of the nanoparticles is associated with the corresponding increase in the surface area of the delivery systems. This, in turn, increases the solubility of the drug molecules within the physiological fluids and hence the bioavailability of the drugs when formulated as nanomedicine as compared to the conventional drug delivery systems (e.g. ointments, elixers, solutions, suspensions, emulsions, tablets and capsules) (Mudshinge, Deore, Patil, & Bhalgat, 2011). Further, the loading and the delivery efficiency of the drugs is also high due to the same reason. The residence time of nanoparticles within the human body may be improved by suitable physical or chemical modifications (Mudshinge, et al., 2011). Due to the aforementioned reasons, the efficacy of the drug improves many- folds, thereby, reducing the requirement of the amount of the drug for the treatment of the disease. The nanoparticulate drug delivery systems are broadly classified either as nanospheres or nanocapsules (Parveen, et al., 2012). Nanospheres are the delivery systems where the continuous matrix is formed by the polymer and the drugs molecules are dispersed homogeneously throughout the matrix. On the other hand, nanocapsules are the delivery systems where the polymer encapsulates the drug (either solid or solubilized in oil). In other words, the drug phase forms the core of the delivery system and the polymer phase forms a thin layer over the drug phase. Further, it is possible to decorate these nanoparticles with other bioactive molecules. The decoration may be achieved either by adsorption or conjugation of the drugs with the functional group of the polymers present on the surface of the nanoparticles. Nanoparticulate drug delivery systems have been explored successfully for the delivery of anticancer drugs, peptides/ proteins, and genetic materials (Parveen, et al., 2012).

The release of the drug from the nanoparticles may either be via diffusion- mediated, erosion- mediated or a combination of the aforementioned processes (Mudshinge, et al., 2011). The interaction amongst the matrix or shell polymer phase and the drugs play an important role in governing the rate of release of the drugs. If the interactions amongst the polymer and the drug molecules are associative in nature, then the rate of the release of the drug molecules will be slower than the cases where there are no interactions or the interactions are repulsive in nature, respectively. Further, the rate of release of the drugs can be lowered by converting the drug molecules into less soluble forms by interacting them with auxiliary ingredients (Mudshinge, et al., 2011).

Chitosan Based Nanoparticles as Delivery Vehicles

Chitosan is a polyelectrolyte polysaccharide which is a deacetylated form of the polysaccharide, chitin (Younes & Rinaudo, 2015). Chitin is abundantly present in crustaceans like shrimps and crabs (Arbia, et al., 2013). Due to the presence of primary amino group, chitosan easily undergoes dissolution in acidic medium. The primary amino group is protonated under the acidic environment. Hence, the chitosan molecules are regarded as cationic polyelectrolyte (Luo & Wang, 2014). The polysaccharide has gained much importance in recent years due to its inherent biocompatibility, antimicrobial activity, and haemostatic properties (Liu, Wu, & Qiu, 2016). The haemostatic property of chitosan is associated with its ability to agglutinate the red blood cells (Smith, Laird, Porter, & Bloch, 2013). Further, chitosan has been reported to have antibacterial properties. This is attributed to its ability to interact with the bacterial cell membrane, which leads to the damage of the cell membrane (Yan, et al., 2016). The aforementioned properties are essential in wound dressing materials. This explains why chitosan has become a material of choice for fabricating wound dressings (Abdelgawad, Hudson, & Rojas, 2014). Chitosan nanoparticles have been explored as drug delivery vehicles for the delivery of the drugs via oral (Luo, Teng, Li, & Wang, 2015), ophthalmic (Chopra, Marwaha, Kaushik, & Dureja, 2015), transdermal (Jana, Manna, Nayak, Sen, & Basu, 2014), and parenteral (Trapani, et al., 2013) routes of administration. The nanoparticles may also be explored to deliver drug to the central nervous system due to their ability to cross blood- brain- barrier (Malmo, Sandvig, Vårum, & Strand, 2013). The chitosan based nanoparticles may be prepared either by covalently crosslinking method, ionotropic gelation method, or desolvation method.

Chitosan nanoparticles have been used for intra-ocular drug delivery by topical administration to the ocular sac. The intra-ocular drug delivery is promoted by the chitosan nanoparticles due to their mucoadhesive and drug penetration enhancing properties. The drug penetration enhancing property of the chitosan nanoparticles is due to the increase in the internalization of the chitosan nanoparticles by the corneal epithelial cells (Chaiyasan, Praputbut, Kompella, Srinivas, & Tiyaboonchai, 2017). The uptake of the chitosan nanoparticles by the cells is dependent on the molecular weight and the degree of deacetylation of the chitosan used. A reduction in the molecular weight and the degree of deactylation has been reported to significantly reduce not only the cellular uptake but also the cytotoxic effects of the chitosan nanoparticles. On the contrary, the antibacterial efficacy of the nanoparticles is potentiated by the low molecular weight chitosan molecules (Huang, Khor, & Lim, 2004; Huang, Ma, Khor, & Lim, 2002). The bronchial relaxation property of theophylline is potentiated, when delivered using chitosan nanoparticles (C. Chen, et al., 2016). Similarly, the rate of absorption of insulin, after intranasal administration using chitosan nanoparticles, is also significantly increased (Dyer, et al., 2002). Chitosan coated liposomes (Gradauer, et al., 2013) and chitosan- polyacrylic (Sung, Sonaje, Liao, Hsu, & Chuang, 2012) nanoparticles show an improved bioavailability of peptide drugs after oral administration by improving the mucoadhesive property. Chitosan nanoparticles have also been explored in the vaccine development (Rezaei Mokarram & Alonso, 2016). Chitosan nanoparticle based nasal formulations containing hepatitis-B surface antigen (HBsAg) as the immunomodulatory agent have been studied extensively (Lebre, Borchard, Faneca, Pedroso de Lima, & Borges, 2016). Parenteral delivery of tacrine containing chitosan nanoparticles have been proposed for the treatment of the Alzheimer's disease (Silva, González-Mira, Lobo, & Amaral, 2013). Similarly, intranasal delivery of 17β – estradiol using chitosan nanoparticles have also shown good success in the treatment of the Alzheimer's disease (X. Wang, Chi, & Tang, 2008).

Gelatin Based Nanoparticles as Delivery Vehicles

Gelatin is a protein- based biopolymer obtained from the animal source. The main sources of gelatin for the pharmaceutical and the biomedical industries include porcine and bovine (Tseng, Tsou, Wang, & Hsu, 2013). It is a denatured protein, synthesized from collagen by partial hydrolysis (acidic or alkaline) (Elzoghby, 2013). It is a United States Food and Drug Administration approved polymer for human consumption and medical use (Chong, Lim, & Sultana, 2014). The main advantage of the gelatin based implants is their biodegradability without any harmful by-products (Marvizadeh, Nafchi, & Jokar, 2014). The improved biological activity of the gelatin constructs have been explained by the presence of RGD (Arginine-Glycine-Aspartic acid) motifs within the gelatin molecules (Lin, Yu, Chen, & Tsai, 2016). Gelatin nanoparticles can be prepared by desolvation (Park, Vo, Kang, Oh, & Lee, 2015), coacervation phase separation (Patra, Basak, & Tibarewala, 2016), emulsification- solvent evaporation (Staff, Landfester, & Crespy, 2013), reverse phase microemulsion (Biausque, Laveille, Anjum, Caps, & Basset, 2015), nanoprecipitation (Lee, Khan, Park, & Lim, 2012), self- assembly (Park, et al., 2015), and layer-by-layer coating (Elzoghby, 2013) methods. The gelatin nanoparticles are non- toxic in nature even at very high concentrations (Gupta, Gupta, Yarwood, & Curtis, 2004). The inflammatory response is not initiated by the gelatin nanoparticles suggesting its biocompatible nature (Brzoska, et al., 2004). The cellular uptake of the gelatin nanoparticles within murine fibroblast cells (Xu & Amiji, 2012) and murine bone marrow dendritic cells (Conrad Coester, Nayyar, & Samuel, 2006) was found to be very good. The mechanism of the uptake of the nanoparticles was due to endocytosis. Due to their ability to be internalized within the cells, the gelatin nanoparticles have been proposed for site-specific drug delivery to the cancer cells. It has been found to carry a payload of both hydrophilic and hydrophobic anti- cancer drugs. Some of the common anticancer drugs which have been used to formulate gelatin nanoparticle based delivery systems include paclitaxel, methotrexate, doxorubicin, resveratrol, cycloheximide, cytarabine, cisplatin, and noscapine (Elzoghby, 2013; Sahoo, Sahoo, Biswas, Guha, & Kuotsu, 2015). The efficacy of the treatment under both *in vitro* and *in vivo* testing conditions was found to be better when gelatin nanoparticle based formulations were used. Gelatin nanoparticles have also been explored as delivery vehicles for proteins and peptides. Protein drugs like fibroblast growth factor, alkaline phosphatase, insulin, and bone morphogenetic protein-2 have been loaded in gelatin nanoparticles (Elzoghby, 2013). Due to the ability of gelatin to be internalized within dendritic cells, the gelatin nanoparticles have been proposed to be used as immunoadjuvant in vaccine development. The architecture of the gelatin matrix is compromised by lysozyme after the internalization within the cells. This resulted in the release of the proteins (capable of eliciting immune response), which are pre-loaded within the gelatin nanoparticles (Elzoghby, 2013; Klier, et al., 2012). The gelatin nanoparticles have found great success in gene delivery applications. In addition to proteins, gelatin nanoparticles have the potential to deliver genetic material as well. The main advantage of the genetic material delivery via gelatin nanoparticles is their ability to deliver the genetic material without using viral delivery vector (Nitta & Numata, 2013). Gelatin nanoparticles have also been explored for the topical delivery of the drug onto the ocular surface. The mucoadhesive property of the gelatin nanoparticles improves the residence time of the drug, thereby, improving the bioavailability of the drug (Mahor, et al., 2016). Transfection of the genetic material into the ocular epithelial layers has also been studied successfully using gelatin nanoparticles (Elzoghby, 2013). Gelatin nanoparticles do not elicit inflammation of the respiratory tract after inhalation. This property of the gelatin nanoparticles make them suitable for pulmonary drug delivery. Pulmonary delivery of the immunostimulating agents has been achieved successfully using gelatin nanoparticles with considerable success. The efficacy of the lung cancer treatment may be improved by using gelatin nanoparticles as the carrier matrices (Elzoghby, 2013).

FOCUS OF THE CHAPTER

The present chapter discusses about the synthesis and the characterization of the chitosan and gelatin based nanoparticles. The ability of the prepared nanoparticles for the delivery of the drugs has also been studied in-depth.

MATERIALS AND METHODS

Materials

Gelatin (type II) and chitosan (>95% deacetylated) were purchased from Himedia lab Pvt. Ltd, Mumbai. Acetone, glacial acetic acid and SCMC were procured from Rankem RFCL Ltd, New Delhi. 25% (v/v) glutaraldehyde and sodium tripolyphosphate (TPP) were purchased from Merck specialitics Pvt. Ltd, Mumbai and Lobachemi, respectively. Salicylic acid (SA) was procured from Fine chemicals, India. Double distilled water was used in the whole study.

Preparation of Gelatin Nanoparticles

Gelatin nanoparticles were prepared by two-step desolvation method with slight modification of the protocol as described by Coester *et al.* (2000) (C. Coester, et al., 2000). In short, a solution of SCMC and gelatin was prepared so as to have a proportion of 1.0: 2.33 (SCMC: Gelatin). 25 ml of acetone (desolvating agent) was added to the 25 ml of the SCMC-gelatin solution. This resulted in the precipitation of the white colored gelatin-SCMC complex. The complex, so obtained, was collected by centrifugation and was subsequently dissolved in 25 ml of water, whose pH was adjusted to 3.5 using glacial acetic acid. 75 ml of acetone was added drop-wise to the above solution, kept on stirring at 600 rpm, which resulted in the desolvation and subsequent precipitation of the gelatin molecules as nanoparticles. Thereafter, 250 µl of glutaraldehyde was added and stirred for 12 h at 600 rpm at room-temperature. This step resulted in the improvement of the stability of the nanoparticles. The remaining solvent was filtered and the nanoparticles were washed thoroughly and were subsequently redispersed in water and stored at 2-8 °C for further analyses. The samples were regarded as GS. Drug loaded nanoparticles were prepared in the similar manner. 1% (w/v) SA was added to the acidic gelatin solution before the polymer was precipitated. The drug loaded particles were regarded as GSS.

Preparation of Chitosan Nanoparticles

Chitosan nanoparticles were prepared by ionic gelation method with slight modification of the protocol as described by Wang *et al.* (2007) (C. Wang, Fu, & Yang, 2007). In short, 0.5 ml of aqueous TPP solution (1% w/v) was added to 49.5 ml of chitosan solution (0.45% w/v in 10% v/v acetic acid), kept under stirring at 600 rpm at room-temperature, until the solution appears to be turbid indicating the formation of nanoparticles. Nanoparticles, so obtained, were separated by centrifugation and washed thoroughly. The nanoparticles were subsequently redispersed in water and stored at 2-8 °C for further analyses. The particles were regarded as CS. Drug loaded nanoparticles were also prepared in the same fashion. 1% (w/v) SA was added to the chitosan solution used for the preparation of the nanoparticles. The drug loaded particles were regarded as CSS.

Characterization of the Nanoparticles

Scanning Electron Microscopy (SEM)

The surface morphology and shape of the nanoparticles were analyzed under scanning electron microscope (JSM-6390, JEOL, Philips, Japan), which was operated at 15 kV. The nanoparticle dispersions were vacuum dried over a glass slide and subsequently mounted on sample holder. The samples were sputtered with platinum before analyzing under the microscope.

X-Ray Diffraction (XRD)

The x-ray diffraction studies were carried out using X'Pert-PRO diffractometer (PANanalytical, Japan). CuKα was used as the x-ray source. The samples were analyzed from 5° 2θ to 50° 2θ at rate of 2° 2θ / min. The XRD parameters such as peak center, AUC and FWHM were calculated using *fityk* software (Wojdyr, 2010)

Fourier Transform Infrared Spectroscopy (FTIR)

The FTIR analysis of the samples were performed by using a Impact-410 spectrophotometer (Nicolet Instrument Corp., Madison, WI) in the spectral region of 4000 cm^{-1} and 400 cm^{-1} as KBr pellets.

Drug Entrapment Efficiency (DEE)

The drug encapsulation efficiency was determined by calculating the difference between total drug taken and free drug in the supernatant after nanoparticles preparation as per method mentioned previously (D. Jain & Banerjee, 2008; Raj & Prabha, 2016). In gist, supernatant was separated by centrifugation and the amount of free drug present in the supernatant (w) was assayed by UV-spectrophotometer at 294 nm. The amount of drug in the supernatant was subtracted from the total amount of drug added (W). In effect, (W-w) will give the amount of drug entrapped in the nanoparticles. The percentage of drug entrapment was calculated using the formula:

$$\% \ Drug \ Entrapment = \frac{\left(\mathrm{W} - \mathrm{w}\right)}{\mathrm{W}} \mathrm{x100}$$

In Vitro Drug Release Studies

In vitro drug release study across a dialysis tube was performed at 37 °C under constant magnetic stirring at 100 rpm at 37 °C. Prior to the experiment the dialysis tube was washed with warm distilled water (80 °C) for 30 min. 2 g of the vaccum dried drug-loaded nanoparticles were dispersed in 5 ml of phosphate buffer (pH 7.2). The dispersion was then placed in the dialysis tube tied at both ends and immersed in receiver chamber filled with 500 ml of phosphate buffer (pH 7.2). 3 ml of the dissolution medium was sampled at an interval of 30 min. 3 ml of fresh phosphate buffer was replaced after each sampling. The study was conducted for 8h. The collected samples were analyzed at 294 nm using UV-visible spectrophotometer (UV-3200, LabIndia, India).

RESULTS AND DISCUSSION

Morphological Analysis (SEM Studies)

The gelatin and chitosan nanoparticles were subjected to scanning electron microscopy to figure out the shape and size of the particulate systems. The micrographs of the particulate systems indicated that the particles were spherical in shape and had sizes of 400 and 600 nm (approx.) for gelatin and chitosan systems, respectively (Figure 1). Based on the size obtained from the micrographs, the particles can be regarded as nanoparticles, since; the size of the organic nanoparticles tends to be in the range of 10 nm to 1000 nm (Soppimath, Aminabhavi, Kulkarni, & Rudzinski, 2001). The GS nanoparticles seem to be slightly distorted or not spherical as compared to CS and CSS. Gelatin-based nanoparticles couldn't retain the spherical shape after drying due to the amorphous nature of the gelatin (evident from x-ray diffraction studies).

X-Ray Diffraction Studies

The x-ray diffractograms of SA and the nanoparticles have been shown in Figure 2. The x-ray diffractogram of SA showed 3 major peaks at 10.86°, 17.18° and 25.26° 2θ. The nature of the diffractogram indicated that SA is crystalline in nature. The x-ray diffractogram of CS nanoparticles showed a peak at 19.7° 2θ and a hump at 21.56° 2θ whereas GS nanoparticles showed a broad peak at 20° 2θ. The full width at half maximum (FWHM, Table 1) indicated that the CS nanoparticles are more crystalline than the GS nanoparticles. The CSS nanoparticles showed 3 sharp peaks at 10.82°, 17.14° and 25.00°

Figure 1. Scanning electron micrographs of nanoparticles (A) GS (B) CS and (c) CSS

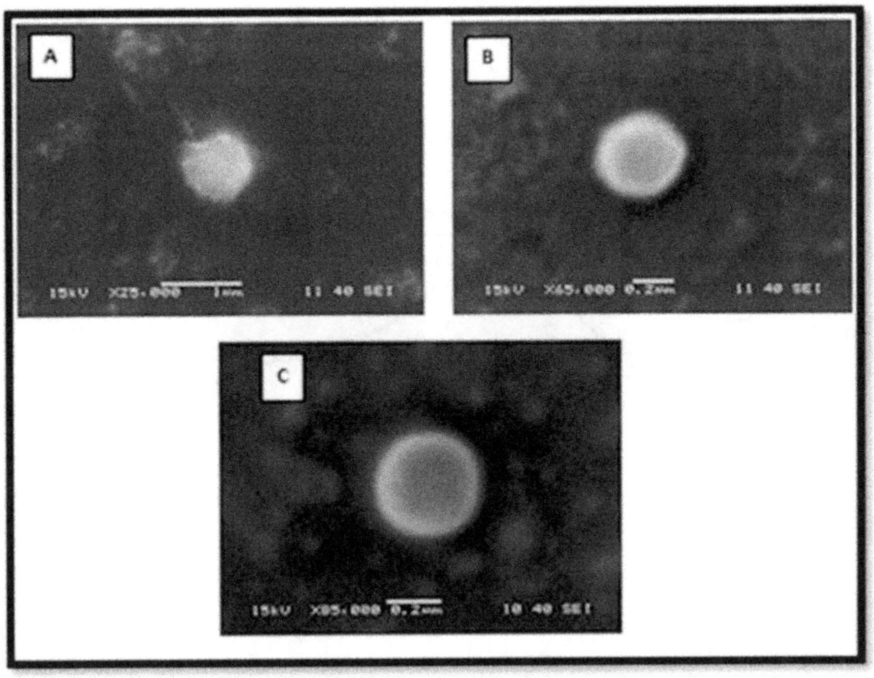

2θ which are the characteristic peaks of SA indicating the presence of SA in the CSS nanoparticles. CSS nanoparticles also showed a peak at 22.0° 2θ and a hump at 24.1° 2θ which may be attributed to the presence of chitosan. A slight shift of the chitosan peaks in CSS from the CS nanoparticles may be attributed to the change in the crystallite orientation of the polymer matrix due to the presence of SA in CSS. Also, with the incorporation of SA in the nanoparticles, there was a decrease in the area under the curve (AUC) of CSS as compared to CS nanoparticles (Table 1). This indicated that there was an increase in the crystallinity of the CSS nanoparticles.

Unlike CSS nanoparticles, x-ray diffractogram of GSS nanoparticles showed only one broad peak at 23.24° 2θ. This may be accounted to the amorphous nature of the nanoparticles (Rejinold, Chennazhi, Nair, Tamura, & Jayakumar, 2011). The absence of peaks for SA may be due to the fact that the SA molecules were entrapped within the amorphous structure of the gelatin and there was no SA present on the surface of the nanoparticles. The incorporation of SA decreased the crystallinity of the gelatin nanoparticles. This might be attributed to the decreased intermolecular hydrogen bonding amongst the amino and carboxylic groups of gelatin molecules due to the increased ionic interactions amongst the amino group of gelatin and carboxylic groups of the salicylic acid.

FTIR Studies

The FTIR spectra for GS (curve a), GSS (curve b), CS (curve c), CSS nanoparticles (curve d) and SA (curve e) are shown in Figure 3.

FTIR spectra of the nanoparticles have been shown in Figure 3. The FTIR spectra of GS nanoparticles showed absorption peaks at 3410 cm^{-1} (NH stretching), 1629 cm^{-1}(amide I, CO and CN stretching) and

Figure 2. XRD profiles. (a) Chitosan (b) Gelatin (c) SCMC (d) G-SCMC physical mixture (e) CS-SA nanoparticles (f) G-SCMC-SA nanoparticles, and (g) SA drug

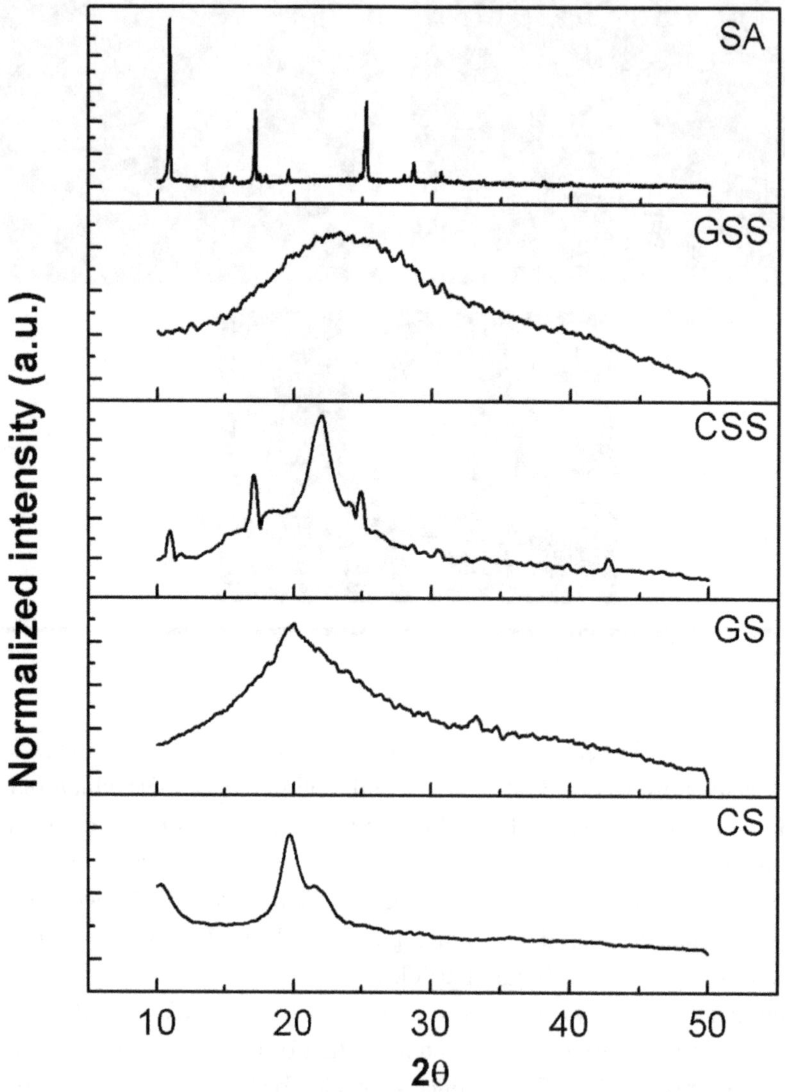

Table 1. The XRD parameters of the nanoparticles

Particulars	CS	CSS	GS	GSS
Peak center	19.8	22.22	19.38	23.24
AUC	2350.56	1841.95	5338.09	4577.21
FWHM	3.62	3.36	11.94	17.2

1534 cm^{-1} (amide-II). The absorption peak at 1630 cm^{-1} indicated the presence of amide-I due to the secondary structure of proteins (gelatin A) (Muyonga, Cole, & Duodu, 2004). Slight shift of the absorption band from 1,630 cm^{-1} to 1,629 cm^{-1} in GSS might be due to the interaction between the negatively

Figure 3. FTIR spectra of: (a) GS (b) GSS (c) CS, and (d) CSS nanoparticles, and (e) SA drug

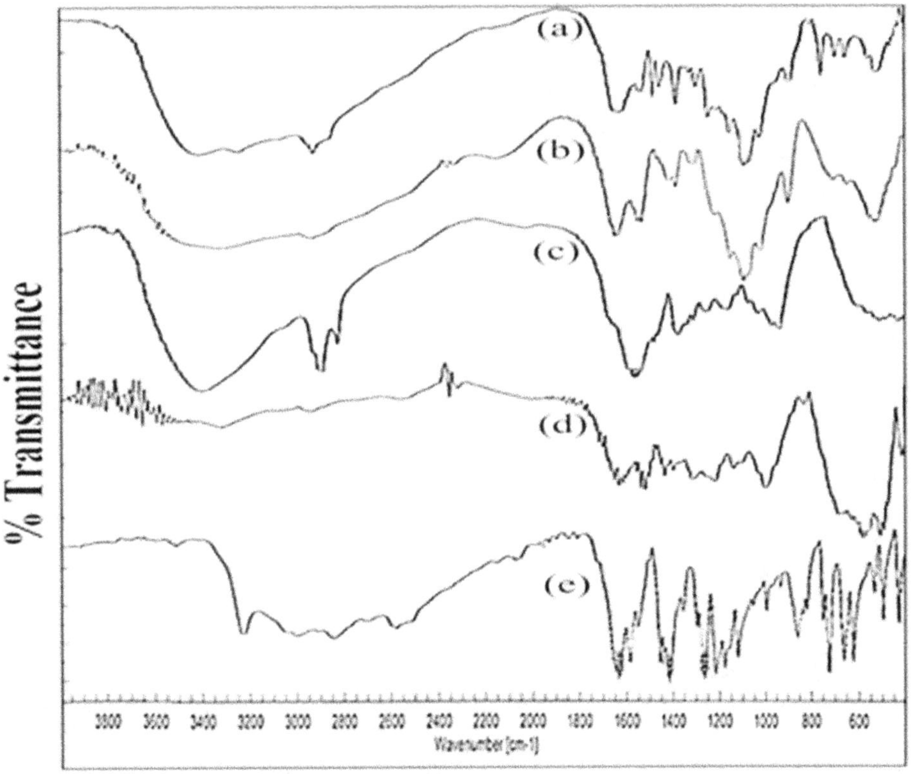

charged carboxy methyl groups and positively charged gelatin (Devi & Maji, 2009). CS showed peaks at 1320 cm^{-1} (corresponding to the acetyl group of chitosan) and in the region between 1420 cm^{-1} and 1435 cm^{-1} (characteristic fingerprints of carbohydrate polymers) (MIN, 2004; Mohammad R, 2008). GSS and CSS nanoparticles showed peaks at 1447 cm^{-1} and 1386 cm^{-1}, which are typical characteristic peaks for the aromatic ring of SA (Alvarez-Ros, Sánchez-Cortés, & Garcla-Ramos, 2000). This gave the confirmation for the presence of SA within the nanoparticles.

Drug Entrapment Efficiency

1% (w/v) SA was used in the preparation of CSS and GSS nanoparticles. Spectrophotometric observation of the supernatant after centrifugation of the prepared nanoparticles confirmed 65±3% and 58±4% drug-loading efficiency for the CSS and GSS nanoparticles, respectively. Since gelatin is amorphous in nature, the GSS nanoparticles couldn't retain the drug within the nanoparticles as effectively as that of CSS.

In Vitro Drug Release Studies

In vitro drug release studies were performed for 8h to study the release pattern of SA from the nanoparticles. The release patterns of SA from the CSS and GSS nanoparticles have been shown in Figure 4. The study showed that the rate of release of SA from GSS nanoparticles was higher as compared to CSS

Figure 4. In vitro release of salicylic acid from the drug loaded nanoparticles

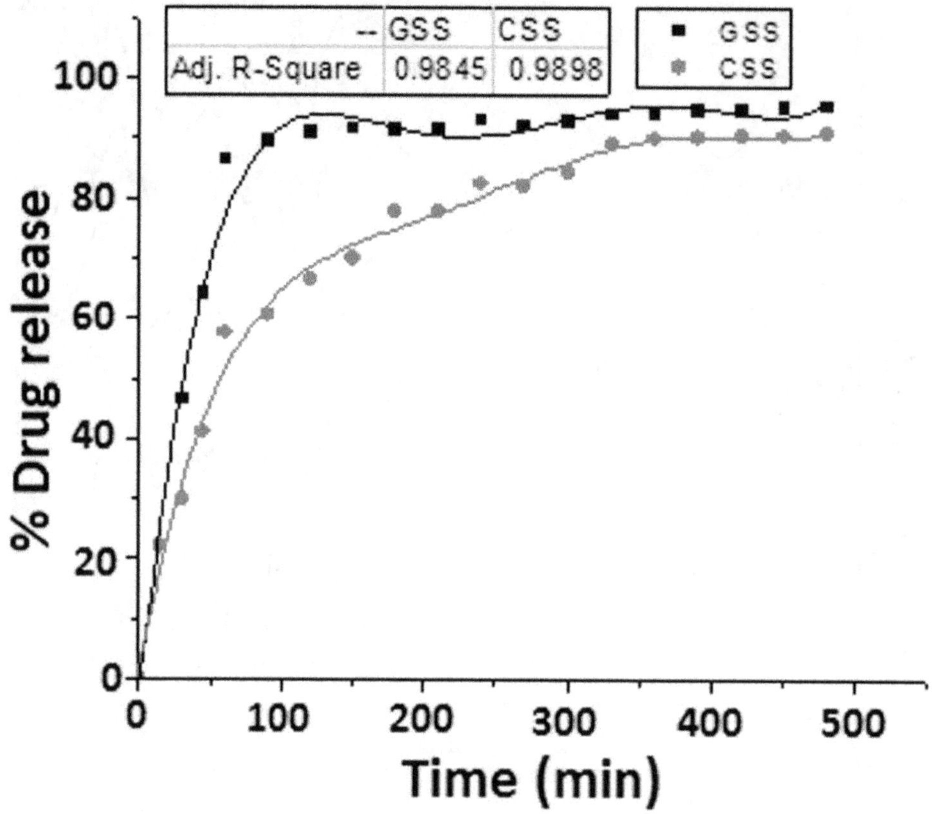

nanoparticles. This observation can be explained by the more amorphous nature of GSS nanoparticles than CSS nanoparticles, which promoted the easy wetting of the polymer chain and subsequent diffusion of water into the core structure (Leuner & Dressman, 2000). This, in turn, enhanced the rate of dissolution and migration of SA molecules from the core of the matrices. Also, the % release of SA was higher in GSS nanoparticles as compared to CSS nanoparticles. The higher crystallinity, the higher is the crystallite sites and hence more are the physical crosslink points (J. K. Li, Wang, & Wu, 1998). This might have resulted in the entrapment of SA molecules within the CSS polymer matrices.

CONCLUSION

The study describes the preparation of chitosan and gelatin based nanoparticles. The chitosan based nanoparticles were prepared by ionic gelation while the gelatin based nanoparticles were prepared by desolvation technique. An attempt was made to characterize the prepared nanoparticles and to study their suitability as a carrier for bioactive agents. The results indicated that the prepared nanoparticles were spherical in shape, irrespective of their method of preparation, though there was a difference in size of the nanoparticles, gelatin nanoparticles being smaller (400 nm) than the chitosan nanoparticles (600 nm). The gelatin based nanoparticles were more amorphous than chitosan nanoparticles and FTIR studies indicated the presence of SA in the SA-loaded nanoparticles. The drug release studies indicated that >90% of the drug was released after a span of 8 h. Hence the developed nanoparticles may be tried as vehicles for bioactive agent delivery.

FUTURE RESEARCH DIRECTIONS

Prior to the biological applications, further characterization of the nanoparticles is needed. The zeta-potential of the nanoparticles needs to be determined. If necessary, suitable surface modification of the nanoparticles needs to be carried out to improve the stability of the nanoparticle dispersion. Lastly, the *in vitro* and *in vivo* biocompatibility of the nanoparticles has to be ascertained for physiological applications.

REFERENCES

Abdelgawad, A. M., Hudson, S. M., & Rojas, O. J. (2014). Antimicrobial wound dressing nanofiber mats from multicomponent (chitosan/silver-NPs/polyvinyl alcohol) systems. *Carbohydrate Polymers*, *100*, 166–178. doi:10.1016/j.carbpol.2012.12.043 PMID:24188851

Abeer, M. M., Amin, M., Iqbal, M. C., & Martin, C. (2014). A review of bacterial cellulose-based drug delivery systems: Their biochemistry, current approaches and future prospects. *The Journal of Pharmacy and Pharmacology*, *66*(8), 1047–1061. PMID:24628270

Alvarez-Ros, M. C., Sánchez-Cortés, S., & García-Ramos, J. V. (2000). Vibrational study of the salicylate interaction with metallic ions and surfaces. *Spectrochimica Acta. Part A: Molecular and Biomolecular Spectroscopy*, *56*(12), 2471–2477. doi:10.1016/S1386-1425(00)00328-0 PMID:11075690

Amoozgar, Z., & Yeo, Y. (2012). Recent advances in stealth coating of nanoparticle drug delivery systems. *Wiley Interdisciplinary Reviews: Nanomedicine and Nanobiotechnology*, *4*(2), 219–233. PMID:22231928

Anand, M., Maruthupandy, M., Kumar, S. P., Ravikumar, S., Jeyaraj, N., & Kumaran, R. (2016). Antimicrobial Potential of Chitosan from Marine Biowastes—A Wealth from Waste Approach. *Journal of Chitin and Chitosan Science*, *4*(1), 55–58. doi:10.1166/jcc.2016.1095

Anselmo, A. C., & Mitragotri, S. (2014). An overview of clinical and commercial impact of drug delivery systems. *Journal of Controlled Release*, *190*, 15–28. doi:10.1016/j.jconrel.2014.03.053 PMID:24747160

Araki, J., Yamanaka, Y., & Ohkawa, K. (2012). Chitin-chitosan nanocomposite gels: Reinforcement of chitosan hydrogels with rod-like chitin nanowhiskers. *Polymer Journal*, *44*(7), 713–717. doi:10.1038/pj.2012.11

Arbia, W., Arbia, L., Adour, L., & Amrane, A. (2013). Chitin extraction from crustacean shells using biological methods-a review. *Food Technology and Biotechnology*, *51*(1), 12.

Aryal, S., Hu, C.-M. J., Fu, V., & Zhang, L. (2012). Nanoparticle drug delivery enhances the cytotoxicity of hydrophobic–hydrophilic drug conjugates. *Journal of Materials Chemistry*, *22*(3), 994–999. doi:10.1039/C1JM13834K

Bamrungsap, S., Zhao, Z., Chen, T., Wang, L., Li, C., Fu, T., & Tan, W. (2012). Nanotechnology in therapeutics: A focus on nanoparticles as a drug delivery system. *Nanomedicine (London)*, *7*(8), 1253–1271. doi:10.2217/nnm.12.87 PMID:22931450

Bernkop-Schnürch, A., & Dünnhaupt, S. (2012). Chitosan-based drug delivery systems. *European Journal of Pharmaceutics and Biopharmaceutics*, *81*(3), 463–469. doi:10.1016/j.ejpb.2012.04.007 PMID:22561955

Biausque, G., Laveille, P., Anjum, D. H., Caps, V., & Basset, J.-M. (2015). *Methods to synthesize NiPt bimetallic nanoparticles by a reversed-phase microemulsion, deposition of NiPt bimetallic nanoparticles on a support, and application of the supported catalyst for CO2 reforming of methane*. Google Patents.

Brzoska, M., Langer, K., Coester, C., Loitsch, S., Wagner, T. O. F., & Mallinckrodt, C. (2004). Incorporation of biodegradable nanoparticles into human airway epithelium cells—in vitro study of the suitability as a vehicle for drug or gene delivery in pulmonary diseases. *Biochemical and Biophysical Research Communications*, *318*(2), 562–570. doi:10.1016/j.bbrc.2004.04.067 PMID:15120637

Chaiyasan, W., Praputbut, S., Kompella, U. B., Srinivas, S. P., & Tiyaboonchai, W. (2017). Penetration of mucoadhesive chitosan-dextran sulfate nanoparticles into the porcine cornea. *Colloids and Surfaces. B, Biointerfaces*, *149*, 288–296. doi:10.1016/j.colsurfb.2016.10.032 PMID:27771493

Chen, C., Zhong, M., Li, G., Yang, F., Huang, R., Xiao, W., & Ding, P. (2016). Optimization on preparation conditions of calcium-crosslinked modified chitosan as potential matrix material for theophylline sustained-release beads and its evaluation of release kinetics. *Journal of Alloys and Compounds*, *658*, 348–355. doi:10.1016/j.jallcom.2015.10.219

Chen, M.-M., Huang, Y.-Q., Cao, H., Liu, Y., Guo, H., Chen, L. S., & Zhang, Q.-Q. et al. (2015). Collagen/chitosan film containing biotinylated glycol chitosan nanoparticles for localized drug delivery. *Colloids and Surfaces. B, Biointerfaces*, *128*, 339–346. doi:10.1016/j.colsurfb.2015.02.024 PMID:25784300

Chen, Z., Li, Z., Lin, Y., Yin, M., Ren, J., & Qu, X. (2013). Bioresponsive Hyaluronic Acid-Capped Mesoporous Silica Nanoparticles for Targeted Drug Delivery. *Chemistry (Weinheim an der Bergstrasse, Germany)*, *19*(5), 1778–1783. doi:10.1002/chem.201202038 PMID:23303570

Cheng, Z., Al Zaki, A., Hui, J. Z., Muzykantov, V. R., & Tsourkas, A. (2012). Multifunctional nanoparticles: Cost versus benefit of adding targeting and imaging capabilities. *Science*, *338*(6109), 903–910. doi:10.1126/science.1226338 PMID:23161990

Chong, L. H., Lim, M. M., & Sultana, N. (2014). Polycaprolactone (PCL)/Gelatin (Ge)-based Electrospun Nanofibers for Tissue engineering and Drug Delivery Application. *Applied Mechanics & Materials*, (554).

Chopra, A. K., Marwaha, R. K., Kaushik, D., & Dureja, H. (2015). Box-Behnken Designed Fluconazole Loaded Chitosan Nanoparticles for Ocular Delivery. *Pharmaceutics & Drug Delivery Research, 2014*.

Coco, R., Plapied, L., Pourcelle, V., Jérôme, C., Brayden, D. J., Schneider, Y.-J., & Préat, V. (2013). Drug delivery to inflamed colon by nanoparticles: Comparison of different strategies. *International Journal of Pharmaceutics*, *440*(1), 3–12. doi:10.1016/j.ijpharm.2012.07.017 PMID:22820482

Coester, C., Kreuter, J., von Briesen, H., & Langer, K. (2000). Preparation of avidin-labelled gelatin nanoparticles as carriers for biotinylated peptide nucleic acid (PNA). *International Journal of Pharmaceutics, 196*(2), 147-149.

Coester, C., Nayyar, P., & Samuel, J. (2006). In vitro uptake of gelatin nanoparticles by murine dendritic cells and their intracellular localisation. *European Journal of Pharmaceutics and Biopharmaceutics, 62*(3), 306–314. doi:10.1016/j.ejpb.2005.09.009 PMID:16316749

Couvreur, P. (2013). Nanoparticles in drug delivery: Past, present and future. *Advanced Drug Delivery Reviews, 65*(1), 21–23. doi:10.1016/j.addr.2012.04.010 PMID:22580334

Deng, Q., Liu, Y., Wang, S., Xie, M., Wu, S., Chen, A., & Wu, W. (2013). Construction of a novel magnetic targeting anti-tumor drug delivery system: Cytosine arabinoside-loaded bacterial magnetosome. *Materials (Basel), 6*(9), 3755–3763. doi:10.3390/ma6093755

Devi, N., & Maji, T. (2009). Preparation and Evaluation of Gelatin/Sodium Carboxymethyl Cellulose Polyelectrolyte Complex Microparticles for Controlled Delivery of Isoniazid. *AAPS PharmSciTech, 10*(4), 1412–1419. doi:10.1208/s12249-009-9344-9 PMID:19937282

Doane, T. L., & Burda, C. (2012). The unique role of nanoparticles in nanomedicine: Imaging, drug delivery and therapy. *Chemical Society Reviews, 41*(7), 2885–2911. doi:10.1039/c2cs15260f PMID:22286540

Dyer, A., Hinchcliffe, M., Watts, P., Castile, J., Jabbal-Gill, I., Nankervis, R., & Illum, L. et al. (2002). Nasal delivery of insulin using novel chitosan based formulations: A comparative study in two animal models between simple chitosan formulations and chitosan nanoparticles. *Pharmaceutical Research, 19*(7), 998–1008. doi:10.1023/A:1016418523014 PMID:12180553

Elsadek, B., & Kratz, F. (2012). Impact of albumin on drug delivery—New applications on the horizon. *Journal of Controlled Release, 157*(1), 4–28. doi:10.1016/j.jconrel.2011.09.069 PMID:21959118

Elzoghby, A. O. (2013). Gelatin-based nanoparticles as drug and gene delivery systems: Reviewing three decades of research. *Journal of Controlled Release, 172*(3), 1075–1091. doi:10.1016/j.jconrel.2013.09.019 PMID:24096021

Elzoghby, A. O., Samy, W. M., & Elgindy, N. A. (2012a). Albumin-based nanoparticles as potential controlled release drug delivery systems. *Journal of Controlled Release, 157*(2), 168–182. doi:10.1016/j.jconrel.2011.07.031 PMID:21839127

Elzoghby, A. O., Samy, W. M., & Elgindy, N. A. (2012b). Protein-based nanocarriers as promising drug and gene delivery systems. *Journal of Controlled Release, 161*(1), 38–49. doi:10.1016/j.jconrel.2012.04.036 PMID:22564368

Foerster, F., Bamberger, D., Schupp, J., Weilbächer, M., Kaps, L., Strobl, S., & Schuppan, D. et al. (2016). Dextran-based therapeutic nanoparticles for hepatic drug delivery. *Nanomedicine (London), 11*(20), 2663–2677. doi:10.2217/nnm-2016-0156 PMID:27628057

Garrait, G., Beyssac, E., & Subirade, M. (2014). Development of a novel drug delivery system: Chitosan nanoparticles entrapped in alginate microparticles. *Journal of Microencapsulation*, *31*(4), 363–372. doi:10.3109/02652048.2013.858792 PMID:24697173

Gradauer, K., Barthelmes, J., Vonach, C., Almer, G., Mangge, H., Teubl, B., & Prassl, R. et al. (2013). Liposomes coated with thiolated chitosan enhance oral peptide delivery to rats. *Journal of Controlled Release*, *172*(3), 872–878. doi:10.1016/j.jconrel.2013.10.011 PMID:24140721

Gupta, A. K., Gupta, M., Yarwood, S. J., & Curtis, A. S. G. (2004). Effect of cellular uptake of gelatin nanoparticles on adhesion, morphology and cytoskeleton organisation of human fibroblasts. *Journal of Controlled Release*, *95*(2), 197–207. doi:10.1016/j.jconrel.2003.11.006 PMID:14980768

Hu, G., Zhang, H., Zhang, L., Ruan, S., He, Q., & Gao, H. (2015). Integrin-mediated active tumor targeting and tumor microenvironment response dendrimer-gelatin nanoparticles for drug delivery and tumor treatment. *International Journal of Pharmaceutics*, *496*(2), 1057–1068. doi:10.1016/j.ijpharm.2015.11.025 PMID:26598487

Huang, M., Khor, E., & Lim, L.-Y. (2004). Uptake and cytotoxicity of chitosan molecules and nanoparticles: Effects of molecular weight and degree of deacetylation. *Pharmaceutical Research*, *21*(2), 344–353. doi:10.1023/B:PHAM.0000016249.52831.a5 PMID:15032318

Huang, M., Ma, Z., Khor, E., & Lim, L.-Y. (2002). Uptake of FITC-chitosan nanoparticles by A549 cells. *Pharmaceutical Research*, *19*(10), 1488–1494. doi:10.1023/A:1020404615898 PMID:12425466

Jain, A., Singh, S. K., Singh, Y., & Singh, S. (2013). Development of lipid nanoparticles of diacerein, an antiosteoarthritic drug for enhancement in bioavailability and reduction in its side effects. *Journal of Biomedical Nanotechnology*, *9*(5), 891-900.

Jain, D., & Banerjee, R. (2008). Comparison of ciprofloxacin hydrochloride-loaded protein, lipid, and chitosan nanoparticles for drug delivery. *Journal of Biomedical Materials Research. Part B, Applied Biomaterials*, *86*(1), 105–112. doi:10.1002/jbm.b.30994 PMID:18098198

Jain, K. K. (2014). *Drug delivery system*. Springer. doi:10.1007/978-1-4939-0363-4

Jana, S., Manna, S., Nayak, A. K., Sen, K. K., & Basu, S. K. (2014). Carbopol gel containing chitosan-egg albumin nanoparticles for transdermal aceclofenac delivery. *Colloids and Surfaces. B, Biointerfaces*, *114*, 36–44. doi:10.1016/j.colsurfb.2013.09.045 PMID:24161504

Karthik, R., Manigandan, V., Saravanan, R., Rajesh, R. P., & Chandrika, B. (2016). Structural characterization and in vitro biomedical activities of sulfated chitosan from Sepia pharaonis. *International Journal of Biological Macromolecules*, *84*, 319–328. doi:10.1016/j.ijbiomac.2015.12.030 PMID:26708430

Kasoju, N., Ali, S. S., Dubey, V. K., & Bora, U. (2013). Exploiting the potential of Collagen as a Natural Biomaterial in Drug Delivery. *Journal of Proteins & Proteomics*, *1*(1).

Kaya, M., Karaarslan, M., Baran, T., Can, E., Ekemen, G., Bitim, B., & Duman, F. (2014). The quick extraction of chitin from an epizoic crustacean species (Chelonibia patula). *Natural Product Research*, *28*(23), 2186–2190. doi:10.1080/14786419.2014.927469 PMID:24933023

Klier, J., Fuchs, S., May, A., Schillinger, U., Plank, C., Winter, G., & Coester, C. et al. (2012). A nebulized gelatin nanoparticle-based CpG formulation is effective in immunotherapy of allergic horses. *Pharmaceutical Research, 29*(6), 1650–1657. doi:10.1007/s11095-012-0686-8 PMID:22302522

Kraft, J. C., Freeling, J. P., Wang, Z., & Ho, R. J. (2014). Emerging research and clinical development trends of liposome and lipid nanoparticle drug delivery systems. *Journal of Pharmaceutical Sciences, 103*(1), 29–52. doi:10.1002/jps.23773 PMID:24338748

Kreuter, J. (2013). Mechanism of polymeric nanoparticle-based drug transport across the blood-brain barrier (BBB). *Journal of Microencapsulation, 30*(1), 49–54. doi:10.3109/02652048.2012.692491 PMID:22676632

Lebre, F., Borchard, G., Faneca, H., Pedroso de Lima, M., & Borges, O. (2016). Intranasal Administration of Novel Chitosan Nanoparticle/DNA Complexes Induces Antibody Response to Hepatitis B Surface Antigen in Mice. *Molecular Pharmaceutics, 13*(2), 472–482. doi:10.1021/acs.molpharmaceut.5b00707 PMID:26651533

Lee, E. J., Khan, S. A., Park, J. K., & Lim, K.-H. (2012). Studies on the characteristics of drug-loaded gelatin nanoparticles prepared by nanoprecipitation. *Bioprocess and Biosystems Engineering, 35*(1-2), 297–307. doi:10.1007/s00449-011-0591-2 PMID:21909678

Leuner, C., & Dressman, J. (2000). Improving drug solubility for oral delivery using solid dispersions. *European Journal of Pharmaceutics and Biopharmaceutics, 50*(1), 47-60.

Li, J. K., Wang, N., & Wu, X. S. (1998). Poly(vinyl alcohol) nanoparticles prepared by freezing–thawing process for protein/peptide drug delivery. *Journal of Controlled Release, 56*(1-3), 117-126.

Li, L., ten Hagen, T. L., Hossann, M., Süss, R., van Rhoon, G. C., Eggermont, A. M., & Koning, G. A. et al. (2013). Mild hyperthermia triggered doxorubicin release from optimized stealth thermosensitive liposomes improves intratumoral drug delivery and efficacy. *Journal of Controlled Release, 168*(2), 142–150. doi:10.1016/j.jconrel.2013.03.011 PMID:23524188

Liang, Y., & Kiick, K. L. (2014). Heparin-functionalized polymeric biomaterials in tissue engineering and drug delivery applications. *Acta Biomaterialia, 10*(4), 1588–1600. doi:10.1016/j.actbio.2013.07.031 PMID:23911941

Lin, W.-H., Yu, J., Chen, G., & Tsai, W.-B. (2016). Fabrication of multi-biofunctional gelatin-based electrospun fibrous scaffolds for enhancement of osteogenesis of mesenchymal stem cells. *Colloids and Surfaces. B, Biointerfaces, 138*, 26–31. doi:10.1016/j.colsurfb.2015.11.017 PMID:26642073

Liu, T.-M., Wu, X.-Z., & Qiu, Y.-R. (2016). Enhanced biocompatibility and antibacterial property of polyurethane materials modified with citric acid and chitosan. *Journal of Biomaterials Science. Polymer Edition*, 1–21. PMID:27102367

Lohcharoenkal, W., Wang, L., Chen, Y. C., & Rojanasakul, Y. (2014). Protein nanoparticles as drug delivery carriers for cancer therapy. *BioMed Research International, 2014*.

Luo, Y., Teng, Z., Li, Y., & Wang, Q. (2015). Solid lipid nanoparticles for oral drug delivery: Chitosan coating improves stability, controlled delivery, mucoadhesion and cellular uptake. *Carbohydrate Polymers, 122*, 221–229. doi:10.1016/j.carbpol.2014.12.084 PMID:25817662

Luo, Y., & Wang, Q. (2014). Recent development of chitosan-based polyelectrolyte complexes with natural polysaccharides for drug delivery. *International Journal of Biological Macromolecules, 64*, 353–367. doi:10.1016/j.ijbiomac.2013.12.017 PMID:24360899

Mahor, A., Prajapati, S. K., Verma, A., Gupta, R., Iyer, A. K., & Kesharwani, P. (2016). Moxifloxacin loaded gelatin nanoparticles for ocular delivery: Formulation and in-vitro, in-vivo evaluation. *Journal of Colloid and Interface Science, 483*, 132–138. doi:10.1016/j.jcis.2016.08.018 PMID:27552421

Makwana, V., Jain, R., Patel, K., Nivsarkar, M., & Joshi, A. (2015). Solid lipid nanoparticles (SLN) of Efavirenz as lymph targeting drug delivery system: Elucidation of mechanism of uptake using chylomicron flow blocking approach. *International Journal of Pharmaceutics, 495*(1), 439–446. doi:10.1016/j.ijpharm.2015.09.014 PMID:26367780

Malmo, J., Sandvig, A., Vårum, K. M., & Strand, S. P. (2013). Nanoparticle mediated P-glycoprotein silencing for improved drug delivery across the blood-brain barrier: A siRNA-chitosan approach. *PLoS ONE, 8*(1), e54182. doi:10.1371/journal.pone.0054182 PMID:23372682

Manzoor, A. A., Lindner, L. H., Landon, C. D., Park, J.-Y., Simnick, A. J., Dreher, M. R., & Dewhirst, M. W. et al. (2012). Overcoming limitations in nanoparticle drug delivery: Triggered, intravascular release to improve drug penetration into tumors. *Cancer Research, 72*(21), 5566–5575. doi:10.1158/0008-5472.CAN-12-1683 PMID:22952218

Marvizadeh, M. M., Nafchi, A. M., & Jokar, M. (2014). Improved physicochemical properties of tapioca starch/bovine gelatin biodegradable films with zinc oxide nanorod. *Journal of Chemical Health Risks, 4*(4).

Min, H. (2004). *Chitosan: Cellular uptake, cytotoxicity and application in gene delivery*. Academic Press.

Mohammad R, K. (2008). A review of several reported procedures to determine the degree of N-acetylation for chitin and chitosan using infrared spectroscopy. *Carbohydrate Polymers, 71*(4), 497-508.

Mudshinge, S. R., Deore, A. B., Patil, S., & Bhalgat, C. M. (2011). Nanoparticles: Emerging carriers for drug delivery. *Saudi Pharmaceutical Journal, 19*(3), 129–141. doi:10.1016/j.jsps.2011.04.001 PMID:23960751

Muyonga, J. H., Cole, C. G. B., & Duodu, K. G. (2004). Fourier transform infrared (FTIR) spectroscopic study of acid soluble collagen and gelatin from skins and bones of young and adult Nile perch (Lates niloticus). *Food Chemistry, 86*(3), 325–332. doi:10.1016/j.foodchem.2003.09.038

Nesamony, J., Singh, P. R., Nada, S. E., Shah, Z. A., & Kolling, W. M. (2012). Calcium alginate nanoparticles synthesized through a novel interfacial cross-linking method as a potential protein drug delivery system. *Journal of Pharmaceutical Sciences, 101*(6), 2177–2184. doi:10.1002/jps.23104 PMID:22411606

Nitta, S. K., & Numata, K. (2013). Biopolymer-based nanoparticles for drug/gene delivery and tissue engineering. *International Journal of Molecular Sciences*, *14*(1), 1629–1654. doi:10.3390/ijms14011629 PMID:23344060

Pal, K., Behera, B., Roy, S., Sekhar Ray, S., & Thakur, G. (2013). Chitosan based delivery systems on a length scale: Nano to macro. *Soft Matter*, *11*(2), 125–142. doi:10.1080/1539445X.2011.586082

Pan, L., He, Q., Liu, J., Chen, Y., Ma, M., Zhang, L., & Shi, J. (2012). Nuclear-targeted drug delivery of TAT peptide-conjugated monodisperse mesoporous silica nanoparticles. *Journal of the American Chemical Society*, *134*(13), 5722–5725. doi:10.1021/ja211035w PMID:22420312

Pang, H., Braun, G. B., Friman, T., Aza-Blanc, P., Ruidiaz, M. E., Sugahara, K. N., & Ruoslahti, E. et al. (2015). A novel endocytic and intercellular transport pathway for drug delivery across blood vessels and into nutrient-deprived tumor cells. *Cancer Research*, *75*(15Supplement), NG03–NG03. doi:10.1158/1538-7445.AM2015-NG03

Park, C., Vo, C. L.-N., Kang, T., Oh, E., & Lee, B.-J. (2015). New method and characterization of self-assembled gelatin–oleic nanoparticles using a desolvation method via carbodiimide/N-hydroxysuccinimide (EDC/NHS) reaction. *European Journal of Pharmaceutics and Biopharmaceutics*, *89*, 365–373. doi:10.1016/j.ejpb.2014.12.002 PMID:25536111

Parveen, S., Misra, R., & Sahoo, S. K. (2012). Nanoparticles: A boon to drug delivery, therapeutics, diagnostics and imaging. *Nanomedicine; Nanotechnology, Biology, and Medicine*, *8*(2), 147–166. doi:10.1016/j.nano.2011.05.016 PMID:21703993

Pathak, Y., & Thassu, D. (2016). *Drug delivery nanoparticles formulation and characterization* (Vol. 191). CRC Press.

Patra, S., Basak, P., & Tibarewala, D. (2016). Synthesis of gelatin nano/submicron particles by binary nonsolvent aided coacervation (BNAC) method. *Materials Science and Engineering C*, *59*, 310–318. doi:10.1016/j.msec.2015.10.011 PMID:26652378

Paul, W., Shelma, R., & Sharma, C. P. (2013). Alginate encapsulated anacardic acid-chitosan self aggregated nanoparticles for intestinal delivery of protein drugs. *Journal of Nanopharmaceutics and Drug Delivery*, *1*(1), 82–91. doi:10.1166/jnd.2013.1012

Pfeiffer, C., Rehbock, C., Hühn, D., Carrillo-Carrion, C., de Aberasturi, D. J., Merk, V., & Parak, W. J. et al. (2014). Interaction of colloidal nanoparticles with their local environment: The (ionic) nanoenvironment around nanoparticles is different from bulk and determines the physico-chemical properties of the nanoparticles. *Journal of the Royal Society, Interface*, *11*(96), 20130931. doi:10.1098/rsif.2013.0931 PMID:24759541

Prabhakar, U., Maeda, H., Jain, R. K., Sevick-Muraca, E. M., Zamboni, W., Farokhzad, O. C., & Blakey, D. C. et al. (2013). Challenges and key considerations of the enhanced permeability and retention effect for nanomedicine drug delivery in oncology. *Cancer Research*, *73*(8), 2412–2417. doi:10.1158/0008-5472.CAN-12-4561 PMID:23423979

Probst, C. E., Zrazhevskiy, P., Bagalkot, V., & Gao, X. (2013). Quantum dots as a platform for nanoparticle drug delivery vehicle design. *Advanced Drug Delivery Reviews*, *65*(5), 703–718. doi:10.1016/j.addr.2012.09.036 PMID:23000745

Ragelle, H., Danhier, F., Préat, V., Langer, R., & Anderson, D. G. (2016). Nanoparticle-based drug delivery systems: A commercial and regulatory outlook as the field matures. *Expert Opinion on Drug Delivery*, 1–14. PMID:27730820

Raj, V., & Prabha, G. (2016). Synthesis, characterization and in vitro drug release of cisplatin loaded Cassava starch acetate–PEG/gelatin nanocomposites. *Journal of the Association of Arab Universities for Basic and Applied Sciences*, *21*, 10–16. doi:10.1016/j.jaubas.2015.08.001

Rejinold, N. S., Chennazhi, K. P., Nair, S. V., Tamura, H., & Jayakumar, R. (2011). Biodegradable and thermo-sensitive chitosan-g-poly(N-vinylcaprolactam) nanoparticles as a 5-fluorouracil carrier. *Carbohydrate Polymers*, *83*(2), 776–786. doi:10.1016/j.carbpol.2010.08.052

Rescignano, N., Fortunati, E., Montesano, S., Emiliani, C., Kenny, J. M., Martino, S., & Armentano, I. (2014). PVA bio-nanocomposites: A new take-off using cellulose nanocrystals and PLGA nanoparticles. *Carbohydrate Polymers*, *99*, 47–58. doi:10.1016/j.carbpol.2013.08.061 PMID:24274478

Rezaei Mokarram, A., & Alonso, M. (2016). Preparation and evaluation of chitosan nanoparticles containing Diphtheria toxoid as new carriers for nasal vaccine delivery in mice. *Archives of Razi Institute*, *61*(1), 13–25.

Sahoo, N., Sahoo, R. K., Biswas, N., Guha, A., & Kuotsu, K. (2015). Recent advancement of gelatin nanoparticles in drug and vaccine delivery. *International Journal of Biological Macromolecules*, *81*, 317–331. doi:10.1016/j.ijbiomac.2015.08.006 PMID:26277745

She, W., Li, N., Luo, K., Guo, C., Wang, G., Geng, Y., & Gu, Z. (2013). Dendronized heparin– doxorubicin conjugate based nanoparticle as pH-responsive drug delivery system for cancer therapy. *Biomaterials*, *34*(9), 2252–2264. doi:10.1016/j.biomaterials.2012.12.017 PMID:23298778

Silva, A., González-Mira, E., Lobo, J. S., & Amaral, M. (2013). Current progresses on nanodelivery systems for the treatment of neuropsychiatric diseases: Alzheimers and Schizophrenia. *Current Pharmaceutical Design*, *19*(41), 7185–7195. doi:10.2174/138161281941131219123329 PMID:23489198

Smith, A. H., Laird, C., Porter, K., & Bloch, M. (2013). Haemostatic dressings in prehospital care. *Emergency Medicine Journal*, *30*(10), 784–789. doi:10.1136/emermed-2012-201581 PMID:23161808

Soppimath, K. S., Aminabhavi, T. M., Kulkarni, A. R., & Rudzinski, W. E. (2001). Biodegradable polymeric nanoparticles as drug delivery devices. *Journal of Controlled Release*, *70*(1), 1–20. doi:10.1016/S0168-3659(00)00339-4 PMID:11166403

Staff, R. H., Landfester, K., & Crespy, D. (2013). *Recent advances in the emulsion solvent evaporation technique for the preparation of nanoparticles and nanocapsules. In Hierarchical Macromolecular Structures: 60 Years after the Staudinger Nobel Prize II* (pp. 329–344). Springer.

Sundar, S., Kundu, J., & Kundu, S. C. (2016). Biopolymeric nanoparticles. *Science and Technology of Advanced Materials*. PMID:27877319

Sung, H.-W., Sonaje, K., Liao, Z.-X., Hsu, L.-W., & Chuang, E.-Y. (2012). pH-responsive nanoparticles shelled with chitosan for oral delivery of insulin: From mechanism to therapeutic applications. *Accounts of Chemical Research*, *45*(4), 619–629. doi:10.1021/ar200234q PMID:22236133

Trapani, A., Di Gioia, S., Ditaranto, N., Cioffi, N., Goycoolea, F. M., Carbone, A., & Alonso, M. J. et al. (2013). Systemic heparin delivery by the pulmonary route using chitosan and glycol chitosan nanoparticles. *International Journal of Pharmaceutics*, *447*(1), 115–123. doi:10.1016/j.ijpharm.2013.02.035 PMID:23454518

Tseng, H. J., Tsou, T. L., Wang, H. J., & Hsu, S. (2013). Characterization of chitosan–gelatin scaffolds for dermal tissue engineering. *Journal of Tissue Engineering and Regenerative Medicine*, *7*(1), 20–31. doi:10.1002/term.492 PMID:22034441

Wang, C., Fu, X., & Yang, L. (2007). Water-soluble chitosan nanoparticles as a novel carrier system for protein delivery. *Chinese Science Bulletin*, *52*(7), 883–889. doi:10.1007/s11434-007-0127-y

Wang, H., Zou, Q., Boerman, O. C., Nijhuis, A. W., Jansen, J. A., Li, Y., & Leeuwenburgh, S. C. G. (2013). Combined delivery of BMP-2 and bFGF from nanostructured colloidal gelatin gels and its effect on bone regeneration in vivo. *Journal of Controlled Release*, *166*(2), 172–181. doi:10.1016/j.jconrel.2012.12.015 PMID:23266450

Wang, W., Ji, X., Na, H. B., Safi, M., Smith, A., Palui, G., & Mattoussi, H. et al. (2014). Design of a Multi-Dopamine-Modified Polymer Ligand Optimally Suited for Interfacing Magnetic Nanoparticles with Biological Systems. *Langmuir*, *30*(21), 6197–6208. doi:10.1021/la500974r PMID:24805794

Wang, X., Chi, N., & Tang, X. (2008). Preparation of estradiol chitosan nanoparticles for improving nasal absorption and brain targeting. *European Journal of Pharmaceutics and Biopharmaceutics*, *70*(3), 735–740. doi:10.1016/j.ejpb.2008.07.005 PMID:18684400

Wohlfart, S., Gelperina, S., & Kreuter, J. (2012). Transport of drugs across the blood–brain barrier by nanoparticles. *Journal of Controlled Release*, *161*(2), 264–273. doi:10.1016/j.jconrel.2011.08.017 PMID:21872624

Wojdyr, M. (2010). Fityk: A general-purpose peak fitting program. *Journal of Applied Crystallography*, *43*(5 Part 1), 1126–1128. doi:10.1107/S0021889810030499

Xu, J., & Amiji, M. (2012). Therapeutic gene delivery and transfection in human pancreatic cancer cells using epidermal growth factor receptor-targeted gelatin nanoparticles. *Journal of Visualized Experiments*, (59), e3612-e3612.

Yan, F., Dang, Q., Liu, C., Yan, J., Wang, T., Fan, B., & Zhang, Z. et al. (2016). 3, 6-O-[N-(2-Aminoethyl)-acetamide-yl]-chitosan exerts antibacterial activity by a membrane damage mechanism. *Carbohydrate Polymers, 149*, 102–111. doi:10.1016/j.carbpol.2016.04.098 PMID:27261735

Yoon, H. Y., Koo, H., Choi, K. Y., Kwon, I. C., Choi, K., & Park, J. H. et al.. (2013). Photo-crosslinked hyaluronic acid nanoparticles with improved stability for in vivo tumor-targeted drug delivery. *Biomaterials, 34*(21), 5273–5280. doi:10.1016/j.biomaterials.2013.03.050 PMID:23591396

Younes, I., & Rinaudo, M. (2015). Chitin and chitosan preparation from marine sources. Structure, properties and applications. *Marine Drugs, 13*(3), 1133–1174. doi:10.3390/md13031133 PMID:25738328

Zhang, X.-Q., Xu, X., Bertrand, N., Pridgen, E., Swami, A., & Farokhzad, O. C. (2012). Interactions of nanomaterials and biological systems: Implications to personalized nanomedicine. *Advanced Drug Delivery Reviews, 64*(13), 1363–1384. doi:10.1016/j.addr.2012.08.005 PMID:22917779

Zhen, Z., Tang, W., Chen, H., Lin, X., Todd, T., Wang, G., & Xie, J. et al. (2013). RGD-modified apoferritin nanoparticles for efficient drug delivery to tumors. *ACS Nano, 7*(6), 4830–4837. doi:10.1021/nn305791q PMID:23718215

Zikakis, J. (2012). *Chitin, chitosan, and related enzymes*. Elsevier.

Zong, Z., Kimura, Y., Takahashi, M., & Yamane, H. (2000). Characterization of chemical and solid state structures of acylated chitosans. *Polymer, 41*(3), 899-906.

KEY TERMS AND DEFINITIONS

Biopolymer: A polymeric substance occurring in the living organisms.

Chitosan: A linear polysaccharide consisting of arbitrarily spread β-(1→4) linked deacetylated D-glucosamine and N-acetyl-D-glucosamine.

Gelatin: A water soluble protein derived from collagen.

In Vitro **Drug Delivery:** A measure of the discharge of active pharmaceutical ingredient from the drug containing matrix in a laboratory controlled environment.

Nanoparticle: A microscopic particle of matter that is measured on nanoscale.

Nanomedicine: Use of nanotechnology for the prevention and cure of diseases.

Sodium Carboxymethyl Cellulose: A sodium salt of cellulose gum, derivative of cellulose with carboxymethyl groups on some of the hydroxyl groups of glucopyranose monomers of the cellulose backbone.

Chapter 11
Methods of Preparation of Nanoparticle Formulations for the Treatment of COPD:
A Short Review

Suprio R. Paul
Amity University, India

Suraj K. Nayak
National Institute of Technology, India

Arfat Anis
King Saud University, Saudi Arabia

Kunal Pal
National Institute of Technology, India

ABSTRACT

The incidence of chronic obstructive pulmonary disease (COPD) has increased many folds in the last couple of decades. This can be explained by the increased pollution levels and drastic changes in the lifestyle of the people. Taking a note of this, in this review, a conscious attempt was made to understand the physiological changes in the respiratory tract in a COPD patient. Nanoparticulate formulations play an extensive role in the current-day treatment regime of COPD patients. Hence, different methods for developing nanoparticulate formulations for the treatment of COPD are discussed in details.

INTRODUCTION

Chronic obstructive pulmonary disease (COPD) is one of the most common diseases with high mortality rate across the globe (Aryal, Diaz-Guzman, & Mannino, 2013). The patient often becomes a burden to the families (Miravitlles, Peña-Longobardo, Oliva-Moreno, & Hidalgo-Vega, 2015). This may be

DOI: 10.4018/978-1-5225-3023-7.ch011

attributed to the high cost associated with the treatment of COPD (Lee & Goldstein, 2015). Additionally, the requirement for an extensive care becomes a necessity for these patients (Perera, Armstrong, Sherrill, & Skrepnek, 2012). This further provides an additional burden to the families. The families, who are economically sound, can hire the services of caregivers. Due to the high cost associated with the hiring of the services of caregivers, most of the families, not only in the developing world but also in the developed countries, are unable to hire the services of the caregivers (Mansfield, et al., 2016). We have to bear in mind that the COPD patients, in their early and mid stages of the disease, are often unable to do the day-to-day activities with relative ease (Monjazebi, et al., 2016). This is due to the disability of the patients (Braido, et al., 2015). Though the name of the disease COPD suggests a chronic obstruction of the air flow to the lungs, the efficiency of the alveolar ventilation and gas exchange also play an important part in COPD (Antonaglia, Ferluga, & Lucangelo, 2016). The abnormalities present in the gaseous exchange quite often lead to sleep disorder, dyspnoea and also a relative increase in the strain experienced while breathing (McNicholas, Verbraecken, & Marin, 2013; Miki, et al., 2013). This, eventually, may further lead to hypoxia (a clinical condition where the oxygen concentration in the tissue is very low) and hypercapnia (a clinical condition where the concentration of carbon dioxide gas in blood is very high) (Kuklisova, Tkacova, Joppa, Wouters, & Sastry, 2016; Littleton, 2015; Zheng, et al., 2016). There is also a marked reduction in the respiratory reserve volume of the COPD patients (Guenette, Webb, & O'Donnell, 2012). Many scientists have predicted that there is an increase in the number of COPD patients due to the increased frequency of smoking and ageing population (Chiappa, Winn, Viñuela, Tipney, & Spector, 2013; Lahousse, et al., 2016). An increase in the air pollution is also playing a major role in the deterioration of the functioning capability of the lungs, thereby, promoting the progression of the disease (Bloemsma, Hoek, & Smit, 2016).

In the present study, we will discuss about the pathophysiological changes in the lungs of the COPD patients and the drug delivery/ formulation technologies employed in the treatment of the COPD patients.

Anatomy and Physiology of Human Respiratory Tract

The respiratory and the cardiovascular systems form an interconnected system (Gimeno-Santos, et al., 2014). The deoxygenated blood is pumped into the lungs for oxygenation by the right ventricle of the heart. After oxygenation of the blood, the blood is drained into the left atrium. The oxygenated blood then enters into the left ventricle from where it is pumped into the body through the aorta through the aortic valve (Hall, 2015). The lungs are the parts of the respiratory system (Figure 1). The respiratory system is classified into upper and lower respiratory tracts (Charlson, et al., 2012). The upper respiratory tract consists of nose, nasal cavity and pharynx, whereas, the lower respiratory tract consists of larynx, trachea, bronchi and the lungs (Beachey, 2013). The trachea bifurcates into two bronchi. Each of these bronchi gains entry into the right and the left lungs, respectively. As the bronchi enter into the lungs, they form smaller bronchioles. The bronchioles then terminate into alveoli. The alveoli are highly vascularised tissue, which promotes the exchange of oxygen and carbon dioxide across the alveolar membrane (Reece, 2015). This exchange of gases allows the maintenance of oxygen and carbon dioxide balance in the blood. During the respiration process, the air is inhaled through the nose which subsequently gains entry into the nasal cavity, where the conditioning of the inhaled air takes place. The conditioned air then passes through the pharynx, larynx, trachea and bronchi to reach the alveoli. After the exchange of the gases have taken place in the alveoli, the air moves in the reverse direction and is exhaled from the nose. A healthy adult lung has the capacity to accommodate 1 pint (~500 ml) of air during each inhala-

Figure 1. Schematic diagram of the respiratory system

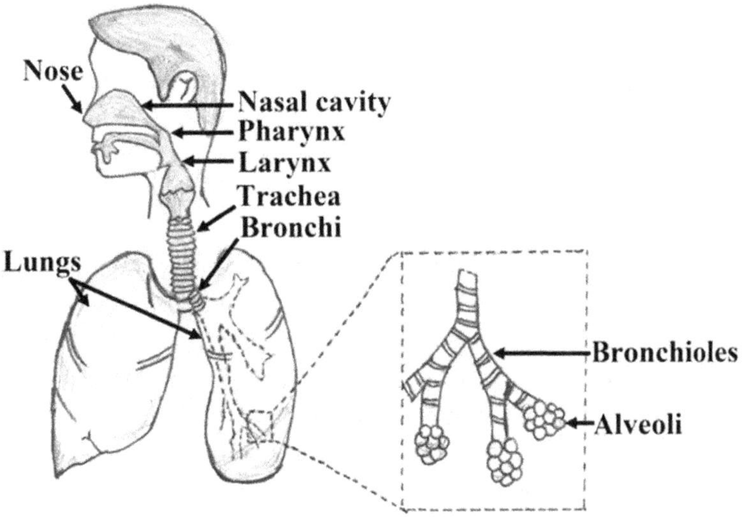

tion stage of the respiration process ("Respiratory System - Pulmonary Ventilation," 2016). The average rate of the respiration in a healthy adult is 15 times per minute (Naftali, Rosenfeld, Wolf, & Elad, 2005). The lungs are delicate organs of the human body and have been reported to be susceptible to different disorders governed by genetic factors, infections and pollution (Bansal, Khatri, & Taneja, 2013; Mathew, et al., 2016; To, et al., 2016). Some of the common respiratory diseases include common cold, cough, pulmonary hypertension, asthma, bronchitis and COPD (Chapman, 2013; Kim & Criner, 2013; Suveer Singh, 2016; Sivtseva, et al., 2015).

MAIN FOCUS OF THE CHAPTER

This chapter focuses on realizing the physiological alterations in the respiratory tract of a COPD patient. In the present-day scenario, the nanoparticulate formulations are playing an important role in the treatment of COPD patients. The methods employed in the development of nanoparticulate formulations for the treatment of COPD are also discussed in detail.

Pathophysiological Changes

Clinically, the COPD is characterised by the hindrance set to the airflow within the respiratory tract (Kruis & Chavannes, 2016). The COPD may be precipitated by a number of factors, which include irritation in the respiratory tract by the irritants and the diseases of the respiratory tracts (Ionescu, 2013), which may lead to significant alteration in the physiological and structural characteristics of the lungs (Tan, et al., 2015). In this section, we will discuss about the pathophysiological changes occurring in the respiratory tract at the lungs which lead to COPD. It is important to note over here that, in many cases, the obstruction to the airflow within the respiratory tract and lungs may not be fully reversible in nature

(Takeyama, Tagaya, Kondo, Tamaoki, & Orimo, 2016). Depending on the severity of the obstruction, clinical intervention may become a necessity.

In recent years, smoking of cigarettes has become a matter of status for middle and upper-class societies (Jacques & Wright, 2015). But the fact remains that the smoke inhaled during the smoking of cigarettes and other related smoking products produces chemical irritants, which gain entry into the respiratory tract (Millqvist, Ternesten-Hasséus, & Bende, 2013). These chemical irritants have been found to induce inflammation of the respiratory tract, thereby, resulting in the swelling of the respiratory tract (Muala, et al., 2014). Thus, swelling results in the narrowing of the respiratory path for respiration. Similarly, with the advent of the modernisation of the human society, there has been an increase in the air pollution levels due to the increase in the industries, vehicles and aircrafts (Sharma, Jain, Khirwadkar, & Kulkarni, 2013). This has resulted in the release of noxious chemicals, capable of irritating the airway passage of the respiratory system (Bae, et al., 2014). The above-mentioned irritants, if exposed to the airway passage, progressively deteriorate the normogenic health of the respiratory system. If proper care is not taken in the early stages of the disease, the disease may become quite severe.

As mentioned in the previous paragraph, the inflamed airway passage causes an obstruction. This obstruction leads to entrapment of air within the lungs. In other words, during the exhalation stage of the respiration process, the patient is unable to exhale the required volume of air (Ionescu, 2013). Due to this reason, the lungs undergo hyperinflation (Sally Singh, et al., 2015). As the disease progresses, the volume of the air entrapped within the lungs is increased significantly, thereby, increasing the residual functional capacity (a parameter used to measure the volume of air remaining after the exhalation stage) (Farkhooy, et al., 2015). Hence, during exercise the COPD patients suffer from shortness of breath. This can be explained by the fact that the entrapped air limits the inhalation of the fresh air into the lungs during the inhalation stage of the respiration process. This phenomenon results in the inhalation of a lower proportion of oxygen during the inhalation process, which is just sufficient to maintain the normal day-to-day activities.

In addition to the inflammatory response, it has been found that the noxious stimuli (smoking and air pollutants) can destroy the functional unit of oxygen and carbon dioxide exchange (alveoli). The process of destruction of the alveoli is clinically regarded as emphysema (Arlbrandt, et al., 2016). Hence, there are lower number alveoli present in the COPD patients. This further reduces the availability of oxygen in the blood and simultaneously reduces the excretion of carbon dioxide from the blood. Due to the above-mentioned abnormalities, COPD patients suffer from hypoxemia and simultaneous hypercapnia (Walker, et al., 2015). The submucosal glands in the COPD patients are generally larger in size. The submucosal glands help in the production of mucous via goblet and mucous cells (Ramos, Krahnke, & Kim, 2014). In a healthy human being, the number of goblet cells has been found to be higher in large bronchi. The number of these cells decreases with the decrease in the airway passage in the smaller bronchioles. The submucosal glands produce mucous as a protective agent, which is responsible for the lubrication of the lungs and entrapment of foreign debris from the respiratory airway passage. Since, in COPD patients, the airway passage of the lungs is irritated, there is an increase in the number of goblet cells with the concurrent increase in the size of the submucosal gland in the respiratory tract (Kim, et al., 2015). Due to this reason, there is an increase in the mucus secretion (Viby, et al., 2015). The lungs of the COPD patients fail to clear out the mucous from the lungs, which in turn, restricts the airway passage. Additionally, the mucous also serves as a good media for the pathogenic bacteria to proliferate and hence, causing infection of the lungs (Ma, et al., 2015). The infection causing bacteria may secrete toxins, which have the capability to damage the alveoli. This can further exacerbate the breathing problem of the COPD patients. From the afore-mentioned discussions, it is quite clear that COPD is a complex clinical condition. Hence, the effective treatment of COPD should concentrate on minimising all the causative propositions.

Methods for Producing Nanoparticulate Pulmonary Drug Delivery Systems for the Treatment of COPD

The delivery of drugs locally to the lungs has gained much attention in the recent years. Such drug delivery systems are commonly categorised as pulmonary drug delivery systems. Pulmonary drug delivery systems have also been successfully tested for systemic delivery of drugs (Ruge, Kirch, & Lehr, 2013). Such delivery systems have been used for the treatment of various clinical conditions associated with the lungs including COPD, cystic fibrosis, asthma, bronchitis, tuberculosis, respiratory tract infections and lung cancer (Hodson & Sherrington, 2014; Ponte, Santos, Macedo, & Souto, 2016; Sin, Sze, & Hogg, 2014; Zhou, et al., 2015). The popularity of these delivery systems can be attributed to the non-invasive mode of drug delivery (Grassin-Delyle, et al., 2012). This allows the patients to administer the drug without the help of any medical professional. In this article, we shall discuss about the pulmonary drug delivery systems based on nanoparlicles. The use of nanoparticulate formulations as drug delivery systems has found many applications in pharmaceutical industry. In fact, nanoparticle- based formulations have shown the properties which cannot be achieved using the conventional drug delivery systems. As for example, nanoparticles of cytosine molecules have shown great potential to cross blood-brain-barrier, thereby, allowing the delivery of the drug into the central nervous system (Bancel, 2013). All these have been possible due to the advancement in the processing technology of the nanoparticles.

In general, the nanoparticle-based pulmonary drug formulations can be prepared both by the bottom-up and the top-down approach (Nigam, et al., 2014). The bottom-up approach of nanoparticles deals with the precipitation of the drug particles from a drug solution. On the other hand, top-down approach of nanoparticle preparation deals with the size reduction (milling) of bigger drug particles to obtain particulate matter having sizes less than 100 nm. In general, the preparation of nanoparticles is carried out in the presence of surfactants. The surfactants help reducing the surface tension of the particle surface and hence, reduce the free energy of the particles. These are necessary to retard Ostwald ripening and agglomeration of the nanoparticles during the preparation stage, which effectively increase the size of the particulates (Gindy, et al., 2014). Ostwald ripening and agglomeration undoes the advantages of the nanoparticle-based drug formulation. Though many nanoparticle preparation methods are available, only some of the methods have been explored for formulating pulmonary drug delivery systems. These methods include wet milling, supercritical fluid extraction, spray drying, electro spray, high-pressure homogenization and recrystallization. In this section, the afore-mentioned methodologies for the production of nanoparticle formulation for pulmonary drug delivery will be discussed.

Wet-Milling

Wet-milling is a process for reduction of size of the drug particles formulated as suspensions. The suspension is poured into a suitable container. The container usually contains balls or rods, which are usually made up of glass, zirconia and high molecular weight polyester derivatives. As the container, which contains the balls or the rods along with the drug suspension, is rotated across the horizontal axis, the drug size is reduced due to the attrition force exerted by the movement of the balls or the rods(Breit, et al., 2013; Vartiainen, Bimbo, Hirvonen, Kauppinen, & Raula, 2016).

Spray Drying

Spray drying is the process for making fine particles in bulk (Figure 3) (Trotta, Haghi, Scalia, Young, & Traini, 2015). In this method, a drug containing solution is atomised (either using a rotary nozzle, fluid nozzle or an ultrasonic nozzle; to form fine droplets), in a closed chamber. Simultaneously, a warm air current is inserted into the chamber. The temperature of the warm air current may be at a temperature of >100 °C. This warm air results in the evaporation of the solvent (aqueous or ethanolic solution) from the fine droplets, thereby, resulting in the formation of fine particles of the drug molecules. From the afore-mentioned process, it may appear that there are chances of thermal degradation of the thermolabile drugs. Hence, such drugs may not be used for the production of the particles. But the fact is that, even though the temperature is maintained at a temperature higher than 100 °C, the temperature sensed by the drug molecule is considerably lower. This can be explained by the cooling effect exerted by the evaporating solvent due to the latent heat of vaporisation (Amstad, et al., 2015; Leung, et al., 2016). Hence, spray drying technique has been used for making particles of both thermolabile and thermostable drugs.

Spray Freeze Drying

The spray freeze drying method is similar to spray drying method in certain aspects. The method includes spraying of drug solutions into the freezing medium (Figure 4). In general, liquid nitrogen is used as the freezing medium. The freezing medium instantly converts the atomised solution into drug containing ice particles. These frozen droplets are subsequently lyophilized. During the lyophilization process, the water molecules are sublimed. This leads to the formation of particles. The particles formed by spray freeze drying method are relatively light and porous in nature as compared to the particles formed by spray drying method (Ali & Lamprecht, 2014; D'Addio, et al., 2013).

Figure 2. Schematic diagram of the wet milling method for preparing nanoparticle formulations

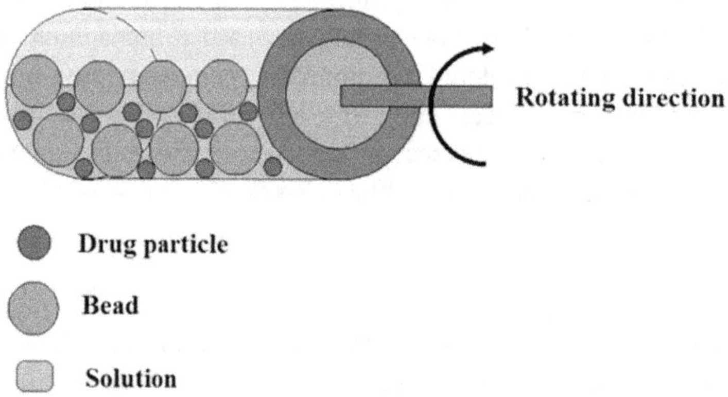

Figure 3. Schematic diagram of the spray drying method for preparing nanoparticle formulations

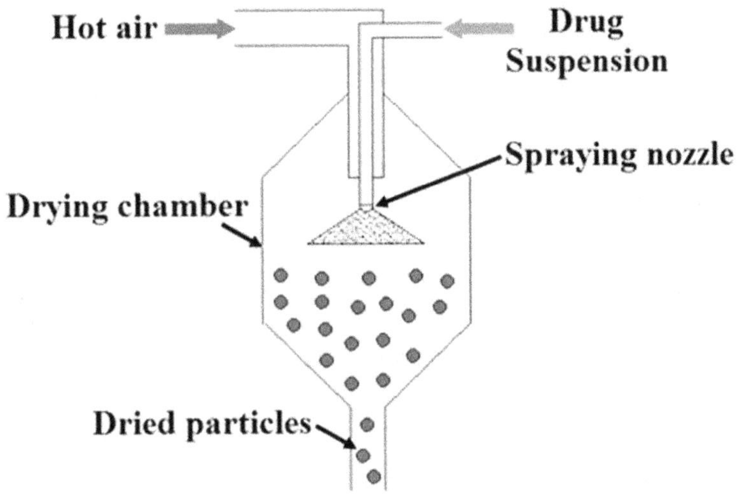

Solvent Precipitation Method

In solvent precipitation method, the particles are formed by aqueous solution of the drug (Figure 5). The solvent molecules are extracted from the drug solution using an anti-solvent. During the precipitation of the particles, ultrasonic probes may be used to improve the efficiency of the drug particle formation. The ultrasonic waves impart physical energy to the particles. It also prevents the formation of agglomerates during the preparation of the particles. This process is often regarded as sono-crystallization method. Another method of the preparation of the particulate drugs includes microprecipitation by opposing liquid jets. In this method, the drug solution is converted into a jet and is allowed to mix with the jet of anti-solvent. As both the jets mix with each other, there is a formation of a region of extreme turbulence. This results in the intense mixing of the two jet components, i.e., drug solution and the anti-solvent. This phenomenon results in the extraction of the aqueous solvent from the drug solution and subsequent precipitation of the drug molecules to form particles (Chan & Kwok, 2013; Patil & Sarasija, 2012).

Figure 4. Schematic diagram of the spray freeze drying method for preparing nanoparticle formulations

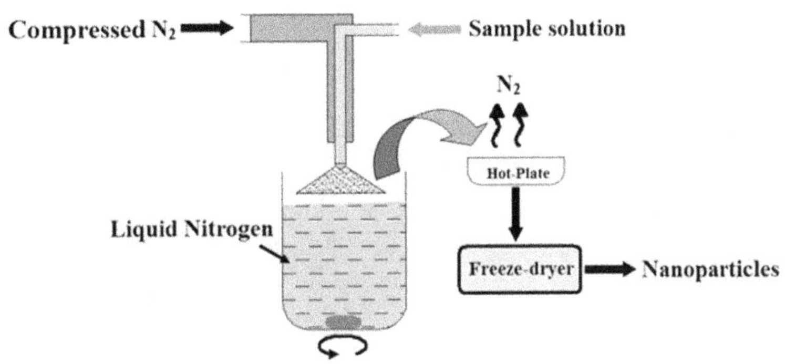

Figure 5. Schematic diagram of the solvent extraction method for preparing nanoparticle formulations

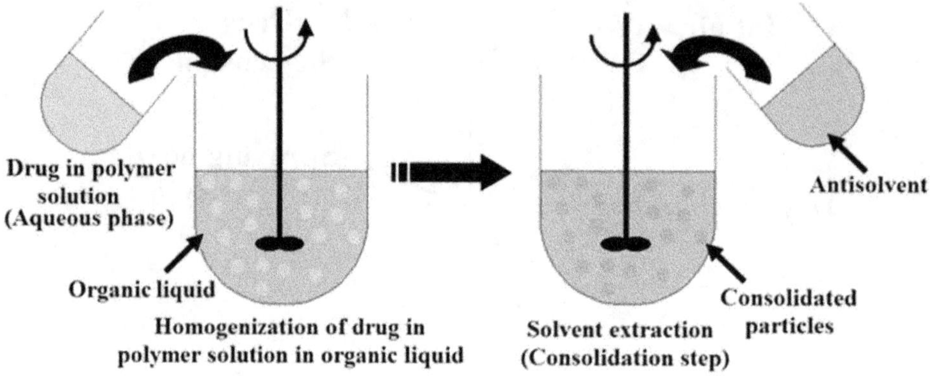

Supercritical Fluid Extraction

The use of supercritical fluid for the preparation of particles has gained much attention in recent years (Figure 6). The main reason for this may be accounted to the low-cost production of particles in the non-toxic environment. This has allowed mass production of the particulate formulations. Drug particles may be prepared using supercritical fluids mainly by two methods. The first method is regarded as Supercritical Anti-solvent Technique (SAS). The second method is known as Rapid Expansion of Supercritical Solutions (RESS) technique. In the SAS technique, a mixture of the solute and the solvent is spread into a precipitator. The precipitator is also supplied with supercritical fluids (usually supercritical CO_2). The drug and the solvent are chosen such that the drug is insoluble in the supercritical fluid, whereas, the solvent is highly soluble in the supercritical CO_2. Hence, the supercritical CO_2 acts as an anti-solvent and extracts out the solvent from the drug solution droplets within the precipitator. The precipitator particles are collected, whereas, the mixture of supercritical CO_2 and solvent is directed towards a separator unit. The separator unit separates the CO_2 from the solvent, which can be collected and reused for producing

Figure 6. Schematic diagram of the super critical fluid extraction method for preparing nanoparticle formulations

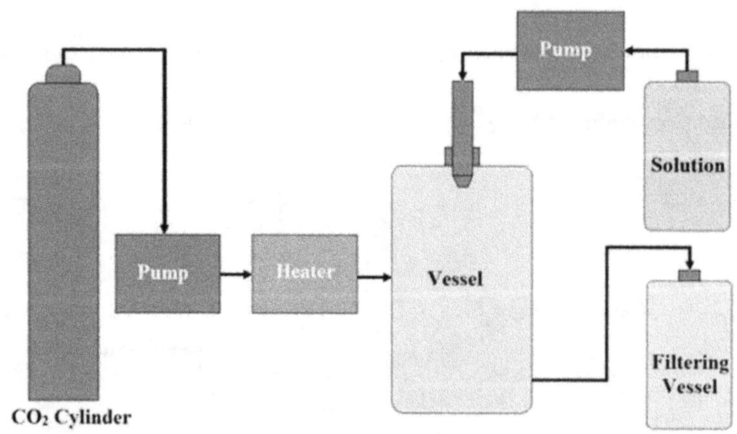

a fresh batch of particle formulation. Unlike the SAS process, the drug is dissolves in supercritical CO_2. The drug solution in the supercritical CO_2 is subsequently depressurised in a depressing cell. Under this condition, there is a formation of supersaturated drug solution which initiates very rapid nucleation of the drug molecules as particulate formulation. The environmental conditions within the depressurizing cell play an important role in grooming the size of the drug particles (Campardelli, et al., 2013; Sheth, et al., 2012).

Electrospray

Electrospraying method of the particle formulation is based on the atomization of the drug solution under electrical forces (Figure 7). The droplets formed during the atomization are electrically active and possess same electrical charge. This phenomenon helps in promoting the self-dispersion of the drug particles by preventing coagulation of the atomised droplets. The manipulation of the size of the drug particles can be achieved by varying the voltage applied to the electrodes and the flow rate of the drug solution. The drug solution is allowed to pass through a capillary, which is supplied with a high voltage potential. Usually, DC potential in the range of 2.2 to 10 KV is used. The capillary is surrounded by another capillary (co-axial capillary). CO_2 is continuously injected through this capillary. The injected CO_2 helps preventing corona discharge due to the static electric field (Eltayeb, Stride, Edirisinghe, & Harker, 2016; Sridhar & Ramakrishna, 2013).

High-Pressure Homogenization

High-pressure homogenization is a kind of disintegration method for the production of nanoparticles (Figure 8). The high-pressure homogenization method can be achieved by two processes, namely, microfluidization and piston-gap homogenizer. In the microfluidization technique, suspension of bigger particles is passed through either a Z-type or Y-type microchannel (Figure 9). In the Z-type microchannels, the particles undergo collision during their passage through the microchannels. This results in the reduction in the size of the particles. On the contrary, Y-type microchannel employs injection of the

Figure 7. Schematic diagram of the electrosparying method for preparing nanoparticle formulations

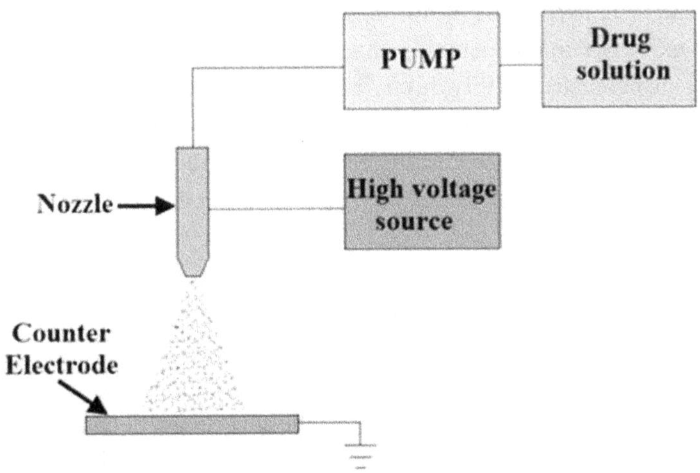

Figure 8. Schematic diagram of the high- pressure homogenizer (piston type) for preparing nanoparticle formulations

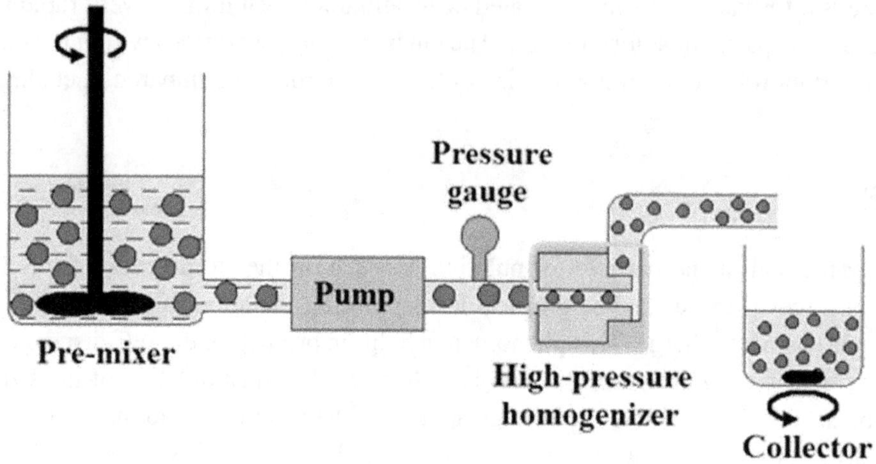

Figure 9. Schematic diagram of the high- pressure homogenizer (Y-shaped microfluidics based) for preparing nanoparticle formulations

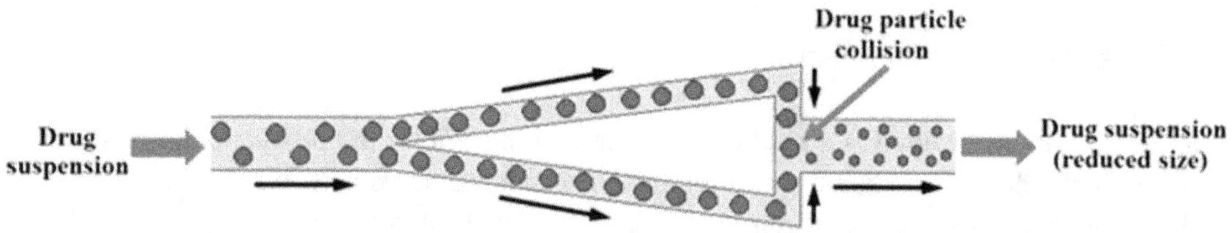

suspension to the Y-channel such that there is a frontal collision between the suspension streams. This collision results in the size reduction of the drug particles. The advancement in the field of piston-gap homogenizers has left microfluidization technique as a least favoured method for the production of nanoparticulate formulation. This is due to the fact that micron-sized particles are also obtained apart from the nanoparticulate drug particles. Furthermore, the production of nanoparticulate formulation by microfluidization technique is a time-consuming and tedious process. The piston-gap homogenizers use a piston to pass the drug suspensions forcibly through a narrow channel whose diameter can be controlled by the user. During the passage of the suspension under high shear rate (induced by the piston pump), the particles undergo collision, thereby, resulting in the reduction in the size (Durán-Lobato, Enguix-González, Fernández-Arévalo, & Martín-Banderas, 2013; Kumar & Randhawa, 2013).

CONCLUSION

The present review discussed about the anatomy and the physiology of the respiratory system. Discussions were also made to have an understanding on the pathophysiological changes in the respiratory tract in patients suffering from COPD. In recent years, with the advancements in the production of nanopar-

Table 1. Nanoparticle preparation methods employed for the preparation of various drug formulations

Process	Drugs	Reference
Spray drying	Salbutamol	(Ahmad, et al., 2009)
	Budesonide microparticles	(Willis, Hayes Jr, & Mansour, 2012)
	Levofloxacin	(Cheow & Hadinoto, 2010; Kho, Cheow, Lie, & Hadinoto, 2010)
Spray freeze drying	Ciprofloxacin nanoplex	(Yu, Teo, Chew, & Hadinoto, 2016)
	Salbutamol sulphate	(Pritchard, 2001)
	Curcumin-chitosan nanoparticle complex	(Yu, Tran, Teo, & Hadinoto, 2016)
Sonocrystallization	Fluticasone or triamcinolone	(Cavanagh, Barman, Hao, Leland, & Thekkedath, 2013)
	Salbutamol sulphate	(Carvalho, 2013)
Supercritical fluid methods	5-fluorouracil	(Kaur, Narang, Rath, & Goyal, 2012)
	Budesonide, triamcinolone acetonide, rifampicin, albuterol sulfate	(A. Singh, Malviya, & K Sharma, 2011)
	Piroxicam (PX) and diclophenac sodium	(Della Porta, Falco, & Reverchon, 2010)
Electrospray	Albuterol	(Gomez, 2002)
High pressure homogenization	Amikacin loaded solid-lipid nanoparticles	(Varshosaz, et al., 2013)
	S-ketoprofen	(Kluge & Mazzotti, 2012)
Wetmilling	Fluticasone	(Chiang, et al., 2009)
	Budesonide	(Mansour, Rhee, & Wu, 2009)

ticulate delivery systems, the use of nanoparticle based pulmonary delivery vehicles has gained importance. The present review also discussed about the different methods of nanoparticulate drug delivery systems, which are commonly used for the treatment of COPD patients. A table briefly summarizing the nanoparticle preparation methods employed for the preparation of various drug formulations has been tabulated in Table 1.

REFERENCES

Ahmad, F. J., Mittal, G., Jain, G. K., Malhotra, G., Khar, R. K., & Bhatnagar, A. (2009). Nano-salbutamol dry powder inhalation: A new approach for treating broncho-constrictive conditions. *European Journal of Pharmaceutics and Biopharmaceutics*, *71*(2), 282–291. doi:10.1016/j.ejpb.2008.09.018 PMID:18984050

Ali, M. E., & Lamprecht, A. (2014). Spray freeze drying for dry powder inhalation of nanoparticles. *European Journal of Pharmaceutics and Biopharmaceutics*, *87*(3), 510–517. doi:10.1016/j.ejpb.2014.03.009 PMID:24657824

Amstad, E., Gopinadhan, M., Holtze, C., Osuji, C. O., Brenner, M. P., Spaepen, F., & Weitz, D. A. (2015). Production of amorphous nanoparticles by supersonic spray-drying with a microfluidic nebulator. *Science*, *349*(6251), 956–960. doi:10.1126/science.aac9582 PMID:26315432

Antonaglia, V., Ferluga, M., & Lucangelo, U. (2016). *Respiratory Mechanics in COPD Patients Who Failed Noninvasive Ventilation. In Noninvasive Mechanical Ventilation* (pp. 287–292). Springer.

Arlbrandt, S., Bold, P., Butcher, A., Carlsson, J. F., Cygan, P., & Ehnebom, J. et al.. (2016). *Highly Selective Inhibitors Of PI3Kγ For The Oral Treatment Of COPD C35. In Innate Immune Mechanisms in COPD* (pp. A4901–A4901). Am Thoracic Soc.

Aryal, S., Diaz-Guzman, E., & Mannino, D. M. (2013). COPD and gender differences: An update. *Translational Research; the Journal of Laboratory and Clinical Medicine, 162*(4), 208–218. doi:10.1016/j.trsl.2013.04.003 PMID:23684710

Bae, H., Jung, K.-H., Kang, G.-H., Kim, W. S., Kim, H. S., Hong, M., et al. (2014). Pyranopyran-1, 8-dione, an active compound from Viticis Fructus ameliorates cigarette smoke induced lung inflammatory response in a murine model (CAM5P. 239). *The Journal of Immunology, 192*(1 Supplement), 180.110-180.110.

Bancel, S. (2013). *Methods and compositions for targeting agents into and across the blood-brain barrier*. Google Patents.

Bansal, M., Khatri, M., & Taneja, V. (2013). Potential role of periodontal infection in respiratory diseases-a review. *Journal of Medicine and Life, 6*(3), 244.

Beachey, W. (2013). *Respiratory Care Anatomy and Physiology, Foundations for Clinical Practice, 3: Respiratory Care Anatomy and Physiology*. Elsevier Health Sciences.

Bloemsma, L. D., Hoek, G., & Smit, L. (2016). Panel Studies On Acute Effects Of Air Pollution. In *Patients With COPD: A Systematic Review And Meta-Analysis C55. Cardiopulmonary Effects of Air Pollution* (pp. A5420–A5420). Am Thoracic Soc.

Braido, F., Baiardini, I., Scichilone, N., Sorino, C., Di Marco, F., Corsico, A., & Canonica, G. W. et al. (2015). Disability in moderate chronic obstructive pulmonary disease: Prevalence, burden and assessment-results from a real-life study. *Respiration, 89*(2), 100–106. doi:10.1159/000368365 PMID:25612914

Breit, P. B. J., Lembke, F., Dubose, D., Burke, M., Muskie-Dukes-Driggs, A., Dobry, D., ... (2013). *A Dry-Powder Inhalable Formulation of Nanocrystalline Camptothecin for the Treatment of Lung Cancer.* Paper presented at the Journal of Aerosol Medicine and Pulmonary Drug Delivery.

Campardelli, R., Cherain, M., Perfetti, C., Iorio, C., Scognamiglio, M., Reverchon, E., & Della Porta, G. (2013). Lipid nanoparticles production by supercritical fluid assisted emulsion–diffusion. *The Journal of Supercritical Fluids, 82*, 34–40. doi:10.1016/j.supflu.2013.05.020

Carvalho, S. R. (2013). *Improved inhalation therapies of brittle powders.* Academic Press.

Cavanagh, T., Barman, S. P., Hao, T., Leland, T. B., & Thekkedath, R. V. (2013). *Preparations of hydrophobic therapeutic agents, methods of manufacture and use thereof.* Google Patents.

Chan, H. K., & Kwok, P. C. L. (2013). Novel particle production technologies for inhalation products. *Inhalation Drug Delivery: Techniques and Products*, 47-62.

Chapman, J. (2013). Therapeutic insight: Chronic obstructive pulmonary disease: Insidious, debilitating and costly. *AJP: The Australian Journal of Pharmacy*, *94*(1116), 78.

Charlson, E. S., Bittinger, K., Chen, J., Diamond, J. M., Li, H., Collman, R. G., & Bushman, F. D. (2012). Assessing bacterial populations in the lung by replicate analysis of samples from the upper and lower respiratory tracts. *PLoS ONE*, *7*(9), e42786. doi:10.1371/journal.pone.0042786 PMID:22970118

Cheow, W. S., & Hadinoto, K. (2010). Enhancing encapsulation efficiency of highly water-soluble antibiotic in poly (lactic-co-glycolic acid) nanoparticles: Modifications of standard nanoparticle preparation methods. *Colloids and Surfaces. A, Physicochemical and Engineering Aspects*, *370*(1), 79–86. doi:10.1016/j.colsurfa.2010.08.050

Chiang, P.-C., Alsup, J. W., Lai, Y., Hu, Y., Heyde, B. R., & Tung, D. (2009). Evaluation of aerosol delivery of nanosuspension for pre-clinical pulmonary drug delivery. *Nanoscale Research Letters*, *4*(3), 254–261. doi:10.1007/s11671-008-9234-1 PMID:20596335

Chiappa, S., Winn, J., Viñuela, A., Tipney, H., & Spector, T. D. (2013). A probabilistic model of biological ageing of the lungs for analysing the effects of smoking, asthma and COPD. *Respiratory Research*, *14*(1), 1. PMID:23721360

DAddio, S. M., Chan, J. G. Y., Kwok, P. C. L., Benson, B. R., Prudhomme, R. K., & Chan, H.-K. (2013). Aerosol delivery of nanoparticles in uniform mannitol carriers formulated by ultrasonic spray freeze drying. *Pharmaceutical Research*, *30*(11), 2891–2901. doi:10.1007/s11095-013-1120-6 PMID:23893019

Della Porta, G., Falco, N., & Reverchon, E. (2010). NSAID Drugs Release from Injectable Microspheres Produced by Supercritical Fluid Emulsion Extraction. *Journal of Pharmaceutical Sciences*, *99*(3), 1484–1499. doi:10.1002/jps.21920 PMID:19780130

Durán-Lobato, M., Enguix-González, A., Fernández-Arévalo, M., & Martín-Banderas, L. (2013). Statistical analysis of solid lipid nanoparticles produced by high-pressure homogenization: A practical prediction approach. *Journal of Nanoparticle Research*, *15*(2), 1–14. doi:10.1007/s11051-013-1443-6

Eltayeb, M., Stride, E., Edirisinghe, M., & Harker, A. (2016). Electrosprayed nanoparticle delivery system for controlled release. *Materials Science and Engineering C*, *66*, 138–146. doi:10.1016/j.msec.2016.04.001 PMID:27207047

Farkhooy, A., Janson, C., Arnardóttir, R. H., Emtner, M., Hedenström, H., & Malinovschi, A. (2015). Impaired Carbon Monoxide Diffusing Capacity is the strongest lung function predictor of decline in 12 minute-walking distance in COPD; a 5-year follow-up study. *COPD: Journal of Chronic Obstructive Pulmonary Disease*, *12*(3), 243–251. doi:10.3109/15412555.2014.948991 PMID:25208266

Gimeno-Santos, E., Rodriguez, D. A., Barberan-Garcia, A., Blanco, I., Vilaró, J., Torralba, Y., & Garcia-Aymerich, J. et al. (2014). Endurance exercise training improves heart rate recovery in patients with COPD. *COPD: Journal of Chronic Obstructive Pulmonary Disease*, *11*(2), 190–196. doi:10.3109/15412555.2013.831401 PMID:24377907

Gindy, M. E., Feuston, B., Glass, A., Arrington, L., Haas, R. M., Schariter, J., & Stirdivant, S. M. (2014). Stabilization of ostwald ripening in low molecular weight amino lipid nanoparticles for systemic delivery of siRNA therapeutics. *Molecular Pharmaceutics, 11*(11), 4143–4153. doi:10.1021/mp500367k PMID:25317715

Gomez, A. (2002). The electrospray and its application to targeted drug inhalation. *Respiratory Care, 47*(12), 1419-1431.

Grassin-Delyle, S., Buenestado, A., Naline, E., Faisy, C., Blouquit-Laye, S., Couderc, L.-J., & Devillier, P. et al. (2012). Intranasal drug delivery: an efficient and non-invasive route for systemic administration: focus on opioids. *Pharmacology & Therapeutics, 134*(3), 366–379. doi:10.1016/j.pharmthera.2012.03.003 PMID:22465159

Guenette, J. A., Webb, K. A., & ODonnell, D. E. (2012). Does dynamic hyperinflation contribute to dyspnoea during exercise in patients with COPD? *The European Respiratory Journal, 40*(2), 322–329. doi:10.1183/09031936.00157711 PMID:22183485

Hall, J. E. (2015). *Guyton and Hall textbook of medical physiology*. Elsevier Health Sciences.

Hodson, M., & Sherrington, R. (2014). Treating patients with chronic obstructive pulmonary disease. *Nursing Standard, 29*(9), 50–58. doi:10.7748/ns.29.9.50.e9061 PMID:25351094

Ionescu, C. M. (2013). *The human respiratory system. In The Human Respiratory System* (pp. 13–22). Springer. doi:10.1007/978-1-4471-5388-7_2

Jacques, S., & Wright, R. (2015). *Code of the suburb: Inside the world of young middle-class drug dealers*. University of Chicago Press. doi:10.7208/chicago/9780226164250.001.0001

Kaur, G., Narang, R., Rath, G., & Goyal, A. K. (2012). Advances in pulmonary delivery of nanoparticles. *Artificial Cells, Blood Substitutes, and Biotechnology, 40*(1-2), 75–96. doi:10.3109/10731199.2011.592494 PMID:21806501

Kho, K., Cheow, W. S., Lie, R. H., & Hadinoto, K. (2010). Aqueous re-dispersibility of spray-dried antibiotic-loaded polycaprolactone nanoparticle aggregates for inhaled anti-biofilm therapy. *Powder Technology, 203*(3), 432–439. doi:10.1016/j.powtec.2010.06.003

Kim, V., & Criner, G. J. (2013). Chronic bronchitis and chronic obstructive pulmonary disease. *American Journal of Respiratory and Critical Care Medicine, 187*(3), 228–237. doi:10.1164/rccm.201210-1843CI PMID:23204254

Kim, V., Oros, M., Durra, H., Kelsen, S., Aksoy, M., Cornwell, W. D., & Criner, G. J. et al. (2015). Chronic bronchitis and current smoking are associated with more goblet cells in moderate to severe COPD and smokers without airflow obstruction. *PLoS ONE, 10*(2), e0116108. doi:10.1371/journal.pone.0116108 PMID:25646735

Kluge, J., & Mazzotti, M. (2012). CO2-assisted high pressure homogenization: A solvent-free process for polymeric microspheres and drug–polymer composites. *International Journal of Pharmaceutics, 436*(1–2), 394–402. doi:10.1016/j.ijpharm.2012.06.048 PMID:22750408

Kruis, A. L., & Chavannes, N. H. (2016). Potential benefits of integrated COPD management in primary care. *Monaldi Archives for Chest Disease*, *73*(3). doi:10.4081/monaldi.2010.297 PMID:21214043

Kuklisova, Z., Tkacova, R., Joppa, P., Wouters, E., & Sastry, M. (2016). Severity of nocturnal hypoxia and daytime hypercapnia predict CPAP failure in patients with COPD and obstructive sleep apnea overlap syndrome. *Sleep Medicine*. PMID:28215237

Kumar, S., & Randhawa, J. K. (2013). High melting lipid based approach for drug delivery: Solid lipid nanoparticles. *Materials Science and Engineering C*, *33*(4), 1842–1852. doi:10.1016/j.msec.2013.01.037 PMID:23498204

Lahousse, L., Ziere, G., Verlinden, V. J., Zillikens, M. C., Uitterlinden, A. G., Rivadeneira, F., & Stricker, B. H. et al. (2016). Risk of frailty in elderly with COPD: A population-based study. *The Journals of Gerontology. Series A, Biological Sciences and Medical Sciences*, *71*(5), 689–695. doi:10.1093/gerona/glv154 PMID:26355016

Lee, A. L., & Goldstein, R. S. (2015). The role of telemedicine. *Controversies in COPD*, *69*, 269–296. doi:10.1183/2312508X.10019014

Leung, S. S., Parumasivam, T., Gao, F. G., Carrigy, N. B., Vehring, R., Finlay, W. H., & Chan, H.-K. et al. (2016). Production of inhalation phage powders using spray freeze drying and spray drying techniques for treatment of respiratory infections. *Pharmaceutical Research*, *33*(6), 1486–1496. doi:10.1007/s11095-016-1892-6 PMID:26928668

Littleton, S. W. (2015). Hypercapnia From Hyperoxia in COPD: Another Piece of the Puzzle or Another Puzzle Entirely? *Respiratory Care*, *60*(3), 473–475. doi:10.4187/respcare.03979 PMID:25737575

Ma, X., Cui, J., Wang, J., Chang, Y., Fang, Q., & Bai, C. et al.. (2015). Multicentre investigation of pathogenic bacteria and antibiotic resistance genes in Chinese patients with acute exacerbation of chronic obstructive pulmonary disease. *The Journal of International Medical Research*. PMID:26152913

Mansfield, E., Bryant, J., Regan, T., Waller, A., Boyes, A., & Sanson-Fisher, R. (2016). Burden and Unmet Needs of Caregivers of Chronic Obstructive Pulmonary Disease Patients: A Systematic Review of the Volume and Focus of Research Output. *COPD: Journal of Chronic Obstructive Pulmonary Disease*, 1-6.

Mansour, H. M., Rhee, Y.-S., & Wu, X. (2009). Nanomedicine in pulmonary delivery. *International Journal of Nanomedicine*, *4*, 299–319. doi:10.2147/IJN.S4937 PMID:20054434

Mathew, N., Xavier, Z., Deshmukh, A. S., Deshmukh, H., Jadhav, S., & Kasat, S. (2016). The Study of Aetiology of Chronic Obstructive Pulmonary Disease (COPD) in Non-Smokers. *Journal of Pulmonary & Respiratory Medicine, 2015.*

McNicholas, W. T., Verbraecken, J., & Marin, J. M. (2013). Sleep disorders in COPD: The forgotten dimension. *European Respiratory Review*, *22*(129), 365–375. doi:10.1183/09059180.00003213 PMID:23997063

Miki, K., Maekura, R., Nagaya, N., Kitada, S., Miki, M., & Yoshimura, K. et al.. (2013). Effects of ghrelin on the dyspnea pattern in cachectic COPD: Exploratory analysis of a substudy of a multicenter, randomized, double-blind, placebo-controlled trial of ghrelin treatment. *The European Respiratory Journal, 42*(Suppl 57), 2499.

Millqvist, E., Ternesten-Hasséus, E., & Bende, M. (2013). Inhalation of menthol reduces capsaicin cough sensitivity and influences inspiratory flows in chronic cough. *Respiratory Medicine, 107*(3), 433–438. doi:10.1016/j.rmed.2012.11.017 PMID:23266255

Miravitlles, M., Peña-Longobardo, L. M., Oliva-Moreno, J., & Hidalgo-Vega, Á. (2015). Caregivers burden in patients with COPD. *International Journal of Chronic Obstructive Pulmonary Disease, 10*, 347. doi:10.2147/COPD.S76091 PMID:25709429

Monjazebi, F., Dalvandi, A., Ebadi, A., Khankeh, H. R., Rahgozar, M., & Richter, J. (2016). Functional Status Assessment of COPD Based on Ability to Perform Daily Living Activities: A Systematic Review of Paper and Pencil Instruments. *Global Journal of Health Science, 8*(3), 210.

Muala, A., Nicklasson, H., Boman, C., Swietlicki, E., Nyström, R., Pettersson, E., et al. (2014). *Respiratory tract deposition of inhaled wood smoke particles in healthy volunteers and subjects with COPD.* Academic Press.

Naftali, S., Rosenfeld, M., Wolf, M., & Elad, D. (2005). The air-conditioning capacity of the human nose. *Annals of Biomedical Engineering, 33*(4), 545–553. doi:10.1007/s10439-005-2513-4 PMID:15909660

Nigam, P., Waghmode, S., Louis, M., Wangnoo, S., Chavan, P., & Sarkar, D. (2014). Graphene quantum dots conjugated albumin nanoparticles for targeted drug delivery and imaging of pancreatic cancer. *Journal of Materials Chemistry B, 2*(21), 3190–3195. doi:10.1039/C4TB00015C

Patil, J., & Sarasija, S. (2012). Pulmonary drug delivery strategies: A concise, systematic review. *Lung India, 29*(1), 44. doi:10.4103/0970-2113.92361 PMID:22345913

Perera, P. N., Armstrong, E. P., Sherrill, D. L., & Skrepnek, G. H. (2012). Acute exacerbations of COPD in the United States: Inpatient burden and predictors of costs and mortality. *COPD: Journal of Chronic Obstructive Pulmonary Disease, 9*(2), 131–141. doi:10.3109/15412555.2011.650239 PMID:22409371

Ponte, B. M., Santos, I. S., Macedo, A. S., & Souto, E. B. (2016). Lipid Nanoplatforms for Pulmonary Drug Delivery. *Nanotechnology and Drug Delivery, Volume Two: Nano-Engineering Strategies and Nanomedicines against Severe Diseases*, 146.

Pritchard, J. (2001). The influence of lung deposition on clinical response. *Journal of Aerosol Medicine, 14*(Supplement 1), 19–26. doi:10.1089/08942680150506303 PMID:11424889

Ramos, F. L., Krahnke, J. S., & Kim, V. (2014). Clinical issues of mucus accumulation in COPD. *International Journal of Chronic Obstructive Pulmonary Disease, 9*, 139–150. PMID:24493923

Reece, W. O. (2015). Overview of the Respiratory System. *Dukes' Physiology of Domestic Animals*, 203.

Respiratory System - Pulmonary Ventilation. (2016). *Science Encyclopedia*. Retrieved from http://science.jrank.org/pages/5846/Respiratory-System-Pulmonary-ventilation.html

Ruge, C. A., Kirch, J., & Lehr, C.-M. (2013). Pulmonary drug delivery: From generating aerosols to overcoming biological barriers—therapeutic possibilities and technological challenges. *The Lancet Respiratory Medicine*, *1*(5), 402–413. doi:10.1016/S2213-2600(13)70072-9 PMID:24429205

Sharma, S. B., Jain, S., Khirwadkar, P., & Kulkarni, S. (2013). The effects of air pollution on the environment and human health. *Indian Journal of Research in Pharmacy and Biotechnology*, *1*(3), 391.

Sheth, P., Sandhu, H., Singhal, D., Malick, W., Shah, N., & Serpil Kislalioglu, M. (2012). Nanoparticles in the pharmaceutical industry and the use of supercritical fluid technologies for nanoparticle production. *Current Drug Delivery*, *9*(3), 269–284. doi:10.2174/156720112800389052 PMID:22283656

Sin, D., Sze, M., & Hogg, J. (2014). Bacterial microbiome of lungs in COPD. *International Journal of Chronic Obstructive Pulmonary Disease*, *9*(1), 229–238. doi:10.2147/COPD.S38932 PMID:24591822

Singh, A., Malviya, R., & Sharma, K, P. (2011). Pulmonary drug delivery system: A novel approach for drug delivery. *Current Drug Therapy*, *6*(2), 137–151. doi:10.2174/157488511795304930

Singh, S. (2016). Respiratory symptoms and signs. *Medicine*, *44*(4), 205–212. doi:10.1016/j.mpmed.2016.02.001

Singh, S., Maltais, F., Tombs, L., Church, A., Iqbal, A., & Riley, J. H. (2015). Effect of umeclidinium/vilanterol (UMEC/VI) on inspiratory capacity/total lung capacity ratio in hyperinflated COPD patients. *The European Respiratory Journal*, *46*(suppl 59), PA1493.

Sivtseva, A. I., Strutynsky, A. V., Krivoshapkin, V. G., Sivtseva, E. N., Ivanova, M. A., & Timofeev, L. F. (2015). Systolic and Diastolic Dysfunction of the Right Heart Ventricle in Patients With Chronic Obstructive Pulmonary Disease in Extremely Cold Climate. *Global Journal of Health Science*, *7*(3), 283.

Sridhar, R., & Ramakrishna, S. (2013). Electrosprayed nanoparticles for drug delivery and pharmaceutical applications. *Biomatter*, *3*(3), e24281. doi:10.4161/biom.24281 PMID:23512013

Takeyama, K., Tagaya, E., Kondo, M., Tamaoki, J., & Orimo, K. (2016). Impact Of Irreversible Airflow Obstruction. In *Clinical Course Of Patients With Severe Atopic Asthma A36. Clinical Problems in Asthma and Allergy* (pp. A1397–A1397). Am Thoracic Soc.

Tan, W., Sin, D., Bourbeau, J., Hernandez, P., Chapman, K., Cowie, R., & Aaron, S. D. et al. (2015). Characteristics of COPD in never-smokers and ever-smokers in the general population: Results from the CanCOLD study. *Thorax*, *70*(9), 822–829. doi:10.1136/thoraxjnl-2015-206938 PMID:26048404

To, T., Zhu, J., Larsen, K., Simatovic, J., Feldman, L., Ryckman, K., & Carlsten, C. et al. (2016). Progression from Asthma to Chronic Obstructive Pulmonary Disease (COPD): Is Air Pollution a Risk Factor?. *American Journal of Respiratory and Critical Care Medicine*, *194*(4), 429–438. doi:10.1164/rccm.201510-1932OC PMID:26950751

Trotta, V., Haghi, M., Scalia, S., Young, P. M., & Traini, D. (2015). *Resveratrol, a Novel Spray Dried Inhalation Powder for the Treatment of COPD and other Inflammatory Lung Diseases.* Paper presented at the Journal of aerosoal medicine and pulmonary drug delivery.

Varshosaz, J., Ghaffari, S., Mirshojaei, S., Jafarian, A., Atyabi, F., Kobarfard, F., et al. (2013). Biodistribution of amikacin solid lipid nanoparticles after pulmonary delivery. *BioMed Research International, 2013*.

Vartiainen, V., Bimbo, L. M., Hirvonen, J., Kauppinen, E. I., & Raula, J. (2016). Drug permeation and cellular interaction of amino acid-coated drug combination powders for pulmonary delivery. *International Journal of Pharmaceutics*, *504*(1), 89–97. doi:10.1016/j.ijpharm.2016.03.052 PMID:27034001

Viby, N.-E., Nexø, E., Kissow, H., Andreassen, H., Clementsen, P., Thim, L., & Poulsen, S. S. (2015). Trefoil factors (TFFs) are increased in bronchioalveolar lavage fluid from patients with chronic obstructive lung disease (COPD). *Peptides*, *63*, 90–95. doi:10.1016/j.peptides.2014.09.026 PMID:25445610

Walker, D. J., Walterspacher, S., Ekkernkamp, E., Storre, J. H., Windisch, W., & Dreher, M. (2015). Walking with non-invasive ventilation does not prevent exercise-induced hypoxaemia in stable hypercapnic COPD patients. *COPD: Journal of Chronic Obstructive Pulmonary Disease*, *12*(5), 546–551. doi:10.3109/15412555.2015.1008690 PMID:26263033

Willis, L., Hayes, D. Jr, & Mansour, H. M. (2012). Therapeutic liposomal dry powder inhalation aerosols for targeted lung delivery. *Lung*, *190*(3), 251–262. doi:10.1007/s00408-011-9360-x PMID:22274758

Yu, H., Teo, J., Chew, J. W., & Hadinoto, K. (2016). Dry powder inhaler formulation of high-payload antibiotic nanoparticle complex intended for bronchiectasis therapy: Spray drying versus spray freeze drying preparation. *International Journal of Pharmaceutics*, *499*(1–2), 38–46. doi:10.1016/j.ijpharm.2015.12.072 PMID:26757148

Yu, H., Tran, T.-T., Teo, J., & Hadinoto, K. (2016). Dry powder aerosols of curcumin-chitosan nanoparticle complex prepared by spray freeze drying and their antimicrobial efficacy against common respiratory bacterial pathogens. *Colloids and Surfaces. A, Physicochemical and Engineering Aspects*, *504*, 34–42. doi:10.1016/j.colsurfa.2016.05.053

Zheng, M., Zhao, M., Tang, L., Zhang, C., Song, L., & Wang, W. (2016). Ginsenoside Rg1 attenuates hypoxia and hypercapnia-induced vasoconstriction in isolated rat pulmonary arterial rings by reducing the expression of p38. *Journal of Thoracic Disease*, *8*(7), 1513–1523. doi:10.21037/jtd.2016.05.71 PMID:27499938

Zhou, Q. T., Leung, S. S. Y., Tang, P., Parumasivam, T., Loh, Z. H., & Chan, H.-K. (2015). Inhaled formulations and pulmonary drug delivery systems for respiratory infections. *Advanced Drug Delivery Reviews*, *85*, 83–99. doi:10.1016/j.addr.2014.10.022 PMID:25451137

ADDITIONAL READING

Celli, B. R., MacNee, W., Agusti, A., Anzueto, A., Berg, B., Buist, A. S., & ZuWallack, R. et al. (2004). Standards for the diagnosis and treatment of patients with COPD: A summary of the ATS/ERS position paper. *The European Respiratory Journal*, *23*(6), 932–946. doi:10.1183/09031936.04.00014304 PMID:15219010

Enright, P. L. (2016). Oxygen desaturation during a 6-min walk identifies a COPD phenotype with an increased risk of morbidity and mortality. *The European Respiratory Journal*, *48*(1), 1–2. doi:10.1183/13993003.00884-2016 PMID:27365500

Ford, E. S., Croft, J. B., Mannino, D. M., Wheaton, A. G., Zhang, X., & Giles, W. H. (2013). COPD surveillance—United States, 19992011. *CHEST Journal*, *144*(1), 284–305. doi:10.1378/chest.13-0809 PMID:23619732

Hardin, M., Hawrylkiewicz, I., Sliwinski, P., Yim, J., Kim, W., & Kim, D. et al.. (2015). Coding Variant Associations With Lung Function And COPD Using An Exome Array. *COPD*, *1*, 1.

Houghton, A. M. (2013). Mechanistic links between COPD and lung cancer. *Nature Reviews. Cancer*, *13*(4), 233–245. doi:10.1038/nrc3477 PMID:23467302

Leidy, N. K., Malley, K. G., Anderson, W. H., Barr, R., Bleeker, E., & Bowler, R. P. et al.. (2016). How'Stable'Is Stable COPD? Daily Symptom Variability Of Subjects Enrolled. In *The SPIROMICS Exacerbation Sub-Study B43. COPD: Phenotypes and Clinical Outcomes* (pp. A3527–A3527). Am Thoracic Soc.

Magnussen, H., Disse, B., Rodriguez-Roisin, R., Kirsten, A., Watz, H., Tetzlaff, K., & Calverley, P. M. A. et al. (2014). Withdrawal of inhaled glucocorticoids and exacerbations of COPD. *The New England Journal of Medicine*, *371*(14), 1285–1294. doi:10.1056/NEJMoa1407154 PMID:25196117

Mukherjee, D., Abbas, S., Minter, J., Whitfield, J., Field, S., & Pearce, S. et al.. (2016). *A Chronic Obstructive Pulmonary Disease (COPD) Service Integrating Community And Hospital Services Can Improve Patient Care And Reduce Hospital Stays A41. The Spectrum COPD Care: From Identification to Policy* (pp. A1523–A1523). Am Thoracic Soc.

Romme, E., McAllister, D., Murchison, J., Van Beek, E., Petrides, G., & Rutten, E. et al.. (2013). Associations between arterial calcification and stiffness, bone density, emphysema and all-cause mortality in COPD patients. *The European Respiratory Journal*, *42*(Suppl 57), 5105.

Shin, J. I., & Brusselle, G. G. (2014). Mechanistic links between COPD and lung cancer: A role of microRNA let-7? *Nature Reviews. Cancer*, *14*(1), 70–70. doi:10.1038/nrc3477-c1 PMID:24505620

Suissa, S., Patenaude, V., Lapi, F., & Ernst, P. (2013). Inhaled corticosteroids in COPD and the risk of serious pneumonia. *Thorax*, *68*(11), 1029–1036. doi:10.1136/thoraxjnl-2012-202872 PMID:24130228

Voelker, H., Albert, R. K., Bailey, W. C., Casaburi, R., Cooper, J., & Curtis, J. L. et al.. (2015). Cardiac Events And Relationship To Rates Of Acute Exacerbation In COPD. *American Journal of Respiratory and Critical Care Medicine*, *191*, A6368.

Wise, R. A., Anzueto, A., Cotton, D., Dahl, R., Devins, T., Disse, B., & Calverley, P. et al. (2013). Tiotropium Respimat inhaler and the risk of death in COPD. *The New England Journal of Medicine*, *369*(16), 1491–1501. doi:10.1056/NEJMoa1303342 PMID:23992515

Zhao, H., Martinez, F. J., Criner, G. J., & Czysz, S. A. (2014). Predictors Of Stability. In *COPD B39. COPD Exacerbations: Precipitating Factors, Prevention and Outcome* (pp. A2891–A2891). Am Thoracic Soc.

KEY TERMS AND DEFINITIONS

Chronic Obstructive Pulmonary Disease (COPD): It refers to a group of diseases such as emphysema, and chronic bronchitis, which makes breathing difficult.

Drug Delivery System: A drug delivery system is a formulation or a device that allows the targeted delivery of a therapeutic substance to the place of action.

High Pressure Homogenization: It is a type of disintegration method for the formation of nanoparticles.

Nanoparticles: Nanoparticles are the particles having diameter in between 1 and 100 nm.

Respiratory System: Respiratory system is a biological system of an organism responsible for the gaseous exchange between the organism and its environment.

Spray Drying: Spray drying refers to the production of a dry powder from a liquid feed through the evaporation of the solvent.

Surfactant: A surfactant is a surface-active substance that reduces the surface tension of a liquid.

Chapter 12
Efficacy of Composites for Fabrication of Orthotic Calipers

Nisha Kumari
Birla Institute of Technology, India

Kaushik Kumar
Birla Institute of Technology, India

ABSTRACT

Composites based materials are becoming more important in the field of aerospace, automobile, medicine, sports, energy etc. The principal quality of composite materials is their higher strength, firmness, resistant to corrosion and comparatively low weight when compared to the metals. The main aim of the work is to replace the presently used orthotic calipers with the proposed polymeric based composites as in last few years composites have generated broad research in engineering and research field due to its smaller density, low cost, recyclable, environment friendly and capable of being decomposed. Here, the authors have proposed two polymeric based composites (thermoplast and thermoset) as an alternative material of construction and hence their various properties were investigated. The authors after performing the experiment unveil that the strength and firmness of thermoset based composites (Epoxy + Carbon) is higher and have lower weight- to- volume ratio than that of the presently used (Aluminum based) orthotic calipers.

INTRODUCTION

Millions of people worldwide are suffering from one or other disability. Disability is an impairment related to speech, vision, locomotion etc. It may be also due to the emergence of new disease or virus transmitted from one person to another. It affects not only their part but also their minds because of their ignorance in the normal society. One of the main disabilities we are here concerned with is polio or often termed as poliomyelitis. Poliomyelitis affects the motor neurons and is the main reason of paralysis and death (Gromeier and Wimmer, 1998). In 1950's it affected lot of patients by its viral infections but thanks to the researchers/ scientist's who discovered vaccine because of which many survived but muscle weak-

DOI: 10.4018/978-1-5225-3023-7.ch012

ness and disfigured joints were the common problems being faced by the patients. Researchers came up with many ideas to make them feel comfortable with the techniques and new treatments so that they can lead their life easily and can work at least or do whatever they wish to do (Thorndike et al.,1956, Cicenia and Hoberman., 1957). Firstly the scientists invented iron lung which worked with the help of the motor. Polio affected patients were covered inside the iron lung fully and sudden expansion and contraction were supplied to the chest with the help of the motor attached to it to maintain the levels of pressure in the body. Patients were covered inside the iron lung for months and even for years as it was dependent on the type of disability of the patients so it was not known by the researchers who will take how much of time to get recovered. Due to this reason iron lung was no longer in use and was forbidden. Secondly, the researchers/scientists thought of passive immunotherapy which consisted of antibody serum which was found to be effective for 80% of the patients but it did not last long as serum duration was only for 5 months and reinjection of the serum was not known by the scientists. And then they developed two vaccines for polio infected patients in 1950 after being tested on them and then they started forwarding to affected patients. But the main problem with the vaccines was that it only restricted paralysis and death it did not help the patients in their walking cycle. Then rehabilitation therapies were started which consisted of electrotherapy and hydrotherapy and created lot of awareness among infected people but it was not possible for every affected individual to join the rehabilitation centers because it demanded a very high cost and not everyone could afford the charges for therapies. Because of the therapies and treatment processes affected children's were unable to go to school for months and years, adults were unable to go to college and their respective jobs and a lot more problem were being faced by them because of acute poliomyelitis. Later, the researchers moved on foot orthotics which mainly included the shoes being designed by the doctor for polio affected survivors or for the patients who were suffering from foot related problems (Sethi et al., 1978, Ferber et al., 2005). After the problems with the SACH (Solid Ankle Cushion Heel) foot, vulcanized rubber foot was designed for the amputees so that they can walk properly, stand and can move without any restriction in movement. This lead to the evolution of Jaipur foot, the intervening space present inside the shoe was filled by providing anextra cushion with the help of layers of sponge rubber (Sethi, 1990).

Polio affected survivors are in need of something which would not limit their movement and it should ensure that they can walk, sit and do whatever they want without any risk of falling and fractures of disfigured joints (Genet et al., 2010). They were in need ofacomponent that is lighter in weight, affordable and aids in correct positioning of joints and weakened muscles. Scientists designed orthotic calipers which helped polio affected survivors in walking with their support (Moreno et al., 2006, Giovacchini et al., 2015, Bregman.,2011). Themain purpose of orthotic calipers (Figure 1) is to provide correct positioning and stability without restricting their movement in walking, sitting and moving from one place to another (Berwick., 1971, Rakib et al., 2015).

In olden days calipers were made up of wood which was made by the villagers. The calipers at that time did not require straps and belts attached to them that why they were affordable to polio affected survivors. The calipers made up of wood were not durable because wood is having a hygroscopic property due to which it absorbs moisture very easily. Then the researchers thought of designing calipers made up of ferrous metals like Iron. Iron was chosen as calipers materials as they possess thehigh strength to weight ratio, they are long lasting material, atmosphere friendly and if required can be recycled too. After little time polio affected survivors raised aproblem with the iron made calipers due to their high weight because of which they were unable to move as it required alot of strength. The another problem with the iron based calipers was its rusting and magnetic property because of which iron calipers were

Figure 1. Commercially available above knee calipers for handicapped
Source: http://www.indiamart.com/proddetail/above-knee-caliper-3092433433.html

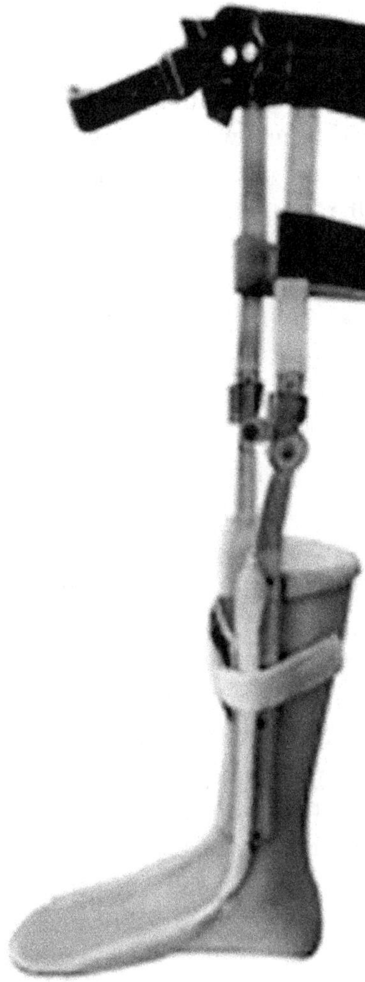

no longer used by the affected patients. And then few decades' later researcher/scientists came up with an idea of designing calipers made up of metals such as Aluminum Alloy and Stainless Steel. Aluminum was selected for designing an orthotic caliper due to its low weight; low cost than previously used materials. Aluminum based calipers are easy to manufacture and are easier to customize for the patients as every patient have different height and shape of the body, metal rods used can be easily customized according to the requirement. Presently aluminum based calipers are used by the polio affected patients still they are not fully satisfied with the weight and cost which is the most important thing to be taken into consideration by the researchers/scientists working on the material of the orthotic calipers. A caliper material now should be chosen by keeping in mind the weight to volume ratio, cost and it should provide required stiffness by the weakened and disfigured joints. For designing lighter weight calipers with improved stiffness and low cost one of the most extensively developed material being used are composite materials. In this work, composite materials are being tried for making calipers.

The literature survey on orthotic calipers indicated that the work has been done mostly on various mechanisms. The material of construction of the braces was initiated with steel and presently using aluminum alloy. No work has been done to replace aluminum alloy with a still lighter material which would reduce the self-weight of the calipers and hence increase the ease and mobility of the users especially polio infected patients.

The main objective of the project is to propose a polymer based polymer composite for fabrication of the braces expecting to reduce the self-weight without affecting the mechanical properties.

Composite materials are made up of one or two additional phases in the primary phase, so there will always be two or three phases in the composite materials (Kröling et al., 2014, Liew et al., 2015, Ashik et al., 2015). Before moving on to the discussion of the material to be chosen for the orthotic calipers, the properties being possessed by the composite materials that have to be taken care of while deciding the material are to be discussed. In the customer driven society, the ultimate requirement is performance and quality. Composite materials are formed by blending up of two macro constituents that have a different physical composition (density, thermal expansion, specific heat, melting point, etc.) and chemical composition (oxidation, corrosion, toxicity, biodegradation, etc.) and are dissolved in each other. Macro- constituents stated here are matrix and reinforcements. In a composite with one matrix, reinforcement can be one or more than one. Matrix provides the bulk form by providing shape and size to the composite by holding the reinforcements when a load is applied. Matrix shares the load with the reinforcement. Reinforcement provides strength to the matrix; they have a prime aim of weight bearing capability. Reinforcement given to the matrix can be in any one of the forms such as fibers, flakes, whiskers, particles. In the same way, matrix can also be divided into three which are polymer matrix composite ceramic matrix composites and metal matrix composites. While preparing a composite material different matrix and different reinforcements can be used but there will be the recognizable interface between the two. The material should have noticeably different physical and chemical composition than the two constituents mixed together so that type of material will be considered as composite materials.

Composite materials are emerging as materials and have gained broad name among researchers and engineering spectrum. Composite materials, due to their inherent properties, can be used for fabrication of orthotic calipers. Composites, generally, are very strong and firm, very light in weight, good mechanical properties, low density, good specific properties when compared to the presently used aluminum calipers (Ku et al., 2011, Powell., 1994). The strength to weight and stiffness to weight ratios respectively are several times greater than steel or presently used aluminum orthotic calipers because of which it can replace conventional materials such as wood and metals. Fatigue properties are better than that of common engineering metals and have higher toughness. They can be so made that they will possess corrosion resistant property and will have the possibility of achieving a combination of properties that are not even attainable by the macro constituents used (Mitra, 2014). Due to such advantages they are finding diverse application in engineering field, automotive, aerospace applications, marine, biomedical, sports, energy and even house-hold applications (Arpitha et al., 2014, Jagannatha and Harish., 2015, Kinsella et al., 2000, Deogonda et al., 2013, Subash and Pillai., 2015, Ansar et al. 2011, Liu et al., 2012, Yu et al., 2016). Apart from the above their recyclability and environment friendliness are added advantages (Gibson, 2015, Suresh et al., 2003). Processing of composite materials is also easier and even complex part can be made very easily and in a simple process in a single shot by manufacturing processes. As less use of composites make the assembly process reduces and leads to the cost reduction and inherent property of providing good surface finish for the polymeric based composites, ease of fabrication and installation is also possible which leads to theapplication of structural components installation in less time.

Polymeric matrix composites are composed of different types of short fibers or continuous fibers blended together by an organic polymeric matrix. They have gained a lot of importance in the market especially in bearing applications (Sayed et al.,1995) and are widely used for several years due to its fatigue properties and lighter weight (Mantia and Morreale., 2011, Shivakumar and Guggari., 2011, Ala-Kleme et al., 2006). Polymeric matrix composites consist of matrix and reinforcements, matrix consists of polymers and reinforcement consists of fibrous material. The properties of apolymeric based matrix are determined by the properties of fibers, resin and the ratio of fiber to resin in composites, geometry and orientation of the composites. Polymers can be of thermoplastic and thermoset nature.

Thermoplastic are materials that convert the raw materials or palettes into solid form by curing. Thermoplastics materials soften when heated above the glass transition temperature or melting point and become hard after cooling. They have linear branched polymeric structures. They inherit properties such as high strength and toughness, better hardness, chemical resistant, durable, self- lubrication, transparency and water proofing. Some of the thermoplastic based composites are polypropylene, liquid crystal polymers, cellulose etc. They can be further blended with other constituents to produce huge family and tangible products and can be recycled (Otheguy et al., 2009). Thermoplastic based composites can be used to manufacture dashboards, toys, electrical products, glass, sheets, etc.

Thermoset based composites cure irreversibly and become permanently hard and rigid after curing. They cannot be remolded and continued heating for a long time leads to degradation and decomposition. They have a cross linked structure of three dimensional network of covalent intermolecular bonds. Different types of thermoset materials used as matrix are epoxies, Bakelite, alkyds, phenolic, polyester, etc.

In the paper work is being done on calipers using thermoplast based polymeric composites (Nylon with 30 percent particulate glass filler). Glass fillers provide better properties such as high strength/ stiffness; low weight ratio and cost when compared to the Aluminum based orthotic calipers (Etcheverry and Barbosa., 2012, Bader., 2002, Yadav et al., 1993, El-Tayeb and Mostafa., 1996).

The authors worked on calipers with thermoset based polymer (carbon fillers with epoxy). Carbon fillers itself has many number of advantages some of which are it will provide excellent strength to weight ratio when compared to any materials. Carbon fillers when bonded with epoxy resin create a composite material that is strong enough to handle the weight of someone and at the same time lighter in weight. It's distinguishable properties such as excellent specific energy absorption (SEA) make it a workable and a versatile and able to use easily product for a variety of commercial product (Meredith et al., 2015). It provides higher strength, stiffness and resistance almost 8 times stronger than Aluminum Alloy and 1.5 times lighter than Aluminum Alloy. Carbon reinforced polymers have a higher strength to weight ratio when compared with the presently used orthotic calipers which are made up of Aluminum (Kastner et al., 2010).

In the present work thermoplast based composite and thermoset based composite are considered and hence their mechanical behavior, including Density, Hardness, Tensile, etc. are evaluated and compared with aluminum in an existing product. For ensuring even distribution of the reinforcements in the matrix, microstructures of the composites are studied using scanning electron microscope.

MATERIALS USED

Epoxy

Low solidifying curing epoxy resin (standard Epoxy AW 106) chemically belonging to the 'epoxide' family and corresponding hardener (HV 953U) were mixed in a quantitative relation (as declared by the manufacturer) for preparation of the thermoset matrix.

Carbon

Carbon in the form of particulate (about 70μ) was taken as filler. The properties of carbon, like high stiffness, high durability, low weight, high chemical resistance, heat tolerance and low thermal growth, make them usable in building components/products in industrial, military, motorsports, alongside totally different competition sports.

Nylon

For thermoplast based composite Nylon is taken as matrix material. Nylon based polymeric composites exhibits predominant physical and mechanical property that provides an outsized verity of end-use performances necessary in many industrial applications. Moreover, self-lubricating nature of Nylon makes it an ideal building material for products like abearing, bush, gear etc.

Glass

Glass in particulate size (about 70μ) has been used as the reinforcing material for the thermoplast composites. It being a moldable filler, can be used with Nylon to manufacture a good vary of structural elements by injection or compression molding.

Conditioning

Nylon being an absorptive material cannot be dried effectively so preheating is needed. Preheating of the Nylon is done in dehumidifier chamber (Figure 2) at 250^0 C for three Hours.

Composite Fabrication (Thermoset Based Composite)

For preparation of Composites, samples were created by intermixture carbon fillers (30%) having a density of (1.15gm/cc) with an Epoxy resin having density of (1.4gm/cc). The composites were fabricated at room temperature. The mixture was poured into mould of 99 X 88 X 4.5 mm. for the formation of the samples the mould was placed in a Carver Press (Figure 3) at a pressure of 5T. By varying the vacuum pressure throughout the heating stage of five metric tons air bubbles and wetness were removed. After hardening of specimens, the samples were taken out of the press for testing.

Figure 2. Dehumidifier for preheating of the nylon
Source: Roy et al., 2016

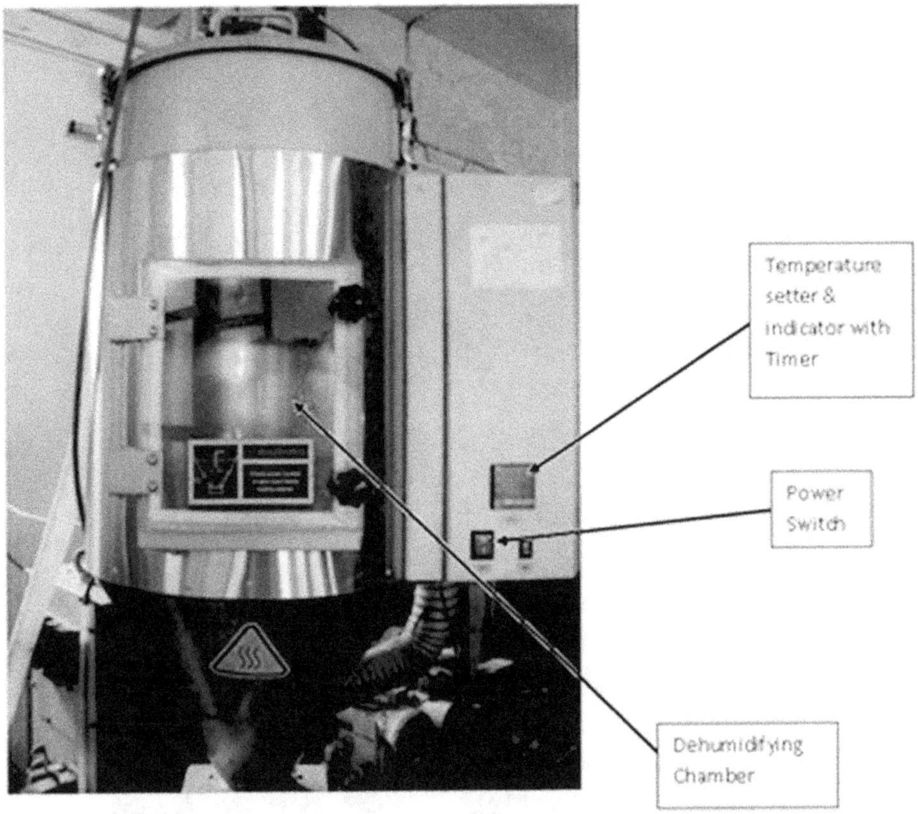

Composite Fabrication (Thermoplast Based Composite)

For preparation of thermoplast based polymeric composite Nylon was taken as the matrix material having a density of 1.14 gm/cc and 30% glass filler having adensity of 2.5 gm/cc. First, Nylon and glass are mixed homogeneously in a single screw extruder (Figure 4) and then the granules are subjected to injection molding process, the process by which the plastic palettes are transformed into molded parts. Composite granules are fed into the hopper of an injection moulding machine (Figure 5). A screw conveys and melts the palettes within the barrel. Typical screw speed setting for the nylon is 60-100 rpm. Melted composite is forced into the closed mold and cooled within the mold. Then the mold is opened and the part is ejected or removed from the mold and the cycle was repeated.

Sample for Testing

The samples prepared using the above processes are trimmed to the exact sizes and testing were performed. One sample each of the three types is shown in Figure 6.

Figure 3. Carver press for thermoset based composite fabrication

Figure 4. Haake single screw extruder

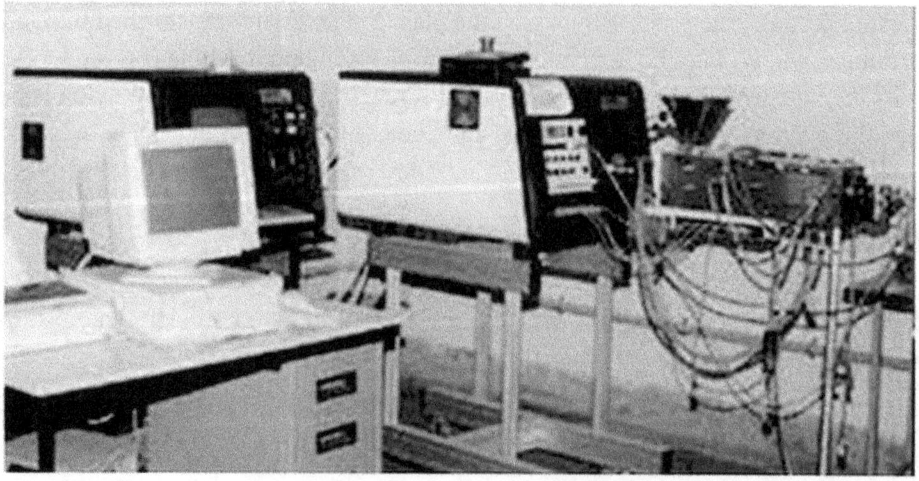

Figure 5. Injection moulding machine

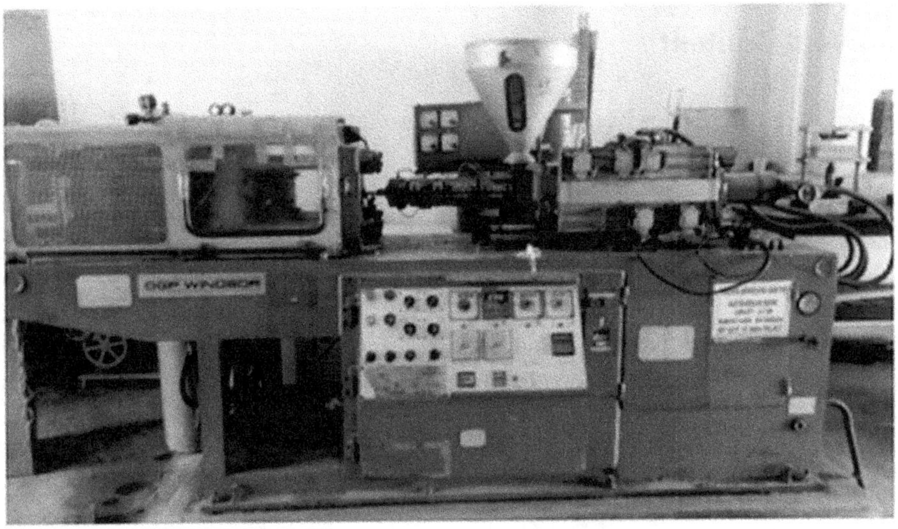

Figure 6. Samples for testing: a: aluminium alloy for an existing calliper; b: composite with nylon and glass filler; c: composite with epoxy and carbon filler

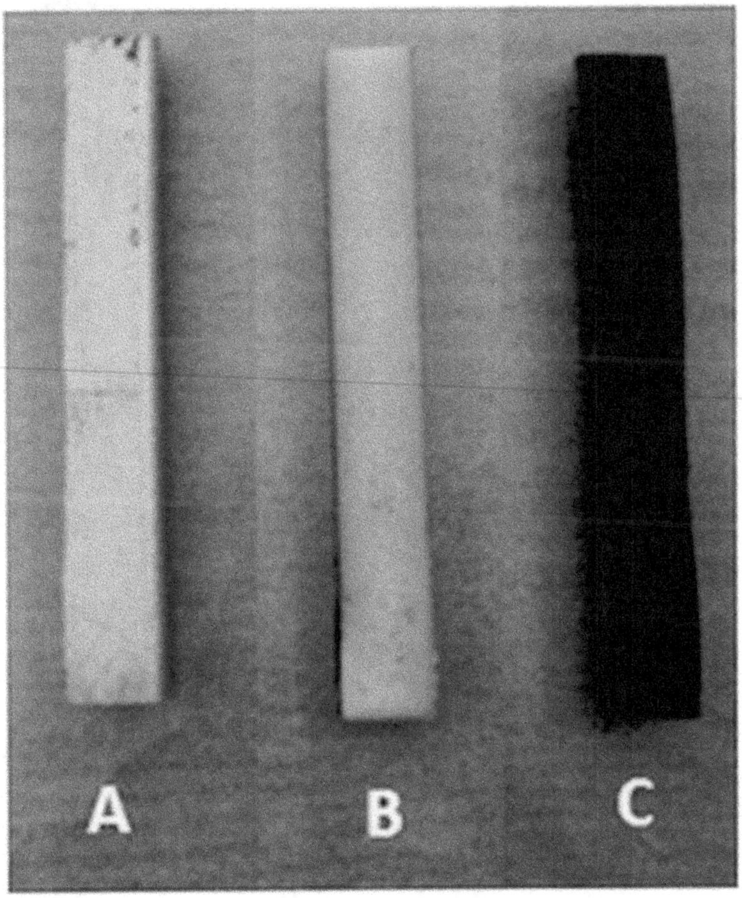

TESTING

Density and Void Content

The density of polymer composites basically depends upon the composition of the composite. It is basically defined as the ratio of mass to the unit volume of a substance. The theoretical densities of composite material would be calculated by using weight additive principle equation and the actual density of composite would be determined experimentally by using simple water immersion method with the help of Mettler Toledo electronic balance (Figure 7). The difference in the theoretical and experimental result would indicate the void. Void content, in the fabricated composites, is an important physical property which can significantly affect the physical and mechanical properties of composite material. Hence estimation of void content is important to evaluate the quality of composites.

Figure 7. Mettler toledo electronic balance

Tensile and Flexural Tests

Universal testing machine (UTM) is used to test the tensile and flexural behavior (Tensile strength, tensile modulus, flexural strength and flexural modulus) of composite materials. The mechanical tests are carried out using Instron universal testing machine (Instron Ltd, UK) (Figure 8). The tensile tests are conducted according to ASTM D 638. The specimens for mechanical tests are of the shape of a rectangular bar of size 100 X 12 X 8 mm. The flexural or 3 point bending tests are undertaken as per ASTM D 790 under the same condition as that of the tensile tests.

Hardness Test

Hardness is the resistance of a material to deformation, indentation or scratching. It is required to ascertain the usage of the composite as a building material for the calipers. It and also reveals the dimensional stability of a composite. The indentation value has high importance for technical applications reflecting

Figure 8. Instron universal testing machine
Source: Roy et al., 2016

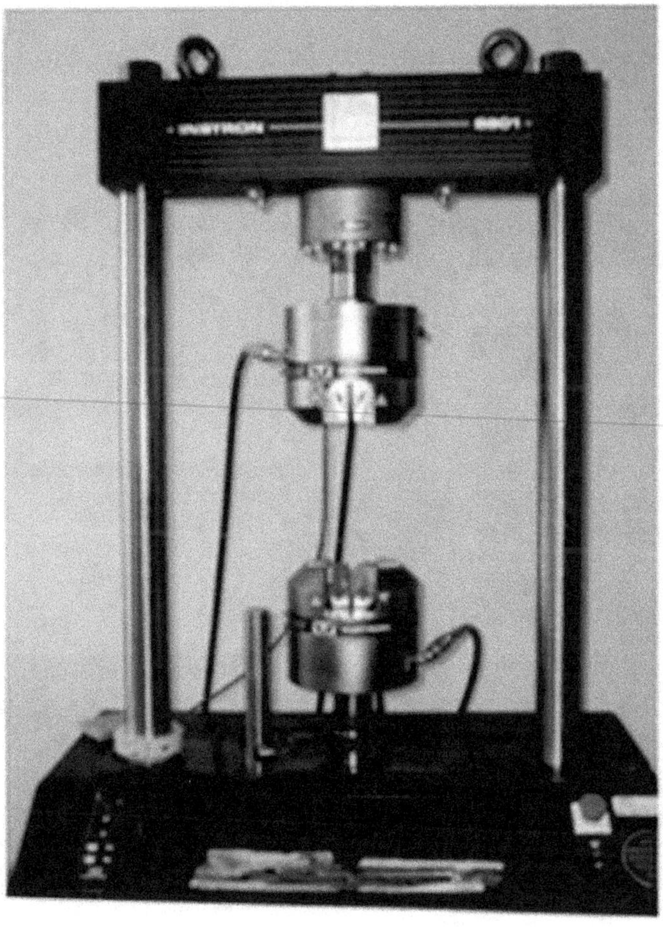

the resistance to permanent deformation, which not only decides modulus, strength, elasticity, plasticity but also provides dimensional stability of a material.

In this study, Vicker's hardness test setup is used to find out the micro-hardness value of polymer composites. The instrument used is a UHL micro hardness tester (Model - VMHT MOT, Sl. No. 1002001, Technische Mikroskopie) (Figure 9) with a Vicker's diamond indenter. The dwell time and speed of indentation are kept constant and a constant load is applied.

Morphological Study

It is important to study a composite material which is newly developed to ensure the structure; chemical composition, distribution etc. are exactly similar as desired. The microstructure characterization of a developed material helps one to quickly understand the reasons for variation of its properties, physical behavior, distribution of fillers into the matrix and performance under certain investigation, hence these techniques are in demand to study and characterize materials. Microstructure characterization of a material is mostly studied by using Scanning Electron Microscopy (SEM) images and it is very useful in order to observe the microstructure in detail.

Scanning electron microscopy examinations are undertaken for the composite material to observe the filler distribution in the matrix. The surfaces of the composite is coated with a thin platinum film on the surface by sputtering and the images are observed in a JEOL (model JSM 6390LV, Japan) microscope (Figure 10).

Figure 9. Vicker's micro hardness tester
Source: Roy et al., 2016

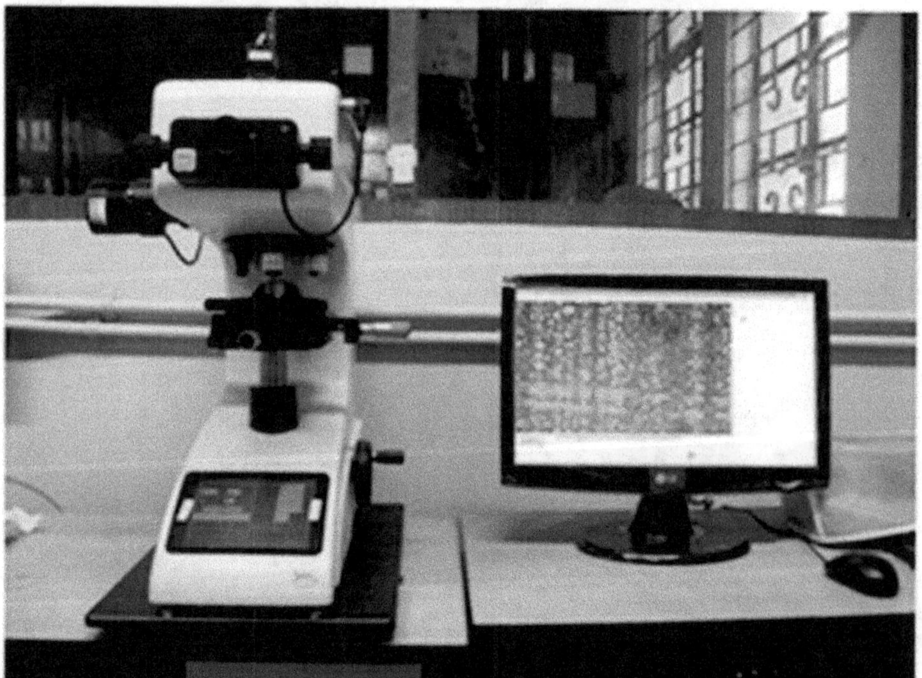

Figure 10. Scanning electron microscope
Source: Roy et al., 2016

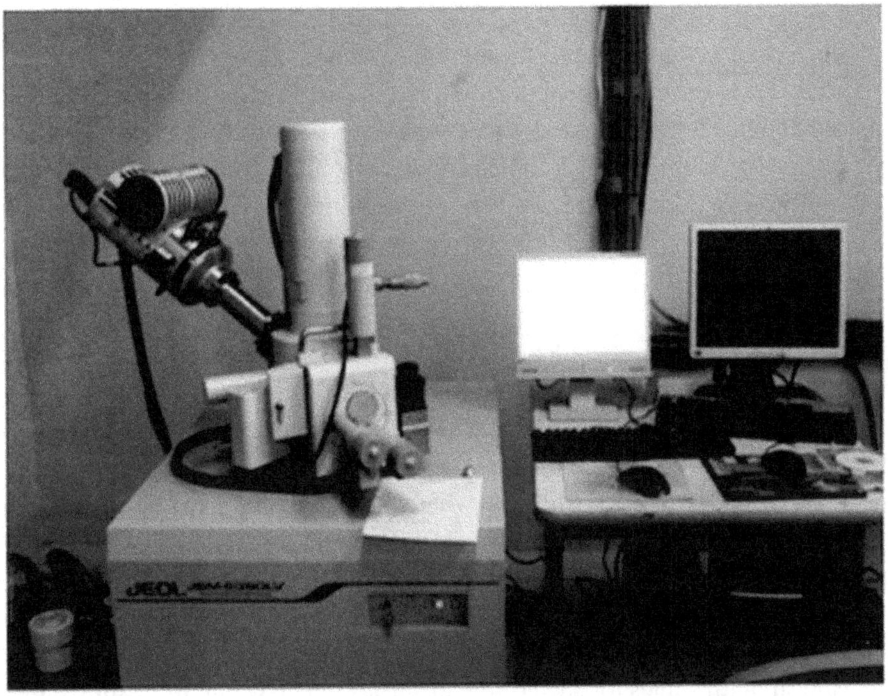

RESULTS AND DISCUSSION

Density and Void Content

Density and void content is very important physical properties of polymeric based composites materials to estimate the characteristic of composites. The experiment is being carried on Mettler Toledo electronic balance by using simple water immersion technique and experimental density is found. Theoretical density is found out by weight additive principle. Void content for the polymeric based composites is determined by taking into consideration the experimental and theoretical data. It is seen that the experimental data is lower than the theoretical indicating presence of voids at the surface forming the common boundary between the composites. It can be seen from the Table 1 that the density of Aluminum alloy is higher than the nylon and epoxy composites. Theoretical value of density of Aluminum is not present because

Table 1. Density and void content of composites and aluminum alloy

Material of Construction	Density (g/cc)		Void fraction (%)
	Experimental	Theoretical	
Aluminium	2.705	--	--
Thermoset Based Composite	1.088	1.144	4.89
Thermoplast Based Composite	1.319	1.378	4.28

it has been taken for reference purpose only. The experimental value obtained is found out after performing EDX of the presently used aluminum based calipers. Here, the only concerned about the void fraction was of the thermoset and thermoplast based composites. The Void content is not zero but less than 5 percent which is an acceptable level for all industrial and household products. The void present in the composites are due to the entrapment of air bubbles during mixing and fabrication of composites. Thermoplast based composites have less void content when compared to thermoset based composite because of the fact that thermoplast based composite was injection molded (processed at high temperature) whereas thermoset based composite were compression molded (processed at high temperature).

Tensile and Flexural Tests

The tensile test was undertaken in accordance with ASTM D 638 where the specimens are placed in the grips and are pulled until failure. The tensile tests are performed with the maximum load capacity of 1 kN with 50 mm gauge length and cross head speed of 2.5 mm / min. The data of tensile modulus and tensile strength are recorded on a computer attached to the machine setup. The flexural or 3 point bending tests was performed as per ASTM D 790 where the test was initiated by applying the load to the specimen at the specified rate until rupture of specimen. The tests are carried out with a span length of 50 mm and cross head speed of 2.5 mm / min and the flexural modulus and flexural strength are recorded from the tests. The mechanical tests are carried out for minimum of three samples for each composition and the average results are reported. The plots of the various values are depicted in the graphs. Figure 11 shows the tensile strength values for each of the materials whereas Figure 12 illustrates avariation in tensile modulus for the same. Figure 13 reveals the flexural strength data whereas Figure 14 indicates the flexural modulus. From all the above graphs it is clear that the mechanical strength of thermoplast based composite (Nylon + Glass) is minimum and for thermoset based composite (Epoxy + Carbon) values are higher than both Aluminum alloy and thermoplast based composite. This can be attributed to the high strength of the carbon reinforcement in epoxy. This indicates that the Aluminum alloy made calipers can be fabricated using thermoset composite.

Figure 11. Comparison of tensile strength with 5% error bars

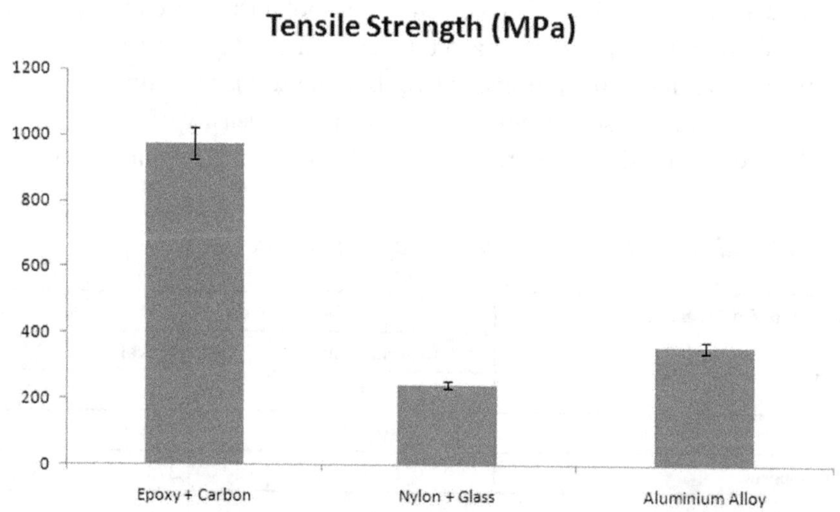

Figure 12. Tensile modulus of proposed and currently used materials with 5% error bars

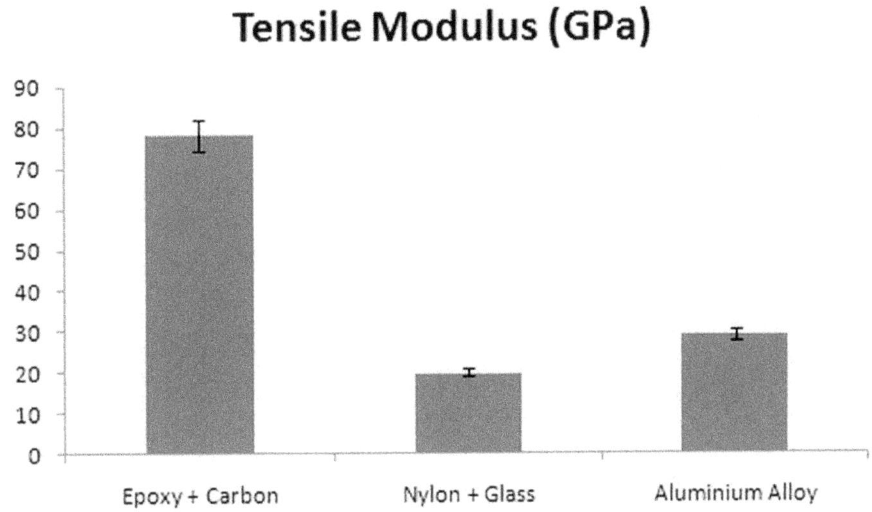

Figure 13. Flexural strength with 5% error bars

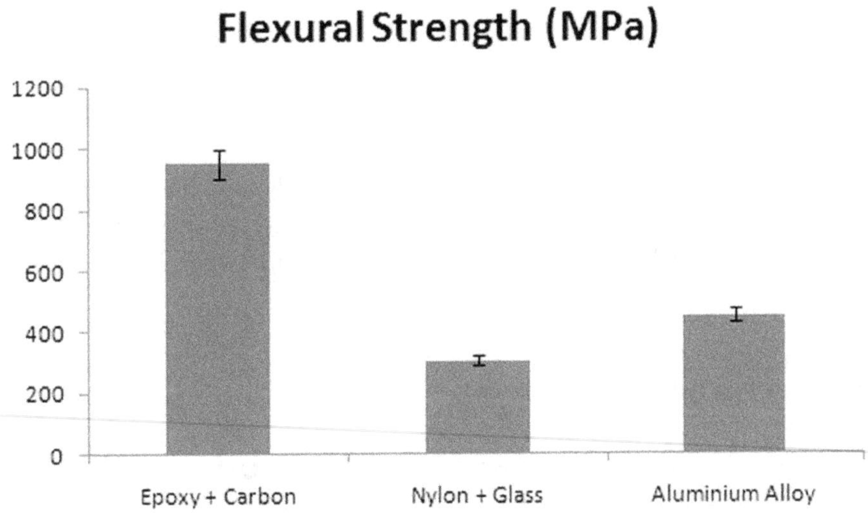

Hardness Test

Hardness is generally classified into three categories with respect to the depth of indentation (d) as follows: (a) Nano-hardness (d <1 μm), (b) Micro-hardness (d = 1-50 μm) and (c) Macro-hardness (d > 50 μm). The test methods commonly used for expressing the relationship between hardness and the size of impression are Brinell, Vicker's and Rockwell hardness tests. In this work, Vicker's hardness test setup is used to find out the micro-hardness value of all three products. The dwell time is kept at 10s while the speed of indentation is set at 50μm/s and indentation load at 100gf. The micro hardness tester is controlled via. a touch screen based system which is part of the tester. The hardness numbers

Figure 14. Flexural modulus of the composites and aluminum alloy with 5% error bars

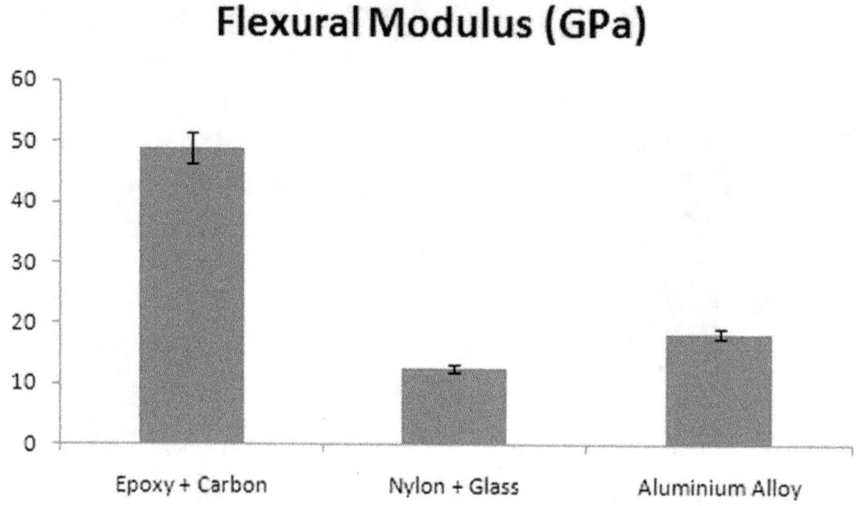

are obtained by processing the indention image. An average of at least three hardness values for each sample is recorded. The plotted graph of recorded data is shown in Figure 15. The trend replicates the tensile and flexural testing results.

Morphological Study

SEM images have been used to investigate the dispersion morphology and the interactions between matrix and filler for both thermoset based and thermoplast based composites. From the figures (Figure 16) it can be visualized that the dispersion and the interaction of filler with the matrix is evenly spread.

Figure 15. Vicker's hardness numbers for the tested samples with 5% error bars

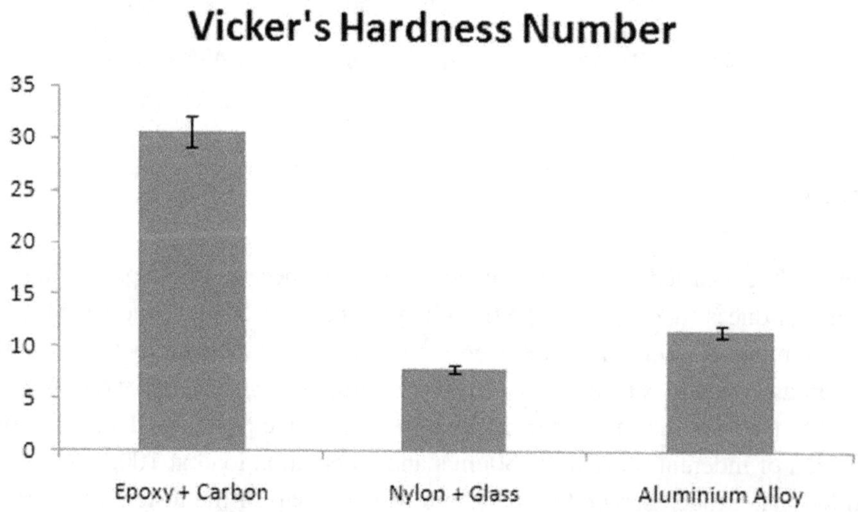

Figure 16. Scanning Electron Microscopy Images: (a) Composite with Nylon and glass filler; (b): Aluminum Alloy for an existing calliper; (c) Composite with Epoxy and carbon filler

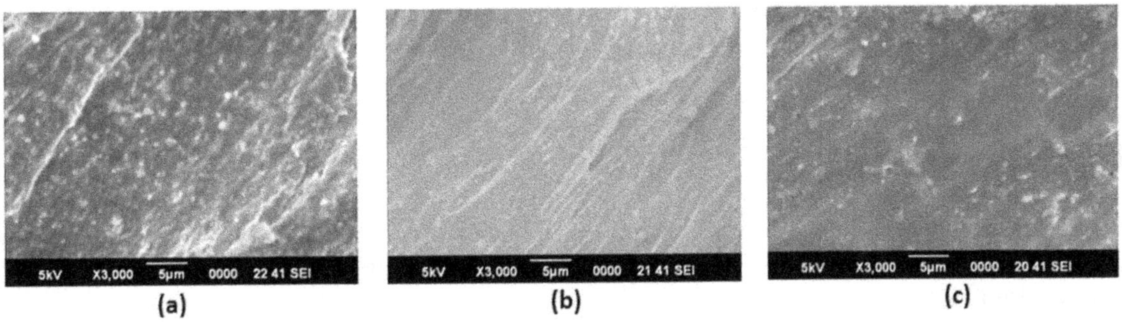

CONCLUSION

In the present work thermoplast based composite and thermoset based composite are considered as an alternative material for the fabrication of an orthotic calipers. Thermoset based Composite samples were made by the intermixture of carbon fillers (30%) with Epoxy resin at room temperature and casting the same in a mould. The mould was placed in a Carver Pressat a pressure of 5T. The thermoplast based composite samples were prepared by mixing Nylon (matrix material) and 30% glass filler in a single screw extruder and then the granules are subjected to injection molding process by melting and forcing composite into the closed mold. The samples were obtained from the mould. The mechanical behavior of the proposed composites are evaluated and compared with Aluminum alloy, existing material of the product. It was observed that mechanical properties such as tensile strength, tensile modulus, flexural strength, flexural modulus and hardness of thermoplast based composite (Nylon + Glass) were minimum and for thermoset based composite (Epoxy + Carbon) values was higher than both Aluminum alloy and thermoplast based composite. In order to ensure the quality of the prepared composites void test and morphological tests were also conducted. They indicated very little presence of voids and uniform distribution of the filler in matrix respectively. It can be concluded that the Aluminium alloy made calipers can be fabricated using thermoset composite. But before going for commercialization it is required that other parameters such as buckling, manufacturability, wear etc. are evaluated.

FUTURE WORK

Orthotic calipers are subjected to adverse climatic conditions so they need to undergo such as tribological testing and dynamic mechanical testing. For actual usage of the calipers it should be completely fabricated and is required to be clinically tested before replacing (aluminum alloy based) with thermoset (Epoxy + Carbon) polymeric based composites.

ACKNOWLEDGMENT

The authors sincerely acknowledge the comments and suggestions of the reviewers that have been instrumental for improving and upgrading the paper in its final form.

REFERENCES

Ala-Kleme, S., Kivikytö-Reponen, P., Liimatainen, J., Hellman, J., & Hannula, S. P. (2006). Abrasive wear properties of metal matrix composites produced by hot isostatic pressing. *Proc. Estonian Acad. Sci. Eng.*, *12*(4), 445–454.

Ansar, M., Xinwei, W., & Chouwei, Z. (2011). Modeling strategies of 3D woven composites: A review. *Composite Structures*, *93*(8), 1947–1963. doi:10.1016/j.compstruct.2011.03.010

Arpitha, G. R., Sanjay, M. R., Naik, L., & Yogesha, B. (2014). Mechanical Properties of Epoxy Based Hybrid Composites Reinforced with Sisal/SIC/Glass Fibers. *International Journal of Engineering Research and General Science*, *2*(5), 398–405.

Ashik, K. P., & Sharma, R. S. (2015). A Review on Mechanical Properties of Natural Fiber Reinforced Hybrid Polymer Composites. *Journal of Minerals and Materials Characterization and Engineering.*, *3*(05), 420–426. doi:10.4236/jmmce.2015.35044

Bader, M. G. (2002). Selection of composite materials and manufacturing routes for cost-effective performance. *Composites. Part A, Applied Science and Manufacturing*, *33*(7), 913–934. doi:10.1016/S1359-835X(02)00044-1

Berwick, A. (1971). The Short Leg Brace For Ankle Foot Stability In Adults. *The Australian Journal of Physiotherapy*, *17*(3), 81–84. doi:10.1016/S0004-9514(14)61110-7 PMID:25025930

Bregman, D. J. J., van der Krogt, M. M., De Groot, V., Harlaar, J., Wisse, M., & Collins, S. H. (2011). The effect of ankle foot orthosis stiffness on the energy cost of walking: A simulation study. *Clinical Biomechanics (Bristol, Avon)*, *26*(9), 955–961. doi:10.1016/j.clinbiomech.2011.05.007 PMID:21723012

Cicenia, E.F., & Hoberman, M. (1957). Braces and Brace Management. *Orthotics and Prosthetics, 11*(44), 43-59.

Deogonda, P., & Chalwa, V. N. (2013). Mechanical Property of Glass Fiber Reinforcement Epoxy Composites. *International Journal of Scientific Engineering and Research.*, *1*(4), 6–9.

El-Sayed, A. A., El-Sherbiny, M. G., Abo-El-Ezz, A. S., & Aggag, G. A. (1995). Friction and wear properties of polymeric composite materials for bearing applications. *Wear*, *184*(1), 45–53. doi:10.1016/0043-1648(94)06546-2

El-Tayeb, N. S. M., & Mostafa, I. M. (1996). The effect of laminate orientations on friction and wear mechanisms of glass reinforced polyester composite. *Wear*, *195*(1-2), 186–191. doi:10.1016/0043-1648(95)06849-X

Etcheverry, M., & Barbosa, E. (2012). Glass Fiber Reinforced Polypropylene Mechanical Properties Enhancement by Adhesion Improvement. *Materials (Basel)*, *5*(12), 1084–1113. doi:10.3390/ma5061084

Ferber, R., Davis, I. M., & Williams, D. S. III. (2005). Effect of foot orthotics on rear foot and tibia joint coupling patterns and variability. *Journal of Biomechanics*, *38*(3), 477–483. doi:10.1016/j.jbiomech.2004.04.019 PMID:15652545

Gene^t, F., Schnitzler, A., Mathieu, S., Autret, K., Théfenne, L., Dizien, O., & Maldjian, A. (2010). Orthotic devices and gait in polio patients. *Annals of Physical and Rehabilitation Medicine*, *53*(1), 51–59. doi:10.1016/j.rehab.2009.11.005 PMID:20022835

Gibson, R. F. (2010). A review of recent research on mechanics of multifunctional composite materials and structures. *Composite Structures*, *92*(12), 2793–2810. doi:10.1016/j.compstruct.2010.05.003

Giovacchini, F., Vannetti, F., Fantozzi, M., & Cempinia, M. (2015). A light-weight active orthosis for hip movement assistance. *Robotics and Autonomous Systems*, *73*, 123–134. doi:10.1016/j.robot.2014.08.015

Gromeier, M., & EckardWimmer, E. (1998). Mechanism of Injury-Provoked Poliomyelitis. *Journal of Virology*, *72*(6), 5056–5060. PMID:9573275

Jagannatha, T. D., & Harish, G. (2015). Mechanical properties of carbon/glass fiber reinforced epoxy hybrid polymer composites. *International Journal of Mechanical Engineering and Robotics Research*, *4*(2), 131 – 137.

Kastner, J., Plank, B., Salaberger, D., & Sekelja, J. (2010). Defect and Porosity Determination of Fibre Reinforced Polymers by X-ray Computed Tomography. *2nd International Symposium on NDT in Aerospace 2010*, 1 – 12

Kinsella, M. (2000). Innovative Automotive Applications Based on ZenTron® High Strength Glass Fiber. *Proc.* 45th*Science of Advanced Materials and Process Engineering Series*, *45*, 1979 – 1994.

Kröling, H., Fleckenstein, J., Nubbo, N., Endres, A., Miletzky, F., & Schabel, S. (2014). Non-Woven and Paper Based Epoxy Composites. *Science and Technology DAS PAPIER*, *6*(7), 2 – 5.

Ku, H., Wang, H., Pattarachaiyakoop, N., & Trada, M. (2011). A review on the tensile properties of natural fiber reinforced polymer composites. *Composites. Part B, Engineering*, *42*(4), 856–873. doi:10.1016/j.compositesb.2011.01.010

La Mantia, F. P., & Morreale, M. (2011). Green composites: A brief review. *Composites. Part A, Applied Science and Manufacturing*, *42*(6), 579–588. doi:10.1016/j.compositesa.2011.01.017

Liew, K. M., Lei, Z. X., & Zhang, L. W. (2015). Mechanical analysis of functionally graded carbon nanotube reinforced composites: A review. *Composite Structures*, *120*, 90–97. doi:10.1016/j.compstruct.2014.09.041

Liu, D., Tang, Y., & Cong, W. L. (2012). A review of mechanical drilling for composite laminates. *Composite Structures*, *94*(4), 1265–1279. doi:10.1016/j.compstruct.2011.11.024

Meredith, J., Bilson, E., & Powe, R. (2015). A performance versus cost analysis of prepreg carbon fibre epoxy energy absorption structures. Composite Structures, 124, 206–213.

Mitra, B. C. (2014). Environment Friendly Composite Materials: Biocomposites and Green Composites. *Defence Science Journal*, *64*(3), 244–261. doi:10.14429/dsj.64.7323

Moreno, J. C., Rocon de Lima, E., & Andres, F. (2006). Design and implementation of an inertial measurement unit for control of artificial limbs: Application on leg orthoses. *Sensors and Actuators. B, Chemical*, *118*(1-2), 333–337. doi:10.1016/j.snb.2006.04.039

Otheguy, M.E., Gibson, A.G., Robinson, M., Findon, E., Cripps, B., Ochoa Mendoz, A., & Castro, A. (2009). Recycling of end of life thermoplastic composite boats. *Plastics, Rubber and Composites: Macromolecular Engineering, 38*(9-10), 406-411.

Powell, R. M. (1994). Utilization of Natural Fibers in Plastic Composites. *Lignocellulosic Plastic Composites*, *4*(27), 23–51.

Rakib, M. I., Choudhury, I. A., Hussain, S., & Osman, N. A. A. (2015). Design and biomechanical performance analysis of a user-friendly orthotic device. *Materials & Design*, *65*, 716–725. doi:10.1016/j.matdes.2014.09.075

Roy, S., Davim, J. P., Bhowmik, S., & Kumar, K. (2016). Estimation of mechanical and tribological properties of epoxy based green composites. In D. Verma, S. Jain, X. Zhang, & P. C. Gope (Eds.), *Green Approaches to Biocomposite Materials Science and Engineering* (pp. 96–124). Hershey, PA: Engineering Science Reference. doi:10.4018/978-1-5225-0424-5.ch005

Sethi, P. K. (1990). Designing Aids for Physically Handicapped in Developing Countries. *Indian Journal of Physical Medicine and Rehabilitation.*, *3*, 1–10.

Sethi, P. K., Udawat, M. P., Kasliwal, S. C., & Chandra, R. (1978). Vulcanized rubber foot for lower limb amputees. *Prosthetics and Orthotics International*, *2*, 125–136. PMID:733459

Shivakumar, S., & Guggari, G. S. (2011). Literature review of fiber reinforced polymer composites. *International Journal of Advances in Engineering and Technology*, *1*(5), 218–226.

Subash, T., & Pillai, S. N. (2015). Bast fibers reinforced green composites for aircraft indoor structures applications [review]. *Journal of Chemical and Pharmaceutical Sciences.*, *7*, 305–307.

Suresh, K. P., Niranjan, H. B., Martin Jebaraj, P., & Chowdiah, M. P. (2003). Tensile and wear properties of aluminum composites. *Wear*, *255*(1-6), 638–642. doi:10.1016/S0043-1648(03)00292-8

Thorndike, , Murphy, & Staros. (1956). Engineering Applied to Orthopedic Bracing. *Orthotics and Prosthetics*, *10*(4), 55–71.

Yadav, S., Mathur, M. K., Dr, S. C., & Kasliwa, S. C. (1993). Saltiel's Design Modification: Short leg brace for knee extensor weakness. *Indian Journal of Physical Medicine and Rehabilitation.*, *6*, 24–29.

Yu, G.-C., Wu, L., Feng, L., & Yang, W. (2016). Thermal and mechanical properties of carbon fiber polymer-matrix composites with a 3D thermal conductive pathway. *Composite Structures*, *149*, 213–219. doi:10.1016/j.compstruct.2016.04.010

Compilation of References

Abbas, N. K., Abbas, L. K., & Muhameed, S. A. (2013). Effect of Thickness on Structural and Optical Properties of $Zn_xCd_{1-x}S$ Thin Films Prepared by Chemical Spray Pyrolysis. *Int. J. Thin Film Sci. Tec.*, *2*, 127–132. doi:10.12785/ijtfst/020208

Abdelgawad, A. M., Hudson, S. M., & Rojas, O. J. (2014). Antimicrobial wound dressing nanofiber mats from multicomponent (chitosan/silver-NPs/polyvinyl alcohol) systems. *Carbohydrate Polymers*, *100*, 166–178. doi:10.1016/j.carbpol.2012.12.043 PMID:24188851

Abeer, M. M., Amin, M., Iqbal, M. C., & Martin, C. (2014). A review of bacterial cellulose-based drug delivery systems: Their biochemistry, current approaches and future prospects. *The Journal of Pharmacy and Pharmacology*, *66*(8), 1047–1061. PMID:24628270

Abughoush, M., Al-Mahasneh, M., Samhouri, M., Al-Holy, M., & Herald, T. (2008). Formulation and Fuzzy Modeling of Viscosity of an Orange-Flavored Carboxymethylcellulose-Whey Protein Isolate Beverage. *International Journal of Food Engineering*, *4*(7), 12. doi:10.2202/1556-3758.1496

Addis, P. B., Ruan, R. R., Keenan, J. M., & Geleva, D. (2011). Medical and nutritional applications of highly refined cellulose. *United States Patent No 8,026,226, 2*(12).

Addis, P. B., Ruan, R. R., Keenan, J. M., & Geleva, D. (2014). Medical and nutritional applications of highly refined cellulose. *United States Patent No 8,623,841, 2*(12).

Adeyeye, M. C., Jain, A. C., Ghorab, M. K. M., & Reilly, W. J. (2002). Viscoelastic Evaluation of Topical Creams Containing Microcrystalline Cellulose/Sodium Carboxymethyl Cellulose as Stabilizer. *AAPS PharmSciTech*, *3*(2), 16–25. doi:10.1208/pt030208 PMID:12916945

Agag, T., & Takeichi, T. (2011). Synthesis and characterization of benzoxazine resin-SiO_2 hybrids by sol-gel process: The role of benzoxazine-functional silane coupling agent. *Polymer*, *52*(13), 2757–2763. doi:10.1016/j.polymer.2011.04.044

Agarwal, T., Gupta, K. A., Alam, S., & Zaidi, M. G. H. (2012). Fabrication and characterization of iron oxide filled polyvinyl pyrrolidone nanocomposites. *International Journal of Composite Materials*, *2*(3), 17–21. doi:10.5923/j.cmaterials.20120203.01

Agool, I. R., Kadhim, K. J., & Hashim, A. (2016). Preparation of (polyvinyl alcohol–polyethylene glycol–polyvinyl pyrrolidinone–titanium oxide nanoparticles) nanocomposites, electrical properties for energy storage and release. *International Journal of Plastics Technology*, *20*(1), 121–127. doi:10.1007/s12588-016-9144-5

Ahmad, F. J., Mittal, G., Jain, G. K., Malhotra, G., Khar, R. K., & Bhatnagar, A. (2009). Nano-salbutamol dry powder inhalation: A new approach for treating broncho-constrictive conditions. *European Journal of Pharmaceutics and Biopharmaceutics*, *71*(2), 282–291. doi:10.1016/j.ejpb.2008.09.018 PMID:18984050

Ahmad, Z., Ansell, M. P., & Smedley, D. (2011). Moisture Absorption Characteristics of Epoxy Based Adhesive Reinforced with CTBN and Ceramic Particles for Bonded-in Timber Connection, Fickian or Non-Fickian Behaviour. *Materials Science and Engineering*, *17*(1), 012011.

Ahmed, N. M., & El-Sabbagh, S. H. (2014). The influence of kaolin and calcined kaolin on SBR composite properties. *Polymer Composites*, *35*(3), 570–580. doi:10.1002/pc.22697

Akinci, A., Sen, S., & Sen, U. (2014). Friction and wear behavior of zirconium oxide reinforced PMMA composites. *Composites. Part B, Engineering*, *56*, 42–47. doi:10.1016/j.compositesb.2013.08.015

Akiyama, M., Yukari, M., Toshihiro, K., Keiko, N., Yasunobu, O., Masayoshi, T., & Naohiro, U. (2007). Preparation of oriented aluminum nitride thin films on polyimide films and piezoelectric response with high thermal stability and flexibility. *Advanced Functional Materials*, *17*(3), 458–462. doi:10.1002/adfm.200600098

Ala-Kleme, S., Kivikytö-Reponen, P., Liimatainen, J., Hellman, J., & Hannula, S. P. (2006). Abrasive wear properties of metal matrix composites produced by hot isostatic pressing. *Proc. Estonian Acad. Sci. Eng.*, *12*(4), 445–454.

Al-bassam, A. A. (1999). Photoconductivity and defect levels in Zn*x*Cd_{1-}*x*Se with (*x*=0.5, 0.55) crystals. *Solar Energy Materials and Solar Cells*, *57*(4), 323–329. doi:10.1016/S0927-0248(98)00154-8

Al-Bayer, R., Zihlif, A., Lahlouh, B., Elimat, Z., & Ragosta, G. (2013). AC electrical and optical characterization of epoxy–Al_2O_3 composites. *Journal of Materials Science Materials in Electronics*, *24*(8), 2866–2872. doi:10.1007/s10854-013-1184-4

Al-Hada, N. M., Saion, E., Talib, Z. A., & Shaari, A. H. (2016). The Impact of Polyvinylpyrrolidone on Properties of Cadmium Oxide Semiconductor Nanoparticles Manufactured by Heat Treatment Technique. *Polymers*, *8*(4), 113. doi:10.3390/polym8040113

Ali, I., & Elleithy, R. (2011). Toughness of HDPE / CaCO. *Composite Structures*, *48*(1), 49–58.

Ali, M. E., & Lamprecht, A. (2014). Spray freeze drying for dry powder inhalation of nanoparticles. *European Journal of Pharmaceutics and Biopharmaceutics*, *87*(3), 510–517. doi:10.1016/j.ejpb.2014.03.009 PMID:24657824

Al-Jubory, A. A. (2012). Study of the effect of copper doping on the structural and optical properties of $Cd_{0.7}Zn_{0.3}S$ nanocrystalline thin films prepared by chemical bath deposition. *International Journal of Science and Technology*, *2*, 707–712.

Al-tabakha, M. M. (2010). HPMC Capsules: Current Status and Future Prospects. *Journal of Pharmacy & Pharmaceutical Sciences*, *13*(3), 428–442. doi:10.18433/J3K881 PMID:21092714

Althues, H., Henle, J., & Kaskel, S. (2007). Functional inorganic nanofillers for transparent polymers. *Chemical Society Reviews*, *36*(9), 1454–1465. doi:10.1039/b608177k PMID:17660878

Alvarez-Ros, M. C., Sánchez-Cortés, S., & García-Ramos, J. V. (2000). Vibrational study of the salicylate interaction with metallic ions and surfaces. *Spectrochimica Acta. Part A: Molecular and Biomolecular Spectroscopy*, *56*(12), 2471–2477. doi:10.1016/S1386-1425(00)00328-0 PMID:11075690

Amoozgar, Z., & Yeo, Y. (2012). Recent advances in stealth coating of nanoparticle drug delivery systems. *Wiley Interdisciplinary Reviews: Nanomedicine and Nanobiotechnology*, *4*(2), 219–233. PMID:22231928

Amstad, E., Gopinadhan, M., Holtze, C., Osuji, C. O., Brenner, M. P., Spaepen, F., & Weitz, D. A. (2015). Production of amorphous nanoparticles by supersonic spray-drying with a microfluidic nebulator. *Science*, *349*(6251), 956–960. doi:10.1126/science.aac9582 PMID:26315432

Anand, M., Maruthupandy, M., Kumar, S. P., Ravikumar, S., Jeyaraj, N., & Kumaran, R. (2016). Antimicrobial Potential of Chitosan from Marine Biowastes—A Wealth from Waste Approach. *Journal of Chitin and Chitosan Science*, *4*(1), 55–58. doi:10.1166/jcc.2016.1095

Anatolii, D. P. (2000). Hybrid polymer-inorganic nanocomposites. *Russian Chemical Reviews*, *69*(0), 53–80.

Anna, B., Milena, Z., Katarzyna, P., Piotr, C., & Joanna, R. (2016). Modification of flexible polyurethane foams by the addition of natural origin fillers. *Polymer Degradation & Stability*, *132*, 32–40. doi:10.1016/j.polymdegradstab.2016.05.002

Annakutty, K. S., & Kishore, K. (1988). Flame retardant polyphosphate esters: 1. Condensation polymers of bisphenols with aryl phosphorodichloridates: Synthesis, characterization and thermal studies. *Polymer*, *29*(4), 756–761. doi:10.1016/0032-3861(88)90096-1

Ansar, M., Xinwei, W., & Chouwei, Z. (2011). Modeling strategies of 3D woven composites: A review. *Composite Structures*, *93*(8), 1947–1963. doi:10.1016/j.compstruct.2011.03.010

Anselmo, A. C., & Mitragotri, S. (2014). An overview of clinical and commercial impact of drug delivery systems. *Journal of Controlled Release*, *190*, 15–28. doi:10.1016/j.jconrel.2014.03.053 PMID:24747160

Antonaglia, V., Ferluga, M., & Lucangelo, U. (2016). *Respiratory Mechanics in COPD Patients Who Failed Noninvasive Ventilation. In Noninvasive Mechanical Ventilation* (pp. 287–292). Springer.

Antonietti, M., & Göltner, C. (1997). Superstructures of functional colloids: Chemistry on the nanometer scale. *Angewandte Chemie International Edition*, *36*(9), 910–928. doi:10.1002/anie.199709101

Anton, Steven, & Henry. (2007). A review of power harvesting using piezoelectric materials (2003–2006). *Smart Materials and Structures*, *16*(3), R1.

Aomar, H., Omar, J., Ahmed, T., Mickael, G., Fouad, M., El Miloud, M., & Lan, T. (2016). Effects of cellulose fiber content on physical properties of polyurethane based composites. *Composite Structures*, *135*, 217–223. doi:10.1016/j.compstruct.2015.09.043

AQUALON. (1999). *Sodium Carboxymetylcellulose Physical and Chemical Properties*. Hercules.

AQUALON. (2002). *Ethylcellulose (EC) Physical and Chemicals Properties*. Retrieved from http://www.ashland.com/industries/pharmaceutical/oral-solid-dose/aqualon-ethylcellulose

Araki, J., Yamanaka, Y., & Ohkawa, K. (2012). Chitin-chitosan nanocomposite gels: Reinforcement of chitosan hydrogels with rod-like chitin nanowhiskers. *Polymer Journal*, *44*(7), 713–717. doi:10.1038/pj.2012.11

Arancibia, C., Costell, E., & Bayarri, S. (2011). Fat replacers in low-fat carboxymethyl cellulose dairy beverages: Color, rheology, and consumer perception. *Journal of Dairy Science*, *94*(5), 2245–2258. doi:10.3168/jds.2010-3989 PMID:21524514

Arbia, W., Arbia, L., Adour, L., & Amrane, A. (2013). Chitin extraction from crustacean shells using biological methods-a review. *Food Technology and Biotechnology*, *51*(1), 12.

Arbuckle, W. S. (1977). Ice cream. Westport, CT: The AVI. Publishing Inc.

Ariffin, A., Mansor, A. S., & Jikan, S. S. (2008). Mechanical, morphological, and thermal properties of polypropylene / kaolin composite. Part I. The effects of surface-treated kaolin and processing enhancement. *Journal of Applied Polymer Science*, *108*(6), 3901–3916. doi:10.1002/app.27950

Ariraman, M., Kumar, R. S., & Alagar, M. (2014). Studies on FMCM-41 reinforced cyanate ester nanocomposites for low k applications. *RSC Advances*, *4*(101), 57759–57767. doi:10.1039/C4RA09399B

Ariraman, M., Kumar, R. S., & Alagar, M. (2015). Design of cyanate ester/azomethine/ZrO_2 nanocomposites high-k dielectric materials by single step sol-gel approach. *Journal of Applied Polymer Science*, *131*(24), 41097–41102.

Arlbrandt, S., Bold, P., Butcher, A., Carlsson, J. F., Cygan, P., & Ehnebom, J. et al.. (2016). *Highly Selective Inhibitors Of PI3Kγ For The Oral Treatment Of COPD C35. In Innate Immune Mechanisms in COPD* (pp. A4901–A4901). Am Thoracic Soc.

Arnott, S., Chandrasekaran, R., Hall, I. H., & Puigjaner, L. C. (1983). Heteronomous DNA. *Nucleic Acids Research*, *11*(12), 4141–4155. doi:10.1093/nar/11.12.4141 PMID:6866768

Arnott, S., Chandrasekaran, R., Hall, I. H., Puigjaner, L. C., Walker, J. K., & Wang, M. (1983, January). DNA secondary structures: Helices, wrinkles, and junctions. *Cold Spring Harbor Symposia on Quantitative Biology*, *47*(0), 53–65. doi:10.1101/SQB.1983.047.01.008 PMID:6574857

Arnott, S., Chandrasekaran, R., Puigjaner, L. C., Walker, J. K., Hall, I. H., Birdsall, D. L., & Ratliff, R. L. (1983). Wrinkled DNA. *Nucleic Acids Research*, *11*(5), 1457–1474. doi:10.1093/nar/11.5.1457 PMID:6572358

Arnott, S., & Wonacott, A. J. (1966). The refinement of the crystal and molecular structures of polymers using X-ray data and stereochemical constraints. *Polymer*, *7*(4), 157–166. doi:10.1016/0032-3861(66)90009-7

Arpitha, G. R., Sanjay, M. R., Naik, L., & Yogesha, B. (2014). Mechanical Properties of Epoxy Based Hybrid Composites Reinforced with Sisal/SIC/Glass Fibers. *International Journal of Engineering Research and General Science*, *2*(5), 398–405.

Aryal, S., Diaz-Guzman, E., & Mannino, D. M. (2013). COPD and gender differences: An update. *Translational Research; the Journal of Laboratory and Clinical Medicine*, *162*(4), 208–218. doi:10.1016/j.trsl.2013.04.003 PMID:23684710

Aryal, S., Hu, C.-M. J., Fu, V., & Zhang, L. (2012). Nanoparticle drug delivery enhances the cytotoxicity of hydrophobic–hydrophilic drug conjugates. *Journal of Materials Chemistry*, *22*(3), 994–999. doi:10.1039/C1JM13834K

Asakura, T., Iwadate, M., Demura, M., & Williamson, M. P. (1999). Structural analysis of silk with 13 C NMR chemical shift contour plots. *International Journal of Biological Macromolecules*, *24*(2), 167–171. doi:10.1016/S0141-8130(98)00082-8 PMID:10342761

Ashik, K. P., & Sharma, R. S. (2015). A Review on Mechanical Properties of Natural Fiber Reinforced Hybrid Polymer Composites. *Journal of Minerals and Materials Characterization and Engineering.*, *3*(05), 420–426. doi:10.4236/jmmce.2015.35044

ASHLAND. (2014). *Improved Mouthfeel in Low-juice and Reduced-calorie Beverages.* Food Technology Report. Retrieved from http://www.ashland.com/Ashland/Static/Documents/ASI/Food/FTR_020_Low_Juice_Reduced_Calorie_Beverages.pdf

Ashokkumar, M., & Muthukumaran, S. (2013). Size Dependent Structural and Optical Properties of $Cd_{0.9}Zn_{0.1}S$ Thin Films. *Physics Procedia*, *49*, 137–144. doi:10.1016/j.phpro.2013.10.020

Asogwa, P. U. (2010). Variation of optical properties with post deposition annealing in chemically deposited CdZnS thin films. *Chalcogenide Letters*, *7*, 501–508.

Asyadi, F., Jawaid, M., Hassan, A., & Wahit, M. U. (2013). Mechanical Properties of Mica-Filled Polycarbonate/Poly (Acrylonitrile-Butadiene-Styrene) Composites. *Polymer-Plastics Technology and Engineering*, *52*(7), 727–736. doi:10.1080/03602559.2012.762672

Atkinson, G. M., Pearson, R. E., Ounaies, Z., Park, C., Harrison, J. S., & Midkiff, J. A. (2003). Piezoelectric polyimide tactile sensors. *Proc. 15th Biennial University/Government/ Industry Microelectronics Symp*, 308-311. doi:10.1109/UGIM.2003.1225750

Atzori, L., Iera, A., & Morabito, G. (2010). The internet of things: A survey. *Computer Networks*, *54*(15), 2787–2805. doi:10.1016/j.comnet.2010.05.010

Auld, B. A. (1973). *Acoustic fields and waves in solids*. Рипол Классик.

Aurrekoetxea, J., Sarrionandia, M., & Gomez, X. (2008). Effects of microstructure on wear behaviour of wood reinforced polypropylene composite. *Wear*, *265*(5), 606–611. doi:10.1016/j.wear.2007.12.013

Axen, N., Hutchings, I. M., & Jacobson, S. (1996). A model for the friction of multiphase materials in abrasion. *Tribological International*, *29*(6), 467–475. doi:10.1016/0301-679X(95)00104-C

Azizah, S., Sian, M. S., Imran, M. I., & Qumrul, A. (2015). Preparation of rubber wood sawdust-based activated carbon and its use as a filler of polyurethane matrix composites for microwave absorption. *New Carbon Materials*, *30*(2), 167–175. doi:10.1016/S1872-5805(15)60182-2

Baccus, J. B., Jr. (1991). Process for makin a low calorie beverage. *United States Patent No 4,986,994*.

Bader, M. G. (2002). Selection of composite materials and manufacturing routes for cost-effective performance. *Composites. Part A, Applied Science and Manufacturing*, *33*(7), 913–934. doi:10.1016/S1359-835X(02)00044-1

Bae, H., Jung, K.-H., Kang, G.-H., Kim, W. S., Kim, H. S., Hong, M., et al. (2014). Pyranopyran-1, 8-dione, an active compound from Viticis Fructus ameliorates cigarette smoke induced lung inflammatory response in a murine model (CAM5P. 239). *The Journal of Immunology*, *192*(1 Supplement), 180.110-180.110.

Baek, S., Rzchowski, M. S., & Aksyuk, V. A. (2012). Giant piezoelectricity in PMN-PT thin films: Beyond PZT. *MRS Bulletin*, *37*(11), 1022–1029. doi:10.1557/mrs.2012.266

Bahurudeen, A., Wani, K., Basit, M. A., & Santhanam, M. (2015)... *Assesment of Pozzolanic Performance of Sugarcane Bagasse Ash. American Society of Civil Engineers*, *28*(2), 1–11.

Bai, Y., Cheng, Z. Y., Bharti, V., Xu, H. S., & Zhang, Q. M. (2000). High-dielectric-constant ceramic-powder polymer composites. *Applied Physics Letters*, *76*(25), 3804–3806. doi:10.1063/1.126787

Balakrishna, N. S., Ismail, H., & Othman, N. (2014). Polypropylene / Rattan Powder / Kaolin Hybrid Composites: Processing, Mechanical and Thermal Properties. *Polymer-Plastics Technology and Engineering*, *53*(5), 451–458. doi:10.1080/03602559.2013.844833

Ballato, A. (1995). Piezoelectricity: Old effect, new thrusts. *IEEE Transactions on Ultrasonics, Ferroelectrics, and Frequency Control*, *42*(5), 916–926. doi:10.1109/58.464826

Balzar, D. (2002). Report on the size-strain Round Robin. *IUCr-Newsletter*, *14*, 228.

Balzar, D., Audebrand, N., Daymond, M. R., Fitch, A., Hewat, A., Langford, J. I., & Toby, B. H. et al. (2004). Size–strain line-broadening analysis of the ceria round-robin sample. *Journal of Applied Crystallography*, *37*(6), 911–924. doi:10.1107/S0021889804022551

Bamrungsap, S., Zhao, Z., Chen, T., Wang, L., Li, C., Fu, T., & Tan, W. (2012). Nanotechnology in therapeutics: A focus on nanoparticles as a drug delivery system. *Nanomedicine (London)*, *7*(8), 1253–1271. doi:10.2217/nnm.12.87 PMID:22931450

Bancel, S. (2013). *Methods and compositions for targeting agents into and across the blood-brain barrier*. Google Patents.

Ban, S. J., Rico, C. W., Um, I. C., & Kang, M. Y. (2012). Antihyperglycemic and antioxidative effects of hydroxyethyl methylcellulose (HEMC) and hydroxypropyl methylcellulose (HPMC) in mice fed with a high fat diet. *International Journal of Molecular Sciences*, *13*(3), 3738–3750. doi:10.3390/ijms13033738 PMID:22489179

Bansal, M., Khatri, M., & Taneja, V. (2013). Potential role of periodontal infection in respiratory diseases-a review. *Journal of Medicine and Life*, *6*(3), 244.

Barra, G. M., Leyva, M. E., Gorelova, M. M., Soares, B. G., & Sens, M. (2001). X-ray photoelectron spectroscopy and electrical conductivity of polyaniline doped with dodecylbenzenesulfonic acid as a function of the synthetic method. *Journal of Applied Polymer Science*, *80*(4), 556–565. doi:10.1002/1097-4628(20010425)80:4<556::AID-APP1130>3.0.CO;2-N

Basu, G and Roy, A.N. (2008). Blending of Jute with Different Natural Fibres. *Journal of Natural Fibres, 4*(4), 13-29.

Battista, O. A. (1962). Food Composition Incorporation Cellulose Crystallite Aggregates. *United States Patent Office No 3,023,104.*

Bauer, S., Bauer-Gogonea, S., Dansachmuller, M., Graz, I., Leonhartsberger, H., Salhofer, H., & Schwodiauer, R. (2003). Modern electrets. *Ultrasonics, IEEE Symposium on, IEEE,* (1), 370-376.

Bayarri, S., Chuliá, I., & Costell, E. (2010). Comparing λ-carrageenan and an inulin blend as fat replacers in carboxymethyl cellulose dairy desserts. Rheological and sensory aspects. *Food Hydrocolloids*, *24*(6–7), 578–587. doi:10.1016/j.foodhyd.2010.02.004

Bayarri, S., & Costell, E. (2011). Flow Properties of Carboxymethylcellulose Dairy Systems With Different Fat Content. *Journal of Food Process Engineering*, *34*(6), 1903–1914. doi:10.1111/j.1745-4530.2009.00383.x

Baykul, M. C., & Orhan, N. (2010). Band alignment of $Cd_{(1-x)}$ Zn_xS produced by spray pyrolysis method. *Thin Solid Films*, *518*(8), 1925–1928. doi:10.1016/j.tsf.2009.07.142

Beachey, W. (2013). *Respiratory Care Anatomy and Physiology, Foundations for Clinical Practice, 3: Respiratory Care Anatomy and Physiology*. Elsevier Health Sciences.

Behera, P. S. (2012). *Dielectric properties of polymer electrolyte* (Doctoral dissertation).

Behniafar, H., & Amirkhalili, S. K. (2014). Poly(4,4`-oxydiphenylene-pyromellitimide)/TiO_2 nanocomposites with surface modified titanium dioxide using methylene diphenyl diisocyante. *Polymer Bulletin*, *71*(4), 775–785. doi:10.1007/s00289-013-1091-0

Ben Difallah, B., Kharrat, M., Dammak, M., & Monteil, G. (2012). Mechanical and tribological response of ABS polymer matrix filled with graphite powder. *Materials & Design*, *34*, 782–787. doi:10.1016/j.matdes.2011.07.001

Bennett, C. J., & Malits, S. J. (1988). Aqueous table syrup with reduced sugar content. *United States Patent No 4,786,521*, 3–7.

Bernd, P., Beatrix, P., Shin, F. G., Chan, H. L. W., & Choy, C. L. (2000). Pyroelectric activity of ferroelectric PT/PVDF-TRFE. *IEEE Transactions on Dielectrics and Electrical Insulation*, *7*(4), 517–522. doi:10.1109/94.868071

Bernkop-Schnürch, A., & Dünnhaupt, S. (2012). Chitosan-based drug delivery systems. *European Journal of Pharmaceutics and Biopharmaceutics*, *81*(3), 463–469. doi:10.1016/j.ejpb.2012.04.007 PMID:22561955

Berwick, A. (1971). The Short Leg Brace For Ankle Foot Stability In Adults. *The Australian Journal of Physiotherapy*, *17*(3), 81–84. doi:10.1016/S0004-9514(14)61110-7 PMID:25025930

Bhowmick, M., Mukhopadhyay, S., & Alagirusamy, R. (2012). Mechanical properties of natural fibre reinforced composites. *Textile Progress*, *44*(2), 85–140. doi:10.1080/00405167.2012.676800

Biausque, G., Laveille, P., Anjum, D. H., Caps, V., & Basset, J.-M. (2015). *Methods to synthesize NiPt bimetallic nanoparticles by a reversed-phase microemulsion, deposition of NiPt bimetallic nanoparticles on a support, and application of the supported catalyst for CO2 reforming of methane*. Google Patents.

Bienias, J., Walczak, M., Surowska, B., & Sobczak, J. (2003). Microstructure and corrosion behaviour of aluminium fly ash composites. *Journal of Optoelectronics and Advanced Materials*, *5*(2), 493–502.

Bloemsma, L. D., Hoek, G., & Smit, L. (2016). Panel Studies On Acute Effects Of Air Pollution. In *Patients With COPD: A Systematic Review And Meta-Analysis C55. Cardiopulmonary Effects of Air Pollution* (pp. A5420–A5420). Am Thoracic Soc.

Bodvik, R., Dedinaite, A., Karlson, L., Bergström, M., Bäverbäck, P., Pedersen, J. S., & Claesson, P. M. et al. (2010). Aggregation and network formation of aqueous methylcellulose and hydroxypropylmethylcellulose solutions. *Colloids and Surfaces. A, Physicochemical and Engineering Aspects*, *354*(1–3), 162–171. doi:10.1016/j.colsurfa.2009.09.040

Bo-Lin, C., Guo-Wei, L., Wei-Tian, W., Dong-Yi, G., Zheng-Hao, C., & Guo-Zhen, Y. (2005). Nonlinear optical properties of Au/PVP composite thin films. *Chinese Physics*, *14*(9).

Bonham, R. A., & Schafer, L. (1974). *International Tables for X-Ray Crystallography*. Birmingham: Kynoch Press.

Bouclé, J., Ravirajan, P., & Nelson, J. (2007). Hybrid polymer–metal oxide thin films for photovoltaic applications. *Journal of Materials Chemistry*, *17*(30), 3141–3153. doi:10.1039/b706547g

Braido, F., Baiardini, I., Scichilone, N., Sorino, C., Di Marco, F., Corsico, A., & Canonica, G. W. et al. (2015). Disability in moderate chronic obstructive pulmonary disease: Prevalence, burden and assessment-results from a real-life study. *Respiration*, *89*(2), 100–106. doi:10.1159/000368365 PMID:25612914

Bregman, D. J. J., van der Krogt, M. M., De Groot, V., Harlaar, J., Wisse, M., & Collins, S. H. (2011). The effect of ankle foot orthosis stiffness on the energy cost of walking: A simulation study. *Clinical Biomechanics (Bristol, Avon)*, *26*(9), 955–961. doi:10.1016/j.clinbiomech.2011.05.007 PMID:21723012

Breit, P. B. J., Lembke, F., Dubose, D., Burke, M., Muskie-Dukes-Driggs, A., Dobry, D., ... (2013). *A Dry-Powder Inhalable Formulation of Nanocrystalline Camptothecin for the Treatment of Lung Cancer.* Paper presented at the Journal of Aerosol Medicine and Pulmonary Drug Delivery.

Brennan, L. B., Isaac, D. H., & Arnold, J. C. (2002). Recycling of acrylonitrile–butadiene–styrene and high-impact polystyrene from waste computer equipment. *Journal of Applied Polymer Science*, *86*(3), 572–578. doi:10.1002/app.10833

Brown, C. J., & Houghton, A. A. (1941). The chemical and physical properties of carboxymethylcellulose and its salts. *Journal of the Society of Chemical Industry*, *60*(10), 254–258. doi:10.1002/jctb.5000601004

Brubaker, C. L., Bourland, F. M., & Wendel, J. E. (1999). The origin and domestication of cotton. In C. W. Smith & J. T. Cothren (Eds.), *Cotton: Origin, History, Technology, and Production* (p. 3). New York: John Wiley and Sons, Inc.

Brzoska, M., Langer, K., Coester, C., Loitsch, S., Wagner, T. O. F., & Mallinckrodt, C. (2004). Incorporation of biodegradable nanoparticles into human airway epithelium cells—in vitro study of the suitability as a vehicle for drug or gene delivery in pulmonary diseases. *Biochemical and Biophysical Research Communications*, *318*(2), 562–570. doi:10.1016/j.bbrc.2004.04.067 PMID:15120637

Buchberger, G., Schwödiauer, R., & Bauer, S. (2008). Flexible large area ferroelectret sensors for location sensitive touchpads. *Applied Physics Letters, 92*(12), 123511. doi:10.1063/1.2903711

Butty, J., Peyghambarian, N., Kao, Y. H., & Mackenzie, J. D. (1996). Room temperature optical gain in sol-gel derived CdS quantum dots. *Applied Physics Letters, 69*(21), 3224–3226. doi:10.1063/1.118017

Cady, W. G. (1946). *Piezoelectricity: An introduction to the theory and application of electromechanical phenomena in crystals.* McGraw-Hill.

Camargo, P., Satyanarayana, K., & Wypych, F. (2009). Nanocomposites, synthesis, structure, properties and new application opportunities. *Materials Research, 12*(1), 1–39. doi:10.1590/S1516-14392009000100002

Campardelli, R., Cherain, M., Perfetti, C., Iorio, C., Scognamiglio, M., Reverchon, E., & Della Porta, G. (2013). Lipid nanoparticles production by supercritical fluid assisted emulsion–diffusion. *The Journal of Supercritical Fluids, 82*, 34–40. doi:10.1016/j.supflu.2013.05.020

Carvalho, S. R. (2013). *Improved inhalation therapies of brittle powders.* Academic Press.

Cary, B., Daniel, J. A., Deepam, M., Shashank, P., & Walter, V. (2014). Advances in Piezoelectric Polymer Composites for Vibrational Energy Harvesting. In Polymer Composites for Energy Harvesting, Conversion, and Storage. American Chemical Society.

Cascone, M. G. (1997). Dynamic–mechanical properties of bioartificial polymeric materials. *Polymer International, 43*(1), 55–69. doi:10.1002/(SICI)1097-0126(199705)43:1<55::AID-PI762>3.0.CO;2-#

Castro, C., Zuluaga, R., Álvarez, C., Putaux, J.-L. L., Caro, G., Rojas, O. J., & Gañán, P. et al. (2012). Bacterial cellulose produced by a new acid-resistant strain of Gluconacetobacter genus. *Carbohydrate Polymers, 89*(4), 1033–1037. doi:10.1016/j.carbpol.2012.03.045 PMID:24750910

Castro, C., Zuluaga, R., Putaux, J. L., Caro, G., Mondragon, I., & Gañán, P. (2011). Structural characterization of bacterial cellulose produced by Gluconacetobacter swingsii sp. from Colombian agroindustrial wastes. *Carbohydrate Polymers, 84*(1), 96–102. doi:10.1016/j.carbpol.2010.10.072

Cavanagh, T., Barman, S. P., Hao, T., Leland, T. B., & Thekkedath, R. V. (2013). *Preparations of hydrophobic therapeutic agents, methods of manufacture and use thereof.* Google Patents.

Central Electricity Authority (CEA). (2014-15). *Report on fly ash generation at coal/lignite based thermal power stations and its utilization in the country for the year 2014-15.* Retrieved from http://cea.nic.in/reports/others/thermal/tcd/flyash_final_1415.pdf

Chaiyasan, W., Praputbut, S., Kompella, U. B., Srinivas, S. P., & Tiyaboonchai, W. (2017). Penetration of mucoadhesive chitosan-dextran sulfate nanoparticles into the porcine cornea. *Colloids and Surfaces. B, Biointerfaces, 149*, 288–296. doi:10.1016/j.colsurfb.2016.10.032 PMID:27771493

Chan, H. K., & Kwok, P. C. L. (2013). Novel particle production technologies for inhalation products. *Inhalation Drug Delivery: Techniques and Products*, 47-62.

Chand, N., Dwivedi, U. K., & Sharma, M. K. (2007). Development and tribological behaviour of UHMWPE filled epoxy gradient composites. *Wear, 262*(1-2), 184–190. doi:10.1016/j.wear.2006.04.012

Chandra, A., Turng, L. S., Gopalan, P., Rowell, R. M., & Gong, S. (2008). Study of utilizing thin polymer surface coating on the nanoparticles for melt compounding of polycarbonate/alumina nanocomposites and their optical properties. *Composites Science and Technology, 68*(3-4), 768–776. doi:10.1016/j.compscitech.2007.08.027

Chandrasekhar. (1999). *Conducting Polymers, Fundamentals and Applications: A Practical Approach*. Kluwer Academic.

Chang, B. P., Akil, H. M., & Nasir, R. B. M., Bandara, & Rajakpase, S. (2014). The effect of ZnO nanoparticles on the mechanical, tribological and antibacterial properties of ultra-high molecular weight polyethylene. *Journal of Reinforced Plastics and Composites*.

Chang, B. P., Akil, H. M., Nasir, R. M., & Nurdijati, S. (2011). Mechanical and antibacterial properties of treated and untreated zinc oxide filled UHMWPE composites. *Journal of Thermoplastic Composite Materials*, *24*(5), 653–667. doi:10.1177/0892705711399848

Chang, B. P., Md Akil, H., & Bt. Md Nasir, R. (2013). Comparative study of micro-and nano-ZnO reinforced UHMWPE composites under dry sliding wear. *Wear*, *297*(1), 1120–1127. doi:10.1016/j.wear.2012.11.083

Chapman, J. (2013). Therapeutic insight: Chronic obstructive pulmonary disease: Insidious, debilitating and costly. *AJP: The Australian Journal of Pharmacy*, *94*(1116), 78.

Charlson, E. S., Bittinger, K., Chen, J., Diamond, J. M., Li, H., Collman, R. G., & Bushman, F. D. (2012). Assessing bacterial populations in the lung by replicate analysis of samples from the upper and lower respiratory tracts. *PLoS ONE*, *7*(9), e42786. doi:10.1371/journal.pone.0042786 PMID:22970118

Chattopadhyay, K. K., & Banerjee, A. N. (2014). *Introduction to Nanoscience and Nanotechnology*. Delhi, India: PHI Learning Private Limited.

Chau, J. L. H., Tung, C. T., Lin, Y. M., & Li, A. K. (2008). Preparation and optical properties of titania/epoxy nanocomposite coatings. *Materials Letters*, *62*(19), 3416–3418. doi:10.1016/j.matlet.2008.02.058

Chen, L., Ruan, R. R., Johnson, J. E., Addis, P. B., Xu, L., & Yi, L. (1998). Cellulose fiber based compositions and film and the process for their manufacture. *United States Patent No 5,817,381*.

Chen, A., Kai-Lun, L., Chien-Chong, H., Tong, M. L., Jiann, S., & Szu, H. C. (2012). Flexible tactile sensors based on nanoimprinted sub-20 NM piezoelectric copolymer nanograssfilms. In *Sensors* (pp. 1–4). IEEE.

Chen, C. H., Chen, W. J., Chen, M. H., & Li, Y. M. (1999). Simultaneous full-inter penetrating polymer networks of blocked polyurethane and vinyl ester. II. Static and dynamic mechanical properties. *Journal of Applied Polymer Science*, *71*(12), 1977–1985. doi:10.1002/(SICI)1097-4628(19990321)71:12<1977::AID-APP7>3.0.CO;2-F

Chen, C., Zhong, M., Li, G., Yang, F., Huang, R., Xiao, W., & Ding, P. (2016). Optimization on preparation conditions of calcium-crosslinked modified chitosan as potential matrix material for theophylline sustained-release beads and its evaluation of release kinetics. *Journal of Alloys and Compounds*, *658*, 348–355. doi:10.1016/j.jallcom.2015.10.219

Chen, D., Qiao, X., Qiu, X., & Chen, J. (2009). Synthesis and electrical properties of uniform silver nanoparticles for electronic applications. *Journal of Materials Science*, *44*(4), 1076–1081. doi:10.1007/s10853-008-3204-y

Cheng, Z., Al Zaki, A., Hui, J. Z., Muzykantov, V. R., & Tsourkas, A. (2012). Multifunctional nanoparticles: Cost versus benefit of adding targeting and imaging capabilities. *Science*, *338*(6109), 903–910. doi:10.1126/science.1226338 PMID:23161990

Chen, H. M., & Liu, R. S. (2011). Architecture of metallic nanostructures: Synthesis strategy and specific applications. *The Journal of Physical Chemistry C*, *115*(9), 3513–3527. doi:10.1021/jp108403r

Chen, J., Feng, Y., Chen, N., Li, B., Chen, F., & Zhang, X. (2013). Fabrication and Properties of Silver-Doped Acrylonitrile Butadiene Styrene (ABS) Composites. *Polymer-Plastics Technology and Engineering*, *52*(11), 1081–1088. doi:10.1080/03602559.2013.769574

Chen, M.-M., Huang, Y.-Q., Cao, H., Liu, Y., Guo, H., Chen, L. S., & Zhang, Q.-Q. et al. (2015). Collagen/chitosan film containing biotinylated glycol chitosan nanoparticles for localized drug delivery. *Colloids and Surfaces. B, Biointerfaces*, *128*, 339–346. doi:10.1016/j.colsurfb.2015.02.024 PMID:25784300

Chen, R., Jakes, K. A., & Foreman, D. W. (2004). Peak-fitting analysis of cotton fiber powder X-ray diffraction spectra. *Journal of Applied Polymer Science*, *93*(5), 2019–2024. doi:10.1002/app.20666

Chen, S. N., Wang, G. J., & Chien, M. C. (2006). Analytical modeling of piezoelectric vibration-induced micro power generator. *Mechatronics*, *16*(7), 379–387. doi:10.1016/j.mechatronics.2006.03.003

Chen, X. D., Wang, X. B., Liu, L., Yang, D. C., & Fan, L. (2005). Functionalized semiconductor nanocrystals for ultra-sensitive detection of peptides. *Analytica Chimica Acta*, *542*(2), 144–150. doi:10.1016/j.aca.2005.03.071

Chen, Y., Ji, X., Jiang, S., Sun, Q., & Jiang, B. (2003). Synthesis and characterization of CdS nanocrystals in poly(styrene-co-maleic anhydride) copolymer. *Colloid & Polymer Science*, *281*(4), 386–389. doi:10.1007/s00396-002-0750-8

Chen, Z., Li, Z., Lin, Y., Yin, M., Ren, J., & Qu, X. (2013). Bioresponsive Hyaluronic Acid-Capped Mesoporous Silica Nanoparticles for Targeted Drug Delivery. *Chemistry (Weinheim an der Bergstrasse, Germany)*, *19*(5), 1778–1783. doi:10.1002/chem.201202038 PMID:23303570

Cheow, W. S., & Hadinoto, K. (2010). Enhancing encapsulation efficiency of highly water-soluble antibiotic in poly (lactic-co-glycolic acid) nanoparticles: Modifications of standard nanoparticle preparation methods. *Colloids and Surfaces. A, Physicochemical and Engineering Aspects*, *370*(1), 79–86. doi:10.1016/j.colsurfa.2010.08.050

Chiang, P.-C., Alsup, J. W., Lai, Y., Hu, Y., Heyde, B. R., & Tung, D. (2009). Evaluation of aerosol delivery of nano-suspension for pre-clinical pulmonary drug delivery. *Nanoscale Research Letters*, *4*(3), 254–261. doi:10.1007/s11671-008-9234-1 PMID:20596335

Chiappa, S., Winn, J., Viñuela, A., Tipney, H., & Spector, T. D. (2013). A probabilistic model of biological ageing of the lungs for analysing the effects of smoking, asthma and COPD. *Respiratory Research*, *14*(1), 1. PMID:23721360

Chisholm, N., Mahfuz, H., Rangari, V. K., & Jeelani, S. (2005). Fabrication and mechanical characterization of carbon/SiC-epoxy nanocomposites. *Composite Structures*, *67*(1), 115–124. doi:10.1016/j.compstruct.2004.01.010

Choi, S., & Kim, J. (2013). Thermal conductivity of epoxy composites with binary-particle system aluminium oxide and aluminium nitride filers. *Composites. Part B, Engineering*, *51*, 140–147. doi:10.1016/j.compositesb.2013.03.002

Chong, L. H., Lim, M. M., & Sultana, N. (2014). Polycaprolactone (PCL)/Gelatin (Ge)-based Electrospun Nanofibers for Tissue engineering and Drug Delivery Application. *Applied Mechanics & Materials,* (554).

Chopra, A. K., Marwaha, R. K., Kaushik, D., & Dureja, H. (2015). Box-Behnken Designed Fluconazole Loaded Chitosan Nanoparticles for Ocular Delivery. *Pharmaceutics & Drug Delivery Research, 2014.*

Chopra, K. L., & Das, S. R. (1983). *Thin Film Solar Cells*. Springer Science & Business Media. doi:10.1007/978-1-4899-0418-8

Christensen, R. M. (2005). *Mechanics of composite materials*. New York: Dover Publications.

Chung, Y., & Shyu, Y. (1999). The effects of pH, salt, heating and freezing on the physical properties of bacterial cellulose-nata. *International Journal of Food Science & Technology*, *34*(1), 23–26. doi:10.1046/j.1365-2621.1999.00231.x

Chu, T. L., Chu, S. S., Britt, J., Ferekides, C., & Wu, C. Q. (1991). Cadmium zinc sulfide films and heterojunctions. *Applied Physics (Berlin)*, *70*(5), 2688–2693. doi:10.1063/1.349384

Chynoweth, T. A., & Bube, R. H. (1980). Electrical Transport in ZnCdS Films Deposited by Spray Pyrolysis. *Journal of Applied Physics*, *51*(3), 1844–1846. doi:10.1063/1.327755

Cicenia, E.F., & Hoberman, M. (1957). Braces and Brace Management. *Orthotics and Prosthetics, 11*(44), 43-59.

Coco, R., Plapied, L., Pourcelle, V., Jérôme, C., Brayden, D. J., Schneider, Y.-J., & Préat, V. (2013). Drug delivery to inflamed colon by nanoparticles: Comparison of different strategies. *International Journal of Pharmaceutics*, *440*(1), 3–12. doi:10.1016/j.ijpharm.2012.07.017 PMID:22820482

Coester, C., Kreuter, J., von Briesen, H., & Langer, K. (2000). Preparation of avidin-labelled gelatin nanoparticles as carriers for biotinylated peptide nucleic acid (PNA). *International Journal of Pharmaceutics, 196*(2), 147-149.

Coester, C., Nayyar, P., & Samuel, J. (2006). In vitro uptake of gelatin nanoparticles by murine dendritic cells and their intracellular localisation. *European Journal of Pharmaceutics and Biopharmaceutics*, *62*(3), 306–314. doi:10.1016/j.ejpb.2005.09.009 PMID:16316749

Commission of the European Unions. (1995). Food Additives Other Than Colours and Sweeteners. *European Parlament*, *1*(1), 1–56.

Company, T. D. C. (2015). *Hydroxypropyl Methylcellulose*. Retrieved from http://www.dow.com/dowwolff/en/industrial_solutions/polymers/hydroxypropyl-methylcellulose/

Cook-Chennault, K. A., Thambi, N., & Sastry, A. M. (2008). Powering MEMS portable devices a review of non-regenerative and regenerative power supply systems with special emphasis on piezoelectric energy harvesting systems. *Smart Materials and Structures*, *17*(4), 043001. doi:10.1088/0964-1726/17/4/043001

Couvreur, P. (2013). Nanoparticles in drug delivery: Past, present and future. *Advanced Drug Delivery Reviews*, *65*(1), 21–23. doi:10.1016/j.addr.2012.04.010 PMID:22580334

Cui, C., Baughman, R. H., Iqbal, Z. T., Kazmar, R., & Dahlstrom, D. K. (1997). Improved piezoelectric ceramic/polymer composites for hydrophone applications. *Synthetic Metals*, *85*(1), 1391–1392. doi:10.1016/S0379-6779(97)80287-1

Cybulska, J., Szymańska-Chargot, M., Zdunek, A., Psonka-Antonczyk, K. M., & Stokke, B. T. (2011). Crystallinity and nanostructure of cellulose from different sources. *11th International Congress on Engineering and Food*, 1–4.

DAddio, S. M., Chan, J. G. Y., Kwok, P. C. L., Benson, B. R., Prudhomme, R. K., & Chan, H.-K. (2013). Aerosol delivery of nanoparticles in uniform mannitol carriers formulated by ultrasonic spray freeze drying. *Pharmaceutical Research*, *30*(11), 2891–2901. doi:10.1007/s11095-013-1120-6 PMID:23893019

Daoud, M. T., Abou El-khair, M., Abdelaziz, M., & Rohatgi, P. (2007). Fabrication, microstructure and compressive behavior of ZC63 Mg–microballoon foam composites. *Composites Science and Technology*, *67*(4), 1842–1853. doi:10.1016/j.compscitech.2006.10.023

Das, M. K., & Rao, K. R. (2006). Evaluation of zidovudine encapsulated ethylcellulose microspheres prepared by water-in-oil-in-oil (w/o/o) double emulsion solvent diffusion technique. *Acta Poloniae Pharmaceutica - Drug Research*, *63*(2), 141–148.

Das, S., Pandey, P., Mohanty, S., & Nayak, S. K. (2016). Effect of nanosilica on the physicochemical, morphological and curing characteristics of transesterified castor oil based polyurethane coatings. *Progress in Organic Coatings*, *97*, 233–243. doi:10.1016/j.porgcoat.2016.04.012

Das, S., Pandey, P., Mohanty, S., & Nayak, S. K. (2017). Evaluation of biodegradability of green polyurethane/nanosilicacomposite synthesized from transesterified castor oil and palm oil based isocyanate. *International Biodeterioration & Biodegradation, 117*, 278–288. doi:10.1016/j.ibiod.2017.01.015

Datta, U. (2009). Application of Jute Geotextiles. *Journal of Natural Fibers, 4*(3), 87–92.

De Keijser, T. H., Langford, J. I., Mittemeijer, E. J., & Vogels, A. B. P. (1982). Use of the Voigt function in a single-line method for the analysis of X-ray diffraction line broadening. *Journal of Applied Crystallography, 15*(3), 308–314. doi:10.1107/S0021889882012035

De Moura, M. R., Aouada, F. A., Monteiro, H., & De Azeredo, C. (2012). Microfluidizer Technique for Improving Microfiber Properties Incorporated Into Edible and Biodegradable Films. In Advances in Microfluidics (pp. 219–240). doi:10.5772/38922

Deanin, R. D., & Manion, M. A. (1999). Article. In C. B. Shonaike & P. G. Simon (Eds.), Polymer Blends and Alloys (p. 1). New York: Marcel Dekker Inc.

Debeaufort, F., Quezada-Gallo, J. A., & Voilley, A. (1998). Edible Films and Coatings: Tomorrow ' s Packagings: A Review. *Critical Reviews in Food Science and Nutrition, 38*(4), 299–313. doi:10.1080/10408699891274219 PMID:9626488

Debeaufort, F., Voilley, A., & Meares, P. (1994). Water vapor permeability and diffusivity through methylcellulose edible films. *Journal of Membrane Science, 91*(1–2), 125–133. doi:10.1016/0376-7388(94)00024-7

Deepak, S., Dinesh, K., Mankaran, S., & Gurmeet, S. (2012). Taste Masking Technologies: A Novel Approach for the Improvement of Organoleptic Property of Pharmaceutical Active Substance. *International Research Journal of Pharmcy, 3*(4), 108–116. Retrieved from http://scholar.google.com/scholar?hl=en&btnG=Search&q=intitle:TASTE+MASKING+TECHNOLOGIES+:+A+NOVEL+APPROACH+FOR+THE+IMPROVEMENT+OF+ORGANOLEPTIC+PROPERTY+OF+PHARMACEUTICAL+ACTIVE+SUBSTANCE#0

Della Porta, G., Falco, N., & Reverchon, E. (2010). NSAID Drugs Release from Injectable Microspheres Produced by Supercritical Fluid Emulsion Extraction. *Journal of Pharmaceutical Sciences, 99*(3), 1484–1499. doi:10.1002/jps.21920 PMID:19780130

De, M., Ghosh, P. S., & Rotello, V. M. (2008). Applications of nanoparticles in biology. *Advanced Materials, 20*(22), 4225–4241. doi:10.1002/adma.200703183

Demura, M., Minami, M., Asakura, T., & Cross, T. A. (1998). Structure of Bombyx mori silk fibroin based on solid-state NMR orientational constraints and fiber diffraction unit cell parameters. *Journal of the American Chemical Society, 120*(6), 1300–1308. doi:10.1021/ja9720691

Deng, Q., Liu, Y., Wang, S., Xie, M., Wu, S., Chen, A., & Wu, W. (2013). Construction of a novel magnetic targeting anti-tumor drug delivery system: Cytosine arabinoside-loaded bacterial magnetosome. *Materials (Basel), 6*(9), 3755–3763. doi:10.3390/ma6093755

Deogonda, P., & Chalwa, V. N. (2013). Mechanical Property of Glass Fiber Reinforcement Epoxy Composites. *International Journal of Scientific Engineering and Research., 1*(4), 6–9.

Deo, S. R., Singh, A. K., Deshmukh, L., Paliwal, L. J., Singh, R. S., & Adhikari, R. (2014). Structural, morphological and optical studies on chemically deposited nanocrystalline CdZnSe thin films. *J. Soudi Chem. Soc., 18*(4), 327–339. doi:10.1016/j.jscs.2014.01.005

Derradji, M., Ramdani, N., Zhang, T., Wang, J., Gong, L. D., Xu, X. D., & Liu, W. B. (2016). Effect of silane surface modified titania nanoparticles on the thermal, mechanical, and corrosion protective properties of a bisphenol-A based phthalonitrile resin. *Progress in Organic Coatings*, *90*, 34–43. doi:10.1016/j.porgcoat.2015.09.021

Derradji, M., Ramdani, N., Zhang, T., Wang, J., Lin, Z. W., Yang, M., & Liu, W. (2015). High thermal and thermomechanical properties obtained by reinforcing a bisphenol-A based phthalonitrile resin with silicon nitride nanoparticles. *Materials Letters*, *149*, 81–84. doi:10.1016/j.matlet.2015.02.122

Desai, S. D., Emanuel, A. L., & Sinha, V. K. (2003). Polyester polyol-based polyurethane adhesive; Effect of treatment on rubber surface, and biomaterial based polyurethane adhesive for bonding rubber and wood joints. *Journal of Polymer Research*, *10*(3), 141–149. doi:10.1023/A:1026099732272

Desborough, I. J., & Hall, I. H. (1977). A comparison of published crystalline structures of poly (tetramethylene terephthalate). *Polymer*, *18*(8), 825–830. doi:10.1016/0032-3861(77)90189-6

Deshmukh, S., & Ounaies, Z. (2010). Active single walled carbon nanotube -polymer composites. *IUTAM Symposium on Multi-Functional Material Structures and System*, 103-110. doi:10.1007/978-90-481-3771-8_11

Desroches, M., Escouvois, M., Auvergne, R., Caillol, S., & Boutevin, B. (2012). From vegetable oils to polyurethanes: Synthetic routes to polyols and main industrial products. *The Polish Review*, *52*(1), 38–79. doi:10.1080/15583724.2011.640443

Devi, N., & Maji, T. (2009). Preparation and Evaluation of Gelatin/Sodium Carboxymethyl Cellulose Polyelectrolyte Complex Microparticles for Controlled Delivery of Isoniazid. *AAPS PharmSciTech*, *10*(4), 1412–1419. doi:10.1208/s12249-009-9344-9 PMID:19937282

Dickinson, E. (1998). Stability and rheological implications of electrostatic milk protein - Polysaccharide interactions. *Trends in Food Science & Technology*, *9*(10), 347–354. doi:10.1016/S0924-2244(98)00057-0

Diftis, N., & Kiosseoglou, V. (2003). Improvement of emulsifying properties of soybean protein isolate by conjugation with carboxymethyl cellulose. *Food Chemistry*, *81*(1), 1–6. doi:10.1016/S0308-8146(02)00236-4

Divakara, S., Niranjana, A. R., & Somashekar, R. (2009). Computation of stacking and twin faults in varieties of cotton fibers using whole powder pattern fitting technique. *Cellulose (London, England)*, *16*(6), 1187–1200. doi:10.1007/s10570-009-9354-5

Divakara, S., Reddy, T., Roy, S., Siddaraju, G. N., & Somashekar, R. (2010). Changes in Stacking Faults and Microstructural Parameters of Dry and Wet Bivoltine Silk Fibers using Whole Powder Pattern Fitting Technique. *Journal of Polymer Materials*, *27*(2), 135–148.

Doane, T. L., & Burda, C. (2012). The unique role of nanoparticles in nanomedicine: Imaging, drug delivery and therapy. *Chemical Society Reviews*, *41*(7), 2885–2911. doi:10.1039/c2cs15260f PMID:22286540

Dong, L., Greco, R., & Orsello, G. (1993). Polycarbonate/acrylonitrilebutadiene styrene blends: 1. Complementary etching techniques for morphology observations. *Polymers*, *34*(7), 1375–1382. doi:10.1016/0032-3861(93)90848-5

Dong, Y. H., & Scardi, P. (2000). MarqX: A new program for whole-powder-pattern fitting. *Journal of Applied Crystallography*, *33*(1), 184–189. doi:10.1107/S002188989901434X

Doring, J., Bovtun, V., Bartusch, J., Erhard, A., Kreutzbruck, M., & Yakimenko, Y. (2010). Nonlinear electromechanical response of the ferroelectret ultrasonic transducers. *Applied Physics. A, Materials Science & Processing, 100*(2), 479–485. doi:10.1007/s00339-010-5752-7

Döring, J., Bovtun, V., Gaal, M., Bartusch, J., Erhard, A., Kreutzbruck, M., & Yakymenko, Y. (2012). Piezoelectric and electrostrictive effects in ferroelectret ultrasonic transducers. *Journal of Applied Physics, 112*(8), 084505. doi:10.1063/1.4759052

Dow Chemical Company. (2013). *Notification of the Determination of Ethylcellulose As Being Generally Recognized As Safe*. Author.

Durán-Lobato, M., Enguix-González, A., Fernández-Arévalo, M., & Martín-Banderas, L. (2013). Statistical analysis of solid lipid nanoparticles produced by high-pressure homogenization: A practical prediction approach. *Journal of Nanoparticle Research, 15*(2), 1–14. doi:10.1007/s11051-013-1443-6

Dutoit, N. E., Wardle, B. L., & Kim, S. G. (2005). Design considerations for MEMS-scale piezoelectric mechanical vibration energy harvesters. *Integrated Ferroelectrics, 71*(1), 121–160. doi:10.1080/10584580590964574

Dyer, A., Hinchcliffe, M., Watts, P., Castile, J., Jabbal-Gill, I., Nankervis, R., & Illum, L. et al. (2002). Nasal delivery of insulin using novel chitosan based formulations: A comparative study in two animal models between simple chitosan formulations and chitosan nanoparticles. *Pharmaceutical Research, 19*(7), 998–1008. doi:10.1023/A:1016418523014 PMID:12180553

Dzunuzovic, E., Jeremic, K., & Nedeljkovic, J. M. (2007). In situ radical polymerization of methyl methacrylate in a solution of surface modified TiO_2 and nanoparticles. *European Polymer Journal, 43*(9), 3719–3726. doi:10.1016/j.eurpolymj.2007.06.026

Eichhorn, S. J., & Davies, G. R. (2006). Modelling the crystalline deformation of native and regenerated cellulose. *Cellulose (London, England), 13*(3), 291–307. doi:10.1007/s10570-006-9046-3

Eiss, N. S. Jr, & Czichos, H. (1986). Tribological studies on rubber-modified epoxies: Influence of material properties and operating conditions. *Wear, 111*(4), 347–361. doi:10.1016/0043-1648(86)90132-8

Elliott, J. H., & Ganz, A. J. (1971). Characterization, I Rheological Property, Organoleptic. *Journal of Texture Studies, 2*(1526), 220–229. doi:10.1111/j.1745-4603.1971.tb00583.x PMID:28371977

Elsadek, B., & Kratz, F. (2012). Impact of albumin on drug delivery—New applications on the horizon. *Journal of Controlled Release, 157*(1), 4–28. doi:10.1016/j.jconrel.2011.09.069 PMID:21959118

El-Sayed, A. A., El-Sherbiny, M. G., Abo-El-Ezz, A. S., & Aggag, G. A. (1995). Friction and wear properties of polymeric composite materials for bearing applications. *Wear, 184*(1), 45–53. doi:10.1016/0043-1648(94)06546-2

Eltayeb, M., Stride, E., Edirisinghe, M., & Harker, A. (2016). Electrosprayed nanoparticle delivery system for controlled release. *Materials Science and Engineering C, 66*, 138–146. doi:10.1016/j.msec.2016.04.001 PMID:27207047

El-Tayeb, N. S. M., & Mostafa, I. M. (1996). The effect of laminate orientations on friction and wear mechanisms of glass reinforced polyester composite. *Wear, 195*(1-2), 186–191. doi:10.1016/0043-1648(95)06849-X

Elzoghby, A. O. (2013). Gelatin-based nanoparticles as drug and gene delivery systems: Reviewing three decades of research. *Journal of Controlled Release, 172*(3), 1075–1091. doi:10.1016/j.jconrel.2013.09.019 PMID:24096021

Elzoghby, A. O., Samy, W. M., & Elgindy, N. A. (2012a). Albumin-based nanoparticles as potential controlled release drug delivery systems. *Journal of Controlled Release, 157*(2), 168–182. doi:10.1016/j.jconrel.2011.07.031 PMID:21839127

Elzoghby, A. O., Samy, W. M., & Elgindy, N. A. (2012b). Protein-based nanocarriers as promising drug and gene delivery systems. *Journal of Controlled Release*, *161*(1), 38–49. doi:10.1016/j.jconrel.2012.04.036 PMID:22564368

Engelen, L., De Wijk, R., Van Der Bilt, A., Prinz, J. F., Janssen, A. M., & Bosman, F. (2005). Relating particles and texture perception. *Physiology & Behavior*, *86*(1–2), 111–117. doi:10.1016/j.physbeh.2005.06.022 PMID:16112152

Erturk, A., & Inman, D. J. (2008). A distributed parameter electromechanical model for cantilevered piezoelectric energy harvesters. *Journal of Vibration and Acoustics*, *130*(4), 041002. doi:10.1115/1.2890402

Erturk, A., & Inman, D. J. (2008). Issues in mathematical modeling of piezoelectric energy harvesters. *Smart Materials and Structures*, *17*(6), 065016. doi:10.1088/0964-1726/17/6/065016

Erturk, A., & Inman, D. J. (2009). An experimentally validated bimorph cantilever model for piezoelectric energy harvesting from base excitations. *Smart Materials and Structures*, *18*(2), 025009. doi:10.1088/0964-1726/18/2/025009

Esthappan, S. K., Kuttappan, S. K., & Joseph, R. (2012). Thermal and mechanical properties of polypropylene/titanium dioxide nanocomposite fibers. *Materials & Design*, *37*, 537–542. doi:10.1016/j.matdes.2012.01.038

Etcheverry, M., & Barbosa, E. (2012). Glass Fiber Reinforced Polypropylene Mechanical Properties Enhancement by Adhesion Improvement. *Materials (Basel)*, *5*(12), 1084–1113. doi:10.3390/ma5061084

Evora, V. M. F., & Shukla, A. (2003). Fabrication, characterization, and dynamic behavior of polyester/TiO_2 nanocomposites. *Materials Science and Engineering A*, *361*(1-2), 358–366. doi:10.1016/S0921-5093(03)00536-7

Fan, H., Wu, Y., Hu, X., Wu, J., & Liao, X. (2011). Characteristics of thin-layer drying and rehydration of nata de coco. *International Journal of Food Science & Technology*, *46*(7), 1438–1444. doi:10.1111/j.1365-2621.2011.02642.x

Farchioni, R., & Grosso, G. (Eds.). (2013). *Organic electronic materials: Conjugated polymers and low molecular weight organic solids* (Vol. 41). Springer Science & Business Media.

Farkhooy, A., Janson, C., Arnardóttir, R. H., Emtner, M., Hedenström, H., & Malinovschi, A. (2015). Impaired Carbon Monoxide Diffusing Capacity is the strongest lung function predictor of decline in 12 minute-walking distance in COPD; a 5-year follow-up study. *COPD: Journal of Chronic Obstructive Pulmonary Disease*, *12*(3), 243–251. doi:10.3109/15412555.2014.948991 PMID:25208266

Fenigstein, M., Eliyahu, S., Gan-Mor, S., & Veierov, D. (2001). Effects of five vegetable oils on the sweetpotato whitefly Bemisia tabaci. *Journal of Phytoparasitica*, *29*(1), 197–206. doi:10.1007/BF02983451

Ferber, R., Davis, I. M., & Williams, D. S. III. (2005). Effect of foot orthotics on rear foot and tibia joint coupling patterns and variability. *Journal of Biomechanics*, *38*(3), 477–483. doi:10.1016/j.jbiomech.2004.04.019 PMID:15652545

Floury, J., Desrumaux, A., Axelos, M., & Legrand, J. (2003). Effect of high pressure homogenisation on methylcellulose as food emulsifier. *Journal of Food Engineering*, *58*(3), 227–238. doi:10.1016/S0260-8774(02)00372-2

Foerster, F., Bamberger, D., Schupp, J., Weilbächer, M., Kaps, L., Strobl, S., & Schuppan, D. et al. (2016). Dextran-based therapeutic nanoparticles for hepatic drug delivery. *Nanomedicine (London)*, *11*(20), 2663–2677. doi:10.2217/nnm-2016-0156 PMID:27628057

Ford, J. L. (1999). Thermal analysis of hydroxypropylmethylcellulose and methylcellulose: Powders, gels and matrix tablets. *International Journal of Pharmaceutics*, *179*(2), 209–228. doi:10.1016/S0378-5173(98)00339-1 PMID:10053214

Fossey, S. A., Némethy, G., Gibson, K. D., & Scheraga, H. A. (1991). Conformational energy studies of β-sheets of model silk fibroin peptides. I. Sheets of poly (Ala-Gly) chains. *Biopolymers, 31*(13), 1529–1541. doi:10.1002/bip.360311309 PMID:1814502

Franke, M. E., Koplin, T. J., & Simon, U. (2006). Metal and metal oxide nanoparticles in chemiresistors: Does the nanoscale matter?. *Small, 2*(1), 36–50. doi:10.1002/smll.200500261 PMID:17193551

Fraser, R. D. B., & MacRue, T. P. (1973). *Conformations in fibrous proteins.* New York: Academic Press.

French, A. D., & Johnson, G. P. (2004). What crystals of small analogs are trying to tell us about cellulose structure. *Cellulose (London, England), 11*(1), 5–22. doi:10.1023/B:CELL.0000014765.94239.fe

Fried Joel, R. (2002). *Polymer science and technology.* Academic Press.

Fultz, B., & Howe, J. (2008). *Transmission Electron Microscopy and Diffractometry of Materials. Springer-Verlag.*

Gaewdang, N., & Gaewdang, T. (2005). Investigations on chemically deposited Cd_1_xZnxS thin films with low Zn content. *Materials Letters, 59*(28), 3577–3584. doi:10.1016/j.matlet.2005.06.031

Galiev, E. E. S., Stakhanov, A. I., Blagodatskikh, I. V., Kobitskaya, E. M., Khokhlov, A. R., Naumkin, A. V., & Pisarev, S. A. et al. (2010). Synthesis of polyvinylpyrrolidone and its nanosilver_based polymer composites in supercritical carbon dioxide. *Polymer Science, Series B, 52*(3–4), 165–173. doi:10.1134/S1560090410030061

Gamay, A. (1996). Process for preparing low fat cheese products. *United States Patent No 5,549,916*, 1–8.

Gamelas, J. A. F., Ferraz, E., & Rocha, F. (2014). An insight into the surface properties of alcined kaolinitic clays: The grinding effect. Colloids and Surfaces A: *Physicochem. Eng. Aspects, 455*, 49–57. doi:10.1016/j.colsurfa.2014.04.038

Garcia, M., van Vliet, G., Cateten, M. G. J., Chavez, F., Norder, B., & Kooi, B. et al.. (2004). Large-scale extrusion processing and characterization of hybrid nylon-6/SiO_2 nanocomposites. *Polymers for Advanced Technologies, 15*(4), 164–172. doi:10.1002/pat.458

Garner, G. M., Shorrocks, N. M., Whatmore, R. W., Goosey, M. T., Seth, P., & Ainger, F. W. (1989). 0–3 piezoelectric composites for large area hydrophones. *Ferroelectrics, 93*(1), 169–176. doi:10.1080/00150198908017341

Garrait, G., Beyssac, E., & Subirade, M. (2014). Development of a novel drug delivery system: Chitosan nanoparticles entrapped in alginate microparticles. *Journal of Microencapsulation, 31*(4), 363–372. doi:10.3109/02652048.2013.858792 PMID:24697173

Gaunø, M. H., Vilhelmsen, T., Larsen, C. C., Boetker, J. P., Wittendorff, J., Rantanen, J., & Østergaard, J. (2013). Real-time in vitro dissolution of 5-aminosalicylic acid from single ethyl cellulose coated extrudates studied by UV imaging. *Journal of Pharmaceutical and Biomedical Analysis, 83*, 49–56. doi:10.1016/j.jpba.2013.04.028 PMID:23708430

Gene^t, F., Schnitzler, A., Mathieu, S., Autret, K., Théfenne, L., Dizien, O., & Maldjian, A. (2010). Orthotic devices and gait in polio patients. *Annals of Physical and Rehabilitation Medicine, 53*(1), 51–59. doi:10.1016/j.rehab.2009.11.005 PMID:20022835

George, J. (1992). *Preparation of Thin Films.* New York: Marcel Dekker, Inc.

Ghezelbash, Z., Ashouri, D., Mousavian, S., Ghandi, A. H., & Rahnama, Y. (2012). Surface modified Al_2O_3 in fluorinated polyimide/Al_2O_3 nanocomposites: Synthesis and characterization. *Bulletin of Materials Science, 35*(6), 925–931. doi:10.1007/s12034-012-0385-4

Ghosh, G., Naskar, M. K., & Patra, A. (2006). Synthesis and characterization of PVP-encapsulated ZnS nanoparticles. *Optical Materials, 28*(8-9), 1047–1053. doi:10.1016/j.optmat.2005.06.003

Gibson, R. F. (2010). A review of recent research on mechanics of multifunctional composite materials and structures. *Composite Structures*, *92*(12), 2793–2810. doi:10.1016/j.compstruct.2010.05.003

Gimeno-Santos, E., Rodriguez, D. A., Barberan-Garcia, A., Blanco, I., Vilaró, J., Torralba, Y., & Garcia-Aymerich, J. et al. (2014). Endurance exercise training improves heart rate recovery in patients with COPD. *COPD: Journal of Chronic Obstructive Pulmonary Disease*, *11*(2), 190–196. doi:10.3109/15412555.2013.831401 PMID:24377907

Gindy, M. E., Feuston, B., Glass, A., Arrington, L., Haas, R. M., Schariter, J., & Stirdivant, S. M. (2014). Stabilization of ostwald ripening in low molecular weight amino lipid nanoparticles for systemic delivery of siRNA therapeutics. *Molecular Pharmaceutics*, *11*(11), 4143–4153. doi:10.1021/mp500367k PMID:25317715

Ginés, R., Libanori, R., Studart, A. R., Bergamini, A., Motavalli, M., & Ermanni, P. (2015). Ceramic–polymer composites with improved dielectric and tribological properties for semi-active damping. *Composites. Part B, Engineering*, *72*, 80–86. doi:10.1016/j.compositesb.2014.11.035

Giovacchini, F., Vannetti, F., Fantozzi, M., & Cempinia, M. (2015). A light-weight active orthosis for hip movement assistance. *Robotics and Autonomous Systems*, *73*, 123–134. doi:10.1016/j.robot.2014.08.015

Gomez, A. (2002). The electrospray and its application to targeted drug inhalation. *Respiratory Care*, *47*(12), 1419-1431.

Gō, N., & Okuyama, K. (1976). A Method for Calculating Dihedral Angles in Helical Polymers with Given Values of Unit Height and Unit Rotation. *Macromolecules*, *9*(5), 867–868. doi:10.1021/ma60053a032

Gradauer, K., Barthelmes, J., Vonach, C., Almer, G., Mangge, H., Teubl, B., & Prassl, R. et al. (2013). Liposomes coated with thiolated chitosan enhance oral peptide delivery to rats. *Journal of Controlled Release*, *172*(3), 872–878. doi:10.1016/j.jconrel.2013.10.011 PMID:24140721

Granstrom, J., Feenstra, J., Sodano, H. A., & Farinholt, K. (2007). Energy harvesting from a backpack instrumented with piezoelectric shoulder straps. *Smart Materials and Structures*, *16*(5), 1810–1820. doi:10.1088/0964-1726/16/5/036

Grassin-Delyle, S., Buenestado, A., Naline, E., Faisy, C., Blouquit-Laye, S., Couderc, L.-J., & Devillier, P. et al. (2012). Intranasal drug delivery: an efficient and non-invasive route for systemic administration: focus on opioids. *Pharmacology & Therapeutics*, *134*(3), 366–379. doi:10.1016/j.pharmthera.2012.03.003 PMID:22465159

Gromeier, M., & EckardWimmer, E. (1998). Mechanism of Injury-Provoked Poliomyelitis. *Journal of Virology*, *72*(6), 5056–5060. PMID:9573275

Guan, C., Lu, C. L., Cheng, Y. R., Song, S. Y., & Yang, B. A. (2009). Facile one-pot route to transparent polymer nanocomposites with high ZnS nanophase contents via in situ bulk polymerization. *Journal of Materials Chemistry*, *19*(5), 617–621. doi:10.1039/B818876A

Guenette, J. A., Webb, K. A., & ODonnell, D. E. (2012). Does dynamic hyperinflation contribute to dyspnoea during exercise in patients with COPD? *The European Respiratory Journal*, *40*(2), 322–329. doi:10.1183/09031936.00157711 PMID:22183485

Gu, J., Wu, G., & Zhang, Q. (2007). Preparation and damping properties of fly ash filled epoxy composites. *Materials Science and Engineering A*, *452*, 614–618. doi:10.1016/j.msea.2006.11.006

Gunduz, O., Ahmad, Z., Stride, E., & Edirisinghe, M. (2012). Continuous Generation of Ethyl Cellulose Drug Delivery Nanocarriers from Microbubbles. *Pharmaceutical Research*, 1–13. doi:10.1007/s11095-012-0865-7 PMID:22956171

Guo, J. H., Liu, Y. C., Chai, T., Jing, S. M., Ma, H., Qin, N., & He, W. M. et al. (2015). Synthesis and properties of nano-silica modified environmentally friendly polyurethane adhesive. *RSC Advances*, *5*(56), 44990–44997. doi:10.1039/C5RA01965F

Guo, Z., Pereira, T., Choi, O., Wang, Y., & Hahn, H. T. (2006). Surface functionalized alumina nanoparticle filled polymeric nanocomposites with enhanced mechanical properties. *Journal of Materials Chemistry*, *16*(27), 2800–2808. doi:10.1039/b603020c

Gupta, A. K., Gupta, M., Yarwood, S. J., & Curtis, A. S. G. (2004). Effect of cellular uptake of gelatin nanoparticles on adhesion, morphology and cytoskeleton organisation of human fibroblasts. *Journal of Controlled Release*, *95*(2), 197–207. doi:10.1016/j.jconrel.2003.11.006 PMID:14980768

Gupta, N., Brar, B. S., & Woldesenbet, E. (2001). Effect of filler addition on the compressive and of glass fiber reinforced epoxies. *Bulletin of Materials Science*, *24*(2), 219–223. doi:10.1007/BF02710105

Hall, I. H., & Pass, M. G. (1976). Chain conformation of poly (tetramethylene terephthalate) and its change with strain. *Polymer*, *17*(9), 807–816. doi:10.1016/0032-3861(76)90036-7

Hall, I. H., & Somashekar, R. (1991). The determination of crystal size and disorder from the X-ray diffraction photograph of polymer fibres. 2. Modelling intensity profiles. *Journal of Applied Crystallography*, *24*(6), 1051–1059. doi:10.1107/S0021889891007707

Hall, J. E. (2015). *Guyton and Hall textbook of medical physiology*. Elsevier Health Sciences.

Hall, N. (2003). Bringing inorganic chemistry to life. *Chemical Communications*, (7): 803–806. PMID:12739626

Hanner, K. A., Safari, A., Newnham, R. E., & Runt, J. (1989). Thin film 0–3 polymer/piezoelectric ceramic composites: Piezoelectric paints. *Ferroelectrics*, *100*(1), 255–260. doi:10.1080/00150198908007920

Han, Y., Chen, Z., Dong, W., & Zhongyin, X. (2014). Improved water resistance, thermal stability, and mechanical properties of waterborne polyurethane nanohybrids reinforced by fumed silica via in situ polymerization. *High Performance Polymers*, *27*(7), 824–832. doi:10.1177/0954008314563058

Haque, A., & Morris, E. R. (1993). Thermogelation of methylcellulose. Part I: Molecular structures and processes. *Carbohydrate Polymers*, *22*(3), 161–173. doi:10.1016/0144-8617(93)90137-S

Harrison, J. S., & Ounaies, Z. (2002). *Polymers, piezoelectric*. John Wiley & Sons, Inc.

Hashimoto, M., Takadama, H., Mizuno, M., & Kokubo, T. (2006). Enhancement of mechanical strength of TiO_2/high-density polyethylene composites for bone repair with silane-coupling treatment. *Materials Research Bulletin*, *41*(3), 515–524. doi:10.1016/j.materresbull.2005.09.014

Hassan, A., Akbari, A., Hing, N. K., & Ratnam, C. T. (2012). Mechanical and thermal properties of ABS / PVC composites: Effect of particles size and surface treatment of ground calcium carbonate. *Polymer-Plastics Technology and Engineering*, *51*(5), 473–479. doi:10.1080/03602559.2011.651242

Hayashi, T. (2006). *The Science and Lore of the Plant Cell Wall: Biosynthesis, Structure and Function*. Universal-Publishers.

Heinze, T., & Koschella, A. (2005). Carboxymethyl ethers of cellulose and starch - A review. *Macromolecular Symposia*, *223*(1), 13–39. doi:10.1002/masy.200550502

Henriksson, M. (2008). *Cellulose Nanofibril Networks and Composites: Preparation, Structure and Properties* (Doctoral dissertation). KTH.

Henry, A. S., Daniel, J. I., & Park, G. (2004). A review of power harvesting from vibration using piezoelectric materials. *Shock and Vibration Digest*, *36*(3), 197–206. doi:10.1177/0583102404043275

Hernandez, S. A. M., Cruz, D., Flores, F., Garcia, S. A., Jimenez, L., Arrocena, M. C., & Santos-Cruz, J. et al. (2014). Optical, Electrical and Photocatalytic Properties of the Ternary Semiconductors $Zn_{xC}d1_{-xS,}$ $CuxC_{d1}-x_{S\,an}dCuxZn1_{-x}S._{Int}$er-*national Journal of Photoenergy, 2014*, *158782*. doi:10.1155/2014/158782

Herrera, M., Matuschek, G., & Kettrup, A. (2003). Thermal degradation of thermoplastic polyurethane elastomers (TPU) based on MDI. Polym*er Degradation & Stability, 78(*2), 323–331. doi:10.1016/S0141-3910(02)00181-7

He, S. J., Valluzzi, R., & Gido, S. P. (1999). Silk I structure in Bombyx mori silk foams. Inte*rnational Journal of Biological Macromolecules, 24*(2*),* 187–195. doi:10.1016/S0141-8130(99)00004-5 PMID:10342764

Hillenbrand, J., & Sessler, G. M. (2000). Piezoelectricity in cellular electret films. IEE*E Transactions on Dielectrics and Electrical Insulation, 7*(4*)*, 537–542. doi:10.1109/94.868074

Hodson, M., & Sherrington, R. (2014). Treating patients with chronic obstructive pulmonary disease. Nu*rsing Standard, 29(9)*, 50–58. doi:10.7748/ns.29.9.50.e9061 PMID:25351094

Hogan, D. T., Bertrand, D. T., Tuazon, M. T., & Tuason, D. C. (1998). Coprocessed microcrystalline cellulose stabilizer compositions, foods containing the same and stabilization method. U*nited States Patent No 5,789,004.*

Hong, F., & Qiu, K. (2008). An alternative carbon source from konjac powder for enhancing production of bacterial cellulose in static cultures by a model strain Acetobacter aceti subsp. xylinus ATCC 23770. *Carbohydrate Polymers*, *72*(3), 545–549. doi:10.1016/j.carbpol.2007.09.015

Hoogesteijn von Reitzenstein, N., Bi, X., Yang, Y., Hristovski, K., & Westerhoff, P. (2016). Morphology, structure, and properties of metal oxide/polymer nanocomposite electrospun mats. *Journal of Applied Polymer Science*, *133*(33), 33. doi:10.1002/app.43811

Houton, K. A., & Wilson, A. J. (2015). Hydrogen-bonded supramolecular polyurethanes. *Polymer International*, *64*(2), 165–173. doi:10.1002/pi.4837

Hsiue, G. H., Kuo, W. J., & Jeng, R. J. (2000). Microstructural and morphological characteristics of PS-SiO_2 nanocomposites. *Polymer*, *41*(8), 2813–2825. doi:10.1016/S0032-3861(99)00478-4

Huang, D. P., & Merolla, T. (1994). Method of making pre-cooked filled pasta products by co-extrusion. *United States Patent No 5,296,247.*

Huang, M., Khor, E., & Lim, L.-Y. (2004). Uptake and cytotoxicity of chitosan molecules and nanoparticles: Effects of molecular weight and degree of deacetylation. *Pharmaceutical Research*, *21*(2), 344–353. doi:10.1023/B:PHAM.0000016249.52831.a5 PMID:15032318

Huang, M., Ma, Z., Khor, E., & Lim, L.-Y. (2002). Uptake of FITC-chitosan nanoparticles by A549 cells. *Pharmaceutical Research*, *19*(10), 1488–1494. doi:10.1023/A:1020404615898 PMID:12425466

Huang, W. S., Humphrey, B. D., & MacDiarmid, A. G. (1986). Polyaniline, a novel conducting polymer. Morphology and chemistry of its oxidation and reduction in aqueous electrolytes. *Journal of the Chemical Society, Faraday Transactions 1*: *Physical Chemistry in Condensed Phases*, *82*(8), 2385–2400.

Hu, G., Zhang, H., Zhang, L., Ruan, S., He, Q., & Gao, H. (2015). Integrin-mediated active tumor targeting and tumor microenvironment response dendrimer-gelatin nanoparticles for drug delivery and tumor treatment. *International Journal of Pharmaceutics*, *496*(2), 1057–1068. doi:10.1016/j.ijpharm.2015.11.025 PMID:26598487

Hughes, J. E., & Girdlestone, S. J. A. (2001). Use of solid waste bagasse from a paper mill as an organic amendment and its effect on some soil properties: A case study. *South African Journal of Plant and Soil, 18*(3), 128–134. doi:10.1080/02571862.2001.10634416

Hult, E. L., Iotti, M., & Lenes, M. (2010). Efficient approach to high barrier packaging using microfibrillar cellulose and shellac. *Cellulose (London, England), 17*(3), 575–586. doi:10.1007/s10570-010-9408-8

Hundiwale, D. G., Kapadi, U. R., Desai, M. C., & Bidkar, S. H. (2002). Mechanical properties of natural rubber filled with flyash. *Journal of Applied Polymer Science, 85*(5), 995–1001. doi:10.1002/app.10465

Iguchi, M., Yamanaka, S., & Budhiono, A. (2000). Bacterial cellulose — a masterpiece of nature ' s arts. *Journal of Materials Science, 35*(2), 261–270. doi:10.1023/A:1004775229149

Ingrid, G., Markus, K., Simona, B. G., Siegfried, B., Stephanie, P. L., Bernd, P., & Sigurd, W. (2009). Flexible active-matrix cells with selectively poled bifunctional polymer-ceramic nanocomposite for pressure and temperature sensing skin. *Journal of Applied Physics, 106*(3), 034503. doi:10.1063/1.3191677

Ionescu, C. M. (2013). *The human respiratory system. In The Human Respiratory System* (pp. 13–22). Springer. doi:10.1007/978-1-4471-5388-7_2

Ivanova, M. I., & Makowski, L. (1998). Iterative low-pass filtering for estimation of the background in fiber diffraction patterns. *Acta Crystallographica. Section A, Foundations of Crystallography, 54*(5), 626–631. doi:10.1107/S0108767398003213

Jacques, S., & Wright, R. (2015). *Code of the suburb: Inside the world of young middle-class drug dealers.* University of Chicago Press. doi:10.7208/chicago/9780226164250.001.0001

Jadidian, B., Hagh, N. M., Winder, A. A., & Ahmad, S. (2009). 25 MHz ultrasonic transducers with lead-free piezoceramic, 13 PZT fiber-epoxy composite, and PVDF polymer active elements. *IEEE Transactions on Ultrasonics, Ferroelectrics, and Frequency Control, 56*(2), 368–378. doi:10.1109/TUFFC.2009.1046 PMID:19251524

Jaffe, B., Cook, W. R., & Jaffe, H. (1971). *Piezoelectric ceramics.* Academic.

Jagannatha, T. D., & Harish, G. (2015). Mechanical properties of carbon/glass fiber reinforced epoxy hybrid polymer composites. *International Journal of Mechanical Engineering and Robotics Research, 4*(2), 131 – 137.

Jain, A., Singh, S. K., Singh, Y., & Singh, S. (2013). Development of lipid nanoparticles of diacerein, an antiosteoarthritic drug for enhancement in bioavailability and reduction in its side effects. *Journal of Biomedical Nanotechnology, 9*(5), 891-900.

Jain, D., & Banerjee, R. (2008). Comparison of ciprofloxacin hydrochloride-loaded protein, lipid, and chitosan nanoparticles for drug delivery. *Journal of Biomedical Materials Research. Part B, Applied Biomaterials, 86*(1), 105–112. doi:10.1002/jbm.b.30994 PMID:18098198

Jain, K. K. (2014). *Drug delivery system.* Springer. doi:10.1007/978-1-4939-0363-4

James, J., Thomas, G. V., Akhina, H., & Thomas, S. (2016). Micro-and Nano-Structured Interpenetrating Polymer Networks: State of the Art, New Challenges, and Opportunities. *Micro-and Nano-Structured Interpenetrating Polymer Networks: From Design to Applications*, 1.

Jana, S., Manna, S., Nayak, A. K., Sen, K. K., & Basu, S. K. (2014). Carbopol gel containing chitosan-egg albumin nanoparticles for transdermal aceclofenac delivery. *Colloids and Surfaces. B, Biointerfaces, 114*, 36–44. doi:10.1016/j.colsurfb.2013.09.045 PMID:24161504

Jaswon, M. A., Gillis, P. P., & Mark, R. E. (1968, September). The elastic constants of crystalline native cellulose. In *Proceedings of the Royal Society of London A: Mathematical, Physical and Engineering Sciences* (Vol. 306, No. 1486, pp. 389-412). The Royal Society. doi:10.1098/rspa.1968.0157

Jean-Mistral, C., Basrour, S., & Chaillout, J. J. (2010). Comparison of electroactive polymers for energy scavenging applications. *Smart Materials and Structures, 19*(8), 085012. doi:10.1088/0964-1726/19/8/085012

JECFA. (2000). *Microcrystalline cellulose. FNP52* (Vol. 4). JECFA.

Jeevananda, T., & Siddaramaiah, . (2001). Thermal and morphological studies on ethylene-vinyl acetate copolymer–polyaniline blends. *Thermochimica Acta, 376*(1), 51–56. doi:10.1016/S0040-6031(01)00530-5

Jellema, R. H., Janssen, A. M., Terpstra, M. E. J., De Wijk, R., & Smilde, A. K. (2005). Relating the sensory sensation creamy mouthfeel in custards to rheological measurements. *Journal of Chemometrics, 19*(3), 191–200. doi:10.1002/cem.922

Jeong, H., Hsu, D. K., Shannon, R. E., & Liaw, P. K. (1994). Characterization of anisotropic elastic constants of silicon-carbide particulate reinforced aluminum metal matrix composites: Part I. Experiment. *Metallurgical and Materials Transactions. A, Physical Metallurgy and Materials Science, 25*(4), 799–809. doi:10.1007/BF02665456

Jia, G., Wang, N. A., Gong, L., & Fei, X. (2009). growth characterization of cdzns thin film prepared by chemical bath deposition. *Chalcogenide Letters, 6*, 463–467.

Jia, G., Wang, N. A., Gong, L., & Fei, X. (2010). optical properties and forming mechanism of cdzns thin film grown by chemical bath deposition. *Chalcogenide Letter, 7*, 349–355.

Jiang, L., Lam, Y. C., Tam, K. C., Chua, T. H., Sim, G. W., & Ang, L. S. (2005). Strengthening acrylonitrile-butadiene-styrene (ABS) with nano-sized and micron-sized calcium carbonate. *Polymer, 46*(1), 243–252. doi:10.1016/j.polymer.2004.11.001

Jia, Q. M., Shan, S., Wang, Y., Gu, L., & Li, J. (2008). Tribological performance and thermal behavior of epoxy resin nanocomposites containing polyurethane and organoclay. *Polymers for Advanced Technologies, 19*(7), 859–864. doi:10.1002/pat.1049

Jia, Q. M., Zheng, M., Chen, H. X., & Shen, R. J. (2005). Synthesis and characterization of polyurethane/epoxy interpenetrating network nanocomposites with organoclays. *Polymer Bulletin, 54*(1-2), 65–73. doi:10.1007/s00289-005-0372-7

Jia, Q. M., Zheng, M., Chen, H. X., & Shen, R. J. (2006a). Synthesis, characterization and properties of organoclay-modified polyurethane/epoxy interpenetrating polymer network nanocomposites. *Polymer International, 55*(3), 257–264. doi:10.1002/pi.1931

Jia, Q. M., Zheng, M., Chen, H. X., & Shen, R. J. (2006b). Morphologies and properties of polyurethane/epoxy resin interpenetrating network nanocomposites modified with organoclay. *Materials Letters, 60*(9-10), 1306–1309. doi:10.1016/j.matlet.2005.11.018

Jia, Q. M., Zheng, M., Chen, H. X., & Shen, R. J. (2006c). The glass transition temperature and microstructure of polyurethane/epoxy resin interpenetrating polymer networks nanocomposites. *Chinese Science Bulletin, 51*(3), 293–298. doi:10.1007/s11434-006-0293-3

Jia, Q. M., Zheng, M., Zhu, Y. C., Li, J. B., & Xu, C. Z. (2007). Effects of organophilic montmorillonite on hydrogen bonding, free volume and glass transition temperature of epoxy resin/polyurethane interpenetrating polymer networks. *European Polymer Journal, 43*(1), 35–42. doi:10.1016/j.eurpolymj.2006.10.016

Jin, H., Zhang, Y. G., Wang, C. S., Sun, Y. F., Yuan, Z. R., Pan, Y. Q., & Cheng, R. et al. (2014). Thermal, mechanical and morphological properties of soybean oil-based polyurethane/epoxy resin interpenetrating polymer networks (IPNs). *Journal of Thermal Analysis and Calorimetry*, *117*(2), 773–781. doi:10.1007/s10973-014-3849-5

Jirapeangtong, K., Siriwatanayothin, S., & Chiewchan, N. (2008). Effects of coconut sugar and stabilizing agents on stability and apparent viscosity of high-fat coconut milk. *Journal of Food Engineering*, *87*(3), 422–427. doi:10.1016/j.jfoodeng.2008.01.001

Jokanovic, U. (2005). Finely dispersed nanoparticle. CRC Press.

Jonas, R., & Farah, L. F. (1998). Production and application of microbial cellulose. *Polymer Degradation & Stability*, *59*(1–3), 101–106. doi:10.1016/S0141-3910(97)00197-3

Joncher, A. K. (1977). The universal dielectric response. *Nature*, *256*(5613), 673–679. doi:10.1038/267673a0

Jonscher, A. K. (1981). A new understanding of the dielectric relaxation of solids. *Journal of Materials Science*, *16*(8), 2037–2060. doi:10.1007/BF00542364

Jo, S. E., Kim, M. S., & Kim, Y. J. (2011). A resonant frequency switching scheme of a cantilever based on polyvinylidene fluoride for vibration energy harvesting. *Smart Materials and Structures*, *21*(1), 015007. doi:10.1088/0964-1726/21/1/015007

Josephson, D. V., & Dahle, C. D. (1949). A new cellulose gum stablizer for ice cream. *Ice Cream Rev.*, *28*, 11–32.

Josephson, D. V., Dahle, C. D., & Patton, S. (1943). A comparison of some ice cream stabilizers. *Southern Dairy Prod. J.*, *33*(4), 34.

Junhee, Loh, & L. (2008). Piezoelectric polymeric thin films tuned by carbon nanotube fillers. In *The 15th International Symposium on: Smart Structures and Materials &Nondestructive Evaluation and Health Monitoring*. International Society for Optics and Photonics.

Junior, C. D. M., Erturk, A., & Inman, D. J. (2009). An electromechanical finite element model for piezoelectric energy harvester plates. *Journal of Sound and Vibration*, *327*(1), 9–25. doi:10.1016/j.jsv.2009.05.015

Jun, J., Kim, J., Bae, Y., & Seo, Y. C. (2011). Enhancement of dispersion and adhesion of B_4C particles in epoxy resin using direct ultrasonic excitation. *Journal of Nuclear Materials*, *416*(3), 293–297. doi:10.1016/j.jnucmat.2011.06.014

Jun, L., Rong, L., Jianming, W., & Hang, S. (1999). Effect of cupric oxide addition on the performance of nickel electrode. *Journal of Power Sources*, *79*(1), 86–90. doi:10.1016/S0378-7753(99)00048-8

Kalia, S., Kaith, B. S., & Kaur, I. (Eds.). (2011). *Cellulose Fibers: Bio- and Nano-Polymer Composites*. Springer Science & Business Media. doi:10.1007/978-3-642-17370-7

Kamada, E., & Mochihara, N. (2001). Stabilizer for meat products and meat product composition. *United States Patent No 6,270,830*, *1*(12), 1–8.

Kamper, S. L., & Fennema, O. (1985). Use of an Edible Film to Maintain Water Vapor Gradients in Foods. *Journal of Food Science*, *50*(2), 382–384. doi:10.1111/j.1365-2621.1985.tb13408.x

Kango, S., Kalia, S., Celli, A., Njuguna, J., Habibi, Y., & Kumar, R. (2013). Surface modification of inorganic nanoparticles for development of organic–inorganic nanocomposites—A review. *Progress in Polymer Science*, *38*(8), 1232–1261. doi:10.1016/j.progpolymsci.2013.02.003

Kang, S., Hong, S. I., Choe, C. R., Park, M., Rim, S., & Kim, J. (2001). Preparation and characterization of epoxy composites filled with functionalized nanosilica particles obtained via sol–gel process. *Polymer*, *42*(3), 879–887. doi:10.1016/S0032-3861(00)00392-X

Kar, M., & Choudhury, P. K. (2007). Formulation and evaluation of ethyl cellulose microspheres prepared by the multiple emulsion technique. *Die Pharmazie*, *62*(2), 122–125. doi:10.1691/ph.2007.2.6062 PMID:17341032

Karthik, R., Manigandan, V., Saravanan, R., Rajesh, R. P., & Chandrika, B. (2016). Structural characterization and in vitro biomedical activities of sulfated chitosan from Sepia pharaonis. *International Journal of Biological Macromolecules*, *84*, 319–328. doi:10.1016/j.ijbiomac.2015.12.030 PMID:26708430

Kaseem, M., Hamad, K., & Deri, F. (2013). On-line rheological measurements and mechanical properties of acrylonitrile-butadiene-styrene / corn starch composite. *International Journal of Polymeric Materials and Polymeric Biomaterials*, *62*(5), 260–264. doi:10.1080/00914037.2011.641698

Kasoju, N., Ali, S. S., Dubey, V. K., & Bora, U. (2013). Exploiting the potential of Collagen as a Natural Biomaterial in Drug Delivery. *Journal of Proteins & Proteomics*, *1*(1).

Kastner, J., Plank, B., Salaberger, D., & Sekelja, J. (2010). Defect and Porosity Determination of Fibre Reinforced Polymers by X-ray Computed Tomography. *2nd International Symposium on NDT in Aerospace 2010*, 1 – 12

Kaur, G., Narang, R., Rath, G., & Goyal, A. K. (2012). Advances in pulmonary delivery of nanoparticles. *Artificial Cells, Blood Substitutes, and Biotechnology*, *40*(1-2), 75–96. doi:10.3109/10731199.2011.592494 PMID:21806501

Kaushik, A., & Kaushik, J. (2010). Solvent absorption characteristics of epoxy-colloidal silica nanocomposites. *Journal of Reinforced Plastics and Composites*, *29*(18), 2821–2833. doi:10.1177/0731684409360995

Kaya, M., Karaarslan, M., Baran, T., Can, E., Ekemen, G., Bitim, B., & Duman, F. (2014). The quick extraction of chitin from an epizoic crustacean species (Chelonibia patula). *Natural Product Research*, *28*(23), 2186–2190. doi:10.1080/14786419.2014.927469 PMID:24933023

Keyser, W. L., & Kinney, D. S. (1983). Low Caliorie Table Syrup Product. *United States Patent No 4,394,399.*

Khanna, P. K., & Singh, N. (2007). Light emitting CdS quantum dots in PMMA: Synthesis and optical studies. *Journal of Luminescence*, *127*(2), 474–482. doi:10.1016/j.jlumin.2007.02.037

Kho, K., Cheow, W. S., Lie, R. H., & Hadinoto, K. (2010). Aqueous re-dispersibility of spray-dried antibiotic-loaded polycaprolactone nanoparticle aggregates for inhaled anti-biofilm therapy. *Powder Technology*, *203*(3), 432–439. doi:10.1016/j.powtec.2010.06.003

Khrenov, V., Klapper, M., Koch, M., & Mullen, K. (2005). Surface functionalized ZnO particles designed for the use in transparent nanocomposites. *Macromolecular Chemistry and Physics*, *206*(1), 95–101. doi:10.1002/macp.200400213

Kim, H. S., Kim, J. H., & Jaehwan, K. (2011). A review of piezoelectric energy harvesting based on vibration. *International Journal of Precision Engineering and Manufacturing*, *12*(6), 1129-1141.

Kim, D. H., Kwon, O. J., Yang, S. R., Park, J. S., & Chun, B. C. (2007). Structural, thermal, and mechanical properties of polyurethane foams prepared with starch as the main component of polyols. *Journal of Fibers and Polymers*, *8*(2), 155–162. doi:10.1007/BF02875785

Kim, K., & Kim, J. (2014). Fabrication of thermally conductive composite with surface modified boron nitride by epoxy wetting method. *Ceramics International*, *40*(4), 5181–5189. doi:10.1016/j.ceramint.2013.10.076

Kim, V., & Criner, G. J. (2013). Chronic bronchitis and chronic obstructive pulmonary disease. *American Journal of Respiratory and Critical Care Medicine, 187*(3), 228–237. doi:10.1164/rccm.201210-1843CI PMID:23204254

Kim, V., Oros, M., Durra, H., Kelsen, S., Aksoy, M., Cornwell, W. D., & Criner, G. J. et al. (2015). Chronic bronchitis and current smoking are associated with more goblet cells in moderate to severe COPD and smokers without airflow obstruction. *PLoS ONE, 10*(2), e0116108. doi:10.1371/journal.pone.0116108 PMID:25646735

Kinsella, M. (2000). Innovative Automotive Applications Based on ZenTron® High Strength Glass Fiber. *Proc.* 45th *Science of Advanced Materials and Process Engineering Series, 45*, 1979 – 1994.

Kishore, K., Kulkarni, S. M., Sharathchandra, S., & Sunil, D. (2002). On the use of an instrumented set-up to characterize the impact behaviour of an epoxy system containing varying fly ash content. *Polymer Testing, 21*(7), 763–771. doi:10.1016/S0142-9418(02)00008-9

Klemm, D., Heublein, B., Fink, H.-P., & Bohn, A. (2005). Cellulose: Fascinating biopolymer and sustainable raw material. *Angewandte Chemie International Edition, 44*(22), 3358–3393. doi:10.1002/anie.200460587 PMID:15861454

Klier, J., Fuchs, S., May, A., Schillinger, U., Plank, C., Winter, G., & Coester, C. et al. (2012). A nebulized gelatin nanoparticle-based CpG formulation is effective in immunotherapy of allergic horses. *Pharmaceutical Research, 29*(6), 1650–1657. doi:10.1007/s11095-012-0686-8 PMID:22302522

Klingshirn, C. F. (1997). *Semiconductor Optics. Springer-Verlag.*

Klose, R. E., & Glicksman, M. (1972). *Handbook of food additives.* CRC Press.

Kluge, J., & Mazzotti, M. (2012). CO2-assisted high pressure homogenization: A solvent-free process for polymeric microspheres and drug–polymer composites. *International Journal of Pharmaceutics, 436*(1–2), 394–402. doi:10.1016/j.ijpharm.2012.06.048 PMID:22750408

Kocserha, I., & Gömze, L. A. (2010). Friction properties of clay compounds. *Applied Clay Science, 48*(3), 425–430. doi:10.1016/j.clay.2010.01.017

Koebel, M. M., Jones, L. C., & Somorjai, G. A. (2008). Preparation of size-tunable, highly monodisperse PVP-protected Pt-nanoparticles by seed-mediated growth. *Journal of Nanoparticle Research, 10*(6), 1063–1069. doi:10.1007/s11051-008-9370-7

Kohajdová, Z., & Karovicová, J. (2008). Influence of hydrocolloids on quality of baked goods. *Acta Sci Pol Technol Aliment, 7*(1), 42–49.

Koide, K., Yoneda, Y., & Musashi, K. (1983). Process for producing a cream cheese-like food. *United States Patent No 4,397,878*, 3–8.

Konishi, T., Kelsey, E., & Singleton, G. T. (1967). Negative potential in scala media during early stage of anoxia. *Acta Oto-Laryngologica, 64*(1-6), 107–118. doi:10.3109/00016486709139097 PMID:6055793

Kraft, J. C., Freeling, J. P., Wang, Z., & Ho, R. J. (2014). Emerging research and clinical development trends of liposome and lipid nanoparticle drug delivery systems. *Journal of Pharmaceutical Sciences, 103*(1), 29–52. doi:10.1002/jps.23773 PMID:24338748

Krässig, H. A. (1993). *Cellulose, structure, accessibility and reactivity.* Philadelphia: Gordon and Breach Publishers.

Kreuter, J. (2013). Mechanism of polymeric nanoparticle-based drug transport across the blood-brain barrier (BBB). *Journal of Microencapsulation, 30*(1), 49–54. doi:10.3109/02652048.2012.692491 PMID:22676632

Krishnan, A. K., George, T. S., Anjana, R., & George, K. E. (2013). Effect of modified kaolin clays on the mechanical properties of polypropylene / polystyrene blends. *Journal of Applied Polymer Science, 127*(2), 1409–1415. doi:10.1002/app.38043

Kröling, H., Fleckenstein, J., Nubbo, N., Endres, A., Miletzky, F., & Schabel, S. (2014). Non-Woven and Paper Based Epoxy Composites. *Science and Technology DAS PAPIER, 6*(7), 2 – 5.

Krol, P. (2007). Synthesis methods, chemical structures and phase structures of linear polyurethanes. Properties and applications of linear polyurethanes in polyurethane elastomers, copolymers and ionomers. *Progress in Materials Science, 52*(6), 915–1015. doi:10.1016/j.pmatsci.2006.11.001

Kruis, A. L., & Chavannes, N. H. (2016). Potential benefits of integrated COPD management in primary care. *Monaldi Archives for Chest Disease, 73*(3). doi:10.4081/monaldi.2010.297 PMID:21214043

Ku, H., Wang, H., Pattarachaiyakoop, N., & Trada, M. (2011). A review on the tensile properties of natural fiber reinforced polymer composites. *Composites. Part B, Engineering, 42*(4), 856–873. doi:10.1016/j.compositesb.2011.01.010

Kuklisova, Z., Tkacova, R., Joppa, P., Wouters, E., & Sastry, M. (2016). Severity of nocturnal hypoxia and daytime hypercapnia predict CPAP failure in patients with COPD and obstructive sleep apnea overlap syndrome. *Sleep Medicine.* PMID:28215237

Kuksal, A., Tiwary, A. K., Jain, N. K., & Jain, S. (2006). Formulation and in vitro, in vivo evaluation of extended- release matrix tablet of zidovudine: Influence of combination of hydrophilic and hydrophobic matrix formers. *AAPS PharmSciTech, 7*(1), E1–E9. doi:10.1208/pt070101 PMID:16584139

Kuliev, R. S., Kuliev, F. A., Mutalibova, A. A., & Kulieva, S. R. (2006). Improving the antioxidant and anticorrosion properties of vegetable oils. *Chemistry and Technology of Fuels and Oils, 42*(1), 55–59. doi:10.1007/s10553-006-0029-8

Kulkarni, S. K. (2009). *Nanotechnology Principles and Practices.* Capital Publishing Company.

Kulkarni, S. M., & Kishore, . (2002). Studies on fly ash-filled epoxy-cast slabs under compression. *Journal of Applied Polymer Science, 84*(13), 2404–2410. doi:10.1002/app.10507

Kumar, A. P., Depan, D., Tomer, N. S., & Singh, R. P. (2009). Nanoscale particles for polymer degradation and stabilization—trends and future perspectives. *Progress in Polymer Science, 34*(6), 479–515. doi:10.1016/j.progpolymsci.2009.01.002

Kumar, B. (2003). Polymer–ceramic composite protonic conductors. *Journal of Power Sources, 123*(2), 132–136. doi:10.1016/S0378-7753(03)00530-5

Kumar, B., Vasekar, P., Pethe, S. A., Dhere, N. G., & Koishiyev, G. T. (2009). Zn*x*Cd$_{1-\square}$*x*S as a heterojunction partner for CuIn$_{1-\square}$*x*Ga$_x$S$_2$ thin film solar cells. *Thin Solid Films, 517*(7), 2295–2299. doi:10.1016/j.tsf.2008.10.108

Kumar, H., Radha, J. C., Ranganathaiah, C., & Siddaramaiah, . (2007). Physico-mechanical and free volume behaviour of guar gum filled polyurethane/polyacrylonitrile biodegradable composites. *European Polymer Journal, 43*(4), 1580–1587. doi:10.1016/j.eurpolymj.2007.01.029

Kumar, H., Siddaramaiah, , Somashekar, R., Mahesh, S. S., & Manjunath, A. (2008). Nanocrystallite regions of guar gum filled PU/PAN composites before and after biodegradation using WAXS. *Journal of Applied Polymer Science, 108*(5), 2762–2771. doi:10.1002/app.27332

Kumar, S., & Randhawa, J. K. (2013). High melting lipid based approach for drug delivery: Solid lipid nanoparticles. *Materials Science and Engineering C, 33*(4), 1842–1852. doi:10.1016/j.msec.2013.01.037 PMID:23498204

Kumar, T. P., & Sankaranarayanan, K. (2009). tunability of structural, surface texture, compositional and optical properties of cdzns thin films by photo assisted chemical bath deposition technique. *Chalcogenide Letters*, *6*, 617–622.

Kunduru, K. R., Basu, A., Haim Zada, M., & Domb, A. J. (2015). Castor oil-based biodegradable polyesters. *Biomacromolecules*, *16*(9), 2572–2587. doi:10.1021/acs.biomac.5b00923 PMID:26301922

Kurosumi, A., Sasaki, C., Yamashita, Y., & Nakamura, Y. (2009). Utilization of various fruit juices as carbon source for production of bacterial cellulose by Acetobacter xylinum NBRC 13693. *Carbohydrate Polymers*, *76*(2), 333–335. doi:10.1016/j.carbpol.2008.11.009

Kwi-Il, P., Minbaek, L., Ying, L., San, M., Geon, T. H., Guang, Z., & Keon, J. L. (2012). Flexible nanocomposite generator made of BaTiO3 nanoparticles and graphitic carbons. *Advanced Materials*, *24*(22), 2999–3004. doi:10.1002/adma.201200105 PMID:22549998

Kwonhoon, H., Ahmad, S., & Richard, E. R. (1991). Colloidal Processing for Improved Piezoelectric Properties of Flexible 0–3 Ceramic–Polymer Composites. *Journal of the American Ceramic Society*, *74*(7), 1699–1702. doi:10.1111/j.1151-2916.1991.tb07165.x

La Mantia, F. P., & Morreale, M. (2011). Green composites: A brief review. *Composites. Part A, Applied Science and Manufacturing*, *42*(6), 579–588. doi:10.1016/j.compositesa.2011.01.017

Lahousse, L., Ziere, G., Verlinden, V. J., Zillikens, M. C., Uitterlinden, A. G., Rivadeneira, F., & Stricker, B. H. et al. (2016). Risk of frailty in elderly with COPD: A population-based study. *The Journals of Gerontology. Series A, Biological Sciences and Medical Sciences*, *71*(5), 689–695. doi:10.1093/gerona/glv154 PMID:26355016

Lakshmikandhan, T., Chandramohan, A., Sethuraman, K., & Alagar, M. (2016). Development and Characterization of Functionalized Al_2O_3 and TiO_2-reinforced polybenzoxazine nanocomposites. *Design of Monomers and Polymers*, *19*(1), 67–76. doi:10.1080/15685551.2015.1092014

Lancaster, J. K. (1972). *Polymer Science: Material Science Handbook* (A. D. Jenkins, Ed.). Amsterdam: North Holland Publishing Company.

Langford, J. I. (1965). Variance as a Measure of Line Broadening: Particle-size Determination. *Nature*, *207*(5000), 966–967. doi:10.1038/207966a0

Langford, J. I. (1968). The variance and other measures of line broadening in powder diffractometry. II. Determination of particle size. *Journal of Applied Crystallography*, *1*(3), 131–138. doi:10.1107/S0021889868005182

Lan, W., & Zhi-Min, D. (2005). Carbon nanotube composites with high dielectric constant at low percolation threshold. *Applied Physics Letters*, *87*(4), 042903. doi:10.1063/1.1996842

Larsen, J. C., Nørby, K. K., Reffstrup, T. K., & Beltoft, V. M. (2004). Opinion of the Scientific Panel on Food Additives, Flavourings, Processing Aids and Materials in Contact with Food (AFC) on a request from the Commission related to Ethyl Cellulose as a food additive Question number EFSA-Q-2003-116. *The EFSA Journal*, *35*, 1–6.

Laukaitish, G., Lindroos, S., Tamulevicius, S., Leskela, M., & Rackitis, M. (2000). SILAR deposition of $Cd_xZn_{1-x}S$ thin films. *Applied Surface Science*, *161*(3-4), 396–405. doi:10.1016/S0169-4332(00)00275-0

Lavoine, N., Desloges, I., & Bras, J. (2014). Microfibrillated cellulose coatings as new release systems for active packaging. *Carbohydrate Polymers*, *103*, 528–537. doi:10.1016/j.carbpol.2013.12.035 PMID:24528763

Lebre, F., Borchard, G., Faneca, H., Pedroso de Lima, M., & Borges, O. (2016). Intranasal Administration of Novel Chitosan Nanoparticle/DNA Complexes Induces Antibody Response to Hepatitis B Surface Antigen in Mice. *Molecular Pharmaceutics*, *13*(2), 472–482. doi:10.1021/acs.molpharmaceut.5b00707 PMID:26651533

Lee, A. L., & Goldstein, R. S. (2015). The role of telemedicine. *Controversies in COPD*, *69*, 269–296. doi:10.1183/2312508X.10019014

Lee, E. J., Khan, S. A., Park, J. K., & Lim, K.-H. (2012). Studies on the characteristics of drug-loaded gelatin nanoparticles prepared by nanoprecipitation. *Bioprocess and Biosystems Engineering*, *35*(1-2), 297–307. doi:10.1007/s00449-011-0591-2 PMID:21909678

Lei, Z., Yang, Q., Shang, W., & Song, X. (2008). Reinforcement of polyurethane/epoxy interpenetrating network nanocomposites with an organically modified palygorskite. *Journal of Applied Polymer Science*, *111*(6), 3150–3162. doi:10.1002/app.29360

Leuner, C., & Dressman, J. (2000). Improving drug solubility for oral delivery using solid dispersions. *European Journal of Pharmaceutics and Biopharmaceutics*, *50*(1), 47-60.

Leung, S. S., Parumasivam, T., Gao, F. G., Carrigy, N. B., Vehring, R., Finlay, W. H., & Chan, H.-K. et al. (2016). Production of inhalation phage powders using spray freeze drying and spray drying techniques for treatment of respiratory infections. *Pharmaceutical Research*, *33*(6), 1486–1496. doi:10.1007/s11095-016-1892-6 PMID:26928668

Li, J. K., Wang, N., & Wu, X. S. (1998). Poly(vinyl alcohol) nanoparticles prepared by freezing–thawing process for protein/peptide drug delivery. *Journal of Controlled Release*, *56*(1-3), 117-126.

Liang, J. Z. (2007). Impact fracture toughness of hollow glass bead-filled polypropylene composites. *Journal of Materials Science*, *42*(3), 841–846. doi:10.1007/s10853-006-0074-z

Liang, J. Z., & Li, F. H. (2006). Measurement of thermal conductivity of hollow glass-bead-filled polypropylene composites. *Polymer Testing*, *25*(4), 527–531. doi:10.1016/j.polymertesting.2006.02.007

Liang, J. Z., Li, R. K. Y., & Tjong, S. C. (2000). Effects of glass bead size and content on the viscoelasticity of filled polypropylene composites. *Polymer Testing*, *19*(2), 213–220. doi:10.1016/S0142-9418(99)00005-7

Liang, X. H., King, D. M., Groner, M. D., Blackson, J. H., Harris, J. D., George, S. M., & Weimer, A. W. (2008). Barrier properties of polymer/alumina nanocomposites membrane fabricated by atomic layer deposition. *Journal of Membrane Science*, *322*(1), 105–112. doi:10.1016/j.memsci.2008.05.037

Liang, Y., & Kiick, K. L. (2014). Heparin-functionalized polymeric biomaterials in tissue engineering and drug delivery applications. *Acta Biomaterialia*, *10*(4), 1588–1600. doi:10.1016/j.actbio.2013.07.031 PMID:23911941

Liew, K. M., Lei, Z. X., & Zhang, L. W. (2015). Mechanical analysis of functionally graded carbon nanotube reinforced composites: A review. *Composite Structures*, *120*, 90–97. doi:10.1016/j.compstruct.2014.09.041

Li, G., & Zhang, Z. (2004). Synthesis of dendritic polyaniline nanofibers in a surfactant gel. *Macromolecules*, *37*(8), 2683–2685. doi:10.1021/ma035891k

Li, H., Liu, G., Liu, B., Chen, W., & Chen, S. (2007). Properties of polyimide/Al_2O_3 hybrids synthesized by in-situ polymerization. *Materials Letters*, *61*(7), 1507–1511. doi:10.1016/j.matlet.2006.07.063

Li, J. (2006). High performance epoxy resin nanocomposites containing both organic montmorillonite and castor oil-polyurethane. *Polymer Bulletin, 56*(4-5), 377–384. doi:10.1007/s00289-005-0492-0

Li, J. P., He, Q. H., Xu, R. F., & Hu, B. X. (2015). Cyanate ester/functionalized silica nanocomposite, synthesis, characterization and properties. *Chinese Journal of Chemistry, 33*(11), 1259–1268. doi:10.1002/cjoc.201500498

Li, L., ten Hagen, T. L., Hossann, M., Süss, R., van Rhoon, G. C., Eggermont, A. M., & Koning, G. A. et al. (2013). Mild hyperthermia triggered doxorubicin release from optimized stealth thermosensitive liposomes improves intratumoral drug delivery and efficacy. *Journal of Controlled Release, 168*(2), 142–150. doi:10.1016/j.jconrel.2013.03.011 PMID:23524188

Li, M., Qiang, X., Xu, W., & Zhang, H. (2015). Synthesis, characterization and application of AFC-based waterborne polyurethane. *Progress in Organic Coatings, 84*, 35–41. doi:10.1016/j.porgcoat.2015.02.009

Lim, S. H., Zeng, K. Y., & He, C. B. (2010). Morphology, tensile and fracture characteristics of epoxy-alumina nanocomposites. *Materials Science and Engineering A, 527*(21-22), 5670–5676. doi:10.1016/j.msea.2010.05.038

Lindahl, O. A., Constantinou, C. E., Anders, E., Yoshinobu, M., & Hallberg, P. (2009). Tactile resonance sensors in medicine. *Journal of Medical Engineering & Technology, 33*(4), 263–273. doi:10.1080/03091900802491188 PMID:19384701

Lin, J. H., Wu, X. M., Ren, T. L., & Liu, L. T. (2007). Modeling and simulation of piezoelectric MEMS energy harvesting device. *Integrated Ferroelectrics, Taylor & Francis, 95*(1), 128–141. doi:10.1080/10584580701756649

Lin, W.-H., Yu, J., Chen, G., & Tsai, W.-B. (2016). Fabrication of multi-biofunctional gelatin-based electrospun fibrous scaffolds for enhancement of osteogenesis of mesenchymal stem cells. *Colloids and Surfaces. B, Biointerfaces, 138*, 26–31. doi:10.1016/j.colsurfb.2015.11.017 PMID:26642073

Lipatov, Y. S., Sergeyeva, L. M., Karabanova, L. V., Rosovitskii, V. F., Skiba, S. I., & Babkina, N. V. (1988). Effect of fillers on viscoelastic properties and compatibility of components in interpenetrating polyurethane-polyacrylate networks. *Polymer Science, USSR., 30*(3), 651–658. doi:10.1016/0032-3950(88)90105-0

Li, S., Lin, M. M., Toprak, M. S., Kim, D. K., & Muhammed, M. (2010). Nanocomposites of polymer and inorganic nanoparticles for optical and magnetic applications. *Nano Review, 1*(1), 5214–5219. doi:10.3402/nano.v1i0.5214 PMID:22110855

Littleton, S. W. (2015). Hypercapnia From Hyperoxia in COPD: Another Piece of the Puzzle or Another Puzzle Entirely? *Respiratory Care, 60*(3), 473–475. doi:10.4187/respcare.03979 PMID:25737575

Liu, D., Tang, Y., & Cong, W. L. (2012). A review of mechanical drilling for composite laminates. *Composite Structures, 94*(4), 1265–1279. doi:10.1016/j.compstruct.2011.11.024

Liu, H., Dong, L., Xie, H., & Xiong, C. (2012). Novel-modified kaolin for enhancing the mechanical and thermal properties of poly (vinyl chloride). *Polymer Engineering and Science, 52*(10), 2071–2077. doi:10.1002/pen.23161

Liu, K., Lin, X., Chen, L., Huang, L., Cao, S., & Wang, H. (2013). Preparation of microfibrillated cellulose/chitosan–benzalkonium chloride biocomposite for enhancing antibacterium and strength of sodium alginate films. *Journal of Agricultural and Food Chemistry, 61*(26), 6562–6567. doi:10.1021/jf4010065 PMID:23750871

Liu, T.-M., Wu, X.-Z., & Qiu, Y.-R. (2016). Enhanced biocompatibility and antibacterial property of polyurethane materials modified with citric acid and chitosan. *Journal of Biomaterials Science. Polymer Edition*, 1–21. PMID:27102367

Liu, Y. L., Hsu, C. Y., Wei, W. L., & Jeong, R. J. (2003). Preparation and thermal properties of epoxy-silica nanocomposites from nanoscale colloidal silica. *Polymer, 44*(18), 5159–5167. doi:10.1016/S0032-3861(03)00519-6

Li, Z. H., & Chen, T. (2012). The mechanical and tribological properties of MAH-grafted HDPE-filled clay / PEI composites. *Polymer-Plastics Technology and Engineering, 51*(8), 776–779. doi:10.1080/03602559.2012.663848

Lligadas, G., Ronda, J., Galia, M., & Cadiz, V. (2010). Plant oils as platform chemicals for polyurethane synthesis: Current state-of-the-art. *Biomacromolecules, 11*(11), 2825–2835. doi:10.1021/bm100839x PMID:20939529

Lohcharoenkal, W., Wang, L., Chen, Y. C., & Rojanasakul, Y. (2014). Protein nanoparticles as drug delivery carriers for cancer therapy. *BioMed Research International, 2014.*

Lokhande, C. D. (1991). Chemical deposition of metal chalcogenide thin films. *Materials Chemistry and Physics, 27*(1), 1–43. doi:10.1016/0254-0584(91)90158-Q

Lometillo, J. E., & Wolcott, J. M. (1983). Process for producing pasta products. *United States Patent No 4,394,397.*

Lotz, B., & Keith, H. D. (1971). Crystal structure of poly (l-Ala-Gly) II: A model for silk I. *Journal of Molecular Biology, 61*(1), 201IN7203-202IN9215.

Lotz, B., & Cesari, F. C. (1979). The chemical structure and the crystalline structures of Bombyx mori silk fibroin. *Biochimie, 61*(2), 205–214. doi:10.1016/S0300-9084(79)80067-X PMID:465571

Lucas, F., Shaw, J. T. B., & Smith, S. G. (1957). The amino acid sequence in a fraction of the fibroin of Bombyx mori. *The Biochemical Journal, 66*(3), 468–479. doi:10.1042/bj0660468 PMID:13459882

Lu, F., Lee, H. P., & Lim, S. P. (2003). Modeling and analysis of micro piezoelectric power generators for micro-electromechanical-systems applications. *Smart Materials and Structures, 13*(1), 57–63. doi:10.1088/0964-1726/13/1/007

Luo, J., Wang, X., Li, J., Zhao, X., & Wang, F. (2007). Conductive hybrid film from polyaniline and polyurethane–silica. *Polymer, 48*(15), 4368–4374. doi:10.1016/j.polymer.2007.05.062

Luo, Y., Teng, Z., Li, Y., & Wang, Q. (2015). Solid lipid nanoparticles for oral drug delivery: Chitosan coating improves stability, controlled delivery, mucoadhesion and cellular uptake. *Carbohydrate Polymers, 122*, 221–229. doi:10.1016/j.carbpol.2014.12.084 PMID:25817662

Luo, Y., & Wang, Q. (2014). Recent development of chitosan-based polyelectrolyte complexes with natural polysaccharides for drug delivery. *International Journal of Biological Macromolecules, 64*, 353–367. doi:10.1016/j.ijbiomac.2013.12.017 PMID:24360899

MacDiarmid, A. G. (1997). Polyaniline and polypyrrole: Where are we headed? *Synthetic Metals, 84*(1-3), 27–34. doi:10.1016/S0379-6779(97)80658-3

MacDiarmid, A. G. (2001). Synthetic metals: A novel role for organic polymers (Nobel lecture). *Angewandte Chemie International Edition, 40*(14), 2581–2590. doi:10.1002/1521-3773(20010716)40:14<2581::AID-ANIE2581>3.0.CO;2-2 PMID:11458347

MacDiarmid, A. G., Chiang, J. C., Halpern, M., Huang, W. S., Mu, S. L., Nanaxakkara, L. D., & Yaniger, S. I. et al. (1985). Polyaniline: Interconversion of metallic and insulating forms. *Molecular Crystals and Liquid Crystals, 121*(1-4), 173–180. doi:10.1080/00268948508074857

Ma, D., Hugener, T. A., Siegel, R. W., Christerson, A., Martensson, E., Onneby, C., & Schadler, L. S. (2005). Influence of nanoparticle surface modification on the electrical behavior of polyethylene nanocomposites. *Nanotechnology, 16*(6), 724–731. doi:10.1088/0957-4484/16/6/016

Madhu, B. J., Ashwini, S. T., Shruthi, B., Divyashree, B. S., Manjunath, A., & Jayanna, H. S. (2014). Structural, dielectric and electromagnetic shielding properties of Ni–Cu nanoferrite/PVP composites. *Material Science and Engineering B. Advanced Functional Solid-State Materials*, *186*, 1–6. doi:10.1016/j.mseb.2014.02.018

Madhu, B. J., Gurusiddesh, M., Kiran, T., Shruthi, B., & Jayanna, H. S. (2016a). Structural, dielectric, AC conductivity and electromagnetic shielding properties of polyaniline/$Ni_{0.5}Zn_{0.5}Fe_2O_4$ composites. *Journal of Materials Science Materials in Electronics*, *27*(8), 7760–7766. doi:10.1007/s10854-016-4764-2

Madhu, B. J., Kavya, M., Razika Banu, S., Shruthi, B., Sowmya, C. P., & Jayanna, H. S. (2012). Frequency Dependence of Dielectric Parameters and AC Conductivity of Mg-Co-Zn Nanoferrites Synthesized Via Combustion Method. *Advanced Materials Research*, *584*, 295–298. doi:10.4028/www.scientific.net/AMR.584.295

Madhu, B. J., Kiran, T., Gurusiddesh, M., & Shruthi, B. (2016b). Synthesis, characterization and dielectric behavior of polyaniline/$Ni_{0.5}Zn_{0.5}Fe_2O_4$ nanocomposites. *Macromolecular Symposia*, *361*(1), 24–29. doi:10.1002/masy.201400265

Maensiri, S., Sangmanee, M., & Wiengmoon, A. (2008). Magnesium ferrite (MgFe 2 O 4) nanostructures fabricated by electrospinning. *Nanoscale Research Letters*, *4*(3), 221–228. doi:10.1007/s11671-008-9229-y PMID:20596441

Mahdi, M. A., & Al-Ani, S. K. J. (2012). Optical characterization of chemical bath deposition $Cd_{1-x}Zn_xS$ thin films. *Int. J. Nanoelectronics and Materials*, *5*, 11–24.

Mahor, A., Prajapati, S. K., Verma, A., Gupta, R., Iyer, A. K., & Kesharwani, P. (2016). Moxifloxacin loaded gelatin nanoparticles for ocular delivery: Formulation and in-vitro, in-vivo evaluation. *Journal of Colloid and Interface Science*, *483*, 132–138. doi:10.1016/j.jcis.2016.08.018 PMID:27552421

Maity, , Nandy, & Mittal. (2014). A brief review on recent advances of extended release technology employed to design the oral dosage forms. *International Journal of Medical and Applied Sciences Research*, *1*(1).

Makwana, V., Jain, R., Patel, K., Nivsarkar, M., & Joshi, A. (2015). Solid lipid nanoparticles (SLN) of Efavirenz as lymph targeting drug delivery system: Elucidation of mechanism of uptake using chylomicron flow blocking approach. *International Journal of Pharmaceutics*, *495*(1), 439–446. doi:10.1016/j.ijpharm.2015.09.014 PMID:26367780

Ma, L., He, H., Jiang, C., Zhou, L., Luo, Y., & Jia, D. (2012). Effect of alkali treatment on structure and mechanical properties of acrylonitrile-butadiene-styrene / bamboo fiber composites. *Journal of Macromolecular Science, Part B: Physics*, *51*(11), 2232–2244. doi:10.1080/00222348.2012.669688

Mallikarjunan, P., Chinnan, M., Balasubramaniam, V., & Phillips, R. (1997). Edible Coatings for Deep-fat Frying of Starchy Products1. *Lebensmittel-Wissenschaft Und-Technologie*, *30*(7), 709–714. doi:10.1006/fstl.1997.0263

Malmo, J., Sandvig, A., Vårum, K. M., & Strand, S. P. (2013). Nanoparticle mediated P-glycoprotein silencing for improved drug delivery across the blood-brain barrier: A siRNA-chitosan approach. *PLoS ONE*, *8*(1), e54182. doi:10.1371/journal.pone.0054182 PMID:23372682

Mansfield, E., Bryant, J., Regan, T., Waller, A., Boyes, A., & Sanson-Fisher, R. (2016). Burden and Unmet Needs of Caregivers of Chronic Obstructive Pulmonary Disease Patients: A Systematic Review of the Volume and Focus of Research Output. *COPD: Journal of Chronic Obstructive Pulmonary Disease*, 1-6.

Mansour, H. M., Rhee, Y.-S., & Wu, X. (2009). Nanomedicine in pulmonary delivery. *International Journal of Nanomedicine*, *4*, 299–319. doi:10.2147/IJN.S4937 PMID:20054434

Manzoor, A. A., Lindner, L. H., Landon, C. D., Park, J.-Y., Simnick, A. J., Dreher, M. R., & Dewhirst, M. W. et al. (2012). Overcoming limitations in nanoparticle drug delivery: Triggered, intravascular release to improve drug penetration into tumors. *Cancer Research*, *72*(21), 5566–5575. doi:10.1158/0008-5472.CAN-12-1683 PMID:22952218

Mariano, M., El Kissi, N., & Dufresne, A. (2014). Cellulose nanocrystals and related nanocomposites: Review of some properties and challenges. *Journal of Polymer Science. Part B, Polymer Physics*, *52*(12), 791–806. doi:10.1002/polb.23490

Marsh, R. E., Corey, R. B., & Pauling, L. (1955). An investigation of the structure of silk fibroin. *Biochimica et Biophysica Acta*, *16*, 1–34. doi:10.1016/0006-3002(55)90178-5 PMID:14363226

Martin, C. R. (1995). Template synthesis of electronically conductive polymer nanostructures. *Accounts of Chemical Research*, *28*(2), 61–68. doi:10.1021/ar00050a002

Martins, I. M. G., Magina, S. P., Oliveira, L., Freire, C. S. R., Silvestre, A. J. D., Neto, C. P., & Gandini, A. (2009). New biocomposites based on thermoplastic starch and bacterial cellulose. *Composites Science and Technology*, *69*(13), 2163–2168. doi:10.1016/j.compscitech.2009.05.012

Marvizadeh, M. M., Nafchi, A. M., & Jokar, M. (2014). Improved physicochemical properties of tapioca starch/bovine gelatin biodegradable films with zinc oxide nanorod. *Journal of Chemical Health Risks*, *4*(4).

Master, J. E. (1997). *Basic failure modes of continuous fiber composites, in engineering materials Handbook* (Vol. 1). ASM Intl.

Mathew, N., Xavier, Z., Deshmukh, A. S., Deshmukh, H., Jadhav, S., & Kasat, S. (2016). The Study of Aetiology of Chronic Obstructive Pulmonary Disease (COPD) in Non-Smokers. *Journal of Pulmonary & Respiratory Medicine, 2015.*

Matt, H. M., & Scalea, F. L. (2007). Macro-fiber composite piezoelectric rosettes for acoustic source location in complex structures. *Smart Materials and Structures*, *16*(4), 1489–1499. doi:10.1088/0964-1726/16/4/064

Ma, X., Cui, J., Wang, J., Chang, Y., Fang, Q., & Bai, C. et al.. (2015). Multicentre investigation of pathogenic bacteria and antibiotic resistance genes in Chinese patients with acute exacerbation of chronic obstructive pulmonary disease. *The Journal of International Medical Research*. PMID:26152913

McNicholas, W. T., Verbraecken, J., & Marin, J. M. (2013). Sleep disorders in COPD: The forgotten dimension. *European Respiratory Review*, *22*(129), 365–375. doi:10.1183/09059180.00003213 PMID:23997063

Meirovitch, L., & Parker, R. G. (2001). Fundamentals of vibrations. *Applied Mechanics Reviews*, *54*(6), B100–B101. doi:10.1115/1.1421112

Melzer, E., Kreuter, J., & Daniels, R. (2003). Ethylcellulose: A new type of emulsion stabilizer. *European Journal of Pharmaceutics and Biopharmaceutics*, *56*(1), 23–27. doi:10.1016/S0939-6411(03)00025-0 PMID:12837477

Meredith, J., Bilson, E., & Powe, R. (2015). A performance versus cost analysis of prepreg carbon fibre epoxy energy absorption structures. Composite Structures, 124, 206–213.

Michaelson, J. C., & McEvoy, A. J. (1994). Interfacial polymerization of aniline. *Journal of the Chemical Society. Chemical Communications*, (1): 79–80. doi:10.1039/c39940000079

Miki, K., Maekura, R., Nagaya, N., Kitada, S., Miki, M., & Yoshimura, K. et al.. (2013). Effects of ghrelin on the dyspnea pattern in cachectic COPD: Exploratory analysis of a substudy of a multicenter, randomized, double-blind, placebo-controlled trial of ghrelin treatment. *The European Respiratory Journal*, *42*(Suppl 57), 2499.

Millqvist, E., Ternesten-Hasséus, E., & Bende, M. (2013). Inhalation of menthol reduces capsaicin cough sensitivity and influences inspiratory flows in chronic cough. *Respiratory Medicine*, *107*(3), 433–438. doi:10.1016/j.rmed.2012.11.017 PMID:23266255

Min, H. (2004). *Chitosan: Cellular uptake, cytotoxicity and application in gene delivery*. Academic Press.

Miravitlles, M., Peña-Longobardo, L. M., Oliva-Moreno, J., & Hidalgo-Vega, Á. (2015). Caregivers burden in patients with COPD. *International Journal of Chronic Obstructive Pulmonary Disease, 10*, 347. doi:10.2147/COPD.S76091 PMID:25709429

Mitchell, K., Ford, J. L., Armstrong, D. J., Elliott, P. N. C., Rostron, C., & Hogan, J. E. (1990). The influence of additives on the cloud point, disintegration and dissolution of hydroxypropylmethylcellulose gels and matrix tables. *International Journal of Pharmaceutics, 66*(1–3), 233–242. doi:10.1016/0378-5173(90)90404-R

Mitra, B. C. (2014). Environment Friendly Composite Materials: Biocomposites and Green Composites. *Defence Science Journal, 64*(3), 244–261. doi:10.14429/dsj.64.7323

Mittemeijer, E. J., & Scardi, P. (Eds.). (2013). *Diffraction analysis of the microstructure of materials* (Vol. 68). Springer Science & Business Media.

Mi, Y. N., Liang, G. Z., Gu, A. J., Zhao, F. P., & Yuan, L. (2013). Thermally conductive aluminum nitride-multiwalled carbon nanotube/cyanate ester composites with high flame retardancy and low dielectric loss. *Industrial & Engineering Chemistry Research, 52*(9), 3342–3353. doi:10.1021/ie3029569

Mohammad R, K. (2008). A review of several reported procedures to determine the degree of N-acetylation for chitin and chitosan using infrared spectroscopy. *Carbohydrate Polymers, 71*(4), 497-508.

Momany, F. A., McGuire, R. F., Burgess, A. W., & Scheraga, H. A. (1975). Energy parameters in polypeptides. VII. Geometric parameters, partial atomic charges, nonbonded interactions, hydrogen bond interactions, and intrinsic torsional potentials for the naturally occurring amino acids. *Journal of Physical Chemistry, 79*(22), 2361–2381. doi:10.1021/j100589a006

Mondal, M. I. H., Hoque, M. A., Latif, M. A., & Islam, M. K. (2014). Improved Characteristics of Acetic Anhydride and Phthalic Anhydride Modified Jute Fibers. *Research Journal of Textile and Apparel, 18*(1), 64–70. doi:10.1108/RJTA-18-01-2014-B008

Monjazebi, F., Dalvandi, A., Ebadi, A., Khankeh, H. R., Rahgozar, M., & Richter, J. (2016). Functional Status Assessment of COPD Based on Ability to Perform Daily Living Activities: A Systematic Review of Paper and Pencil Instruments. *Global Journal of Health Science, 8*(3), 210.

Monteavaro, L. L., DaSilva, E. O., Costa, O. A. P., Samios, D., Gerbase, A. E., & Petzhold, C. L. (2005). Polyurethane networks from formiated soy polyols: Synthesis and mechanical characterization. *Journal of the American Oil Chemists Society, 82*(5), 365–371. doi:10.1007/s11746-005-1079-0

Moore, L. J., & Shoemaker, C. F. (1981). Sensory textural properties of stabilized ice-cream. *Journal of Food Science, 46*(2), 399–402, 409. doi:10.1111/j.1365-2621.1981.tb04870.x

Moreno, J. C., Rocon de Lima, E., & Andres, F. (2006). Design and implementation of an inertial measurement unit for control of artificial limbs: Application on leg orthoses. *Sensors and Actuators. B, Chemical, 118*(1-2), 333–337. doi:10.1016/j.snb.2006.04.039

Mo, T. C., Wang, H. W., Chen, S. Y., & Yeh, Y. C. (2008). Synthesis and dielectric properties of polyaniline/titanium dioxide nanocomposites. *Ceramics International, 34*(7), 1767–1771. doi:10.1016/j.ceramint.2007.06.002

Muala, A., Nicklasson, H., Boman, C., Swietlicki, E., Nyström, R., Pettersson, E., et al. (2014). *Respiratory tract deposition of inhaled wood smoke particles in healthy volunteers and subjects with COPD*. Academic Press.

Mudshinge, S. R., Deore, A. B., Patil, S., & Bhalgat, C. M. (2011). Nanoparticles: Emerging carriers for drug delivery. *Saudi Pharmaceutical Journal, 19*(3), 129–141. doi:10.1016/j.jsps.2011.04.001 PMID:23960751

Mukherjee, A., Fu, M., & Mitra, P. (2015). Influence of Zn incorporation in CdS: Structural and morphological studies. *Journal of Physics and Chemistry of Solids*, *82*, 50–55. doi:10.1016/j.jpcs.2015.03.009

Murthy, N. S., Akkapeddi, M. K., & Orts, W. J. (1998). Analysis of lamellar structure in semicrystalline polymers by studying the absorption of water and ethylene glycol in nylons using small-angle neutron scattering. *Macromolecules*, *31*(1), 142–152. doi:10.1021/ma9707603

Muyonga, J. H., Cole, C. G. B., & Duodu, K. G. (2004). Fourier transform infrared (FTIR) spectroscopic study of acid soluble collagen and gelatin from skins and bones of young and adult Nile perch (Lates niloticus). *Food Chemistry*, *86*(3), 325–332. doi:10.1016/j.foodchem.2003.09.038

Nabil, A. N. (2004). Studies on the effect of titanate coupling agent (2.0%) on the mechanical properties of flyash-filled polybutadiene rubber. *Journal of Applied Polymer Science*, *93*(3), 1293–1298. doi:10.1002/app.20548

Nadeem, M., & Mushtaq, A. (2009). Thermal cycling induced reversibility/irreversibility in the ac electrical properties of La 0.50 Ca 0.50 MnO 3+ δ by impedance spectroscopy. *Journal of Applied Physics*, *106*(7), 073713. doi:10.1063/1.3238301

Naftali, S., Rosenfeld, M., Wolf, M., & Elad, D. (2005). The air-conditioning capacity of the human nose. *Annals of Biomedical Engineering*, *33*(4), 545–553. doi:10.1007/s10439-005-2513-4 PMID:15909660

Nagamani, K., Reddy, M. V., Lingappa, Y., Reddy, K. T. R., & Miles, R. W. (2012). Physical Properties of $Zn_xCd_{1-x}S$ Nanocrytalline Layers Synthesized by Solution Growth Method. *Int. J. Optoelectron. Eng*, *2*, 1–4. doi:10.5923/j.ijoe.20120202.01

Nakajima, Y., Qu, X. M., & Natori, S. (1987). Interaction between liposomes and sarcotoxin IA, a potent antibacterial protein of Sarcophaga peregrina (flesh fly). *The Journal of Biological Chemistry*, *262*(4), 1665–1669. PMID:3543008

Naseri, M. G., Saion, E. B., & Shaari, A. H. (2012) Role of PVP on the Phase Composition and Morphology of Manganese Ferrite Nanoparticles Prepared by Thermal Treatment Method. *2nd International Conference on Environment Science and Biotechnology IPCBEE*, *48*, 32. Doi:10.7763/IPCBEE

Nesamony, J., Singh, P. R., Nada, S. E., Shah, Z. A., & Kolling, W. M. (2012). Calcium alginate nanoparticles synthesized through a novel interfacial cross-linking method as a potential protein drug delivery system. *Journal of Pharmaceutical Sciences*, *101*(6), 2177–2184. doi:10.1002/jps.23104 PMID:22411606

Newnham, R. E., Skinner, D. P., & Cross, L. E. (1978). Connectivity and piezoelectric-pyroelectric composites. *Materials Research Bulletin*, *13*(5), 525–536. doi:10.1016/0025-5408(78)90161-7

Ng, C. B., Schadler, L. S., & Siegel, R. W. (1999). Synthesis and mechanical properties of TiO_2-epoxy nanocomposites. *Nanostructured Materials*, *12*(1-4), 507–510. doi:10.1016/S0965-9773(99)00170-1

Nhuapeng, W., Thamjaree, W., Kumfu, S., Singjai, P., & Tunkasiri, T. (2008). Fabrication and mechanical properties of silicon carbidenanowires/epoxy resin composites. *Current Applied Physics*, *8*(3-4), 295–299. doi:10.1016/j.cap.2007.10.074

Niesen, T. P., & De Guire, M. R. (2002). Review: Deposition of ceramic thin films at low temperatures from aqueous solutions. *Solid State Ionics*, *151*(1-4), 61–68. doi:10.1016/S0167-2738(02)00604-5

Nigam, P., Waghmode, S., Louis, M., Wangnoo, S., Chavan, P., & Sarkar, D. (2014). Graphene quantum dots conjugated albumin nanoparticles for targeted drug delivery and imaging of pancreatic cancer. *Journal of Materials Chemistry B*, *2*(21), 3190–3195. doi:10.1039/C4TB00015C

Nitta, S. K., & Numata, K. (2013). Biopolymer-based nanoparticles for drug/gene delivery and tissue engineering. *International Journal of Molecular Sciences*, *14*(1), 1629–1654. doi:10.3390/ijms14011629 PMID:23344060

Nugroho, D. A., & Aji, P. (2015). Characterization of Nata de Coco Produced by Fermentation of Immobilized Acetobacter xylinum. *Agriculture and Agricultural Science Procedia, 3*, 278–282. doi:10.1016/j.aaspro.2015.01.053

Nunes, S. P., Peinemann, K. V., Ohlrogge, K., Alpers, A., Keller, M., & Pires, A. T. N. (1999). Membranes of poly(ether imide) and nanodispersed silica. *Journal of Membrane Science, 157*(2), 219–226. doi:10.1016/S0376-7388(98)00379-2

O'Sullivan, A. (1997). Cellulose: the structure slowly unravels. *Cellulose, 4*(3), 173–207.

Okiyama, A., Motoki, M., & Yamanaka, S. (1993). Bacterial Cellulose .3. Development of a New Form of Cellulose. *Food Hydrocolloids, 6*(6), 493–501. 10.1016/S0268-005X(09)80073-8

Okiyama, A., Motoki, M., & Yamanaka, S. (1992). Bacterial cellulose II. Processing of the gelatinous cellulose for food materials. *Food Hydrocolloids, 6*(5), 479–487. doi:10.1016/S0268-005X(09)80033-7

Okiyama, A., Motoki, M., & Yamanaka, S. (1993). Bacterial cellulose IV. Application to processed foods. *Food Hydrocolloids, 6*(6), 503–511. doi:10.1016/S0268-005X(09)80074-X

Okuyama, K. (2008). Revisiting the molecular structure of collagen. *Connective Tissue Research, 49*(5), 299–310. doi:10.1080/03008200802325110 PMID:18991083

Okuyama, K., Noguchi, K., Miyazawa, T., Yui, T., & Ogawa, K. (1997). Molecular and crystal structure of hydrated chitosan. *Macromolecules, 30*(19), 5849–5855. doi:10.1021/ma970509n

Okuyama, K., Somashekar, R., Noguchi, K., & Ichimura, S. (2001). Refined molecular and crystal structure of silk I based on Ala–Gly and (Ala–Gly) 2—Ser–Gly peptide sequence. *Biopolymers, 59*(5), 310–319. doi:10.1002/1097-0282(20011015)59:5<310::AID-BIP1028>3.0.CO;2-5 PMID:11514934

Okuyama, K., Takanashi, K., Nakajima, Y., Hasegawa, Y., Hirabayashi, K., & Nishi, N. (1988). Analysis of silk I structure by X-ray and electron diffraction methods. *The Journal of Sericultural Science of Japan, 57*(1), 23–30.

Oliveira, D. (2004). Optimization of enzymatic production of biodiesel from castor oil in organic solvent medium. *Applied Biochemistry and Biotechnology, 115*(1-3), 771–780. doi:10.1385/ABAB:115:1-3:0771 PMID:15054231

Olivia, A. I., Corona, J. E., Patino, R., & Aviles, A. I. (2014). Chemical bath deposition of CdS thin films doped with Zn and Cu. *Bulletin of Materials Science, 37*(2), 247–255. doi:10.1007/s12034-014-0642-9

Olivia, A. I., Rodriguez, R. C., Canto, S., Sosa, V., Quintana, P., & Pena, J. L. (2003). Comparison of properties of CdS thin films grown by two techniques. *Applied Surface Science, 205*(1-4), 56–64. doi:10.1016/S0169-4332(02)01081-4

Omrani, A., Simon, L. C., & Rostami, A. A. (2009). The effects of alumina nanoparticle on the properties of an epoxy resin system. *Materials Chemistry and Physics, 114*(1), 145–150. doi:10.1016/j.matchemphys.2008.08.090

Organization, W. H. (2006). Compendium of Food Additive Specifications. *FAO JECFA Monographs*, (63).

Otheguy, M.E., Gibson, A.G., Robinson, M., Findon, E., Cripps, B., Ochoa Mendoz, A., & Castro, A. (2009). Recycling of end of life thermoplastic composite boats. *Plastics, Rubber and Composites: Macromolecular Engineering, 38*(9-10), 406-411.

Ou, Y., Yang, F., & Chen, J. (1997). Interfacial interaction and mechanical properties of nylon 6 potassium titanate composites prepared by in-situ polymerization. *Journal of Applied Polymer Science, 64*(12), 2317–2322. doi:10.1002/(SICI)1097-4628(19970620)64:12<2317::AID-APP7>3.0.CO;2-O

Ou, Y., Yang, F., & Yu, Z. (1998). A new conception on the toughness of nylon 6/silica nanocomposite prepared via in situ polymerization. *Journal of Polymer Science. Part B, Polymer Physics, 36*(5), 789–795. doi:10.1002/(SICI)1099-0488(19980415)36:5<789::AID-POLB6>3.0.CO;2-G

Padam, G., Malhotra, G. L., & Rao, S. U. M. (1988). Studies on solution-grown thin films of $Zn_xCd_{1-x}S$. *Journal of Applied Physics*, *63*(3), 770–774. doi:10.1063/1.340069

Pal, K., Behera, B., Roy, S., Sekhar Ray, S., & Thakur, G. (2013). Chitosan based delivery systems on a length scale: Nano to macro. *Soft Matter*, *11*(2), 125–142. doi:10.1080/1539445X.2011.586082

Panda, H. (2010). *The Complete Book on Gums and Stabilizers for Food Industry*. Asia Pacific Business Press Inc.

Pang, H., Braun, G. B., Friman, T., Aza-Blanc, P., Ruidiaz, M. E., Sugahara, K. N., & Ruoslahti, E. et al. (2015). A novel endocytic and intercellular transport pathway for drug delivery across blood vessels and into nutrient-deprived tumor cells. *Cancer Research*, *75*(15Supplement), NG03–NG03. doi:10.1158/1538-7445.AM2015-NG03

Pankove, J. I. J. I. (1975). Optical Processes in Semiconductors. Dover Publications Inc.

Pan, L., He, Q., Liu, J., Chen, Y., Ma, M., Zhang, L., & Shi, J. (2012). Nuclear-targeted drug delivery of TAT peptide-conjugated monodisperse mesoporous silica nanoparticles. *Journal of the American Chemical Society*, *134*(13), 5722–5725. doi:10.1021/ja211035w PMID:22420312

Panwar, V., Park, J. O., Park, S. H., Kumar, S., & Mehra, R. M. (2010). Electrical, dielectric, and electromagnetic shielding properties of polypropylene-graphite composites. *Journal of Applied Polymer Science*, *115*(3), 1306–1314. doi:10.1002/app.29702

Park, C., Vo, C. L.-N., Kang, T., Oh, E., & Lee, B.-J. (2015). New method and characterization of self-assembled gelatin–oleic nanoparticles using a desolvation method via carbodiimide/N-hydroxysuccinimide (EDC/NHS) reaction. *European Journal of Pharmaceutics and Biopharmaceutics*, *89*, 365–373. doi:10.1016/j.ejpb.2014.12.002 PMID:25536111

Park, J. S., Chae, H., Chung, H. K., & Lee, S. I. (2011). Thin film encapsulation for flexible AM-OLED: A review. *Semiconductor Science and Technology*, *26*(3), 034001. doi:10.1088/0268-1242/26/3/034001

Parveen, S., Misra, R., & Sahoo, S. K. (2012). Nanoparticles: A boon to drug delivery, therapeutics, diagnostics and imaging. *Nanomedicine; Nanotechnology, Biology, and Medicine*, *8*(2), 147–166. doi:10.1016/j.nano.2011.05.016 PMID:21703993

Patel, I., Siores, E., & Shah, T. (2010). Utilisation of smart polymers and ceramic based piezoelectric materials for scavenging wasted energy. *Sensors and Actuators. A, Physical*, *159*(2), 213–218. doi:10.1016/j.sna.2010.03.022

Pathak, Y., & Thassu, D. (2016). *Drug delivery nanoparticles formulation and characterization* (Vol. 191). CRC Press.

Pathan, H. M., & Lokhande, C. D. (2004). Deposition of metal chalcogenide thin films by successive ionic layer adsorption and reaction (SILAR) method. *Bulletin of Materials Science*, *27*(2), 85–111. doi:10.1007/BF02708491

Patil, J., & Sarasija, S. (2012). Pulmonary drug delivery strategies: A concise, systematic review. *Lung India*, *29*(1), 44. doi:10.4103/0970-2113.92361 PMID:22345913

Patil, K. C. (2008). *Chemistry of nanocrystalline oxide materials: combustion synthesis, properties and applications*. World Scientific. doi:10.1142/6754

Patra, S., Basak, P., & Tibarewala, D. (2016). Synthesis of gelatin nano/submicron particles by binary nonsolvent aided coacervation (BNAC) method. *Materials Science and Engineering C*, *59*, 310–318. doi:10.1016/j.msec.2015.10.011 PMID:26652378

Paul, A. I., Justine, I. N., Alabi, O. O., & Adamu, W. U. (2012). Recent Trends and Developments in Surface Hardening Technology, A Focus on New Ceramics for Surface Modification in Industry. *International Journal of Science & Engineering Research*, *3*(11), 1–22.

Paul, W., Shelma, R., & Sharma, C. P. (2013). Alginate encapsulated anacardic acid-chitosan self aggregated nanoparticles for intestinal delivery of protein drugs. *Journal of Nanopharmaceutics and Drug Delivery*, *1*(1), 82–91. doi:10.1166/jnd.2013.1012

Perera, P. N., Armstrong, E. P., Sherrill, D. L., & Skrepnek, G. H. (2012). Acute exacerbations of COPD in the United States: Inpatient burden and predictors of costs and mortality. *COPD: Journal of Chronic Obstructive Pulmonary Disease*, *9*(2), 131–141. doi:10.3109/15412555.2011.650239 PMID:22409371

Periadurai, T., Vijayakumar, C. T., & Balasubramanian, M. (2010). Thermal decomposition and flame retardant behaviour of SiO_2-phenolic nanocomposite. *Journal of Analytical and Applied Pyrolysis*, *89*(2), 244–249. doi:10.1016/j.jaap.2010.08.010

Petera, A. J., Lee, C. W. (2012). Electronic and optical properties of CdS/CdZnS nanocrystals. *Chin. Phys. B, 21*, 087302 (1-8).

Petroni, S., Guido, F., Torre, B., Falqui, A., Todaro, M., Cingolani, R., &DeVittorio, M. (2012). Tactile multisensing on flexible aluminum nitride. *Royal Society of Chemistry, 137*(22), 5260-5264.

Pfeiffer, C., Rehbock, C., Hühn, D., Carrillo-Carrion, C., de Aberasturi, D. J., Merk, V., & Parak, W. J. et al. (2014). Interaction of colloidal nanoparticles with their local environment: The (ionic) nanoenvironment around nanoparticles is different from bulk and determines the physico-chemical properties of the nanoparticles. *Journal of the Royal Society, Interface*, *11*(96), 20130931. doi:10.1098/rsif.2013.0931 PMID:24759541

Phadke, K. V., Manjeshwar, L. S., & Aminabhavi, T. M. (2014). Microspheres of Gelatin and Poly(ethylene glycol) Coated with Ethyl Cellulose for Controlled Release of Metronidazole. *ACS Publications*, *53*, 6575–6584.

Phisalaphong, M., & Chiaoprakobkij, N. (2012). Applications and Products—Nata de Coco. In Bacterial NanoCellulose: A Sophisticated Multifunctional Material (pp. 143–156). Academic Press.

Ponte, B. M., Santos, I. S., Macedo, A. S., & Souto, E. B. (2016). Lipid Nanoplatforms for Pulmonary Drug Delivery. *Nanotechnology and Drug Delivery, Volume Two: Nano-Engineering Strategies and Nanomedicines against Severe Diseases*, 146.

Popa, N. C., & Balzar, D. (2002). An analytical approximation for a size-broadened profile given by the lognormal and gamma distributions. *Journal of Applied Crystallography*, *35*(3), 338–346. doi:10.1107/S0021889802004156

Porfiri, M., Nguyen, N. Q., & Gupta, N. (2009). Thermal conductivity of multiphase particulate composite materials. *Journal of Materials Science*, *44*(6), 1540–1550. doi:10.1007/s10853-008-3040-0

Powell, R. M. (1994). Utilization of Natural Fibers in Plastic Composites. *Lignocellulosic Plastic Composites*, *4*(27), 23–51.

Prabhakar, U., Maeda, H., Jain, R. K., Sevick-Muraca, E. M., Zamboni, W., Farokhzad, O. C., & Blakey, D. C. et al. (2013). Challenges and key considerations of the enhanced permeability and retention effect for nanomedicine drug delivery in oncology. *Cancer Research*, *73*(8), 2412–2417. doi:10.1158/0008-5472.CAN-12-4561 PMID:23423979

Prasertmanakit, S., Praphairaksit, N., Chiangthong, W., & Muangsin, N. (2009). Ethyl cellulose microcapsules for protecting and controlled release of folic acid. *AAPS PharmSciTech*, *10*(4), 1104–1112. doi:10.1208/s12249-009-9305-3 PMID:19763838

Praveen, R. N., Muruli, D., & Nayak, S. S. (2013). Sliding wear behavior of Acrylonitrile Butadiene Styrene (ABS) thermoplastic composites. *International Journal of Mechanical Engineering and Robotics Research*, *2*(4), 361–372.

Press, W. H., Flannery, B. P., Teukolsky, S. A., & Vetterling, W. T. (1986). *Numerical Recipes*. Cambridge, UK: Cambridge Univ. Press.

Pritchard, J. (2001). The influence of lung deposition on clinical response. *Journal of Aerosol Medicine*, *14*(Supplement 1), 19–26. doi:10.1089/08942680150506303 PMID:11424889

Priya, S., & Inman, D. J. (Eds.). (2009). Energy harvesting technologies. New York: Springer.

Priya, S. (2007). Advances in energy harvesting using low profile piezoelectric transducers. *Journal of Electroceramics*, *19*(1), 167–184. doi:10.1007/s10832-007-9043-4

Probst, C. E., Zrazhevskiy, P., Bagalkot, V., & Gao, X. (2013). Quantum dots as a platform for nanoparticle drug delivery vehicle design. *Advanced Drug Delivery Reviews*, *65*(5), 703–718. doi:10.1016/j.addr.2012.09.036 PMID:23000745

Psomiadou, E., Arvanitoyannis, I., & Yamamoto, N. (1996). Edible films made from natural resources; microcrystalline cellulose (MCC), methylcellulose (MC) and corn starch and polyols - Part 2. *Carbohydrate Polymers*, *31*(4), 193–204. doi:10.1016/S0144-8617(96)00077-X

Qin, D. K., Liang, G. Z., Gu, A. J., & Yuan, L. (2014). Fabrication of SiO_2 covered ZrO_2 electrospun fibres with controllable structure and their novel high-performance composites with outstanding thermal, mechanical and dielectric properties. *Journal of Composites Materials*, *48*(26), 3201–3214. doi:10.1177/0021998313507619

Quintavalla, S., & Vicini, L. (2002). Antimicrobial food packaging in meat industry. *Meat Science*, *62*(3), 373–380. doi:10.1016/S0309-1740(02)00121-3 PMID:22061613

Ragelle, H., Danhier, F., Préat, V., Langer, R., & Anderson, D. G. (2016). Nanoparticle-based drug delivery systems: A commercial and regulatory outlook as the field matures. *Expert Opinion on Drug Delivery*, 1–14. PMID:27730820

Raghavender, A. T., & Jadhav, K. M. (2009). Dielectric properties of Al-substituted Co ferrite nanoparticles. *Bulletin of Materials Science*, *32*(6), 575–578. doi:10.1007/s12034-009-0087-8

Rahman, I. A., & Padavettan, V. (2012). Synthesis of Silica Nanoparticles by Sol-Gel, Size-Dependent Properties, Surface Modification, and Applications in Silica-Polymer Nanocomposites—A Review. *Journal of Nanomaterials*, *2012*, 1–15. doi:10.1155/2012/132424

Rajathi, S., Sankarasubramabian, N., Ramsnathan, K., & Senthamizhselvi, M. (2012). Structural and optoelectronic properties of pyralytically sprayed CdZnS thin films. *ChalcogenideLett.*, *9*, 495–500.

Raj, V., & Prabha, G. (2016). Synthesis, characterization and in vitro drug release of cisplatin loaded Cassava starch acetate–PEG/gelatin nanocomposites. *Journal of the Association of Arab Universities for Basic and Applied Sciences*, *21*, 10–16. doi:10.1016/j.jaubas.2015.08.001

Rakib, M. I., Choudhury, I. A., Hussain, S., & Osman, N. A. A. (2015). Design and biomechanical performance analysis of a user-friendly orthotic device. *Materials & Design*, *65*, 716–725. doi:10.1016/j.matdes.2014.09.075

Ramachandran, G. N., Ramakrishnan, C., & Sasisekharan, V. (1963). Stereochemistry of polypeptide chain configurations. *Journal of Molecular Biology*, *7*(1), 95–99. doi:10.1016/S0022-2836(63)80023-6 PMID:13990617

Ramdani, N., Liu, W., Wang, J., & Derradji, M. (2017). Ceramic Based Polybenzoxazine Micro- and Nanocomposites. In Advanced and Emerging Polybenzoxazine Science and Technology. Amsterdam: Elsevier.

Ramdani, N., Derradji, M., Wang, J., Mokhnache, E. O., & Liu, W. (2016, September). Improvements of Thermal, Mechanical, and Water-Resistance Properties of Polybenzoxazine/Boron Carbide Nanocomposites. *Journal of the Minerals Metals & Materials Society*, *68*(9), 2533–2542. doi:10.1007/s11837-016-2040-9

Ramdani, N., Derradji, M., Wang, J., Mokhnache, E.-O., Liu, W., Liu, Y., & Dong, W. (2016). Experimental and modeling of thermal and dielectric properties of aluminum nitride-reinforced polybenzoxazine hybrids. *Journal of Thermal Analysis and Calorimetry*, *126*(2), 561–570. doi:10.1007/s10973-016-5492-9

Ramdani, N., Wang, J., & Liu, W. (2014). Preparation and thermal properties of polybenzoxazine/TiC hybrids. *Advanced Materials Research*, *887*(1), 49–52. doi:10.4028/www.scientific.net/AMR.887-888.49

Ramdani, N., Wang, J., Wang, H., Feng, T., Derradji, M., & Liu, W. (2014). Mechanical and thermal properties of silicon nitride reinforced polybenzoxazine nanocomposites. *Composites Science and Technology*, *105*, 73–79. doi:10.1016/j.compscitech.2014.10.006

Ramos, F. L., Krahnke, J. S., & Kim, V. (2014). Clinical issues of mucus accumulation in COPD. *International Journal of Chronic Obstructive Pulmonary Disease*, *9*, 139–150. PMID:24493923

Rao, M., Walter, R. H., & Cooley, H. J. (1981). Effect of Heat Treatment on the Flow Properties of Aqueous Guar Gum and Sodium Carboxymethylcellulose (CMC) Solutions. *Journal of Food Science*, *46*(3), 896–899. doi:10.1111/j.1365-2621.1981.tb15374.x

Rasheva, Z., Zhang, G., & Burkhart, T. (2010). A correlation between the tribological and mechanical properties of short carbon fibers reinforced PEEK materials with different fiber orientations. *Tribology International*, *43*(8), 1430–1437. doi:10.1016/j.triboint.2010.01.020

Ravangave, L. S., Biradar, U. V., & Misal, S. D. (2012). The Effect of Ionic Composition on Structural and Optical Properties of $Cd_xZn_{1-x}S$ Thin films Grown by Spray pyrolysis. *Int. J. Sci. Res. Publ.*, *2*, 1–5.

Raviprakash, Y., Bangera, K. V., & Shivakumar, G. K. (2010). Growth, structural and optical properties of $Cd_xZn_{1-x}S$ thin films deposited using spray pyrolysis technique. *Current Applied Physics*, *10*(1), 193–198. doi:10.1016/j.cap.2009.05.020

Ray, S. S., & Okamoto, M. (2003). Polymer/layered silicate nanocomposite, a review from preparation to processing. *Progress in Polymer Science*, *28*(11), 1539–1541. doi:10.1016/j.progpolymsci.2003.08.002

Reece, W. O. (2015). Overview of the Respiratory System. *Dukes' Physiology of Domestic Animals*, 203.

Reetz, M. T., & Helbig, W. (1994). Size-selective synthesis of nanostructured transition metal clusters. *Journal of the American Chemical Society*, *116*(16), 7401–7402. doi:10.1021/ja00095a051

Regand, A., & Goff, H. D. (2003). Structure and ice recrystallization in frozen stabilized ice cream model systems. *Food Hydrocolloids*, *17*(1), 95–102. doi:10.1016/S0268-005X(02)00042-5

Rehman, S., & Ahmed, M. (2002). Effect of carboxy methyl cellulose and carrageenan gum on the shelf life of bread. *Pakistan Journal of Food Science*, *13*(1), 49–52.

Rejinold, N. S., Chennazhi, K. P., Nair, S. V., Tamura, H., & Jayakumar, R. (2011). Biodegradable and thermo-sensitive chitosan-g-poly(N-vinylcaprolactam) nanoparticles as a 5-fluorouracil carrier. *Carbohydrate Polymers*, *83*(2), 776–786. doi:10.1016/j.carbpol.2010.08.052

Rescignano, N., Fortunati, E., Montesano, S., Emiliani, C., Kenny, J. M., Martino, S., & Armentano, I. (2014). PVA bio-nanocomposites: A new take-off using cellulose nanocrystals and PLGA nanoparticles. *Carbohydrate Polymers*, *99*, 47–58. doi:10.1016/j.carbpol.2013.08.061 PMID:24274478

Respiratory System - Pulmonary Ventilation. (2016). *Science Encyclopedia*. Retrieved from http://science.jrank.org/pages/5846/Respiratory-System-Pulmonary-ventilation.html

Rezaei Mokarram, A., & Alonso, M. (2016). Preparation and evaluation of chitosan nanoparticles containing Diphtheria toxoid as new carriers for nasal vaccine delivery in mice. *Archives of Razi Institute*, *61*(1), 13–25.

Rhee, S. H., & Choi, J. Y. (2002). Preparation of a bioactive poly(methyl methacrylate)/silica nanocomposite. *Journal of the American Ceramic Society*, *85*(5), 1318–1320. doi:10.1111/j.1151-2916.2002.tb00269.x

Rhodes, K. H. (1978). Preparation of liquid and frozen yogurt products. *United States Patent No 4,110,476*.

Ribárik, G., Ungár, T., & Gubicza, J. (2001). MWP-fit: A program for multiple whole-profile fitting of diffraction peak profiles by ab initio theoretical functions. *Journal of Applied Crystallography*, *34*(5), 669–676. doi:10.1107/S0021889801011451

Ritzhaupt-Kleissl, E., Bohm, J., Haubelt, J., & Hanemann, T. (2006). Thermoplastic polymer nanocomposites for applications in optical devices. *Materials Science and Engineering C*, *26*(5-7), 1067–1071. doi:10.1016/j.msec.2006.02.004

Robbins, M. D., & Rodriguez, S. S. (1984). Low calorie baked products. *United States Patent No 4,461,782*, 1–5.

Rohatgi, P. K., Kim, J. K., Gupta, N., Alaraj, S., & Daoud, A. (2006). Compressive characteristics of A356/fly ash cenosphere composites synthesized by pressure infiltration technique. *Composites. Part A, Applied Science and Manufacturing*, *37*(3), 430–437. doi:10.1016/j.compositesa.2005.05.047

Rokicki, G., Parzuchowski, P. G., & Mazurek, M. (2015). Non-isocyanate polyurethanes: Synthesis, properties, and applications. *Polymers for Advanced Technologies*, *26*(7), 707–761. doi:10.1002/pat.3522

Rong, M. Z., Zhang, M. Q., Liu, H., Zeng, H. M., Wetzel, B., & Friedrich, K. (2001). Microstructure and tribological behavior of polymeric nanocomposites. *Industrial Lubrication and Tribology*, *53*(2), 72–77. doi:10.1108/00368790110383993

Roosen, A. R., & Carter, W. C. (1998). Simulations of microstructural evolution: anisotropic growth and coarsening. *Physica A: Statistical Mechanics and its Applications, 261*(1), 232-247.

Roselman, I. C., & Tabor, D. (1976). The friction of carbon fibres. *Journal of Physics. D, Applied Physics*, *9*(17), 2517–2532. doi:10.1088/0022-3727/9/17/012

Rossing, T. D. (2007). *Springer Handbook of Acoustics*. Berlin: Springer. doi:10.1007/978-0-387-30425-0

Roundy, S., & Wright, P. K. (2004). A piezoelectric vibration based generator for wireless electronics. *Smart Materials and Structures*, *13*(5), 1131–1142. doi:10.1088/0964-1726/13/5/018

Rowe, R. C., Sheskey, P. J., & Weller, P. J. (Eds.). (2006). Handbook of Pharmaceutical Excipients. APhA Publications. doi:10.1097/PRS.0b013e31822adba3

Rowe, R. C. (1986). The effect of the molecular weight of ethyl cellulose on the drug release properties of mixed films of ethyl cellulose and hydroxypropylmethylcellulose. *International Journal of Pharmaceutics*, *29*(1), 37–41. doi:10.1016/0378-5173(86)90197-3

Roy, S., Davim, J. P., Bhowmik, S., & Kumar, K. (2016). Estimation of mechanical and tribological properties of epoxy based green composites. In D. Verma, S. Jain, X. Zhang, & P. C. Gope (Eds.), *Green Approaches to Biocomposite Materials Science and Engineering* (pp. 96–124). Hershey, PA: Engineering Science Reference. doi:10.4018/978-1-5225-0424-5.ch005

Rozenberg, B. A., & Tenne, R. (2008). Polymer-assisted fabrication of nanoparticles and nanocomposites. *Progress in Polymer Science, 33*(1), 40–112. doi:10.1016/j.progpolymsci.2007.07.004

Ruge, C. A., Kirch, J., & Lehr, C.-M. (2013). Pulmonary drug delivery: From generating aerosols to overcoming biological barriers—therapeutic possibilities and technological challenges. *The Lancet Respiratory Medicine, 1*(5), 402–413. doi:10.1016/S2213-2600(13)70072-9 PMID:24429205

Rutkowski, J. V., & Levint, B. C. (1986). Acrylonitrile-Butadiene-Styrene Copolymers (ABS): Pyrolysis and Combustion Products and their Toxicity-A Review of the Literature. *Fire and Materials, 10*(3-4), 93–105. doi:10.1002/fam.810100303

Saba, N., Tahir, P. M., & Jawaid, M. (2014). A review on potentiality of nano filler/natural fiber filled polymer hybrid composites. *Polymers, 6*(8), 2247–2273. doi:10.3390/polym6082247

Saha, D., & Bhattacharya, S. (2010). Hydrocolloids as thickening and gelling agents in food: A critical review. *Journal of Food Science and Technology, 47*(6), 587–597. doi:10.1007/s13197-010-0162-6 PMID:23572691

Sahoo, N., Sahoo, R. K., Biswas, N., Guha, A., & Kuotsu, K. (2015). Recent advancement of gelatin nanoparticles in drug and vaccine delivery. *International Journal of Biological Macromolecules, 81*, 317–331. doi:10.1016/j.ijbiomac.2015.08.006 PMID:26277745

Sailor, M. J., & Curtis, C. L. (1994). Conductiong polymer connections for molecular devices. *Advanced Materials, 6*(9), 688–692. doi:10.1002/adma.19940060917

Sain, S., Patra, S., & Pradhan, S. K. (2012). Quickest ever single-step mechanosynthesis of $Cd_{0.5}Zn_{0.5}S$ quantum dots: Nanostructure and optical characterizations. *Materials Research Bulletin, 47*(4), 1062–1072. doi:10.1016/j.materresbull.2011.12.057

Sain, S., & Pradhan, S. K. (2012). Mechanochemical solid state synthesis of $(Cd_{0.8}Zn_{0.2})S$ quantum dots: Microstructure and optical characterizations. *Journal of Alloys and Compounds, 509*(10), 4176–4184. doi:10.1016/j.jallcom.2011.01.035

Sajeev, C., Vinay, G., Archna, R., & Saha, R. N. (2002). Oral controlled release formulation of diclofenac sodium by microencapsulation with ethyl cellulose. *Journal of Microencapsulation, 19*(6), 753–760. doi:10.1080/0265204021000022734 PMID:12569023

Salehi-Khojin, A., Hosseini, M. R., & Jalili, N. (2009). Underlying mechanics of active nanocomposites with tunable properties. *Composites Science and Technology, 69*(3), 545–552. doi:10.1016/j.compscitech.2008.11.033

Sameoto, D., See-Ho, T., & Parameswaran, M. (2007). Polymer MEMS processing for multi-user applications. *Sensors and Actuators A: Physical, 134*(2), 457-464.

Samir, O. M., & Somashekar, R. (2007). Intrinsic strain effect on crystal and molecular structure of (dch32) cotton fiber. *Powder Diffraction, 22*(1), 20-26.

Samir, O. M., Manjunath, A., Parameswara, P., Somashekarappa, H., & Somashekar, R. (2008). Micro-structural Parameters of Polymer Materials using WAXS data: Bimodals Crystal Size Distribution. *Journal of Polymer Materials, 25*(4), 529–540.

Sanap, V. B., & Pawar, B. H. (2011). study of chemical bath deposited nanocrystalline cdzns thin films. *Journal of Optoelectronics and Biomedical Materials, 3*, 39–43.

Sanchez, C., Julian, B., Belleville, P., & Popall, M. (2005). Applications of hybrid organic–inorganic nanocomposites. *Journal of Materials Chemistry, 15*(35-36), 3559–3592. doi:10.1039/b509097k

Sangappa, E. Y., Mahesh, S. S., & Somashekar, R. (2005). Crystal structure of raw pure Mysore silk fibre based on (Ala-Gly) 2-Ser-Gly peptide sequence using Linked-Atom-Least-Squares method. *Journal of Biosciences*, *30*(2), 259–268. doi:10.1007/BF02703707 PMID:15933415

Sanjay, M. R., & Yogesha, B. (2016). Studies on Mechanical Properties of Jute/E-Glass Fiber Reinforced Epoxy Hybrid Composites. *Journal of Minerals and Materials Characterization and Engineering*, *4*(01), 15–25. doi:10.4236/jmmce.2016.41002

Sarkar, N. (1979). Thermal Gelation Properties of Methyl and Hydroxypropyl Methylcellulose. *Journal of Applied Polymer Science*, *24*(4), 1073–1087. doi:10.1002/app.1979.070240420

Sarkar, N. (1995). Kinetics of thermal gelation of methylcellulose and hydroxypropylmethylcellulose in aqueous solutions. *Carbohydrate Polymers*, *26*(3), 195–203. doi:10.1016/0144-8617(94)00107-5

Sarkar, S., Guibal, E., Quignard, F., & SenGupta, A. K. (2012). Polymer-supported metals and metal oxide nanoparticles: Synthesis, characterization, and applications. *Journal of Nanoparticle Research*, *14*(2), 1–24. doi:10.1007/s11051-011-0715-2 PMID:22448125

Sarwar, M. I., Zulfiqar, S., & Ahmad, Z. (2008). Polyamide-silica nanocomposites, mechanical, morphological and thermomechanical investigations. *Polymer International*, *57*(2), 292–296. doi:10.1002/pi.2343

Scardi, P., & Leoni, M. (2001). Diffraction line profiles from polydisperse crystalline systems. *Acta Crystallographica. Section A, Foundations of Crystallography*, *57*(5), 604–613. doi:10.1107/S0108767301008881 PMID:11526309

Scardi, P., & Leoni, M. (2002). Whole powder pattern modelling. *Acta Crystallographica. Section A, Foundations of Crystallography*, *58*(2), 190–200. doi:10.1107/S0108767301021298 PMID:11832590

Scardi, P., & Leoni, M. (2006). Line profile analysis: Pattern modelling versus profile fitting. *Journal of Applied Crystallography*, *39*(1), 24–31. doi:10.1107/S0021889805032978

Schön, J. (1996). Coefficient of friction of composite delamination surfaces. *Wear*, *237*(1), 77–89. doi:10.1016/S0043-1648(99)00315-4

Schwartz, C. J., & Bahadur, S. (2000). Studies on the tribological behavior and transfer film-counterface bond strength for polyphenylene sulphide filled with nanoscale alumina particles. *Wear*, *237*(2), 261–273. doi:10.1016/S0043-1648(99)00345-2

Sekhar, H., & Rao, D. N. (2012). Spectroscopic studies on Zn-doped CdS nanopowders prepared by simple coprecipitation method. *Journal of Materials Science*, *47*(4), 1964–1971. doi:10.1007/s10853-011-5991-9

Selvam, B., Marimuthu, P., Narayanasamy, R., Anandakrishnan, V., Tun, K. S., Gupta, M., & Kamaraj, M. (2014). Dry sliding wear behaviour of zinc oxide reinforced magnesium matrix nano-composites. *Materials & Design*, *58*, 475–481. doi:10.1016/j.matdes.2014.02.006

Sengupta, R., Bandyopadhyay, A., Sabharwal, S., Chaki, T. K., & Bhowmick, A. K. (2005). Polyamide 6,6/in situ silica hybrid nanocomposites by sol–gel technique, synthesis, characterization and properties. *Polymer*, *46*(10), 3343–3354. doi:10.1016/j.polymer.2005.02.104

Seshan, K. (Ed.). (2002). *Handbook of thin film deposition processes and techniques*. Noyes Publication.

Sethi, P. K. (1990). Designing Aids for Physically Handicapped in Developing Countries. *Indian Journal of Physical Medicine and Rehabilitation.*, *3*, 1–10.

Sethi, P. K., Udawat, M. P., Kasliwal, S. C., & Chandra, R. (1978). Vulcanized rubber foot for lower limb amputees. *Prosthetics and Orthotics International*, *2*, 125–136. PMID:733459

Sharma, S. B., Jain, S., Khirwadkar, P., & Kulkarni, S. (2013). The effects of air pollution on the environment and human health. *Indian Journal of Research in Pharmacy and Biotechnology*, *1*(3), 391.

Sheth, P., Sandhu, H., Singhal, D., Malick, W., Shah, N., & Serpil Kislalioglu, M. (2012). Nanoparticles in the pharmaceutical industry and the use of supercritical fluid technologies for nanoparticle production. *Current Drug Delivery*, *9*(3), 269–284. doi:10.2174/156720112800389052 PMID:22283656

Sheu, F., Wang, C. L., & Shyu, Y. T. (2000). Fermentation of Monascus purpureus on bacterial cellulose-nata and the color stability of Monascus-nata complex. *Journal of Food Science*, *65*(2), 342–345.

She, W., Li, N., Luo, K., Guo, C., Wang, G., Geng, Y., & Gu, Z. (2013). Dendronized heparin– doxorubicin conjugate based nanoparticle as pH-responsive drug delivery system for cancer therapy. *Biomaterials*, *34*(9), 2252–2264. doi:10.1016/j.biomaterials.2012.12.017 PMID:23298778

Shih, Y. F., Jeng, R. J., & Wei, K. M. (2003). Carbon black containing interpenetrating polymer networks based on unsaturated polyester/epoxy III: Thermal and pyrolysis analysis. *Journal of Analytical and Applied Pyrolysis*, *70*(1), 129–141. doi:10.1016/S0165-2370(02)00102-X

Shimizu, M. (1941). Eine röntgenographische Untersuchung des Sericins. *Bull. Sericult. Exp. Stn*, *10*, 441–474.

Shirakawa, H., Louis, E. J., MacDiarmid, A. G., Chiang, C. K., & Heeger, A. J. (1977). Synthesis of electrically conducting organic polymers: Halogen derivatives of polyacetylene,(CH) x. *Journal of the Chemical Society. Chemical Communications*, (16), 578–580. doi:10.1039/c39770000578

Shirkhani, A. K., & Madadlou, A. (n.d.). *On stabilization of acidified milk drinks – a review*. Academic Press.

Shivakumar, S., & Guggari, G. S. (2011). Literature review of fiber reinforced polymer composites. *International Journal of Advances in Engineering and Technology*, *1*(5), 218–226.

Shi, Z., Zhang, Y., Phillips, G. O., & Yang, G. (2014). Utilization of bacterial cellulose in food. *Food Hydrocolloids*, *35*, 539–545. doi:10.1016/j.foodhyd.2013.07.012

Sieg, J. M. B., & J. (2009). Fiber. In *Ingredients in Meat Products: Properties, Functionality and Applications* (pp. 83–109). <ALIGNMENT.qj></ALIGNMENT>10.1007/978-0-387-71327-4

Siegel, R. W., Chang, S. K., Ash, B. J., Stone, J. A. P. M., Doremus, R. W., & Schadler, L. S. (2001). Mechanical behavior of polymer and ceramic matrix nanocomposites. *Scripta Materialia*, *44*(8-9), 2061–2064. doi:10.1016/S1359-6462(01)00892-2

Silva, A., González-Mira, E., Lobo, J. S., & Amaral, M. (2013). Current progresses on nanodelivery systems for the treatment of neuropsychiatric diseases: Alzheimers and Schizophrenia. *Current Pharmaceutical Design*, *19*(41), 7185–7195. doi:10.2174/13816128194113121912329 PMID:23489198

Sin, D., Sze, M., & Hogg, J. (2014). Bacterial microbiome of lungs in COPD. *International Journal of Chronic Obstructive Pulmonary Disease*, *9*(1), 229–238. doi:10.2147/COPD.S38932 PMID:24591822

Singer, N. S., Chang, H. H., Tang, P., & Dunn, J. M. (1990). Carbohydrate cream substitute. *United States Patent No 4,911,946*, 2–4.

Singh, A., Malviya, R., & Sharma, K, P. (2011). Pulmonary drug delivery system: A novel approach for drug delivery. *Current Drug Therapy*, *6*(2), 137–151. doi:10.2174/157488511795304930

Singha, S., & Thomas, M. J. (2008). Dielectric properties of epoxy nanocomposites. *IEEE Transactions on Dielectrics and Electrical Insulation, 15*(1), 12–23. doi:10.1109/T-DEI.2008.4446732

Singh, S. (2016). Respiratory symptoms and signs. *Medicine, 44*(4), 205–212. doi:10.1016/j.mpmed.2016.02.001

Singh, S., Maltais, F., Tombs, L., Church, A., Iqbal, A., & Riley, J. H. (2015). Effect of umeclidinium/vilanterol (UMEC/VI) on inspiratory capacity/total lung capacity ratio in hyperinflated COPD patients. *The European Respiratory Journal, 46*(suppl 59), PA1493.

Singh, V. P., & Singh, S. (1976). Some physical properties of ZnCdS solid solutions. *Czechoslovak Journal of Physics B, 26*(10), 1161–1166. doi:10.1007/BF01591089

Sivtseva, A. I., Strutynsky, A. V., Krivoshapkin, V. G., Sivtseva, E. N., Ivanova, M. A., & Timofeev, L. F. (2015). Systolic and Diastolic Dysfunction of the Right Heart Ventricle in Patients With Chronic Obstructive Pulmonary Disease in Extremely Cold Climate. *Global Journal of Health Science, 7*(3), 283.

Skilling, J., & Bryan, R. K. (1984). Maximum entropy image reconstruction: General algorithm. *Monthly Notices of the Royal Astronomical Society, 211*(1), 111–124. doi:10.1093/mnras/211.1.111

Skotheim, T. A., & Reynolds, J. (Eds.). (2007). *Handbook of Conducting Polymers* (Vol. 1-2). CRC Press.

Smith, & Wallace, A. (1989). The role of piezocomposites in ultrasonic transducers. *Ultrasonics Symposium, 1989. Proceedings, IEEE*, 755-766.

Smith, A. H., Laird, C., Porter, K., & Bloch, M. (2013). Haemostatic dressings in prehospital care. *Emergency Medicine Journal, 30*(10), 784–789. doi:10.1136/emermed-2012-201581 PMID:23161808

Smith, P. C., & Arnott, S. (1978). LALS: A linked-atom least-squares reciprocal-space refinement system incorporating stereochemical restraints to supplement sparse diffraction data. *Acta Crystallographica. Section A, Crystal Physics, Diffraction, Theoretical and General Crystallography, 34*(1), 3–11. doi:10.1107/S0567739478000029

Snyder, R. L., Bunge, H. J., & Fiala, J. (1999). *Defect and microstructure analysis by diffraction.* Oxford Univ. Press.

Soares, B. G., Souza, F. G. Jr, Manjunath, A., Somashekarappa, H., Somashekar, R., & Siddaramaiah, . (2007). Variation of long periodicity in blends of styrene butadiene, styrene copolymer/polyaniline using small angle X-ray scattering data. *Pramana, 69*(3), 435–443. doi:10.1007/s12043-007-0144-z

Somashekar, R., & Gopalkrishne Urs, R. (1993). Simulation of X-ray diffraction patterns for silk fibres using paracrystalline statistics. *Pramana, 40*(5), 335–343. doi:10.1007/BF02847494

Somashekar, R., Hall, I. H., & Carr, P. D. (1989). The determination of crystal size and disorder from X-ray diffraction photographs of polymer fibres. 1. The accuracy of determination of Fourier coefficients of the intensity profile of a reflection. *Journal of Applied Crystallography, 22*(4), 363–371. doi:10.1107/S0021889889004085

Song, H. J., Zhang, Z. Z., Men, X. H., & Luo, Z. Z. (2010). A study of the tribological behavior of nano-ZnO-filled polyurethane composite coatings. *Wear, 269*(1), 79–85. doi:10.1016/j.wear.2010.03.011

Song, W. C., & Lee, J. H. (2009). Growth and characterization of $Zn_xCd_{1-x}S$ films prepared by using chemical bath deposition for photovoltaic devices. *Journal of the Korean Physical Society, 54*(4), 1660–1665. doi:10.3938/jkps.54.1660

Soppimath, K. S., Aminabhavi, T. M., Kulkarni, A. R., & Rudzinski, W. E. (2001). Biodegradable polymeric nanoparticles as drug delivery devices. *Journal of Controlled Release, 70*(1), 1–20. doi:10.1016/S0168-3659(00)00339-4 PMID:11166403

Souza, F. G. Jr, Soares, B. G., Mantovani, G. L., Manjunath, A., Somashekarappa, H., Somashekar, R., & Siddaramaiah, . (2006). Blends of styrene butadiene styrene TRI block copolymer/polyaniline-Characterization by WAXS. *Polymer, 47*(6), 2163–2171. doi:10.1016/j.polymer.2006.01.033

Spek, A. L. (2003). Single-crystal structure validation with the program PLATON. *Journal of Applied Crystallography, 36*(1), 7–13. doi:10.1107/S0021889802022112

Spence, J. C. H. (2003). *High-resolution electron microscopy, Monograph on the Physics and Chemistry of materials* (3rd ed.). New York: Oxford Science Publications.

Sperling, L. H. (1981). *Interpenetrating Polymer Networks and Related Materials*. New York: Plenum Press. doi:10.1007/978-1-4684-3830-7

Sreekumar, T. V., Tao, L. M., Guo, B. G., Kumar, H., Hauge, S., & Smalley, R. H. (2004). R.E. *Advanced Materials, 16*(58).

Sridhar, R., & Ramakrishna, S. (2013). Electrosprayed nanoparticles for drug delivery and pharmaceutical applications. *Biomatter, 3*(3), e24281. doi:10.4161/biom.24281 PMID:23512013

Staff, R. H., Landfester, K., & Crespy, D. (2013). *Recent advances in the emulsion solvent evaporation technique for the preparation of nanoparticles and nanocapsules. In Hierarchical Macromolecular Structures: 60 Years after the Staudinger Nobel Prize II* (pp. 329–344). Springer.

Stephen, A. M. (2006). *Food Polysaccharides and Their Applications*. CRC Press. http://doi.org/doi:10.1201/9781420015164.ch6

Stokes, A. R. (1948). A numerical Fourier-analysis method for the correction of widths and shapes of lines on X-ray powder photographs. *Proceedings of the Physical Society, 61*(4), 382–391. doi:10.1088/0959-5309/61/4/311

Strom, G., Ohgren, C., & Ankerfors, M. (2013). Nanocellulose as an additive in foodstuff. *Innventia Report*, (403), 1–25.

Strydom, D. J., Haylett, T., & Stead, R. H. (1977). The amino-terminal sequence of silk fibroin peptide Cp—a reinvestigation. *Biochemical and Biophysical Research Communications, 79*(3), 932–938. doi:10.1016/0006-291X(77)91200-1 PMID:597316

Subash, T., & Pillai, S. N. (2015). Bast fibers reinforced green composites for aircraft indoor structures applications [review]. *Journal of Chemical and Pharmaceutical Sciences., 7*, 305–307.

Su, J. F., Huang, Z., Yuan, X. Y., Wang, X. Y., & Li, M. (2010). Structure and properties of carboxymethyl cellulose/soy protein isolate blend edible films crosslinked by Maillard reactions. *Carbohydrate Polymers, 79*(1), 145–153. doi:10.1016/j.carbpol.2009.07.035

Sundar, S., Kundu, J., & Kundu, S. C. (2016). Biopolymeric nanoparticles. *Science and Technology of Advanced Materials*. PMID:27877319

Sung, H.-W., Sonaje, K., Liao, Z.-X., Hsu, L.-W., & Chuang, E.-Y. (2012). pH-responsive nanoparticles shelled with chitosan for oral delivery of insulin: From mechanism to therapeutic applications. *Accounts of Chemical Research, 45*(4), 619–629. doi:10.1021/ar200234q PMID:22236133

Sun-Waterhouse, D., Bekkour, K., Wadhwa, S. S., & Waterhouse, G. I. N. (2013). Rheological and Chemical Characterization of Smoothie Beverages Containing High Concentrations of Fibre and Polyphenols from Apple. *Food and Bioprocess Technology, 7*(2), 409–423. doi:10.1007/s11947-013-1091-y

Suresh, K. P., Niranjan, H. B., Martin Jebaraj, P., & Chowdiah, M. P. (2003). Tensile and wear properties of aluminum composites. *Wear, 255*(1-6), 638–642. doi:10.1016/S0043-1648(03)00292-8

Su, S. J., & Kuramoto, N. (2000). Processable polyaniline-titanium dioxide nanocomposites, effect of titanium dioxide on the conductivity. *Synthetic Metals*, *114*(2), 147–153. doi:10.1016/S0379-6779(00)00238-1

Suzuki, F., Onozato, K., & Kurokawa, Y. (1990). A Formation of compatible poly (vinyl alcohol)/alumina gel composite and its properties. *Journal of Applied Polymer Science*, *39*(2), 371–381. doi:10.1002/app.1990.070390214

Tabuchi, M. (2007). Nanobiotech versus synthetic nanotech? *Nature Biotechnology*, *25*(4), 389–390. doi:10.1038/nbt0407-389 PMID:17420738

Tagliavia, G., Porfiri, M., & Gupta, N. (2009). Vinyl ester—glass hollow particle composites: Dynamic mechanical properties at high inclusion volume fraction. *Journal of Composite Materials*, *43*(5), 561–582. doi:10.1177/0021998308097683

Tainter, M. L. (1943). Methyl Cellulose as a Colloid Laxative. *Experimental Biology and Medicine*, *54*(1), 77–79. doi:10.3181/00379727-54-14309

Takahashi, Y., Gehoh, M., & Yuzuriha, K. (1999). Structure refinement and diffuse streak scattering of silk (Bombyx mori). *International Journal of Biological Macromolecules*, *24*(2), 127–138. doi:10.1016/S0141-8130(98)00080-4 PMID:10342756

Takeshi, Y., Toshinobu, U., & Toyoki, K. (1982). Piezoelectricity of a high-content lead zirconatetitanate/polymer composite. *Journal of Applied Physics*, *53*(6), 4328–4332. doi:10.1063/1.331211

Takeyama, K., Tagaya, E., Kondo, M., Tamaoki, J., & Orimo, K. (2016). Impact Of Irreversible Airflow Obstruction. In *Clinical Course Of Patients With Severe Atopic Asthma A36. Clinical Problems in Asthma and Allergy* (pp. A1397–A1397). Am Thoracic Soc.

Tang, D. Y., Cai, W., Zhou, D., Zhao, L., & Gongyi, C. K. Y. (2002a). A morphological study of modified PU/PMMA semi-IPNs and nanocomposites with $BaTiO_3$. *Materials Science and Technology*, *10*, 260–262.

Tang, D. Y., Hong, L., & Weimin, C. (2002b). Synthesis and application studies of castor oil PU/PMMA IPNs with $BaTiO_3$ fiber nanocomposites. *Ferroelectrics*, *265*(1), 259–264. doi:10.1080/00150190208260627

Tang, D. Y., Jin, Z., & Qiang, L. S. (2007). Electric and damping properties of PU based IPNs-$BaTiO_3$ nanocomposites. *Key Engineering Materials*, *118*, 336–338.

Tang, D. Y., Liancheng, Z., Zheng, J., & Weimin, C. (2002c). Preparation, morphology, and thermoelectric property studies of $BaTiO_3$ superfine fiber/castor oil polyurethane-based IPN nanocomposites. *Journal of Applied Polymer Science*, *84*(4), 709–715. doi:10.1002/app.10024

Tang, D. Y., Lili, L., Jusheng, Z., Zheng, J., & Liancheng, Z. (2004). Influence of barium titanate on the damping properties of interpenetrating polymer networks based nanocomposites. *Journal of Aeronautical Materials*, *5*, 40–43.

Tang, D. Y., Qin, C., Cai, W., & Zhao, L. (2003a). Preparation, morphology, and mechanical properties of modified-PU/UPR graft-IPN nanocomposites with $BaTiO_3$ fiber. *Materials Chemistry and Physics*, *82*(1), 73–77. doi:10.1016/S0254-0584(03)00162-7

Tang, D. Y., Xu, Q. Y., Cai, W. M., Zhou, D. R., & Zhao, L. H. (2003b). Preparation and mechanical properties of $BaTiO_3$ nanometer powder/COPU based IPNs nanocomposites. *Materials Science and Technology*, *11*, 181–184.

Tan, W., Sin, D., Bourbeau, J., Hernandez, P., Chapman, K., Cowie, R., & Aaron, S. D. et al. (2015). Characteristics of COPD in never-smokers and ever-smokers in the general population: Results from the CanCOLD study. *Thorax*, *70*(9), 822–829. doi:10.1136/thoraxjnl-2015-206938 PMID:26048404

Tarantino, L. M. (2006). *Notification of the Determination of Hydroxypropyl Methylcellulose As Being Generally Recognized As Safe.* Keller and Heckman LLP.

Tauc, J. (1974). *Amorphous and Liquid Semiconductor.* New York: Plenum Press. doi:10.1007/978-1-4615-8705-7

Tejayadi, S. (2004). Flavored milk manufacturing processes and compositions. *United States Patent Application No 10/437,317, 1*(19).

Teterycz, H., Rac, O., Wozniak, P. S., & Oles, D. (2013). the formation mechanism of colloidal spheres of zno in ethylene glycol. *Digest J. Nanomaterials and Biostructures, 8*, 1157–1167.

Thakur, I. S., Ray, A. K., & Singhal, A. (2015). Ethanol from Sugar Cane Bagasse of Pulp and Paper Mill Effluent by Cryptococcus albidus and Saccharomyces cerevisiae. *Energy Sources, Part A: Recovery, Utilization, and Environmental Effects, 37*(11), 1172–1179. doi:10.1080/15567036.2011.605431

Thakur, S., Katyal, S. C., & Singh, M. (2007). Improvement in electric and dielectric properties of nanoferrite synthesized via reverse micelle technique. *Applied Physics Letters, 91*(26), 262501. doi:10.1063/1.2824454

Thakur, V. K., Singha, A. S., & Thakur, M. K. (2012). Green Composites from Natural Fibers: Mechanical and chemical aging properties. *International Journal of Polymer Analysis and. Characterization, 17*, 401–407.

Thomas, G. M., Paquita, E. M., & Thomas, J. P. (2002). *Cellulose Ethers.* Encyclopedia Of Polymer Science and Technology.

Thorndike, , Murphy, & Staros. (1956). Engineering Applied to Orthopedic Bracing. *Orthotics and Prosthetics, 10*(4), 55–71.

Tiemstra, P. J. (1971). Microcrystalline cellulose compositions co-dried with hydrocolloids. *United States Patent No 3,573,058*, 1–3.

Tiwana, M., Redmond, S. J., & Lovell, N. H. (2012). A review of tactile sensing technologies with applications in biomedical engineering. *Sensors and Actuators. A, Physical, 179*, 17–31. doi:10.1016/j.sna.2012.02.051

Tiwari, R., & Garcia, E. (2011). The state of understanding of ionic polymer metal composite architecture: A review. *Smart Materials and Structures, 20*(8), 083001. doi:10.1088/0964-1726/20/8/083001

Tiwari, S. B., Murthy, T. K., Pai, M. R., Mehta, P. R., & Chowdary, P. B. (2003). Controlled release formulation of tramadol hydrochloride using hydrophilic and hydrophobic matrix system. *AAPS PharmSciTech, 4*(3), 1–6. doi:10.1208/pt040331 PMID:14621963

Tomé, L. C., Freire, M. G., Rebelo, L. P. N., Silvestre, A. J. D., Neto, C. P., Marrucho, I. M., & Freire, C. S. R. (2011). Surface hydrophobization of bacterial and vegetable cellulose fibers using ionic liquids as solvent media and catalysts. *Green Chemistry, 13*(9), 2464. doi:10.1039/c1gc15432j

Tosun, B. S., Pettit, C., Campbell, S. A., & Aydil, E. S. (2012). Structure and Composition of $Zn_xCd_{1-x}S$ Films Synthesized through Chemical Bath Deposition. *ACS Applied Materials & Interfaces, 4*(7), 3676–3684. doi:10.1021/am300771k PMID:22732000

To, T., Zhu, J., Larsen, K., Simatovic, J., Feldman, L., Ryckman, K., & Carlsten, C. et al. (2016). Progression from Asthma to Chronic Obstructive Pulmonary Disease (COPD): Is Air Pollution a Risk Factor?. *American Journal of Respiratory and Critical Care Medicine, 194*(4), 429–438. doi:10.1164/rccm.201510-1932OC PMID:26950751

Trakulsujaritchok, T., & Hourston, D. J. (2006). Damping characteristics and mechanical properties of silica filled PUR/PEMA simultaneous inter-penetrating polymer networks. *European Polymer Journal, 42*(11), 2968–2976. doi:10.1016/j.eurpolymj.2006.07.028

Trapani, A., Di Gioia, S., Ditaranto, N., Cioffi, N., Goycoolea, F. M., Carbone, A., & Alonso, M. J. et al. (2013). Systemic heparin delivery by the pulmonary route using chitosan and glycol chitosan nanoparticles. *International Journal of Pharmaceutics*, *447*(1), 115–123. doi:10.1016/j.ijpharm.2013.02.035 PMID:23454518

Trotta, V., Haghi, M., Scalia, S., Young, P. M., & Traini, D. (2015). *Resveratrol, a Novel Spray Dried Inhalation Powder for the Treatment of COPD and other Inflammatory Lung Diseases.* Paper presented at the Journal of aerosoal medicine and pulmonary drug delivery.

Tsekmes, I. A., Kochetov, R., Morshuis, P. H. F., & Smit, J. J. (2015). The role of particle distribution in the dielectric response of epoxy–boron nitride nanocomposites. *Journal of Materials Science*, *50*(3), 1175–1186. doi:10.1007/s10853-014-8674-5

Tseng, H. J., Tsou, T. L., Wang, H. J., & Hsu, S. (2013). Characterization of chitosan–gelatin scaffolds for dermal tissue engineering. *Journal of Tissue Engineering and Regenerative Medicine*, *7*(1), 20–31. doi:10.1002/term.492 PMID:22034441

Tsuji, M., Hashimoto, M., Nishizawa, Y., & Tsuji, T. (2004). Synthesis of gold nanorods and nanowires by a microwave–polyol method. *Materials Letters*, *58*(17), 2326–2330. doi:10.1016/j.matlet.2004.02.020

Tuason, D. C., Jr., Ruszkay, T. A., & Heese, S. (2004). Fat-like agents for low calorie food compositions. *United States Patent No 6,689,405, 1*(12), 4–9.

Uddin, M. S., Hawlader, M. N. A., & Zhu, H. J. (2001). Microencapsulation of ascorbic acid: Effect of process variables on product characteristics. *Journal of Microencapsulation, 18*(2), 199–209. doi:10.1080/02652040010000352 PMID:11253937

Ulus, H., Üstün, T., Eskizeybek, V., & Sinan, Ö. (2014). Boron nitride-MWCNT/epoxy hybrid nanocomposites, Preparation and mechanical properties. *Applied Surface Science*, *318*, 37–42. doi:10.1016/j.apsusc.2013.12.070

Upadhye, S. B., & Rajabi-Siahboomi, A. R. (2013). Properties andApplications of Polyethylene Oxide and Ethylcellulose for Tamper Resistance and Controlled Drug Delivery. In *Advances in the Pharmaceutical Sciences* (Vol. 9, pp. 145–158). Springer New York. doi:10.1007/978-1-4614-8432-5

Valencia-Chamorro, S. A., Palou, L., Del Rio, M. A., & Pérez-Gago, M. B. (2008). Inhibition of Penicillium digitatum and Penicillium italicum by Hydroxypropyl Methylcellulose-Lipid Edible Composite Films Containing Food Additives with Antifungal Properties. *Journal of Agricultural and Food Chemistry*, *56*(23), 11270–11278. doi:10.1021/jf802384m PMID:19012404

Valley, L. (2002). *United States Patent 6495177.* Washington, DC: US Patent Office.

Van den Ende, D. A., Van de Wiel, H. J., Groen, W. A., & Van der Zwaag, S. (2011). Direct strain energy harvesting in automobile tires using piezoelectric PZT–polymer composites. *Smart Materials and Structures*, *21*(1), 015011. doi:10.1088/0964-1726/21/1/015011

van Ruth, S. M., de Witte, L., & Uriarte, A. R. (2004). Volatile Flavor Analysis and Sensory Evaluation of Custard Desserts Varying in Type and Concentration of Carboxymethyl Cellulose. *Journal of Agricultural and Food Chemistry*, *52*(26), 8105–8110. doi:10.1021/jf048647g PMID:15612803

Varshosaz, J., Ghaffari, S., Mirshojaei, S., Jafarian, A., Atyabi, F., Kobarfard, F., et al. (2013). Biodistribution of amikacin solid lipid nanoparticles after pulmonary delivery. *BioMed Research International, 2013.*

Vartiainen, V., Bimbo, L. M., Hirvonen, J., Kauppinen, E. I., & Raula, J. (2016). Drug permeation and cellular interaction of amino acid-coated drug combination powders for pulmonary delivery. *International Journal of Pharmaceutics*, *504*(1), 89–97. doi:10.1016/j.ijpharm.2016.03.052 PMID:27034001

Velterop, L., Delhez, R., Keijser, T. H. D., Mittemeijer, E. J., & Reefman, D. (2000). X-ray diffraction analysis of stacking and twin faults in fcc metals: A revision and allowance for texture and non-uniform fault probabilities. *Journal of Applied Crystallography*, *33*(2), 296–306. doi:10.1107/S0021889800000133

Venkatragavaraj, E., Satish, B., Vinod, P. R., & Vijaya, M. S. (2001). Piezoelectric properties of ferroelectric PZT-polymer composites. *Journal of Physics. D, Applied Physics*, *34*(4), 487–492. doi:10.1088/0022-3727/34/4/308

Viby, N.-E., Nexø, E., Kissow, H., Andreassen, H., Clementsen, P., Thim, L., & Poulsen, S. S. (2015). Trefoil factors (TFFs) are increased in bronchioalveolar lavage fluid from patients with chronic obstructive lung disease (COPD). *Peptides*, *63*, 90–95. doi:10.1016/j.peptides.2014.09.026 PMID:25445610

Villagran, F. V., Dria, G. J., Butterbaugh, J. L., & Small, L. E. (2004). Creamy mouthfeel agent for foods and beverages. *United States Patent No 6,673,384.*

Villalobos-Carvajal, R., Hernández-Muñoz, P., Albors, A., & Chiralt, A. (2009). Barrier and optical properties of edible hydroxypropyl methylcellulose coatings containing surfactants applied to fresh cut carrot slices. *Food Hydrocolloids*, *23*(2), 526–535. doi:10.1016/j.foodhyd.2008.02.008

Vojdani, F., & Torres, J. A. (1990). Potassium Sorbate Permeability of Methylcellulose and Hydroxypropyl Methylcellulose Coatings: Effect of Fatty Acids. *Journal of Food Science*, *55*(3), 841–846. doi:10.1111/j.1365-2621.1990.tb05244.x

Vollath, D., & Szabo, D. V. (2004). Synthesis and properties of nanocomposites. *Advanced Engineering Materials*, *6*(3), 117–127. doi:10.1002/adem.200300568

Vollath, D., Szabo, D. V., & Schlabach, S. (2004). Oxide/polymer nanocomposites as new luminescent materials. *Journal of Nanoparticle Research*, *6*(2), 181–191. doi:10.1023/B:NANO.0000034624.56366.ac

Walker, D. J., Walterspacher, S., Ekkernkamp, E., Storre, J. H., Windisch, W., & Dreher, M. (2015). Walking with noninvasive ventilation does not prevent exercise-induced hypoxaemia in stable hypercapnic COPD patients. *COPD: Journal of Chronic Obstructive Pulmonary Disease*, *12*(5), 546–551. doi:10.3109/15412555.2015.1008690 PMID:26263033

Wanga, S. T., Barringer, S. A., & Hansenbpt, P. M. T. (1998). Effects of carboxymethylcellulose and guar gum on ice crystal propagation in a sucrose-lactose solution. *Food Hydrocolloids*, *12*(2), 211–215. doi:10.1016/S0268-005X(98)00032-0

Wang, C. S., Wang, Y. T., Liu, W. J., Yin, H. Y., Yuan, Z. R., Wang, Q. J., & Cheng, R. et al. (2012). Natural fibrous nanoclay reinforced soy polyol-based polyurethane. *Materials Letters*, *78*, 85–87. doi:10.1016/j.matlet.2012.03.067

Wang, C., Fu, X., & Yang, L. (2007). Water-soluble chitosan nanoparticles as a novel carrier system for protein delivery. *Chinese Science Bulletin*, *52*(7), 883–889. doi:10.1007/s11434-007-0127-y

Wang, G.-X., Zhang, B.-L., Yu, Z.-L., & Qu, M.-Z. (2005). Manganese oxide/MWNTs composite electrodes for supercapacitors. *Solid State Ionics*, *176*(11-12), 1169–1174. doi:10.1016/j.ssi.2005.02.005

Wang, H., Xu, P., Zhong, W., Shen, L., & Du, Q. (2005). Transparent poly(methyl methacrylate)/silica/zirconia nanocomposites with excellent thermal stabilities. *Polymer Degradation & Stability*, *87*(2), 319–327. doi:10.1016/j.polymdegradstab.2004.08.015

Wang, H., Zou, Q., Boerman, O. C., Nijhuis, A. W., Jansen, J. A., Li, Y., & Leeuwenburgh, S. C. G. (2013). Combined delivery of BMP-2 and bFGF from nanostructured colloidal gelatin gels and its effect on bone regeneration in vivo. *Journal of Controlled Release*, *166*(2), 172–181. doi:10.1016/j.jconrel.2012.12.015 PMID:23266450

Wang, J.-J., Hsu, T.-H., Yeh, C.-N., Tsai, J.-W., & Su, Y.-C. (2011). Piezoelectric polydimethylsiloxane films for MEMS transducers. *Journal of Micromechanics and Microengineering*, *22*(1), 015013. doi:10.1088/0960-1317/22/1/015013

Wang, K., Chen, L., Wu, J., Toh, M. L., He, C., & Yee, A. F. (2005). Epoxy nanocomposites with highly exfoliated clay, mechanical properties and fracture mechanisms. *Macromolecules, 38*(3), 788–800. doi:10.1021/ma048465n

Wang, L. L., Zhang, L. Q., & Tian, M. (2012). Mechanical and tribological properties of acrylonitrile–butadiene rubber filled with graphite and carbon black. *Materials & Design, 39*, 450–457. doi:10.1016/j.matdes.2012.02.051

Wang, S. F., Wang, Y. R., Cheng, K. C., & Hsaio, Y. P. (2009). Characteristics of polyimide/barium titanate composite film. *Ceramics International, 35*(1), 265–268. doi:10.1016/j.ceramint.2007.10.004

Wang, T., Chen, S., Wang, Q., & Pei, X. (2010). Damping analysis of polyurethane/ epoxy graft interpenetrating polymer network composites filled with short carbon fiber and micro hollow glass bead. *Materials & Design, 31*(8), 3810–3815. doi:10.1016/j.matdes.2010.03.029

Wang, W., Ji, X., Na, H. B., Safi, M., Smith, A., Palui, G., & Mattoussi, H. et al. (2014). Design of a Multi-Dopamine-Modified Polymer Ligand Optimally Suited for Interfacing Magnetic Nanoparticles with Biological Systems. *Langmuir, 30*(21), 6197–6208. doi:10.1021/la500974r PMID:24805794

Wang, X., Chi, N., & Tang, X. (2008). Preparation of estradiol chitosan nanoparticles for improving nasal absorption and brain targeting. *European Journal of Pharmaceutics and Biopharmaceutics, 70*(3), 735–740. doi:10.1016/j.ejpb.2008.07.005 PMID:18684400

Wang, Y., Lim, S., Luo, J. L., & Xu, Z. H. (2006). Tribological and corrosion behaviors of Al_2O_3/polymer nanocomposite coatings. *Wear, 260*(9-10), 976–983. doi:10.1016/j.wear.2005.06.013

Wang, Z. G., Zu, X. T., Xiang, X., & Yu, H. J. (2006). Photoluminescence from TiO_2/PMMA nanocomposite prepared by gamma radiation. *Journal of Nanoparticle Research, 8*(1), 137–139. doi:10.1007/s11051-005-7527-1

Wan, Y. Z., Chen, G. C., Raman, S., Xin, J. Y., Li, Q. Y., Huang, H., & Luo, H. L. et al. (2006). Friction and wear behavior of three-dimensional braided carbon fiber/epoxy composites under dry sliding conditions. *Wear, 260*(9-10), 933–941. doi:10.1016/j.wear.2005.06.007

Warren, B. E. (1955). A generalized treatment of cold work in powder patterns. *Acta Crystallographica, 8*(8), 483–486. doi:10.1107/S0365110X55001503

Warren, B. E. (1959). X-ray studies of deformed metals. *Progress in Metal Physics, 8*, 147–202. doi:10.1016/0502-8205(59)90015-2

Warren, B. E., & Averbach, B. L. (1950). The effect of cold-work distortion on X-ray patterns. *Journal of Applied Physics, 21*(6), 595–599. doi:10.1063/1.1699713

Warren, B. E., & Averbach, B. L. (1952). The separation of cold-work distortion and particle size broadening in X-ray patterns. *Journal of Applied Physics, 23*(4), 497–497. doi:10.1063/1.1702234

Weijing, L., Ke, X., Chengshuang, W., Bin, Q., Yifan, S., Yuge, Z., & Rongshi, C. et al. (2015). Carbon nanofibers reinforced soy polyol-based polyurethane Nanocomposites- Thermal and mechanical characterization. *Journal of Thermal Analysis and Calorimetry, 123*(3), 2459–2468.

Welberry, T. R., & Butler, B. D. (1994). Interpretation of diffuse X-ray scattering via models of disorder. *Journal of Applied Crystallography, 27*(3), 205–231. doi:10.1107/S0021889893011392

Williams, C. B., & Rob, B. Y. (1996). Analysis of a micro-electric generator for microsystems. *Sensors and Actuators. A, Physical, 52*(1-3), 8–11. doi:10.1016/0924-4247(96)80118-X

Willis, L., Hayes, D. Jr, & Mansour, H. M. (2012). Therapeutic liposomal dry powder inhalation aerosols for targeted lung delivery. *Lung*, *190*(3), 251–262. doi:10.1007/s00408-011-9360-x PMID:22274758

Wohlfart, S., Gelperina, S., & Kreuter, J. (2012). Transport of drugs across the blood–brain barrier by nanoparticles. *Journal of Controlled Release*, *161*(2), 264–273. doi:10.1016/j.jconrel.2011.08.017 PMID:21872624

Wojdyr, M. (2010). Fityk: A general-purpose peak fitting program. *Journal of Applied Crystallography*, *43*(5 Part 1), 1126–1128. doi:10.1107/S0021889810030499

Wren, M., & Mullen, J. (1974). Process for producing low calorie pasta. *United States Patent No 3,843,818*.

Wu, C. S. (2004). *In*-situ polymerization of titanium isopropoxideinpoly caprolactone, properties and characterization of the hybrid nanocomposites. *Journal of Applied Polymer Science*, *92*(3), 1749–1757. doi:10.1002/app.20135

Wu, H. C., & Kessler, M. R. (2015). Multifunctional Cyanate Ester Nanocomposites Reinforced by Hexagonal Boron Nitride after Noncovalent Biomimetic Functionalization. *ACS Applied Materials & Interfaces*, *7*(10), 5915–5926. doi:10.1021/acsami.5b00147 PMID:25726956

Wu, J., Yang, S., Gao, S., Hu, A., Liu, J., & Fan, L. (2005). Preparation, morphology and properties of nano-sized Al_2O_3/polyimide hybrid films. *European Polymer Journal*, *41*(1), 73–81. doi:10.1016/j.eurpolymj.2004.08.014

Wu, T., & Ke, Y. (2007). Melting crystallization and optical behaviors of poly (ethylene terephthalate)-silica/polystyrene nanocomposite films. *Thin Solid Films*, *515*(13), 5220–5226. doi:10.1016/j.tsf.2006.12.029

Wu, X., He, G., Gu, S., Hu, Z., & Yao, P. (2007). Novel interpenetrating polymer network sulfonated poly (phthalazinone ether sulfone ketone)/polyacrylic acid proton exchange membranes for fuel cell. *Journal of Membrane Science*, *295*(1-2), 80–87. doi:10.1016/j.memsci.2007.02.039

Xia, D., Caijuan, T., Rongzhe, T., Wei, L., Lianghuan, F., Jingquan, Z., & Zhi, L. et al. (2011). Structural and optical properties of $Cd_{0.8}Zn_{0.2}S$ thin films. *J. Semicond.*, *32*(2), 022003. doi:10.1088/1674-4926/32/2/022003

Xia, Z. Y., Yu, J., Cheng, L., Liu, L., & Wang, F. (2010). Study of the Tensile Properties of Jute / Cotton Blended Yarns Using Weibull Distributions. *Research Journal of Textile and Apparel*, *14*(1), 18–25. doi:10.1108/RJTA-14-01-2010-B002

Xie, G. Y., Zhuang, G. S., Sui, G. X., & Yang, R. (2010). Tribological behavior of PEEK / PTFE composites reinforced with potassium titanate whiskers. *Wear*, *268*(3), 424–430. doi:10.1016/j.wear.2009.08.032

Xu, J., & Amiji, M. (2012). Therapeutic gene delivery and transfection in human pancreatic cancer cells using epidermal growth factor receptor-targeted gelatin nanoparticles. *Journal of Visualized Experiments*, (59), e3612-e3612.

Yadav, S., Mathur, M. K., Dr, S. C., & Kasliwa, S. C. (1993). Saltiel's Design Modification: Short leg brace for knee extensor weakness. *Indian Journal of Physical Medicine and Rehabilitation.*, *6*, 24–29.

Yamaguchi, T., Matsufusa, J., & Oshida, A. Y. (1992). Optical Transitions in RF Sputtered $CuIn_xGa_{1-x}Se_2$ Thin Films. *Japanese Journal of Applied Physics*, *31*(Part 2, No. 6A), L703–L706. doi:10.1143/JJAP.31.L703

Yan, F., Dang, Q., Liu, C., Yan, J., Wang, T., Fan, B., & Zhang, Z. et al. (2016). 3, 6-O-[N-(2-Aminoethyl)-acetamide-yl]-chitosan exerts antibacterial activity by a membrane damage mechanism. *Carbohydrate Polymers*, *149*, 102–111. doi:10.1016/j.carbpol.2016.04.098 PMID:27261735

Yang, Sharma, & Sheu. (1989). *United States Patent 4882152*. Washington, DC: US Patent Office.

Yan, H. X., Zhang, M. M., Feng, S. Y., Wang, Q. Q., & Zhang, J. P. (2014). Effects of macromolecular silane coupling agent on the mechanical properties of nano-SiO_2/cyanate ester nanocomposites. *Journal of Reinforced Plastics and Composites*, *33*(2), 127–136. doi:10.1177/0731684413509701

Yan, H. X., Zhang, M. M., Liu, C., & Zhang, J. P. (2013). Nano-Si_3N_4/epoxidized silane/cyanate ester composites for electronic packaging. *Polymer Bulletin, 70*(11), 2923–2933. doi:10.1007/s00289-013-0997-x

Yan, X. H., Ning, R. C., Liang, G. Z., Huang, Y., & Lu, T. L. (2007). The effect of silane coupling agent on the sliding wear behavior of nanometer ZrO_2/bismaleimide composites. *Journal of Materials Science, 42*(3), 958–965. doi:10.1007/s10853-006-0008-9

Yeganeh, H., & Pejman, H. T. (2007). Preparation and properties of novel biodegradable polyurethane networks based on castor oil and poly (ethyleneglycol). *Polymer Degradation & Stability, 92*(3), 480–489. doi:10.1016/j.polymdegradstab.2006.10.011

Yilmaz, B., & Olgun, A. (2008). Studies on cement and mortar containing low-calcium fly ash, limestone and dolomitic limestone. *Cement and Concrete Composites, 30*(3), 194–201. doi:10.1016/j.cemconcomp.2007.07.002

Yoon, H. Y., Koo, H., Choi, K. Y., Kwon, I. C., Choi, K., & Park, J. H. et al.. (2013). Photo-crosslinked hyaluronic acid nanoparticles with improved stability for in vivo tumor-targeted drug delivery. *Biomaterials, 34*(21), 5273–5280. doi:10.1016/j.biomaterials.2013.03.050 PMID:23591396

You, Y. L., Li, D. X., Si, G. J., Lv, R. Y., Deng, X. (2014). Improvement in the tribological properties of polyamide 6: Talc, glass fiber, graphite, and ultrahigh-molecular-weight polyethylene. *Journal of Thermoplastic Composite Materials.*

Younas, M., Nadeem, M., Atif, M., & Grossinger, R. (2011). Metal-semiconductor transition in NiFe2O4 nanoparticles due to reverse cationic distribution by impedance spectroscopy. *Journal of Applied Physics, 109*(9), 093704. doi:10.1063/1.3582142

Younes, I., & Rinaudo, M. (2015). Chitin and chitosan preparation from marine sources. Structure, properties and applications. *Marine Drugs, 13*(3), 1133–1174. doi:10.3390/md13031133 PMID:25738328

Yu, G.-C., Wu, L., Feng, L., & Yang, W. (2016). Thermal and mechanical properties of carbon fiber polymer-matrix composites with a 3D thermal conductive pathway. *Composite Structures, 149*, 213–219. doi:10.1016/j.compstruct.2016.04.010

Yu, H., Teo, J., Chew, J. W., & Hadinoto, K. (2016). Dry powder inhaler formulation of high-payload antibiotic nanoparticle complex intended for bronchiectasis therapy: Spray drying versus spray freeze drying preparation. *International Journal of Pharmaceutics, 499*(1–2), 38–46. doi:10.1016/j.ijpharm.2015.12.072 PMID:26757148

Yu, H., Tran, T.-T., Teo, J., & Hadinoto, K. (2016). Dry powder aerosols of curcumin-chitosan nanoparticle complex prepared by spray freeze drying and their antimicrobial efficacy against common respiratory bacterial pathogens. *Colloids and Surfaces. A, Physicochemical and Engineering Aspects, 504*, 34–42. doi:10.1016/j.colsurfa.2016.05.053

Yu, L., Lee, J. I., Shin, K. W., Park, C. E., & Holze, R. (2003). Preparation of aqueous polyaniline dispersions by micellar-aided polymerization. *Journal of Applied Polymer Science, 88*(6), 1550–1555. doi:10.1002/app.12127

Yu, S., Hing, P., & Hu, X. (2002). Thermal conductivity of polystyrene-Aluminium nitride composites. *Composites. Part A, Applied Science and Manufacturing, 33*(2), 289–292. doi:10.1016/S1359-835X(01)00107-5

Yuwono, A. H., Liu, B., Xue, J., Wang, J., Elim, H. I., Ji, W., & White, T. J. et al. (2004). Controlling the crystallinity and nonlinear optical properties of transparent TiO_2-PMMA nanohybrids. *Journal of Materials Chemistry, 14*(20), 2978–2987. doi:10.1039/B403530E

Yu, Y., Zhang, X. N., & Xie, S. L. (2009). Optimal shape control of a beam using piezoelectric actuators with low control voltage. *Smart Materials and Structures, 18*(9), 095006. doi:10.1088/0964-1726/18/9/095006

Zecher, D., & Van Coilie, R. (1992). Cellulose derivatives. In Thickening and gelling agents for food (pp. 40–66). http://doi.org/ doi:10.1007/978-1-4615-3552-2_3

Zeng, A., Zheng, Y., Guo, Y., Qiu, S., & Cheng, L. (2012). Effect of tetra-needle-shaped zinc oxide whisker (T-ZnOw) on mechanical properties and crystallization behavior of isotactic polypropylene. *Materials & Design*, *34*, 691–698. doi:10.1016/j.matdes.2011.05.028

Zhang, X., & Manohar, S. K. (2004). Polyaniline nanofibers: chemical synthesis using surfactants. *Chemical Communications*, (20), 2360-2361.

Zhang, C. Q., Madbouly, S. A., & Kessler, M. R. (2015). Biobased polyurethanes prepared from different vegetable oils. *ACS Applied Materials & Interfaces*, *7*(2), 1226–1233. doi:10.1021/am5071333 PMID:25541678

Zhang, H., Wang, B., Li, H., Jiang, Y., & Wang, J. (2003). Synthesis and characterization of nanocomposites of silicon dioxide and polyurethane and epoxy resin interpenetrating network. *Polymer International*, *52*(9), 1493–1497. doi:10.1002/pi.1247

Zhang, H., Xu, R., Wu, Z., & Zhou, C. (2003). The synthesis and characterization of polyurethane/clay nanocomposites. *Polymer International*, *52*(5), 790–794. doi:10.1002/pi.1152

Zhang, J., Tu, W., & Dai, Z. (2012). Synthesis and characterization of transparent and high impact resistance polyurethane coatings based on polyester polyols and isocyanate trimers. *Progress in Organic Coatings*, *75*(4), 579–583. doi:10.1016/j.porgcoat.2012.05.005

Zhang, M. C., Gu, A. J., Liang, G. Z., & Yuan, L. (2012). Preparation of high thermal conductive aluminum nitride/cyanate ester nanocomposite using a new macromolecular coupling agent. *Polymers for Advanced Technologies*, *23*(11), 1503–1510. doi:10.1002/pat.2073

Zhang, M., & Singh, R. (2004). Mechanical reinforcement of unsaturated polyester by Al_2O_3 nanoparticles. *Materials Letters*, *58*(3-4), 408–412. doi:10.1016/S0167-577X(03)00512-3

Zhang, W. G., & Zhang, X. Y. (2013). Study on the wear resistance of nm-SiC modified cyanate ester resins. *Thermosetting Resin*, *28*, 46–48.

Zhang, X.-Q., Xu, X., Bertrand, N., Pridgen, E., Swami, A., & Farokhzad, O. C. (2012). Interactions of nanomaterials and biological systems: Implications to personalized nanomedicine. *Advanced Drug Delivery Reviews*, *64*(13), 1363–1384. doi:10.1016/j.addr.2012.08.005 PMID:22917779

Zhang, X., & Simon, L. C. (2005). In situ polymerization of hybrid polyethylene alumina nanocomposites. *Macromolecular Materials and Engineering*, *290*(6), 573–583. doi:10.1002/mame.200500075

Zhang, Y., Wang, S., Liu, D., Zhang, Q., Wang, W., Ren, B., & Luo, H. et al. (2011). Fabrication of angle beam two-element ultrasonic transducers with PMN–PT single crystal and PMN–PT/epoxy 1–3 composite for NDE applications. *Sensors and Actuators. A, Physical*, *168*(1), 223–228. doi:10.1016/j.sna.2011.04.004

Zhang, Z., Yang, J. L., & Friedrich, K. (2004). Creep resistant polymeric nanocomposites. *Polymer*, *45*(10), 3481–3485. doi:10.1016/j.polymer.2004.03.004

Zhao, C. B., Su, L., Xu, S. C., & Yang, X. J. (2014). Study on KH550 modified boron ntride/cyanate ester resin thermal conductive composites. *New Chemical Materials*, *42*, 119–121.

Zheng, L., Ding, Z., Zhang, M., & Sun, J. (2011). Microencapsulation of bayberry polyphenols by ethyl cellulose: Preparation and characterization. *Journal of Food Engineering*, *104*(1), 89–95. doi:10.1016/j.jfoodeng.2010.11.031

Zheng, M., Zhao, M., Tang, L., Zhang, C., Song, L., & Wang, W. (2016). Ginsenoside Rg1 attenuates hypoxia and hypercapnia-induced vasoconstriction in isolated rat pulmonary arterial rings by reducing the expression of p38. *Journal of Thoracic Disease*, *8*(7), 1513–1523. doi:10.21037/jtd.2016.05.71 PMID:27499938

Zhen, Z., Tang, W., Chen, H., Lin, X., Todd, T., Wang, G., & Xie, J. et al. (2013). RGD-modified apoferritin nanoparticles for efficient drug delivery to tumors. *ACS Nano*, *7*(6), 4830–4837. doi:10.1021/nn305791q PMID:23718215

Zhou, Q. T., Leung, S. S. Y., Tang, P., Parumasivam, T., Loh, Z. H., & Chan, H.-K. (2015). Inhaled formulations and pulmonary drug delivery systems for respiratory infections. *Advanced Drug Delivery Reviews*, *85*, 83–99. doi:10.1016/j.addr.2014.10.022 PMID:25451137

Zhou, Q., Lau, S., Wu, D., & Shung, K. K. (2011). Piezoelectric films for high frequency ultrasonic transducers in biomedical applications. *Progress in Materials Science*, *56*(2), 139–174. doi:10.1016/j.pmatsci.2010.09.001 PMID:21720451

Zhu, B. L., & Wang, J. L. (2015). Grafting modification onto anatase TiO_2 by multistep and its effect on disoersion properties of CE composite materials the properties test of static mechanical, dynamic mechanical and heat resistance. *Applied Chemical Industry*, *44*, 1661–1667.

Zikakis, J. (2012). *Chitin, chitosan, and related enzymes*. Elsevier.

Zong, Z., Kimura, Y., Takahashi, M., & Yamane, H. (2000). Characterization of chemical and solid state structures of acylated chitosans. *Polymer, 41*(3), 899-906.

Zorba, M., & Ova, G. (1999). An improved method for the quantitative determination of carboxymethyl cellulose in food products. *Food Hydrocolloids*, *13*(1), 73–76. doi:10.1016/S0268-005X(98)00062-9

Index

A

ABS 78-82, 92-94, 105
AC conductivity 209, 214-215, 220-222

B

bacterial cellulose 38, 40, 61-64, 67-68
bagasse 80-82
band gap 187-188, 191-193, 202-204, 210
biopolymer 226, 229, 246
biopolymeric nanoparticles 225-226

C

carbon 53-54, 79-80, 144-145, 160-161, 163, 171, 198, 248, 250, 267, 271-272, 275, 280, 283
Carboxymethylcellulose 43, 45
CBD 186, 188-193, 198, 203
cellulose 24, 26, 38-45, 47, 49-68, 79, 92, 160, 226, 246, 271
cellulose derivatives 38, 42, 67-68
cenosphere 158, 161-179
ceramics 99-100, 107, 111, 142-143, 211
chitosan 51, 225-226, 228-236, 246
Chronic Obstructive Pulmonary Disease (COPD) 247, 266
clay 81-82, 92, 113, 144, 160
composites 78-82, 85, 87-94, 100, 106, 109, 111, 122-123, 139, 144-146, 148-150, 152, 158-162, 164-179, 211, 213-214, 216-217, 219, 221, 267, 270-272, 276, 278-280, 282-283
conductivity 105-108, 113, 123-127, 136-137, 159-160, 209, 214-215, 217, 219-222

D

dielectric 100, 106, 108-109, 111, 113, 133, 141-142, 144, 149, 209-211, 214-215, 217-220, 222
dielectric constant 100, 108, 113, 141-142, 209, 214-215, 218-219, 222
dielectric loss tangent 209, 214-215, 218-220, 222
drug delivery system 266
drug delivery vehicles 225-226, 228
drug release 54, 58, 225, 232, 235, 237

E

elastically yielding 142-143
electron hopping model 209, 221-222
energy harvesters 139, 147-148, 150
epoxy 80, 101, 103-105, 107-108, 110-111, 150, 160-162, 267, 271-272, 275, 279-280, 283

F

fibres 1-2, 17, 24-26, 29, 31-32, 125-126
food 38, 41-43, 45, 47-48, 50-57, 59-61, 63, 65, 67-68, 82, 229
FT-IR 160, 164-165, 209, 213, 215-217, 225, 231, 233, 235-236

G

glass 79-80, 106, 160, 164, 174, 191-193, 202, 219, 231, 251, 271-273, 275, 280, 283
green composites 79, 158

H

hardness 56-57, 63, 86-87, 92, 94, 100, 105, 159, 164, 166, 187, 271, 277-278, 281-283
High Pressure Homogenization 266

I

In Vitro Drug Delivery 246
IPN 158-162, 164-178

J

jute 79, 81

L

LPA 129

M

Magnesium Oxide 209, 211, 214
matrix 3, 7, 52, 55, 57-59, 78-82, 88-89, 93-94, 101, 103, 105-108, 110-111, 124, 130, 136, 144, 159-161, 166-167, 169, 174, 176, 178-179, 211, 213-214, 218, 221-222, 227-229, 233, 246, 270-273, 278, 282-283
mechanical properties 41, 47, 56, 62, 79-80, 85, 92, 99-100, 105, 108, 112, 122, 124-126, 159-161, 164, 270, 276, 283
methylcellulose 38, 50-53, 59-60
Morphological features 158
Morphological Study 88, 93, 278, 282

N

nanocomposites 99-108, 110-113, 122-123, 139, 152, 160-161, 209, 212-214, 218, 222
Nanomedicine 227, 246
nanoparticles 99-100, 102-103, 105-108, 110-111, 113, 187, 197, 199, 209-214, 218, 222, 225-237, 251, 255, 266
Nylon 105, 271-273, 275, 279-280, 283

O

optical properties 108, 110-111, 113, 186-188, 193, 199, 212

P

pattern fitting 1, 18, 28-30
piezoelectricity 140, 142-143
polyaniline 108, 122-126, 128, 160
polymers 1, 28, 48, 51, 54, 57, 59, 79-80, 99-100, 102, 105-106, 111, 113, 122-127, 132, 136-137, 139, 141-144, 149, 152, 159, 161, 211-213, 226-228, 235, 271
pulmonary drug delivery 230, 251
PVDF 139, 141-146, 149-150
PVP 209, 212-219, 221-222

R

respiratory system 248-250, 256, 266

S

SAXS 122, 127-129, 133, 135-136, 160
SEM 89, 93-94, 158, 160-161, 164, 177-178, 213, 225, 231-232, 278, 282
Sodium carboxymethyl cellulose 43, 45, 226, 246
spray drying 251-253, 266
surface-treatment 100, 110, 113
surfactant 55, 123, 266

T

TEM 110, 160, 186, 198-199, 204
tensile behaviors 160
tensile properties 80, 166
Thermal characteristics 158
thermal properties 58, 79, 99-100, 106, 112, 160-161
thermoset and thermoplast based composites 280
thin film 186-194, 200, 202-203

V

void 85, 89-90, 94, 276, 279-280, 283
void content 85, 89-90, 94, 276, 279-280

W

WAXS 122, 127-130, 132, 136, 160

X

XRD 160, 186, 191, 193-194, 198-200, 204, 209, 213, 216, 225, 231, 234

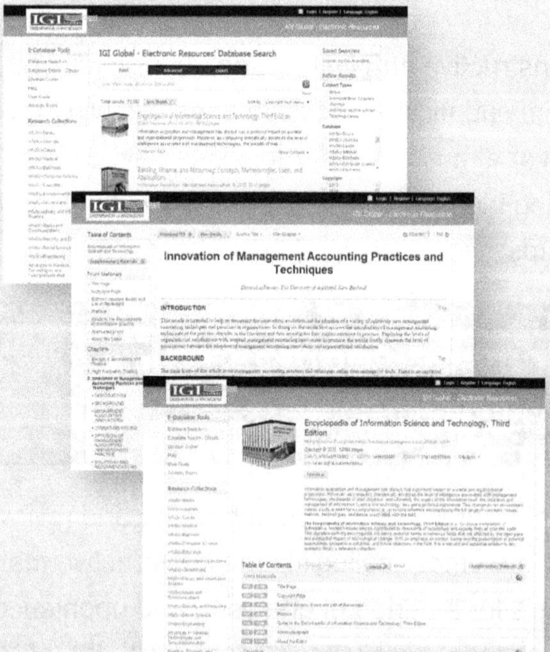

www.ingramcontent.com/pod-product-compliance
Lightning Source LLC
LaVergne TN
LVHW062137300926
849161LV00010B/2554